Encyclopedia of Molecular Dynamics

Encyclopedia of Molecular Dynamics

Edited by **Nick Wu**

Callisto Reference

New York

Published by Callisto Reference,
106 Park Avenue, Suite 200,
New York, NY 10016, USA
www.callistoreference.com

Encyclopedia of Molecular Dynamics
Edited by Nick Wu

International Standard Book Number: 978-1-63239-275-6 (Hardback)

Printed in China.

Contents

Preface

This book aims to highlight the current researches and provides a platform to further the scope of innovations in this area. This book is a product of the combined efforts of many researchers and scientists, after going through thorough studies and analysis from different parts of the world. The objective of this book is to provide the readers with the latest information of the field.

Molecular systems consist of a vast number of particles whose physical movements can be studied under molecular dynamics. This book is a compilation of the productive functions of molecular dynamics simulations to resolve a broad variety of technical and production complications. The data in this book presents the progress of molecular dynamics recreations in various fields of science and technology, such as nanotechnology, energy research, and environmental science, due to the progress of new dynamics conjectures and the astonishing power of today's technology. This book begins by introducing molecular dynamics simulations, and then, demonstrates the computer research done by means of molecular dynamics simulations in the researches of artificial and natural macromolecules. This book also deals with a lot of other related topics, such as difficult development and dynamics of polymers like DNA, RNA, etc.

I would like to express my sincere thanks to the authors for their dedicated efforts in the completion of this book. I acknowledge the efforts of the publisher for providing constant support. Lastly, I would like to thank my family for their support in all academic endeavors.

Editor

Part 1

Dynamics of Polymers

1

Introduction to Molecular Dynamics Simulations: Applications in Hard and Soft Condensed Matter Physics

Martin Oliver Steinhauser
Research Group Shock Waves in Soft Biological Matter, Department of Composite Materials
Fraunhofer Institute for High-Speed Dynamics, Ernst-Mach-Institut,
EMI, Eckerstrasse 4, Freiburg
Germany

1. Introduction

Today, computer experiments play a very important role in science. In the past, physical sciences were characterized by an interplay between experiment and theory. In theory, a model of the system is constructed, usually in the form of a set of mathematical equations. This model is then validated by its ability to describe the system behavior in a few selected cases, simple enough to allow a solution to be computed from the equations. One might wonder why one does not simply derive all physical behavior of matter from an as small as possible set of fundamental equations, e.g. the Dirac equation of relativistic quantum theory. However, the quest for the fundamental principles of physics is not yet finished; thus, the appropriate starting point for such a strategy still remains unclear. But even if we knew all fundamental laws of nature, there is another reason, why this strategy does not work for ultimately predicting the behavior of matter on any length scale, and this reason is the growing complexity of fundamental theories – which are based on the dynamics of particles – when they are applied to systems of macroscopic (or even microscopic) dimensions. In almost all cases, even for academic problems involving only a few particles[1], a strict analytical solution is not possible and solving the problem very often implies a considerable amount of simplification. In contrast to this, in experiments, a system is subject to measurements, and results are collected, very often in the form of large data sets of numbers from which one strives to find mathematical equations describing the data by generalization, imagination and by thorough investigation. Very rarely, normally based on symmetries which allow inherent simplifications of the original problem, is an analytical solution at hand which describes exactly the evidence of the experiment given by the obtained data sets. Unfortunately, many academic and practical physical problems of interest do not fall under this category of "simple" problems, e.g. disordered systems, where there is no symmetry which helps to simplify the treatment.

The advent of modern computers which basically arose from the Manhattan project in the United States in the 1950s added another element to (classical) experiment and theory, namely

[1] In fact, already the three-body problem which involves three coupled ordinary differential equations is not solvable analytically.

the *computer experiment*. Some traditionalists working in theory, who have not followed the modern developments of computer science and its applications in the realm of physics, biology, chemistry and many more scientific fields, still repudiate the term "experiment" in the context of computer simulations. However, this term is most certainly fully justified!

In a computer experiment, a model is still provided by theory, but the calculations are carried out by the machine by following a series of instructions (the *algorithm*) usually coded in some high-level language and translated (*compiled*) into assembler commands which provide instructions how to manipulate the contents of processor registers. The results of computer simulations are just numbers, data which have to be interpreted by humans, either in the form of graphical output, as tables or as function plots. By using a machine to carry out the calculations necessary for solving a model, more complexity can be introduced and more realistic systems can be investigated.

Simulation is seen sometimes as theory, sometimes as experiment. On the one side, one is still dealing with models, not with "real systems"[2] On the other side, the procedure of verifying a model by computer simulation resembles an experiment very closely: One performs a run, then analyzes the results in pretty much the same way as an experimental physicist does. Simulations can come very close to experimental conditions which allows for interpreting and understanding the experiments at the microscopic level, but also for studying regions of systems which are not accessible in "real" experiments[3], too expensive to perform, or too dangerous. In addition, computer simulations allow for performing *thought experiments*, which are impossible to do in reality, but whose outcome greatly increases our understanding of fundamental processes or phenomena. Imagination and creativity, just like in mathematics[4], physics and other scientific areas, are very important qualities of a computer simulator!

From a principal point of view, theory is traditionally based on the reductionist approach: one deals with complexity by reducing a system to simpler subsystems, continuing until the subsystems are simple enough to be represented with solvable models. From this perspective one can regard simulation as a convenient tool to verify and test theories and the models associated with them in situations which are too complex to be handled analytically[5]. Here, one implies that the model represents the level of the theory where the interest is focused.

However, it is important to notice that simulation can play a more important role than just being a tool to be used as an aid to reductionism because it can be considered as an *alternative* to it. Simulation increases considerably the threshold of complexity which separates solvable und unsolvable models. One can take advantage of this threshold shift and move up several levels in our description of physical systems. Thanks to the presence of simulation, we do not need to work with models as simple as those used in the past. This gives the researcher an additional freedom for exploration. As an example, one could mention the interatomic potentials which, in the past, were obtained by two-body potentials with simple

[2] In this context one has to realize that often, in experiments, too, considerable simplifications of the investigated "real system" are done, e.g. when preparing it in a particular state in terms of pressure, temperature or other degrees of freedom.

[3] For example, systems at a pressure comparable to that in the interior of the sun.

[4] The famous David Hilbert once commented the question of what became of one of his students: "He became a writer - he didn't have enough imagination."

[5] For example, when computing the phase diagram of a substance modeled by a certain force law.

analytical form, such as Morse or Lennard-Jones. Today, the most accurate potentials contain many-body terms and are determined numerically by reproducing as closely as possible the forces predicted by ab-initio methods. These new potentials could not exist without simulation, so simulation is not only a connecting link between theory and experiment, but it is also a powerful tool to make progress in new directions. Readers, interested in these more "philosophical" aspects of computational science will be able to find appropriate discussions in the first chapters of refs. (Haile, 1992; Steinhauser, 2008; 2012). In the following, we focus on *classical* molecular dynamics (MD) simulations, i.e. a variant of MD which neglects any wave character of discrete atomic particles, describing them as classical in the Newtonian sense and not referring to any quantum mechanical considerations.

2. The objective of molecular dynamics simulations

Molecular dynamics computer experiments are done in an attempt of understanding the properties of assemblies of molecules in terms of their structure and the microscopic interactions between them. We provide a guess at the interactions between molecules, and obtain exact predictions of bulk properties. The predictions are "exact" in the sense that they can be made as accurate as we like, subject to the limitations imposed by our computer budget. At the same time, the hidden dynamic details behind bulk measurements can be revealed. An example is the link between the diffusion coefficient and the velocity autocorrelation function, with the latter being very hard to measure experimentally, but the former being very easy to measure. Ultimately one wants to make direct comparison with experimental measurements made on specific materials, in which case a good model of molecular interactions is essential. The aim of so-called ab-inito MD is to reduce the amount of guesswork and fitting in this process to a minimum. On the other hand, sometimes one is merely interested in phenomena of a rather generic nature, or one wants to discriminate between bad and good theories. In this case it is not necessary to have a perfectly realistic molecular model, but one that contains the essential physics may be quite suitable.

2.1 Molecular interactions

Classical MD simulation consists of the numerical, step-by-step solution of the classical Newtonian equations of motion, which for a simple atomic system may be written as

$$m_i \ddot{\vec{r}}_i = \vec{F}_i = -\frac{\partial}{\partial \vec{r}_i}\Phi, \tag{1}$$

where the vector symbol in the partial derivative is a physicsts' abbreviatory notation for the derivate of each individual component. To solve Eq. 1 one needs to calculate the forces $\vec{F}_i$ acting on the atoms, which are usually derived from a potential energy function $\Phi(\vec{r}^N)$, where $\vec{r}^N = (\vec{r}_1, \vec{r}_2, ..., \vec{r}_N)$ represents the complete set of $3N$ atomic coordinates.

2.1.1 Bonded interactions

Using the notion of intermolecular potentials acting between the particles of a system one cannot only model fluids made of simple spherically symmetric particles but also more complex molecules with internal degrees of freedom (due to their specific monomer connectivity). If one intends to incorporate all aspects of the chemical bond in complex molecules one has to treat the system with quantum chemical methods. Usually, one considers the internal degrees of freedom of polymers and biomacromolecules by using generic potentials that describe bond lengths l_i, bond angles θ and torsion angles ϕ. When neglecting the fast electronic degrees of freedom, often bond angles and bond lengths can be assumed to be constants. In this case, the potential includes lengths l_0 and angles θ_0, ϕ_0 at equilibrium about which the molecules are allowed to oscillate, and restoring forces which ensure that the system attains these equilibrium values on average. Hence the bonded interactions Φ_{bonded} for polymeric macromolecular systems with internal degrees of freedom can be treated by using some or all parts of the following potential term:

$$\Phi_{bonded}(r,\theta,\phi) = \frac{\kappa}{2}\sum_i (\mid \vec{r}_i - \vec{r}_{i-1} - l_0 \mid)^2 + \frac{k_\theta}{2}\sum_k (\theta_k - \theta_0)^2 + \frac{\beta}{2}\sum_m (\phi_m - \phi_0)^2. \tag{2}$$

Here, the summation indices sum up the number of bonds i at positions $\vec{r}_i$, the number of bond angles k between consecutive monomers along a macromolecular chain and the number of torsion angles m along the polymer chain. A typical value of $\kappa = 5000$ ensures that the fluctuations of bond angles are very small (below 1%). The terms l_0, θ_0 and ϕ_0 are the equilibrium distance, bond angle and torsion angle, respectively.

In particular in polymer physics, very often a Finitely Extensible Non-linear Elastic (FENE) potential is used which - in contrast to a harmonic potential - restricts the maximum bond length of a polymer bond to a prefixed value R_0 (Steinhauser, 2005):

$$\Phi_{FENE}(r) = \begin{cases} -\frac{1}{2}\kappa R_0^2 \ \ln(1 - \frac{r^2}{R_0^2}) & r < R_0, \\ \infty & \text{otherwise.} \end{cases} \tag{3}$$

The FENE potential can be used instead of the first term on the right hand side of the bonded potential in Eq. 2. Figure 1 illustrates the different parameters which are used in the description of bonded interactions in Eq. 2. Further details on the use of potentials in macromolecular biology and polymer physics may be found in (Feller, 2000; Schlenkrich et al., 1996; Siu et al., 2008; Steinhauser, 2008).

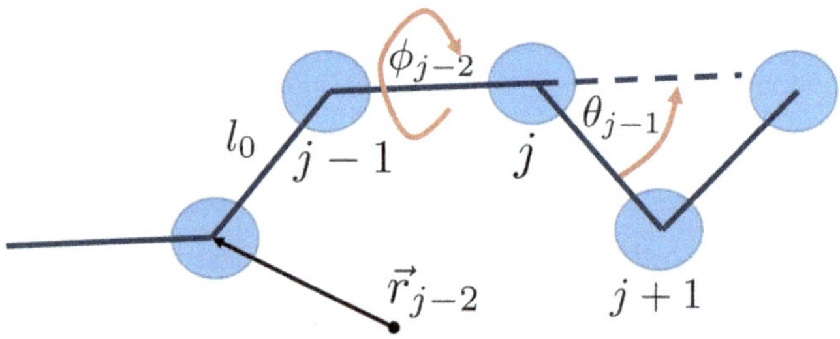

Fig. 1. Illustration of the potential parameters used for modeling bonded interactions.

2.1.2 Non-bonded interactions

Various physical properties are determined by different regions of the potential hypersurface of interacting particles. Thus, for a complete determination of potential curves, widespread experiments are necessary. For a $N-$body system the total energy Φ_{nb}, i.e. the potential hypersurface of the non-bonded interactions can be written as (Allen & Tildesly, 1991)

$$\Phi_{nb}(\vec{r}) = \sum_{i}^{N} \phi_1(\vec{r}_i) + \sum_{i}^{N}\sum_{j>i}^{N} \phi_2(\vec{r}_i, \vec{r}_j) + \sum_{i}^{N}\sum_{j>i}^{N}\sum_{k>j>i}^{N} \phi_3(\vec{r}_i, \vec{r}_j, \vec{r}_k) + \cdots, \quad (4)$$

where $\phi_1, \phi_2, \phi_3, \ldots$ are the interaction contributions due to external fields (e.g. the effect of container walls) and due to pair, triple and higher order interactions of particles. In classical MD one often simplifies the potential by the hypothesis that all interactions can be described by pairwise additive potentials. Despite this reduction of complexity, the efficiency of a MD algorithm taking into account only pair interactions of particles is rather low (of order $\mathcal{O}(N^2)$) and several optimization techniques are needed in order to improve the runtime behavior to $\mathcal{O}(N)$.

The simplest general Ansatz for a non-bonded potential for spherically symmetric systems, i.e. $\Phi(\vec{r}) = \Phi(r)$ with $r = |\vec{r}_i - \vec{r}_j|$ is a potential of the following form:

$$\Phi_{nb}(r) = \Phi_{Coulomb}(r) + \left(\frac{C_1}{r}\right)^{12} + \left(\frac{C_2}{r}\right)^{6}. \quad (5)$$

Parameters C_1 and C_2 are parameters of the attractive and repulsive interaction and the electrostatic energy $\Phi_{Coulomb}(r)$ between the particles with position vectors $\vec{r}_i$ and $\vec{r}_j$ is given by:

$$\Phi_{Coulomb}(r) = \frac{1}{\epsilon} k \cdot \sum_{i}\sum_{j>i} \frac{z_i z_j e^2}{|\vec{r}_i - \vec{r}_j|}. \quad (6)$$

The constant $k = 1$ in the cgs–system of units and ϵ is the dielectric constant of the medium, for example $\epsilon_{air} = 1$ for air, $\epsilon_{prot} = 4$ for proteins or $\epsilon_{H_20} = 82$ for water. The variables z_i and z_j denote the charge of individual monomers in the macromolecule and e is the electric charge of an electron.

The probably most commonly used form of the potential of two neutral atoms which are only bound by Van-der-Waals interactions, is the *Lennard-Jones (LJ), or (a-b) potential* which has the form (Haberland et al., 1995)

$$\Phi_{a,b}(r) = \alpha\varepsilon \left[\left(\frac{\sigma_0}{r}\right)^{a} + \left(\frac{\sigma_0}{r}\right)^{b}\right], \quad (7)$$

where

$$\alpha = \frac{1}{a-b}\left(\frac{a^a}{b^b}\right)^{\frac{1}{a-b}}, \; \Phi_{\min} = \varepsilon \text{ and } \Phi(\sigma) = 0. \quad (8)$$

The most often used LJ-(6-12) potential for the interaction between two particles with a distance $r = |\vec{r}_i - \vec{r}_j|$ then reads (cf. Eq. 5):

$$\Phi_{LJ}(r) = 4\varepsilon \left[\left(\frac{\sigma_0}{r}\right)^{12} + \left(\frac{\sigma_0}{r}\right)^{6}\right]. \quad (9)$$

Parameter ε determines the energy scale and σ_0 the length scale. In simulations one uses dimensionless *reduced units* which tend to avoid numerical errors when processing very small numbers, arising e.g. from physical constants such as the Boltzmann constant $k_B = 1.38 \times$

10^{-23}J/K. In these reduced (simulation) units, one MD timestep is measured in units of $\hat{\tau} = (m\sigma^2/\varepsilon)^{1/2}$, where m is the mass of a particle and ε and σ_0 are often simply set to $\sigma_0 = \varepsilon = k_B T = 1$. Applied to real molecules, for example to Argon with $m = 6.63 \times 10^{-23}kg, \sigma_0 \approx 3.4 \times 10^{-10}m$ and $\varepsilon/k_B \approx 120K$ one obtains a typical MD timestep of $\hat{\tau} \approx 3.1 \times 10^{-13}$s.

Using an exponential function instead of the repulsive r^{-12} term, one obtains the *Buckingham potential* (Buckingham, 1938):

$$\Phi(r) = b\exp(-ar) - \frac{c}{r^6} - \frac{d}{r^8}. \tag{10}$$

This potential however has the disadvantage of using a numerically very expensive exponential function and it is known to be unrealistic for many substances at small distances r where it has to be modified accordingly.

For reasons of efficiency, a classical MD potential should be short-ranged in order to keep the number of force calculations between interacting particles at a minimum. Therefore, instead of using the original form of the potential in Eq. 9, which approaches 0 at infinity, it is common to use a modified form, where the potential is simply cut off at its minimum value $r = r_{\text{min}} = \sqrt[6]{2}$ and shifted to positive values by ε such that it is purely repulsive and smooth at $r = r_{\text{cut}} = \sqrt[6]{2}$:

$$\Phi_{LJ}^{\text{cut}}(r) = \begin{cases} 4\,\varepsilon\left\{\left(\frac{\sigma_0}{r}\right)^{12} - \left(\frac{\sigma_0}{r}\right)^{6}\right\} + \varepsilon & r \le 2^{1/6}\sigma_0, \\ 0 & \text{otherwise.} \end{cases} \tag{11}$$

Another extension of the potential in Eq. 9 is proposed in (Steinhauser, 2005) where a smooth attractive part is introduced again, in order to allow for including different solvent qualities of the solvent surrounding the polymer:

$$\Phi_{cos}(r) = \left[\frac{1}{2}\cdot\cos(\alpha r^2 + \beta) + \gamma\right]\varepsilon. \tag{12}$$

This additional term adds an attractive part to the potential of Eq. 11 and at the same time – by appropriately choosing parameters α, β and γ – keeps the potential cutoff at r_{cut} smooth. The parameters α, β and γ are determined analytically such that the potential tail of Φ_{cos} has zero derivative at $r = 2^{1/6}$ and at $r = r_{\text{cut}}$, while it is zero at $r = r_{\text{cut}}$ and has value γ at $r = 2^{1/6}$, where γ is the depth of the attractive part. Further details can be found in (Steinhauser, 2005). When setting $r_{\text{cut}} = 1.5$ one sets $\gamma = -$ and obtains α and β as solutions of the linear set of equations

$$2^{1/3}\alpha + \beta = \pi, \tag{13}$$

$$2.25\alpha + \beta = 2\pi. \tag{14}$$

The total unbounded potential can then be written as:

$$\Phi_{Total}(r,\lambda) = \begin{cases} \Phi_{LJ}^{\text{cut}}(r) - \lambda\varepsilon & 0 < r < 2^{1/6}\,\sigma_0, \\ \lambda\,\Phi_{cos}(r) & 2^{1/6}\,\sigma_0 \le r < r_{\text{cut}}, \\ \infty & \text{otherwise,} \end{cases} \tag{15}$$

where λ is a new parameter of the potential which determines the depth of the attractive part. Instead of varying the solvent quality in the simulation by changing temperature T directly (and having to equilibrate the particle velocities accordingly), one can achieve a phase transition in polymer behavior by changing λ accordingly, cf. Fig. 2.

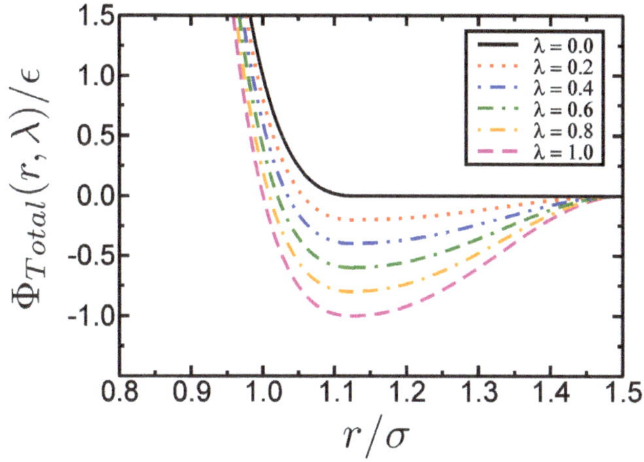

Fig. 2. Graph of the total unbounded potential of Eq. 15 which allows for modeling the effects of different solvent qualities.

Using coarse-grained models in the context of lipids and proteins, where each amino acid of the protein is represented by two coarse-grained beads, it has become possible to simulate lipoprotein assemblies and protein-lipid complexes for several microseconds (Shih et al., 2006).

The assumption of a short ranged interaction is usually fulfilled very well for all (uncharged) polymeric fluids. However, as soon as charged systems are involved this assumption breaks down and the calculation of the Coulomb force requires special numerical treatment due to its infinite range.

2.2 Calculation of forces

The most crucial part of a MD simulation is the force calculation. At least 95% of a MD code is spent with the force calculation routine which uses a search algorithm that determines interacting particle pairs. Therefore this is the task of a MD program which has to be optimized first and foremost. We will review a few techniques that have become standard in MD simulations which enhance the speed of force calculations considerably and speed up the algorithm from $\mathcal{O}(N^2)$ run time to $\mathcal{O}(N)$ run time. Starting from the original LJ potential between two particles i and j with distance $r = |\vec{r}_i - \vec{r}_j|$ of Eq. 7, one obtains the potential function for N interacting particles as the following double sum over all particles:

$$\Phi(\vec{r}_1, ..., \vec{r}_N) = \sum_{i=1}^{N} \sum_{j=i+1}^{N} \Phi_{LJ}(r) = 4\varepsilon \sum_{i=1}^{N} \sum_{j=i+1}^{N} \left(\frac{\sigma_0}{r}\right)^6 \times \left(\left(\frac{\sigma_0}{r}\right)^6 - 1\right). \quad (16)$$

The corresponding force $\vec{F}_i$ exerted on particle i by particle j is given by the gradient with respect to $\vec{r}_i$ as:

$$\vec{F}_i = -\nabla_{\vec{r}_i}\Phi_{LJ}(\vec{r}_1, ..., \vec{r}_N) = -24 \times \varepsilon \sum_{j=1, j\neq i}^{N} \frac{1}{r^2} \times \left(\frac{\sigma_0}{r}\right)^6 \times \left(1 - 2 \times \left(\frac{\sigma_0}{r}\right)^6\right) \vec{r}_{ij}, \quad (17)$$

where $\vec{r}_{ij} = (\vec{r}_i - \vec{r}_j)$ is the direction vector between particles i and j at positions $\vec{r}_i$ and $\vec{r}_j$, and $r = |\vec{r}_i - \vec{r}_j|$. Hence, in general, the force $\vec{F}_i$ on particle i is the sum over all forces $\vec{F}_{ij} := -\nabla_{\vec{r}_i}\Phi$ between particle i and all other particles j:

$$\vec{F}_i = \sum_{i=1, j\neq i}^{N} \vec{F}_{ij}. \quad (18)$$

The least favorable method of looking for interacting pairs of particles and for calculating the double sum in Eqs. 16 and 17 is the "brute force" method that simply involves taking a double loop over all particles in the (usually) cubic simulation box, thus calculating $\frac{1}{2}N(N-1)$ interactions with a N^2 efficiency. This algorithm becomes extremely inefficient for systems of more than a few thousand particles, cf. Fig. 3(a).

2.3 The MD algorithm

The last decade has seen a rapid development in our understanding of numerical algorithms which have been summarized in a recent book (Steinhauser, 2008) that presents the current state of the field.

When introducing an N-dimensional position vector $\vec{r}^N = (\vec{r}_1, \vec{r}_2, ..., \vec{r}_N)$, the potential energy $\Phi(\vec{r}^N)$ and the momenta $\vec{p}^N = (\vec{p}_1, \vec{p}_2, ..., \vec{p}_N)$, in terms of which the kinetic energy may be written as $K(\vec{p}^N)$, then the total energy H of a classical conservative system is given by $H = \Phi + K$. The equations of motion determining all dynamics of the particles can be written as

$$\dot{\vec{r}}_i = \vec{p}_i / m_i \quad and \quad \dot{\vec{p}}_i = \vec{F}_i. \quad (19)$$

This is a system of coupled ordinary differential equations. Many methods exist to solve this set of equations numerically, among which the so-called *velcity Verlet-algorithm* is the one that is the most used. This algorithm integrates the equations of motion by performing the following four steps, where $\vec{r}_i$, $\vec{v}_i$, $\vec{a}_i = \vec{T}_i / m_i$ are the position, velocity and acceleration of the i-th particle, respectively:

$$\text{Calculate } \vec{v}_i(t + \frac{1}{2}\delta t) = \vec{v}_i(t) + \frac{1}{2}\delta t \vec{F}_i(t), \quad (20)$$

$$\text{Calculate } \vec{r}_i(t + \delta t) = \vec{r}_i(t) + \delta t \vec{v}_i(t + \frac{1}{2}\delta t), \quad (21)$$

$$\text{Derive } \vec{a}_i(t + \delta t) \quad \text{from the interaction potential using } \vec{r}(t + \delta t), \quad (22)$$

$$\text{Calculate } \vec{v}_i(t + \delta t) = \vec{v}_i(t + \frac{1}{2}\delta t) + + \frac{1}{2}\delta t \vec{a}_i(t + \delta t). \quad (23)$$

Further details about this standard algorithm can be found elsewhere (Steinhauser, 2008). It is exactly time reversible, symplectic, low order in time (hence permitting large timesteps), and it requires only one expensive force calculation per timestep.

2.3.1 Neighbor lists

In general, in molecular systems, the potential as well as the corresponding force decays very fast with the distance r between the particles. Thus, for reasons of efficiency, in molecular simulations one often uses the modified LJ potential of Eq. 11 which introduces a cutoff r_{cut} for the potential. The idea here is to neglect all contributions in the sums in Eqs. 16 and 17 that are smaller than the threshold r_{cut} which characterizes the range of the interaction. Thus, in this case the force $\vec{F}_i$ on particle i is approximated by

$$\vec{F}_i \approx -24 \times \varepsilon \sum_{\substack{j=1,j\neq i \\ 0<r\leq r_{\text{cut}}}}^{N} \frac{1}{r^2} \times \left(\frac{\sigma_0}{r}\right)^6 \times \left(1 - 2 \times \left(\frac{\sigma_0}{r}\right)^6\right) \vec{r}_{ij}. \tag{24}$$

Contributions to the force on particle i that stem from particles j with $r \leq r_{\text{cut}}$ are neglected. This introduces a small error in the computation of the forces and the total energy of the system, but it reduces the overall computational effort from $\mathcal{O}(N^2)$ to $\mathcal{O}(N)$. For systems with short-ranged or rapidly decaying potentials, a very efficient algorithm for the search of potentially interacting particles, i.e. those particles that are within the cutoff distance r_{cut} of a particle i, has been developed (Hockney, 1970). In MD this algorithm can be implemented most efficiently by geometrically dividing the volume of the (usually cubic) simulation box into small cubic cells whose sizes are slightly larger than the interaction range r_{cut} of particles, cf. Fig. 3b. The particles are then sorted into these cells using the linked-cell algorithm (LCA). The LCA owes its name to the way in which the particle data are arranged in computer memory, namely as linked list for each cell. For the calculation of the interactions it is then sufficient to calculate the distances between particles in neighboring cells only, since cells which are further than one cell apart are by construction beyond the interaction range. Thus, the number of distance calculations is restricted to those particle pairs of neighboring cells only which means that the sums in Eq. 18 are now split into partial sums corresponding to the decomposition of the simulation domain into cells. For the force $\vec{F}_i$ on particle i in cell number n one obtains a sum of the form

$$\vec{F}_i = \sum_{\substack{\text{cell } m \\ m\in\Omega(n)}} \sum_{\substack{j\in\{\text{all particles in cell } m\} \\ j\neq i}} \vec{F}_{ij}, \tag{25}$$

where $\Omega(n)$ denotes cell n itself together with all cells that are direct neighbors of cell n. The linked-cell algorithm is a simple loop over all cells of the simulation box. For each cell there is a linked list which contains a root pointer that points to the first particle in the respective cell which then points to the next particle within this particular cell, until the last particle is reached which points to zero, indicating that all particles in this cell have been considered. Then the algorithm switches to the root pointer of the next cell and the procedure is repeated until all interacting cells have been considered, cf. Fig. 3.

Assuming the average particle density in the simulation box as $\langle\rho\rangle$ then the number of particles in each one of the subcells is $\langle\rho\rangle\, r_{\text{cut}}^3$. The total number of subcells is $N/\,\langle\rho\rangle\, r_{\text{cut}}^3$ and

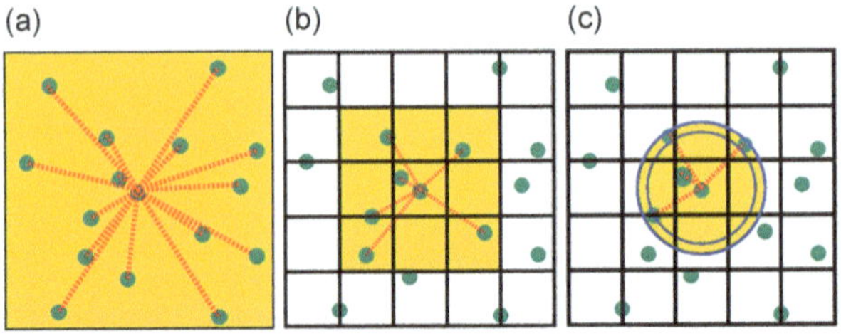

Fig. 3. MD Optimization schemes for the search of potentially interacting particles. (a) The least efficient all particle "brute force" approach with run time $\mathcal{O}(N^2)$ (b) The linked-cell algorithm which reduces the search effort to $\mathcal{O}(N)$. (c) The linked-cell algorithm combined with neighbor lists which further reduces the search effort by using a list of potentially interacting neighbor particles which can be used for several timesteps before it has to be updated. In this 2D representation, the radius of the larger circle is $r_{\text{cut}} + r_{skin}$ and the inner circle, which contains actually interacting particles, has radius r_{cut}.

the total number of neighbor cells of each subcell is 26 in a cubic lattice in three dimensions (3D). Due to Newton's third law only half of the neighbors actually need to be considered. Hence, the order to which the linked-cell algorithm reduces the search effort is given by:

$$\frac{26}{2}\left(\langle\rho\rangle\, r_{\text{cut}}^2\right)\frac{N}{\langle\rho\rangle\, r_{\text{cut}}^3} = 13(\langle\rho\rangle\, r_{\text{cut}}^3)^2 N. \tag{26}$$

For this method to function, the size of the simulation box has to be at least $3r_{\text{cut}}$, cf. Fig. 3. For simulations of dense melts with many particles, this requirement is usually met. Consequently, by this method, the search-loop effort is reduced to $\mathcal{O}(N)$, but with a pre-factor that still can be very large, depending on the density of particles $\langle\rho\rangle$ and the interaction range r_{cut}.

2.3.2 Boundary conditions

In a MD simulation only a very small number of particles can be considered. To avoid the (usually) undesired artificial effects of surface particles which are not surrounded by neighboring particles in all directions and thus are exerted to non-isotropic forces, one introduces *periodic boundary conditions*. Using this technique, one measures the "bulk" properties of the system, due to particles which are located far away from surfaces. As a rule, one uses a cubic simulation box were the particles are located. This cubic box is periodically repeated in all directions. If, during a simulation run, a particle leaves the central simulation box, then one of its image particles enters the central box from the opposite direction. Each of the image particles in the neighboring boxes moves in exactly the same way, cf. Fig. 4 for a two dimensional visualization.

The cubic box is used almost exclusively in simulations with periodic boundaries, mainly due to its simplicity, however also spherical boundary conditions have been investigated were the three-dimensional surface of the sphere induces a non-Euclidean metric. The use of periodic boundary conditions allows for the simulation of bulk properties of systems with a relatively small number of particles.

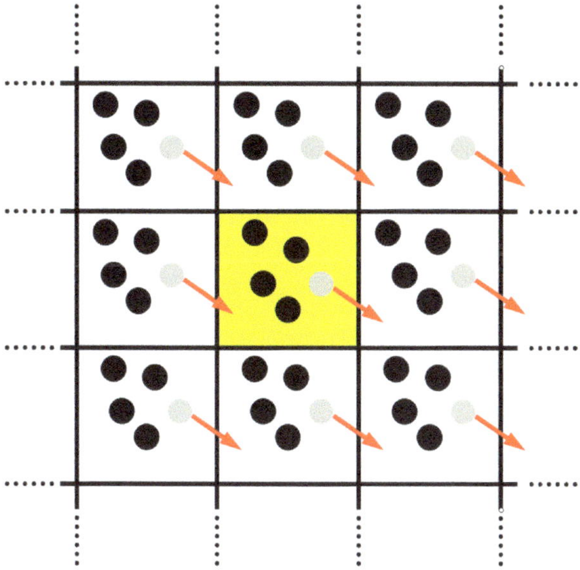

Fig. 4. Two-dimensional schematic of periodic boundary conditions. The particle trajectories in the central simulation box are copied in every direction.

2.3.3 Minimum image convention

The question whether the measured properties with a small, periodically extended system are to be regarded as representative for the modeled system depends on the specific observable that is investigated and on the range of the intermolecular potential. For a LJ potential with cut-off as in Eq. 11 no particle can interact with one of its images and thus be exposed to the artificial periodic box structure which is imposed upon the system. For long range forces, also interactions of far away particles have to be included, thus for such systems the periodic box structure is superimposed although they are actually isotropic. Therefore, one only takes into account those contributions to the energy of each one of the particles which is contributed by a particles that lies within a cut-off radius that is at the most $1/2L_B$ with boxlenth L_B. This procedure is called *minimum image convention*. Using the minimum image convention, each particle interacts with at the most $(N-1)$ particles. Particularly for ionic systems a cut-off has to be chosen such that the electro-neutrality is not violated. Otherwise, particles would start interacting with their periodic images which would render all calculations of forces and energies erroneous.

3. Complex formation of charged macromolecules

A large variety of synthetic and biological macromolecules are polyelectrolytes (Manning, 1969). The most well-known polyelectrolyte biopolymers, proteins, DNA and RNA, are responsible for functions in living systems which are incomparably more complex and diverse than the functions usually discussed for synthetic polymers present in the chemical industry. For example, polyacrylic acid is the main ingredient for diapers and dispersions of copolymers of acrylamide or methacrylamide and methacrylic acid are fundamental for cleaning water. In retrospect, during the past 30 years, despite the tremendous interest in polyelectrolytes, unlike neutral polymers (de Gennes, 1979; Flory, 1969), the general understanding of the behavior of electrically charged macromolecules is still rather poor. The contrast between our understanding of neutral and charged polymers results form the long range nature of the electrostatic interactions which introduce new length and time scales that render an analytical treatment beyond the Debye-Hückel approximation very complicated (Barrat & Joanny, 2007; Debye & Hückel, 1923). Here, the traditional separation of scales, which allows one to understand properties in terms of simple scaling arguments, does not work in many cases. Experimentally, often a direct test of theoretical concepts is not possible due to idealizing assumptions in the theory, but also because of a lack of detailed control over the experimental system, e.g. in terms of the molecular weight. Quite recently, there has been increased interest in hydrophobic polyelectrolytes which are water soluble, covalently bonded sequences of polar (ionizable) groups and hydrophobic groups which are not (Khoklov & Khalatur, 2005). Many solution properties are known to be due to a complex interplay between short-ranged hydrophobic attraction, long-range Coulomb effects, and the entropic degrees of freedom. Hence, such polymers can be considered as highly simplified models of biologically important molecules, e.g. proteins or lipid bilayers in cell membranes. In this context, computer simulations are a very important tool for the detailed investigation of charged macromolecular systems. A comprehensive review of recent advances which have been achieved in the theoretical description and understanding of polyelectrolyte solutions can be found in (Holm et al., 2004).

3.1 Two oppositely charged macromolecules

The investigation of aggregates between oppositely charged macromolecules plays an important role in technical applications, particularly in biological systems. For example, DNA is associated with histone proteins to form the chromatin. Aggregates of DNA with cationic polymers or dendrimers are discussed in the context of their possible application as DNA vectors in gene therapies (Gössl et al., 2002; Yamasaki et al., 2001). Here, we present MD simulations of two flexible, oppositely charged polymer chains and illustrate the universal scaling properties of the resulting polyelectrolyte complexes that are formed when the chains collapse and build compact, cluster-like structures which are constrained to a small region in space (Steinauser, 1998; Winkler et al., 2002). The properties are investigated as a function of chain length N and interaction strength ξ. Starting with Eq. 5 and using $k = 1$ (cgs-system of units) the dimensionless interaction parameter

$$\xi = \xi_B k_B T / \epsilon\sigma \tag{27}$$

can be introduced, where the Bjerrum length ξ_B is given by:

$$\xi_B = e^2 / \epsilon k_B T, \tag{28}$$

where k_B is the Boltzmann constant, T is temperature, ϵ is the energy scale from the Lennard-Jones potential of Eq. 11, σ defines the length scale (size of one monomer) and e is the electronic charge.

The interaction parameter for the here presented study is chosen in the range of $\xi = 0, ..., 100$ which covers electrically neutral chains ($\xi = 0$) in good solvent as well as highly charged chain systems ($\xi = 100$). The monomers in the chains are connected by harmonic bonds

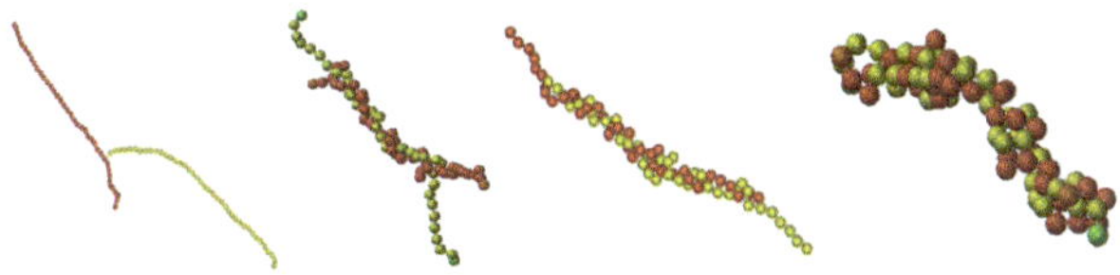

Fig. 5. Twisted, DNA-like polyelectrolyte complexes formed by electrostatic attraction of two oppositely charged linear macromolecules with $N = 40$ at different time intervals $\tau = 0$ (a), $\tau = 10500$ (b), $\tau = 60000$ (c) and $\tau = 120000$ (d), where τ is given in reduced Lennard-Jones units (Allen & Tildesly, 1991). The interaction strength is $\xi = 8$ (Steinauser, 1998; Winkler et al., 2002).

using the first term of the bonded potential of Eq. 2. The interaction with the solvent is taken into account by a stochastic force ($\vec{\Gamma}_i$) and a friction force with a damping constant χ, acting on each mass point. The equations of motion of the system are thus given by the Langevin equations

$$m\ddot{\vec{r}}_i = \vec{F}_i - \chi m \dot{\vec{r}}_i + \vec{\Gamma}_i. \tag{29}$$

The force $\vec{F}_i$ comprises the force due to the sum of the potentials of Eq. 11 with cutoff $r_{\text{cut}} = 1.5$, Eq. 6 with $k = 1$, $z_{i/j} = \pm 1$, and the first term on the right-hand side of the bonded potential in Eq. 2 with $\kappa = 5000\varepsilon/\sigma$ and bond length $l_0 = \sigma_0 = 1.0$. The stochastic force $\vec{\Gamma}_i$ is assumed to be stationary, random, and Gaussian (white noise). The electrically neutral system is placed in a cubic simulation box and periodic boundary conditions are applied for the intermolecular Lennard-Jones interaction according to Eq. 11, thereby keeping the density $\rho = N/V = 2.1 \times 10^{-7}/\sigma^3$ constant when changing the chain length N. The number of monomers N per chain was chosen as $N = 10, 20, 40, 80$ and 160 so as to cover at least one order of magnitude. For the Coulomb interaction a cutoff that is half the boxlength $r_{\text{cut}} = 1/2L_B$ was chosen. This can be done as the eventually collapsed polyelectolyte complexes which are analyzed are confined to a small region in space which is much smaller than r_{cut}. In the following we discuss exemplarily some scaling properties of charged linear macromolecules in the collapsed state. The simulations are started with two well separated and equilibrated chains in the simulation box. After turning on the Coulomb interactions with opposite charges $z_{i/j} = \pm 1$ in the monomers of both chains, the chains start to attract each other. In a first step during the aggregation process the chains start to twist around each other and form helical like structures as presented in Fig. 5. In a second step, the chains start to form a compact globular structure because of the attractive interactions between dipoles formed by oppositely charged monomers, see the snapshots in Fig. 6(a).

Figure 6(a) exhibits the universal scaling regime of R_g obtained for intermediate interaction strengths ξ and scaled by $(N-1)^{2/3}$. Here, the data of various chain lengths fall nicely on top of each other. This scaling corresponds to the scaling behavior of flexible chains in a bad solvent and is also in accordance with what was reported by Shrivastava and Muthukumar (Srivastava & Muthukumar, 1994). The change of R_g is connected with a change of the density ρ of the polyelectrolyte aggregate. However, in Fig. 6(b), which presents an example of ρ for $\xi = 4$, only a slight dependence of the density on the chain length N can be observed. ρ measures the radial monomer density with repsect to the center of mass of the total system. For longer chains, there is a plateau while for short chains there is a pronounced maximum of the density for small distances from the center of mass. While this maximum

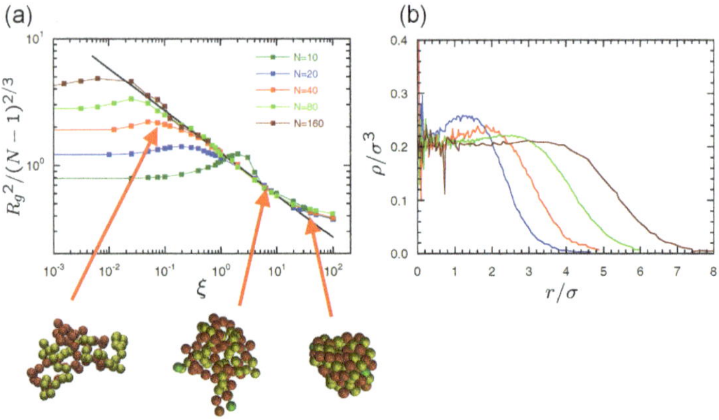

Fig. 6. (a) Radii of gyration as a function of the interaction strength ξ for various chain lengths according to (Steinhauser, 2008; Winkler et al., 2002). The radius of gyration $R_g{}^2$ which is measure for the size of a polymer chain is scaled by $(N-1)^{2/3}$, where $(N-1)$ is the number of bonds of a single chain. Also displayed are sample snapshots of the collapsed globules with $N = 40$ and interaction strengths $\xi = 0.4, 4, 40$. (b) Radial density of monomers with respect to the center of mass of a globule and interaction strength $\xi = 4$ for different chain lengths, $N = 20$ (blue), $N = 40$ (red), $N = 80$ (green) and $N = 160$ (brown).

vanished with decreasing ξ it appears also at higher interaction strengths for longer chains. Monomers on the outer part of the polyelectrolyte complex experience a stronger attraction by the inner part of the cluster than the monomers inside of it, and for smaller ξ, chains of different lengths are deformed to different degrees which leads to a chain length dependence of the density profile.

4. Equilibrium and Non-Equilibrium Molecular Dynamics (NEMD)

An understanding of the behavior of fully flexible linear polymers in dilute solutions and of dense melts has been achieved decades ago by the fundamental works of Rouse and Zimm (Zimm, 1956), as well as of Doi and Edwards (Doi & Edwards, 1986) and deGennes (de Gennes, 1979). In contrast to the well understood behavior of fully flexible linear polymers in terms of the Rouse (Prince E. Rouse, 1953) and the reptation model (de Gennes, 1979; Doi & Edwards, 1986), semiflexible polymer models have received increasing attention recently, as on the one hand they can be applied to many biopolymers like actin filaments, proteins or DNA (Käs et al., 1996; Ober, 2000) and on the other hand even for polymers considered flexible, like Polyethylene, the Rouse model fails as soon as the local chemical structure can no longer be neglected. One of the effects of this structure is a certain stiffness in the polymer chain due to the valence angles of the polymer backbone (Paul et al., 1997). Thus, semiflexible polymers are considerably more difficult to treat theoretically, as they require the fulfillment of additional constraints, such as keeping the total chain length fixed, which render these models more complex and often nonlinear. A deeper understanding of the rheological, dynamical and structural properties of semiflexible or even rod-like polymers is thus of great practical and fundamental interest.

4.1 Theoretical framework

The Kratky-Porod chain model (or worm-like chain model, WLC) (Kratky & Porod, 1949) provides a simple description of inextensible semiflexible polymers with positional fluctuations that are not purely entropic but governed by their bending energy Φ_{bend} and characterized e.g. by their persistence length L_p. The corresponding elastic energy

$$\Phi_{\text{bend}} = \frac{\kappa}{2}\int_0^L ds \left(\frac{\partial^2 \mathbf{r}}{\partial s^2}\right)^2 \tag{30}$$

of the inextensible chain of length L depends on the local curvature of the chain contour s, where $\vec{r}(s)$ is the position vector of a mass point (a monomer) on the chain and κ is a constant (Doi & Edwards, 1986).

Harris and Hearst formulated an equation of motion for the WLC model by applying Hamilton's principle with the constraint that the second moment of the total chain length be fixed and obtained the following expressions for the bending $\vec{F}_{\text{bend}}$ and tension forces $\vec{F}_{\text{tens}}$

$$\vec{F}_{\text{bend}} = \mu \frac{\partial^4 \vec{r}}{\partial s^4}, \tag{31}$$

$$\vec{F}_{\text{tens}} = \omega \frac{\partial^2 \vec{r}}{\partial s^2}. \tag{32}$$

Applying this result to elastic light scattering, this model yields correct results in the flexible coil limit (Harris & Hearst, 1966), but it fails at high stiffness, where it deviates from the solution obtained for rigid rods (Harris & Hearst, 1967).

A different model was proposed by Soda (Soda, 1973), where the segmental tension forces are modeled by stiff harmonic springs. This approach avoids large fluctuations in the contour length but has the disadvantage that an analytic treatment of the model is possible only for few limiting cases. Under the assumption that the longitudinal tension relaxes quickly, the bending dynamics can be investigated using a normal mode analysis (Aragón & Pecora, 1985; Soda, 1991). However, this approach cannot account for the flexible chain behavior which is observed on large length scales in the case $L_p \ll L$.

Winkler, Harnau and Reineker (Harnau et al., 1996; R.G. Winkler, 1994) considered a Gaussian chain model and used a Langevin equation similar to the equation employed by Harris and Hearst, but introduced separate Lagrangian multipliers for the end points of the chain, thus avoiding the problems of the Harris and Hearst equation in the rod-like limit. Thus, the equation used in (Harris & Hearst, 1966) is contained in the model used by Winkler, Harnau and Reineker and can be regained by setting all Lagrangian multipliers equal along the chain contour and at the end points. The expansion of the position vector $\vec{r}(s)$ in normal coordinates in the approach used in (R.G. Winkler, 1994) and resolving the obtained equations for the relaxation times τ_p or the normal mode amplitudes $X_p(t)$ leads to a set of transcendental equations, the solution of which cannot be given in closed form. For some limiting cases, Harnau, Winkler and Reineker showed the agreement of the approximate solution of the transcendental equations with atomistic simulation results of a n-$C_{100}H_{202}$ polymer melt (Harnau et al., 1999) that were performed by Paul, Yoon and Smith (Paul et al., 1997).

In contrast to fully flexible polymers, the modeling of *semiflexible* and *stiff* macromolecules has received recent attention, because such models can be successfully applied to biopolymers such as proteins, DNA, actin filaments or rodlike viruses (Bustamante et al., 1994; Ober, 2000). Biopolymers are wormlike chains with persistence lengths l_p (or Kuhn segment lengths l_K) comparable to or larger than their contour length L and their rigidity and relaxation behavior are essential for their biological functions.

4.2 Modeling and simulation of semiflexible macromolecules

Molecular Dynamics simulations were performed using the MD simulation package "MD-Cube", which was originally developed by Steinhauser (Steinhauser, 2005). A coarse-grained bead-spring model with excluded volume interactions as a model for dilute solutions of polymers in solvents of varying quality, respectively for polymer melts, is employed (Steinhauser, 2008; Steinhauser & et. al., 2005). A compiler switch allows for turning on and off the interaction between different chains. Thus, one can easily switch the type of simulation from single polymers in solvent to polymer melts. The excluded volume for each monomer is taken into account through the potential of Eq. 11.

Neighboring mass points along the chains are connected by harmonic bonds with the following potential for the bonded interactions

$$\Phi_{\text{bonded}}(r) = \frac{K}{2}(r - d_0)^2 \,, \tag{33}$$

which is often used in polymer simulations of charged, DNA-like biopolymers, see e.g. (Steinauser, 1998; Winkler et al., 2002). Note, that the potential in Eq. 33 corresponds to the first term on the right-hand side of Eq. 2. In order to keep fluctuations of the bond lengths and thus the fluctuations of the overall chain length L small (below 1%), a large value for the force constant $K = 10000\varepsilon/\sigma^2$ is chosen, where ε and σ are parameters of the truncated LJ-potential in Eq. 11. The average bondlength d_0 is taken to be 0.97σ which is the equilibrium distance of a potential that is composed of the FENE (Finite Extensible Non-Linear Elastic) potential – which is frequently employed in polymer simulations (Steinhauser, 2005) – and the truncated LJ-potential of Eq. 11. In combination with the LJ-potential this particle distance keeps the chain segments from artificially crossing each other (Steinhauser, 2008). The FENE potential exhibits very large fluctuations of bond lengths which are unrealistic for the investigation of semiflexible or stiff polymers. This is the reason for choosing the simple harmonic potential in Eq. 33). It is noted, that in principle the exact analytic form of the bonded potential when using a coarse-grained polymer model is actually irrelevant as long as it ensures that the the basic properties of polymers are modeled correctly such as the specific connectivity of monomers in a chain, the non-crossability of monomer segments (topological constraints), or the flexibility/stiffness of a chain. Thus, very often, a simple potential that can be quickly calculated in a numerical approach is used.

The stiffness, i.e. the bending rigidity of the chains composed of N mass points, is introduced into the coarse-grained model by the following bending potential

$$\Phi_{\text{bend}} = \frac{\kappa}{2}\sum_{i=1}^{N-1}(\vec{u}_{i+1} - \vec{u}_i)^2 = \kappa\sum_{i=1}^{N-1}(1 - \vec{u}_{i+1}\cdot\vec{u}_i) \,, \tag{34}$$

where $\vec{u}$ is the unit bond vector $\vec{u}_i = (\vec{r}_{i+1} - \vec{r}_i)/|\,\vec{r}_{i+1} - \vec{r}_i\,|$ connecting consecutive monomers, and $\vec{r}_i$ is the position vector to the i-th monomer. The total force acting on monomer i is thus given by

$$\vec{F}_i = -\partial/\partial\vec{r}_i\Phi_{\text{total}} = -\partial/\partial\vec{r}_i\left(\Phi_{\text{LJ}}^{\text{cut}} + \Phi_{\text{bonded}} + \Phi_{\text{bend}}\right) . \tag{35}$$

Assuming a NVT ensemble at temperature $k_B T = 1\varepsilon$, the trajectories of all particles $i = (1, ..., N)$ are generated by integrating the stochastic equations of motion

$$\frac{d^2\vec{r}_i}{dt^2} = -\xi\frac{d\vec{r}_i}{dt} + \frac{1}{m}\vec{F}_i + \frac{1}{m}\vec{F}_i^S , \tag{36}$$

with a particle friction coefficient ξ and a Gaussian stochastic force $\vec{F}_i^S$ that satisfies

$$< \vec{F}_i^S(t) >= 0 , \tag{37}$$

and the fluctuation-dissipation-theorem

$$< F_i^S(t)F_j^S(t') >= 2kT\xi\delta_{ij}\delta(t - t') . \tag{38}$$

The equations of motion are integrated using the Brownian dynamics algorithm proposed by van Gunsteren and Berendsen for a canonical ensemble (van Gunsteren & Berendsen, 1982) which – for vanishing particle friction ξ – changes into the velocity Verlet algorithm for a microcanonical ensemble. The algorithm is used with a constant timestep of $\Delta t = 5 \times 10^{-3}\tau$, where τ is the time unit of the simulation.

4.2.1 Results

The crossover scaling from coil-like, flexible structures on large length scales to stretched conformations at smaller scales can be seen in the scaling of the structure function $S(q)$ when performing simulations with different values of k_θ (Steinhauser, Schneider & Blumen, 2009).

In Fig. 7(a) the structure functions of the simulated linear polymer chains of length $N = 700$ are displayed for different persistence lengths. The chains show a scaling according to q^ν. The stiffest chains exhibit a q^{-1}–scaling which is characteristic for stiff rods. The dotted and dashed lines display the expected theoretical scaling behavior.

Thus, by varying parameter k_θ, the whole range of bending stiffness of chains from fully flexible chains to stiff rods can be covered. The range of q–values for which the crossover from flexible to semiflexible and stiff occurs, shifts to smaller q–values with increasing stiffness k_θ of the chains. The scaling plot in Fig. 7(b) shows that the transition occurs for $q \approx 1/l_K$, i.e. on a length scale of the order of the statistical Kuhn length. In essence, only the fully flexible chains (red data points) exhibit a deviation from the master curve on large length scales (i.e. small q–values), which corresponds to their different global structure compared with semi-flexible macromolecules. Examples for snapshots of stiff and semiflexible chains are displayed in Fig. 8.

For a theoretical treatment, following Doi and Edwards (Doi & Edwards, 1986), we expand the position vector $\vec{r}(s,t)$ of a polymer chain, parameterized with time t and contour length s

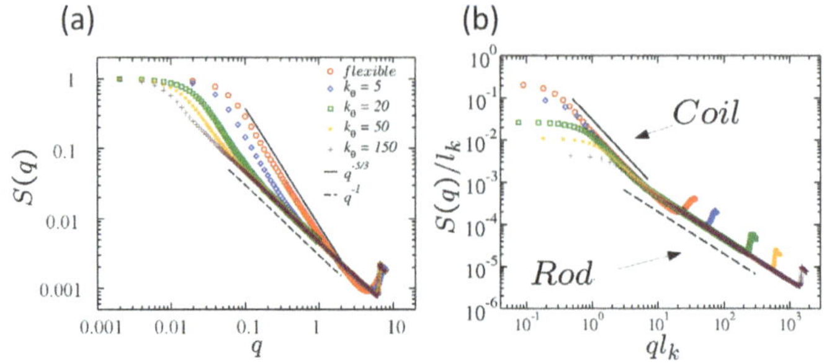

Fig. 7. (a) Structure function $S(q)$ of single linear chains with $N = 700$ and varying stiffness k_θ. The scaling regimes (fully flexible and stiff rod) are indicated by a straight and dashed line, respectively. (b) Scaling plot of $S(q)/l_K$ versus $q \cdot l_K$ using the statistical segment length l_K, adapted from (Steinhauser, Schneider & Blumen, 2009).

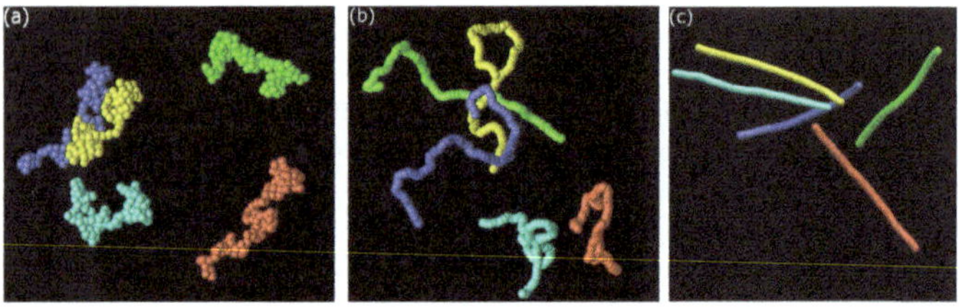

Fig. 8. Simulation snapshots of (a) flexible chains ($k_\theta = 0$), (b) semiflexible chains ($k_\theta = 20$), (c) stiff, rod-like chains ($k_\theta = 50$).

in normal modes $\vec{X}_p(t)$ as follows:

$$\vec{r}(s,t) = \vec{X}_0(t) + 2\sum_{p=1}^{\infty}(t)\cos\left(\frac{p\pi}{L}s\right). \tag{39}$$

Resolving Eq. 39 for the normal modes and inserting the result into the Langevin equations of motion for $\vec{r}(s,t)$ one obtains after some algebraic manipulations for the relaxation time τ_p

$$\tau_p = \frac{3k_B T\pi^2 p^2}{2L^2\xi}\left[\frac{1}{L_p} + \frac{L_p\pi^2}{NL}p^2\right], \tag{40}$$

which can be interpreted physically as arising from contributions due to an entropic force term $\propto p^2$ and a bending force term $\propto p^4$. In Fig. 9 we show that our simulation results for semiflexible chains scale according to Eq. 40.

Figure 10 exhibits the results of a NEMD step-shear simulation, from which the shear modulus $G(t)$ has been determined. The NEMD scheme produces the same results for the shear modulus $G(t)$ as conventional methods at equilibrium, which are based on the Green-Kubo equation.

Finally, in Fig. 11 we illustrate our step-shear simulation scheme for polymer melts with two corresponding NEMD simulation snapshots.

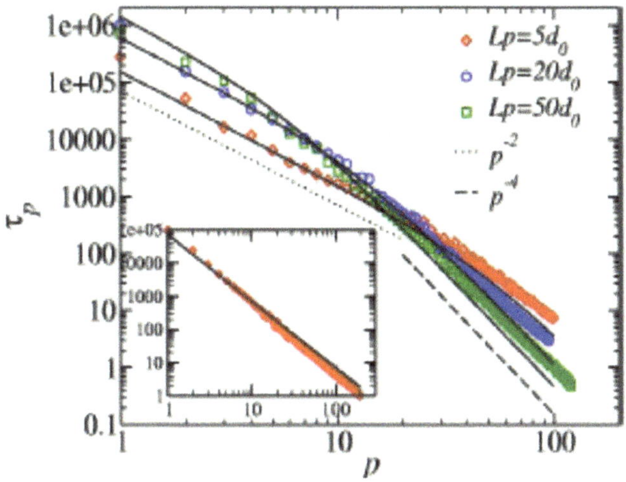

Fig. 9. Scaling of relaxation time τ_p for semiflexible chains ($N = 700$) with different L_p. The dotted and dashed lines show the scaling behavior according to Eq. 40. The inset shows τ_p of our simulated flexible chains compared with the Rouse model.

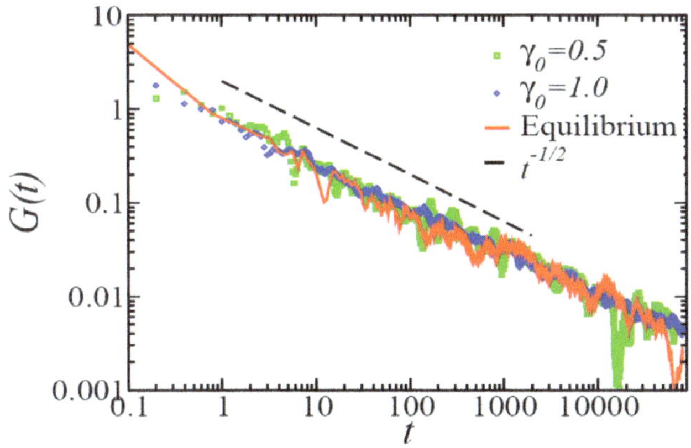

Fig. 10. Shear modulus $G(t)$ obtained from NEMD simulations compared with conventional (red line) equilibrium methods. The dashed line displays the expected Rouse scaling behavior and γ_0 displays the respective shear rates of the systems.

5. Shock wave failure of granular materials

In the following we discuss a recently proposed concurrent multiscale approach for the simulation of failure and cracks in brittle materials which is based on mesoscopic particle dynamics, the Discrete Element Method (DEM), but which allows for simulating macroscopic properties of solids by fitting only a few model parameters (Steinhauser, Grass, Strassburger & Blumen, 2009).

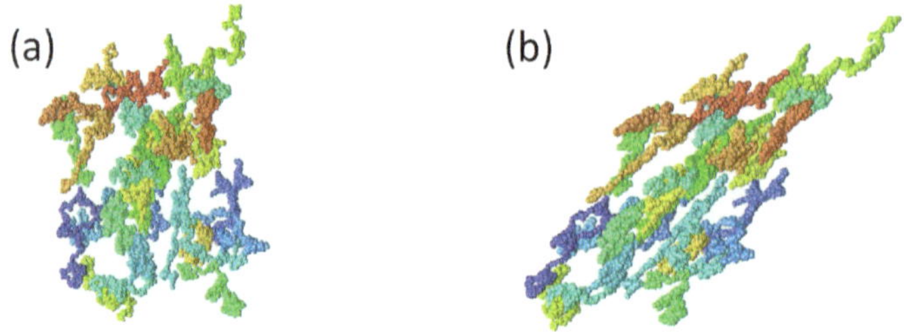

Fig. 11. Non-equilibrium shear-simulation of $N = 100$ polymer chains with $N = 100$ particles each (a) before and (b) after shearing the system. For reasons of clarity only 30 different chains of the system are displayed.

5.1 Introduction: Multiscale modeling of granular materials using MD

Instead of trying to reproduce the geometrical shape of grains on the microscale as seen in two-dimensional (2D) micrographs, in the proposed approach one models the macroscopic solid state with soft particles, which, in the initial configuration, are allowed to overlap, cf. Fig. 12(a). The overall system configuration, see Fig. 12(b), can be visualized as a network of links that connect the centers of overlapping particles, cf. Fig. 12(c).

The degree of particle overlap in the model is a measure of the force that is needed to detach particles from each other. The force is imposed on the particles by elastic springs. This simple model can easily be extended to incorporate irreversible changes of state such as plastic flow in metals on the macro scale. However, for brittle materials, where catastrophic failure occurs after a short elastic strain, in general, plastic flow behavior can be completely neglected. Additionally, a failure threshold is introduced for both, extension and compression of the springs that connect the initial particle network. By adjusting only two model parameters for the strain part of the potential, the correct stress-strain relationship of a specific brittle material as observed in (macroscopic) experiments can be obtained. The model is then applied to other types of external loading, e.g. shear and high-speed impact, with no further model adjustments, and the results are compared with experiments performed at EMI.

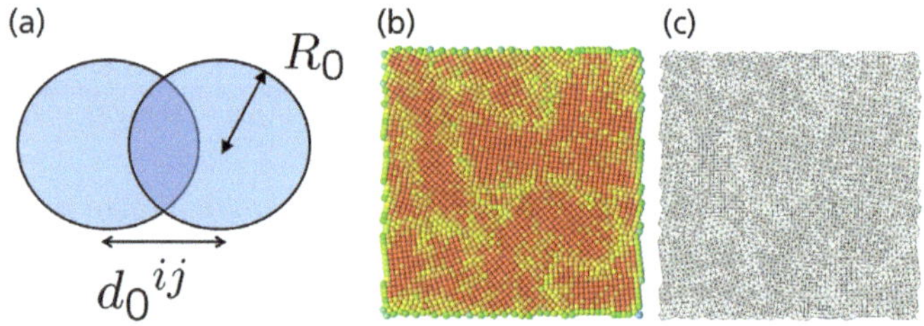

Fig. 12. The particle model as suggested in (Steinhauser, Grass, Strassburger & Blumen, 2009). (a) Overlapping particles with radii R_0 and the initial (randomly generated) degree of overlap indicated by $d_0{}^{ij}$. Here, only two particles are displayed. In the model the number of overlapping particles is unlimited and *each* individual particle pair contributes to the overall pressure and tensile strength of the solid. (b) Sample initial configuration of overlapping particles ($N = 2500$) with the color code displaying the coordination number: red (8), yellow (6), and green (4). (c) The same system displayed as an unordered network.

5.2 Model potentials

The main features of a coarse-grained model in the spirit of Occam's razor with only few parameters, are the repulsive forces which determine the materials resistance against pressure and the cohesive forces that keep the material together. A material resistance against pressure load is introduced by a simple Lennard Jones type repulsive potential Φ_{rep}^{ij} which acts on every pair of particles $\{ij\}$ once the degree of overlap d^{ij} decreases compared to the initial overlap ${d_0}^{ij}$:

$$\phi_{rep}^{ij}\left(\gamma, d^{ij}\right) = \begin{cases} \gamma {R_0}^3 \left(\left(\frac{{d_0}^{ij}}{d^{ij}} \right)^{12} - 2 \left(\frac{{d_0}^{ij}}{d^{ij}} \right)^{6} + 1 \right) & : \quad 0 < d^{ij} < {d_0}^{ij} \\ 0 & : \quad d^{ij} \geq {d_0}^{ij} \end{cases} . \tag{41}$$

Parameter γ scales the energy density of the potential and prefactor ${R_0}^3$ ensures the correct scaling behavior of the calculated total stress $\Sigma_{ij}\sigma^{ij} = \Sigma_{ij}F^{ij}/A$ which, as a result, is independent of N. Figure 13 shows that systems with all parameters kept constant, but only N varied, lead to the same slope (Young's modulus) in a stress-strain diagram. In Eq. 41 R_0 is the constant radius of the particles, $d^{ij} = d^{ij}(t)$ is the instantaneous mutual distance of each interacting pairs $\{ij\}$ of particles, and ${d_0}^{ij} = d^{ij}(t=0)$ is the initial separation which the pair $\{ij\}$ had in the starting configuration. Every single pair $\{ij\}$ of overlapping particles is associated with a different initial separation ${d_0}^{ij}$ and hence with a different force. The minimum of each individual particle pair $\{ij\}$ is chosen such that the body is force-free at the start of the simulation.

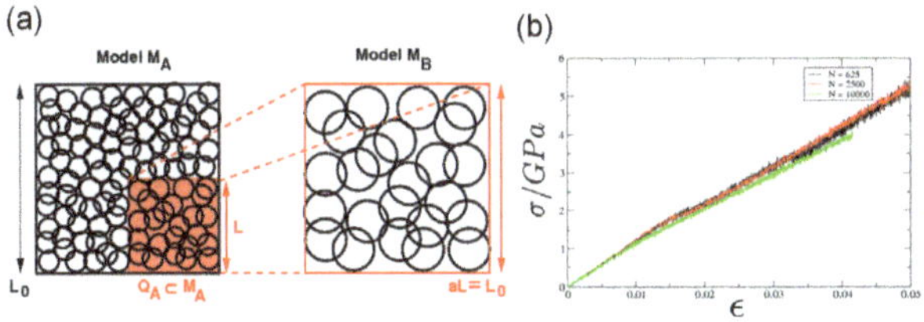

Fig. 13. (a) Schematic of the intrinsic scaling property of the proposed material model. Here, only the 2D case is shown for simplicity. The original system (Model M_a) with edge length L_0 and particle radii R_0 is downscaled by a factor of $1/a$ into the subsystem Q_A of M_A (shaded area) with edge length L, while the particle radii are upscaled by factor a. As a result, model M_B of size $aL = L_0$ is obtained containing much fewer particles, but representing the same macroscopic solid, since the stress-strain relation (and hence, Young's modulus E) upon uni-axial tensile load is the same in both models. (b) Young's modulus E of systems with different number of particles N in a stress-strain ($\sigma - \varepsilon$) diagram. In essence, E is indeed independent of N.

When the material is put under a low tension the small deviations of particle positions from equilibrium will vanish as soon as the external force is released. Each individual pair of overlapping particles can thus be visualized as being connected by a spring, the equilibrium length of which equals the initial distance ${d_0}^{ij}$. This property is expressed in the cohesive potential by the following equation:

$$\Phi_{coh}^{ij}\left(\lambda, d^{ij}\right) = \lambda R_0 \left(d^{ij} - {d_0}^{ij}\right)^2, \; d^{ij} > 0. \tag{42}$$

In this equation, λ (which has dimension [energy/length]) determines the strength of the potential and prefactor R_0 again ensures a proper intrinsic scaling behavior of the material response. The total potential is the following sum:

$$\Phi_{tot} = \Sigma_{ij} \left(\phi^{ij}_{rep} + \phi^{ij}_{coh} \right) . \tag{43}$$

The repulsive part of Φ_{tot} acts only on particle pairs that are closer together than their mutual initial distance ${d_0}^{ij}$, whereas the harmonic potential Φ_{coh} either acts repulsively or cohesively, depending on the actual distance d^{ij}. Failure is included in the model by introducing two breaking thresholds for the springs with respect to compressive and to tensile failure, respectively. If either of these thresholds is exceeded, the respective spring is considered to be broken and is removed from the system. A tensile failure criterium is reached when the overlap between two particles vanishes, i.e. when:

$$d^{ij} > (2R_0). \tag{44}$$

Failure under pressure load occurs when the actual mutual particle distance is less by a factor α (with $\alpha \in (0,1)$) than the initial mutual particle distance, i.e. when

$$d^{ij} < \alpha \cdot {d_0}^{ij}. \tag{45}$$

Particle pairs without a spring after pressure or tensile failure still interact via the repulsive potential and cannot move through each other.

An appealing feature of this model, as opposed to many other material models used for the description of brittle materials, see e.g. (Cundall & Strack, 1979; Leszczynski, 2003; Walton & Braun, 1986), is its simplicity. The proposed model has a total of only three free parameters: γ and λ for the interaction potentials and α for failure. These model parameters can be adjusted to mimic the behavior of specific materials.

5.3 Shock wave simulations and comparison with experiments

Finally, in Fig. 14, non-equilibrium MD simulation (NEMD) results for systems with varying shock impact velocities are presented and compared with high-speed impact experiments performed at EMI with different ceramic materials (Al_2O_3 and SiC) in the so-called edge-on-impact (EOI) configuration. These oxide and non-oxide ceramics represent two major classes of ceramics that have many important applications. The impactor hits the target at the left edge. This leads to a local compression of the particles in the impact area.

The top series of snapshots in Fig. 14(a) shows the propagation of a shock wave through the material. The shape of the shock front and also the distance traveled by it correspond very well to the high-speed photographs in the middle of Fig. 14(a). These snapshots were taken at comparable times after the impact had occurred in the experiment and in the simulation, respectively. In the experiments which are used for comparison, specimens of dimensions $(100 \times 100 \times 10) mm$ were impacted by a cylindrical blunt steel projectile of length $23\,mm$, mass $m = 126\,g$ and a diameter of $29.96\,mm$ (Steinhauser et al., 2006). After a reflection of the pressure wave at the free end of the material sample, and its propagation back into the material, the energy stored in the shock wave front finally disperses in the material. One can study in great detail the physics of shock waves traversing the material and easily

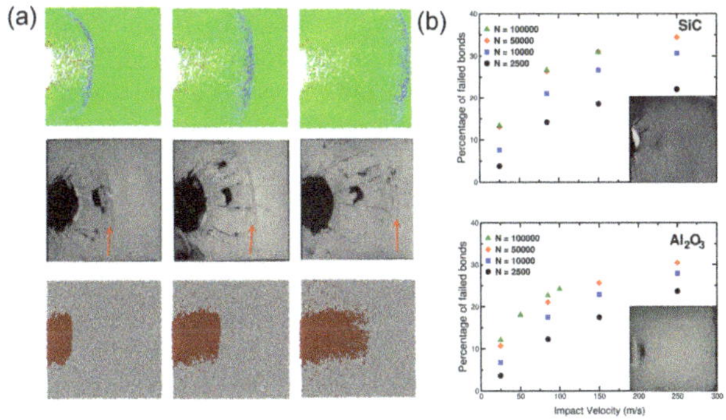

Fig. 14. Results of a simulation of the edge-on-impact (EOI) geometry, except this time, the whole macroscopic geometry of the experiment can be simulated while still including a microscopic resolution of the system. The impactor is not modeled explicitly, but rather its energy is transformed into kinetic energy of the particle bonds at the impact site. (a) Top row: A pressure wave propagates through the material and is reflected at the free end as a tensile wave (not shown). Middle row: The actual EOI experiment with a SiC specimen. The time interval between the high-speed photographs is comparable with the simulation snapshots above. The red arrows indicate the propagating shock wave front. Bottom row: The same simulation run but this time only the occurring damage in the material with respect to the number of broken bonds is shown. (b) Number of broken bonds displayed for different system sizes N, showing the convergence of the numerical scheme. Simulation parameters $(\alpha, \gamma, \lambda)$ are chosen such that the correct stress-strain relations of two different materials (Al_2O_3 and SiC) are recovered in the simulation of uniaxial tensile load. The insets show high-speed photographs of SiC and Al_2O_3, respectively, $4\mu s$ after impact.

identify strained or compressed regions by plotting the potential energies of the individual pair bonds. Also failure in the material can conveniently be visualized by plotting only the failed bonds as a function of time, cf. the bottom series of snapshots in Fig. 14(a). A simple measure of the degree of damage is the number of broken bonds with respect to the their total initial number. This quantity is calculated from impact simulations of Al_2O_3 and SiC, after previously adjusting the simulation parameters γ, λ and α accordingly. Figure 14(b) exhibits the results of this analysis. For all impact speeds the damage in the SiC-model is consistently larger than in the one for Al_2O_3 which is also seen in the experiments.

The impactor is not modeled explicitly, but rather its total kinetic energy is transformed into kinetic energy of the particles in the impact region. Irreversible deformations of the particles such as plasticity or heat are not considered in the model, i.e. energy is only removed from the system by broken bonds. Therefore, the development of damage in the material is slightly overestimated.

6. Conclusions

In summary, we presented an introduction into the molecular dynamics method, discussing the choice of potentials and fundamental algorithms for the implementation of the MD method. We then discussed proto-typical applications of MD, namely the collapse of two oppositely charged macromolecules (polyelectrolytes) and the simulation of semiflexible bio-macromolecules[6]. We demonstrated how semiflexibility, or stiffness of polymers can be included in the potentials describing the interactions of particles. We finally showed a somewhat unusual application of MD in the field of solid state physics where we modeled the brittle failure behavior of a typical ceramic and simulated explicitly the set-up of corresponding high-speed impact experiments. We showed that the discussed multiscale particle model reproduces the macroscopic physics of shock wave propagation in brittle materials very well while at the same time allowing for a resolution of the material on the micro scale and avoiding typical problems (large element distortions, element-size dependent results) of Finite Elements, which constitutes a different type of discretization for simulation problems that are closely connected with macroscopic experiments. The observed failure and crack pattern in impact MD simulations can be attributed to the random initial distribution of particle overlaps which generates differences in the local strength of the material. By generating many realizations of systems with different random initial overlap distributions of particles, the average values obtained from these many simulations lead to the presented fairly accurate results when compared with experimental high-speed impact studies.

7. References

Allen, M. & Tildesly, D. (1991). *Computer Simulation of Liquids*, Oxford University Press.

Aragón, S. & Pecora, R. (1985). Dynamics of wormlike chains, *Macromolecules* 18: 1868–1875.

Barrat, J.-L. & Joanny, F. (2007). Theory of Polyelectrolyte Solutions, *Adv. Chem. Phys.* 94: 1–66.

Buckingham, R. (1938). The Classical Equation of State of Gaseous Helium, Neon and Argon, *Proc. Roy. Soc* A106: 264.

Bustamante, C., Marko, J., Siggia, E. & Smith, S. (1994). Entropic Elasticity of Lambda-Phage DNA, *Science* 265: 1599–1600.

Cundall, P. & Strack, O. (1979). A Discrete Numerical Model for Granular Assemblies, *Geotechnique* 29: 47–65.

de Gennes, P. G. (1979). *Scaling Concepts in Polymer Physics*, Cornell University Press, Ithaca, London.

Debye, P. & Hückel, E. (1923). The Theory of Electrolytes. I. Lowering of Freezing Point and Related Phenomena, *Physikalische Zeitschrift* 24: 185–206.

Doi, M. & Edwards, S. (1986). *The Theory of Polymer Dynamics*, Clarendon Press, Oxford.

Feller, S. (2000). An Improved Empirical Potential Energy Function for Molecular Simulations of Phospholipids, *J. Phys. Chem. B* 104: 7510–7515.

Flory, P. (1969). *Statistical Mechanics of Chain Molecules*, Wiley, New York.

Gössl, I., Shu, L., Schlüter, A. & Rabe, J. (2002). Complexes of DNA and positively charged dendronized polymers, *J. Am. Chem. Soc.* 124: 6860.

Haberland, R., Fritsche, S., Peinel, G. & Heinzinger, K. (1995). *Molekulardynamik - Grundlagen und Anwendungen*, Friedrich VIeweg & Sohn Verlagsgesellschaft mbH, Braunschweig, Wiesbaden.

Haile, J. (1992). *Molecular dynamics simulation: Elementary Methods*, Wiley, New York.

Harnau, L., Winkler, R. & Reineker, P. (1996). Dynamic structure factor of semiflexible macromolecules in dilute solution, *J. Chem. Phys.* 104(16): 6255–6258.

[6] Semiflexibility is a characteristic feature of bio-macromolecules such as DNA, F-actin, intermediate filaments or microtubuli which are present in the cytoplasm of cells.

Harnau, L., WInkler, R. & Reineker, P. (1999). Comment on "Chain Motion in an unentangled polyethylene melt: A critical test of the Rouse model by molecular dynamics simulations anhd neutron spin echo spectroscopy, *Phys. Rev. Lett.* 82: 2408.

Harris, R. A. & Hearst, J. E. (1966). On polymer dynamics, *J. Chem. Phys.* 44: 2595–2602.

Harris, R. & Hearst, J. (1967). On Polymer Dynamics. III. Elastic Light Scattering, *J. Chem. Phys.* 46: 398.

Hockney, R. (1970). The Potential Calculation And Some Applications, *Methods Comp. Phys.* 9: 136–211.

Holm, C., Joanny, J.-P., Kremer, K., Netz, R., Reineker, P., Seidel, C., Vilgis, T. A. & Winkler, R. G. (2004). Polyelectrolyte Theory, *Adv. Polym. Sci.* 166: 67–111.

Käs, J., Strey, H., Tang, J., Finger, D., Ezzell, R., Sackmann, E. & Janmey, P. (1996). F-actin, a model polymer for semiflexible chains in dilute, semidilute, and liquid crystalline solutions, *Biophys. J.* 70: 609–625.

Khoklov, A. R. & Khalatur, P. G. (2005). Solution Properties of Charged Hydrophobic/Hydrophilic Copolymers, *Curr. Opin. Colloid Interface Sci.* 10: 22–29.

Kratky, O. & Porod, G. (1949). Röntgenuntersuchung gelöster Fadenmoleküle, *Rec. Trv. Chim* 68: 1106–1115.

Leszczynski, J. S. (2003). A Discrete Model of a Two-Particle Contact Applied to Cohesive Granular Materials, *Granular Matter* 5: 91.

Manning, G. (1969). Limiting Laws and Counterion Condensation in Polyelectrolyte Solutions I. Colligative Properties, *J. Chem. Phys.* 51: 924–933.

Ober, C. (2000). Shape Persisitence of Synthetic Polymers, *Science* 288(5465): 448–449.

Paul, W., Smith, G. & Yoon, D. (1997). Static and dynamic properties of a n - $C_{100}H_{202}$ melt from molecular dynamics simulations, *Macromolecules* 30: 7772–7780.

Prince E. Rouse, J. (1953). A Theory of the Linear Viscoelastic Properties of Dilute Solutions of Coiling Polymers, *J. Chem. Phys.* 21: 1281–1286.

R.G. Winkler, P. Reineker, L. H. (1994). Models and equilibrium properties of stiff molecular chains, *J. Chem. Phys.* 101(9): 8119–8129.

Schlenkrich, M., Brinckmann, J., MacKerell, A. & Karplus, M. (1996). *Empirical Potential Energy Function for Phospholipids: Criteria for Parameter Optimization and Applications*, Birkhäuser, Boston.

Shih, A. Y., Arkhipov, A., Freddolino, P. L. & Schulten, K. (2006). Coarse Grained Protein-Lipid Model With Application to Lipoprotein Particles, *J. Phys. Chem. B* 110: 3674–3684.

Siu, S. W., Vácha, R., Jungwirth, P. & Böckmann, R. A. (2008). Biomolecular Simulations of Membranes: Physical Properties from Different Force Fields, *J. Chem. Phys.* 125: 125103.

Soda, K. (1973). Dynamics of Stiff Chains. I. Equation of Motion, *J. Phys. Soc. Jpn.* 35: 866–870.

Soda, K. (1991). Differences in thermal conformation fluctuation between two models for semiflexible chains, *J. Chem. Phys.* 95: 9337–9347.

Srivastava, D. & Muthukumar, M. (1994). Interpenetration Of Interacting Polyelectrolytes, *Macromolecules* 27: 1461–1465.

Steinauser, M. O. (1998). *Molekulardynamik-Simulationen von Polyelectrolyten unterschiedlicher Ladungsdichte: Polyelektrolyt-Komplexbildung*, Master's thesis, University of Ulm, Germany.

Steinhauser, M. O. (2005). A Molecular Dynamics Study on Universal Properties of Polymer Chains in Different Solvent Qualitites. Part I: A Review of Linear Chain Properties, *J. Chem. Phys.* 122: 094901.

Steinhauser, M. O. (2008). *Computational Multiscale Modeling of Solids and Fluids – Theory and Applications*, Springer, Berlin, Heidelberg, New York.

Steinhauser, M. O. (2012). *Computer Simulation in Physics and Engineering*, de Gruyter. To be published in October 2012, ISBN 978-3-11-025590-4.

Steinhauser, M. O. & et. al. (2005). MAVO MMM-Tools: Entwicklung von durchgängigen Multiskalen Material Modellierungen, *Technical Report 17/05*, Fraunhofer Ernst-Mach-Institute (EMI), Freiburg, Eckerstrasse 4, 79104 Freiburg i. Br., Germany.

Steinhauser, M. O., Grass, K., Strassburger, E. & Blumen, A. (2009). Impact Failure of Granular Materials – Non-Equilibrium Multiscale Simulations and High-Speed Experiments, *International Journal of Plasticity* 25: 161–182.

Steinhauser, M. O., Grass, K., Thoma, K. & Blumen, A. (2006). A Nonequilibrium Molecular Dynamics Study on Shock Waves, *Europhys. Lett.* 73: 62.

Steinhauser, M. O., Schneider, J. & Blumen, A. (2009). Simulating Dynamic Crossover Behavior of Semiflexible Linear Polymers in Solution and in the Melt, *J. Chem. Phys.* 130: 164902.

van Gunsteren, W. & Berendsen, H. (1982). Algorithms for Brownian Dynamics, *Molec. Phys.* 45: 637.

Walton, O. R. & Braun, R. L. (1986). Viscosity, Granular-Temperature, and Stress Calculation for Shearing Assemblies of Inelastic, Frictional Disks, *J. Rheol.* 39(5): 949.

Winkler, R. G., Steinhauser, M. O. & Reineker, P. (2002). Complex Formation in Systems of Oppositely Charged Polyelectrolytes: A Molecular Dyamics Simulation Study, *Phys. Rev. E* 66: 021802.

Yamasaki, Y., Teramoto, Y. & Yoshikawa, K. (2001). Disappearance Of The Negative Charge in Giant DNA With A Folding transition, *Biophys. J.* 80: 2823Ű2832.

Zimm, B. H. (1956). Dynamics of Polymer Molecules in Dilute Solution: Viscoelasticity, Flow Birefringence and Dielectric Loss, *J. Chem. Phys.* 24: 269–278.

2

Molecular Dynamics Simulation of Synthetic Polymers

Claudia Sandoval
Center for Bioinformatics and Integrative Biology (CBIB), Facultad de Ciencias Biológicas
Universidad Andres Bello, Santiago
Fraunhofer Chile Research Foundation - Center for Systems Biotechnology, FCR-CSB
Chile

1. Introduction

The polymeric systems are great interesting both academic and industrial level, due to interesting applications in several areas. Nowadays, the trends for the development of new materials are focused on the study of macromolecular complex systems. A macromolecular complex system is formed by the interaction between the macromolecules of the same system. There are many macromolecular complex systems; this chapter will refer to complex systems of synthetic polymer blends(Donald R. P., 2000) and synthetic polymer at the air-water interface(Gargallo et al., 2011). The most synthetic polymers are characterized by flexible structure able to adopt multiple and variable forms. The study about molecular behavior in polymers has attracted the attention of many researchers over the years. Studies in dilute solutions of macromolecules have contributed to solving the problem relating to the molecular characterization. Despite, the emergence of characterization methods and theories that have allowed for interpretation of experimental behavior, researchers still have not fully resolved the problem with interactions at the atomic level in polymer systems. The most polymers are amorphous so in some cases are impossible to obtain information about atomic structure through x-ray techniques or atomic force microscopy. At this point, the computer simulations have played an important role in microscopic understanding of dynamical properties of some polymeric systems.

In this chapter a review of molecular simulation methodologies focused on polymer systems will be discussed. This review is based on two case studies. The first one corresponds to a monolayer of amphiphilic polymers and their study at the air-water interface(Gargallo et al., 2009). Three polymers were selected Poly(ethylene oxide) (PEO), Poly(tetrahydrofuran) (PTHF) and Poly(ε-caprolactone) (PEC) and their cyclodextrin inclusion complexes. The second system corresponds to study the motion of side chains of polymers in condensed phase. For this reason, a diblock copolymers of Polystyrene-block-poly(t-butylacrylate) has been considered(Encinar et al., 2008). The methodology to build the models in both systems will be provided, as well as, the methodology to obtain the atomic parameters, necessary to molecular simulations.

To carry out molecular dynamics studies of polymer systems is necessary to know some important physicochemical properties or thermodynamic parameters of polymers. For the other hand, it is important to know the different methodologies employed to carry out molecular dynamic simulation of polymers.

2. Physicochemical properties of polymers

In this chapter we concéntrate efforts on giving the reader a brief description of some important physicochemical properties of the polymers for molecular simulation analysis analysis. Macromolecular conformation is an important point to consider. The macromolecular chains in dilute solution have variables internal rotational angles, Figure 1. The internal rotational angle ϕ can take values that fluctuate within a range so that different minimum energy can be obtained for various values of the angles ϕ. The energy difference associated with the variation of this angle is minimal. This leads to a large number of stable conformations and significant mobility to oscillate around these conformations. Both factors cause that the macromolecule has a significant flexibility. Experimentally the dimensional parameters that characterize a macromolecule can only be characterized in terms of statistical parameters that represents an average over all conformations. The *end-to-end distance* and *radius of gyration* are statistical parameters that characterize the spatial distribution of a chain macromolecule(Flory, 1969, 1995).

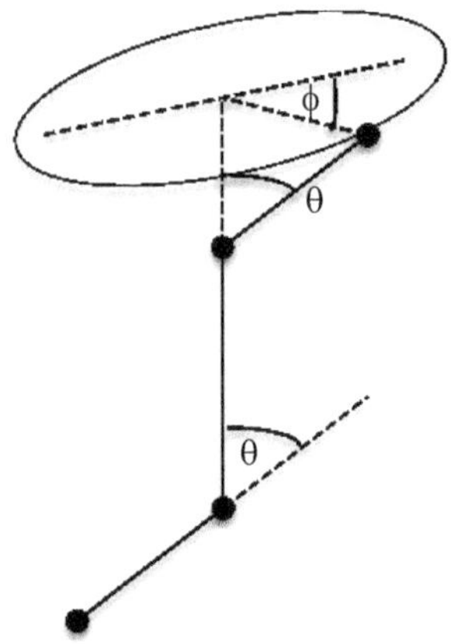

Fig. 1. Three successive bonds for a chain. Where θ and ϕ represents the bond angle and internal rotation angle, respectively.

The *end-to-end distance* $< r^2 >$ represents the distance between end chains. This parameter can vary from a maximum value, where the polymer chain is fully extended to a minimum value corresponding to the sum of the van der Waals radii of the terminal groups, Figure 2.

The end-to-end distance is a statistical parameter that characterizes the spatial distribution of the chains relating the microscopic with macroscopic properties, also allows to relate the hydrodynamic parameters with thermodynamic parameter, and finally allows to obtain information respect to flexibility chain.

a)

b)

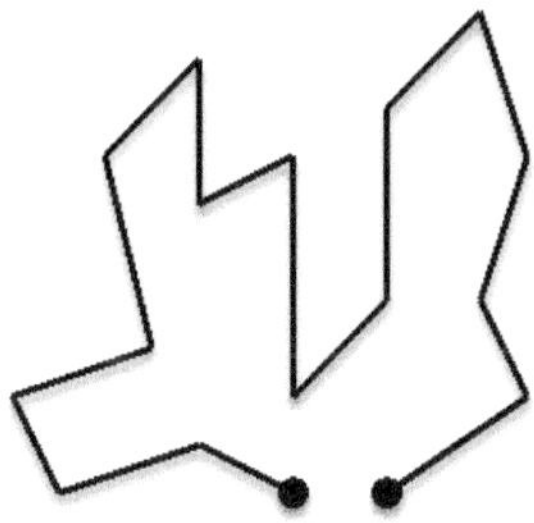

Fig. 2. End-to-end distance $< r^2 >$ representation to a) fully extended chain model, where $< r^2 >$ takes a maximum value and b) $< r^2 >$ takes a minimum value and correspond to the sum of the van der Waals radii.

$< r^2 >$ is defined by the equation 1:

$$< r^2 >= \int_0^{\infty} r^2 W(r) dr \quad (1)$$

where W(r) is probability distribution function, defined by

$$\int_0^{\infty} W(r) dr = 1$$

But the chains may adopt different conformational states depending on the characteristics of the chain, features such as substituents and the solvent. For the different conformational states, there are different chains models. Is not intended to provide a thorough review of statistical mechanics of polymer chains(Flory, 1969), but is important to note that the value of $< r^2 >$ depends on two main factors, "short-range" and "long-range" factors. For these two factors, different chain models were raised in the past. Short-range factors involved characteristics of the bonds and interaction between neighboring atoms or groups along a chain, this chain correspond to a rotation free model. Hence, $< r^2 >$ can be calculated knowing the values of bond angles and bond distances for all the conformations, the value obtained is called dimension mean square of end-to-end distance $\langle r^2 \rangle_{of}$. Take into account factors such as bond angle restrictions and steric hindrance between substituent groups, we can obtain unperturbed dimension $\langle r^2 \rangle_0$, and may be expressed by equation 2:

$$\langle r^2 \rangle_0 = \sigma^2 \langle r^2 \rangle_{of} \quad (2)$$

where σ is the stiffness coefficient or conformational factor. The $\langle r^2 \rangle_0$ is so-called "theta condition".

In long-range factors are involved the interactions between distant segments along the chain, as well as interaction with the solvent. Taking into account that, two distant segments of a chain, cannot occupy the same place at the same time, arise the excluded volume concept. Under effect of these interactions the end-to-end distance to linear chains in dilute solution, may be described by:

$$< r^2 >^{1/2} = < r^2 >_0^{1/2} \alpha \tag{3}$$

where α linear expansion chain factor. In equation 3, the root-mean-square end-to-end distance $< r^2 >^{1/2}$ depends of number of segments, temperature and solvent. When α>1 them the interaction between solvent and chains are favorable, this means that excluded volume effect is important. Instead, if the interactions between segments and solvent are unfavorable them α tend to decrease; this causes folding of the chains to minimize the contact with the solvent. These effects can be balance and them the polymer chain adopts an unperturbed state.

The *radius of gyration* is the parameter more appropriate to describe the dimensions of a macromolecule. The mean-square radius of gyration $< R_g^2 >$ represents the mean square of the distance between each atom in the chain and center of mass of the chain and may be defined to following equation 4:

$$< R_g^2 > = (1/n) \sum_1^n < r_i^2 > \tag{4}$$

where n is the segments numbers of the chain and $< r_i^2 >$ represents the mean-square distance of the *i*-th segment from the center of mass of the chains. Through the random flight model we can express the radius of gyration in terms to mean-square end-to-end distance, indicating their interchangeability. For a freely rotating chain of *n* segments of length *l*, the following relationship is established:

$$< R_g^2 > = \frac{1}{6} < r^2 > = \frac{nl^2}{6} \tag{5}$$

A complete derivation of these equations can be reviewed in various texts of statistical mechanics of polymers(Flory, 1995) (Yamakawa, 2001). We have said that the dimensional parameters are important to know because of their contribution to the characterization of polymers. In addition, these parameters can be calculated both experimentally and theoretically. For example, radius of gyration has been calculated, in several works, from molecular simulation trajectories and compared with the experimental values(Lee et al., 2006; Prosa et al., 2001, 1997).

3. Molecular dynamic simulation of supramolecular structures at the air-water interface

A great interesting in the last year toward to study theoretical of self-assembly of macromolecular systems have been report(Duncan et al., 2011; Korchowiec et al., 2011; Saldias et al., 2009). The self-assembly of macromolecular systems are great interesting due to applications in several areas, as medicine, drug delivery and smart materials(Wick and Cummings, 2011). The polymers at the air-water interface are considering self-assembly systems. The studies of polymers at the air-water interface considering several molecular simulation methodologies, such as atomistic(Gargallo et al., 2009) and coarse-grained(Duncan et al., 2011) models. In this chapter we will focus on atomistic molecular simulation methodology. The aim of studies of polymer at the air-water interface, using molecular simulation, is characterized the orientation of polymer chains at the interface, as well as, calculate the different interaction between amphiphilic polymer and the water subphase.

3.1 Build of the molecular models

In general, the models used to reproduce several polymer chains at air-water interface, consist of the build a bilayer of polymers with an enough thick water slab between of the chains. All the systems were simulated taking into account three experimental surface areas 10, 20 and 30Å^2 per molecules. The water slab should be larger in one axis (i.e. z-axis), to avoid the contact between the polymer chains, see Figure 3. A vacuum space larger enough to avoid the contact between chains of different side of the bilayer, should be considered. To the constructor of different molecular structure (repeating unit of the polymers) any molecular builder is required, the important point is generated the pdb coordinate file. To generate the polymer chains a script was made in Tcl language and them executed in VMD(Humphrey et al., 1996) program and this way, the polymer chain of ten or twelve repeating unit were obtained. The build of water slab was made to using VMD program. For this purpose, was necessary obtained the coordinate of each polymer chain and them located these chains in the correct coordinates in order to build the water slap in the center of the coordinate and this way, obtained the model that represent several polymer chains oriented at the air-water interface, as Figure 3.

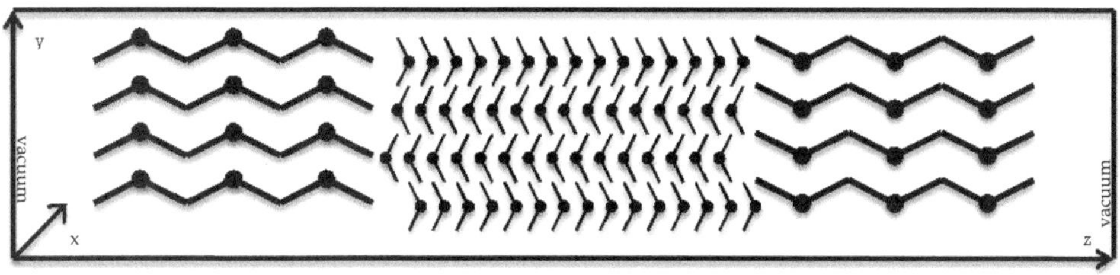

Fig. 3. Drawing in two dimension of polymeric monolayer at the air-water interface. The dimension in x and y-axis are smaller in relation with z-axis.

To the build of the inclusion complexes of cyclodextrin-polymers, α and γ-cyclodextrin were considering. The cyclodextrin structures were obtained from Protein Data Bank, PDB entry 1CXF for α-CD and 1D3C for γ-CD. These structure were fully minimized, using CHARMM force field and a good agreement with the experimental geometry was obtained. To carry out the minimization and molecular dynamic simulation of synthetic polymer was necessary to use a force field, which content the atomic parameter appropriate to a particular polymer. To obtained the atomic parameters to Poly(ethylene oxide) (PEO), Poly(tetrahydrofuran) (PTHF) and Poly(ε-caprolactone) (PEC) we used the CHARMM force field, this force field is open source and it is possible aggregate the parameter because the polymers implicate in the study was necessary calculate the atomic parameter, through quantum mechanics. For this purpose, Paratool package implemented in VMD(Humphrey et al., 1996) software was used. Six models were built, three corresponding to the inclusion complexes: PEC-CD, PEO-CD and PTHF-CD and the polymer respectively at the air water interface, see Figure 4.

3.2 Considerations to molecular dynamic simulations

All Molecular dynamic (MD) simulations were carried out using NAMD (Kalé et al., 1999) program and the CHARMM27(Kuttel et al., 2002) force field. Simulations were performed in

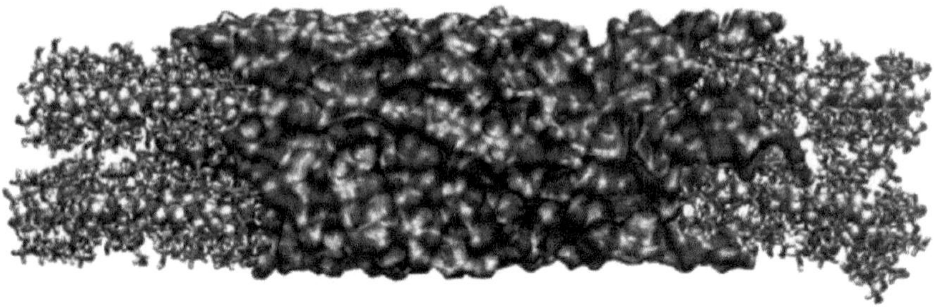

Fig. 4. Snapshot for the CD-PTHF inclusion complex, cyclodextrins are drawn with line style and PTHF with van der Waals style. The drawing method for water molecules is solid surface. Carbon, hydrogen and oxygen atoms are green, white and red respectively.

NPT ensemble; the non-bonded van der Waals and electrostatic interactions were calculated with a cut-off using a switching function starting at a distance of 10Å and reaching zero at 12Å. The TIP3P(Jorgensen et al., 1983) water model was employed for the solvent. Periodic boundary conditions were applied with a rigid cell. The particle mesh Ewald(Darden et al., 1993) (PME) method was employed for computation of electrostatic forces. An integration time step of 1fs was assumed, permitting a multiple time stepping algorithm(Grubmüller et al., 1991) to be employed in which interactions involving covalent bonds were updated every time step. In all simulations, Langevin dynamics were utilized to keep a constant temperature of 298K, likewise the hybrid Nosé-Hoover Langevin piston method(Izrailev et al., 1997) was used to control a constant pressure of 1 atm. The system was first equilibrated for 1ns. The backbone carbons of the polymers in the inclusion complexes were restrained with a 0.5kcal/mol/Å^2 spring constant during relaxation. Then, simulations of 5ns were performed to all the systems. A novel alternative to obtain the charges and potentials to each atoms or groups is through CHARMM General Force Field (CGenFF)(Vanommeslaeghe et al., 2010). This force field is an extension of the CHARMM force field to small molecules and included a wide range of chemical groups. The molecule, in our case the monomer, is required in mol2 format to be uploading in the web site of Paramchem and this way, obtain the charges and potentials atomic. The server provides the topology and parameter files, necessaries to run the MD simulation. Topology file contains the monomer residue, atom types and the charges. Then, is mandatory to make changes to the topology file because we need create a topology for a polymer containing many repeat units. At the moment the topology file only have one residue that corresponding to the monomeric structure, so is require to build three types of residues. A "head" residue that defined the initial monomer chain, the "repeat" residue that is corresponding to the residue in-between chain and "tail" residue that defined the final monomer chain. Another way is to define only one residue corresponding to de monomer and build a path union, where is specified the hydrogen atoms that most be eliminated and the atoms that most be linked. This way is possible obtaining the coordinate and connectivity file to a polymeric chain.

3.3 Properties polymeric systems, calculations and analysis

From MD simulations trajectories several interesting properties cab be obtained and give insight about the behavior of inclusion complexes and polymer chains at the air-water interface. The most important point to know about the supramolecular systems(Komarov et al., 2011; Krishnamoorthy et al., 2011) at the interface is the orientational correlation about the order of monolayer. In the case of study of the inclusion complexes(Almeida et al., 2011; Boonyarattanakalin et al., 2011; Pan et al., 2011; Shi et al., 2011), we calculate the *order parameter* (Sr). This property is defined in terms of second-order Legendre polynomials and is the average of an angle made by a section of the polymer chain in a specified direction. This orientation correlation function S(r) is defined by equation:

$$S(r) = 0.5\left(3cos^2\theta_{ij}(r) - 1\right) \quad (6)$$

where θ is the angle between *i* and *j* vectors, when the center of mass of the associated chains of the vectors *i* and *j* are separated by the distance *r*. For the case of the inclusion complexes, we selected the atoms carbon C-C unit vectors of the different chains from PTHF, PEC and PEO as the vectors *i* and *j*. This parameter is indicative of the tendency towards polymers ordering in systems containing flexible chains. A value of 1.0 for S(r) mean that the chains are parallel to each other and cero value for a disordered distribution.

In the Figure 5, the S(r) variation as a function of time for these polymers is shown. It is clear that PEC is more organized at the air-water interface than PTHF and PEO. These results were in good agreement with the experimental data. Comparing the order parameter for PEC, PTHF and PEO and their respective ICs, can observe that the ICs are more ordered at air-water interface than their precursor polymer, Figure 6 a, b and c.

The order parameter shown differences evident in lose conformational freedom for cyclodextrins-polymers systems. The inclusion complexes are stiffer than single chains explaining the lower values of S(r) of the chains compared to those included in cyclodextrins.

The *radial distribution function*, g(r), is another important property that describe the probability to find an atom in a distance r from another atom and can be represent by equation 7, where R_i is the distance of atoms *i* from the center of mass and N the total number of atoms. By this way, the estimation of g(r) between atoms belonging to hydrophilic groups of the precursor polymer and water, using computational methods, give a view of the orientation of polymer chains at the air-water interface.

$$g^2_{(r)} = \langle \frac{1}{N} \sum_{i=1}^{N} (R_i^2) \rangle \quad (7)$$

Figure 7 and 8 presents g(r) between the hydrophobic groups of the different polymer chains and also g(r) between hydrophilic groups of the different polymer chains and water molecules. The electrostatic interactions between hydrophilic groups of the precursor polymers and water molecules are established. Nevertheless, some differences were seen in g(r) plots to PEC, PEO and PTHF. In the case of, PEC and PTHF (Figure 7a,b) the distance between the hydrophilic groups and water molecules are slightly lower in compare to the distances in PEC.

The $< r^2 >$ *and* $< R_g^2 >$ parameters can be estimate from molecular dynamics simulation trajectories, in the same way that the radial distribution function and order parameter. Scripts for each of the parameter were written in Tcl language and were ran in VMD program. The scripts can be to provide the reader through a mail to the author.

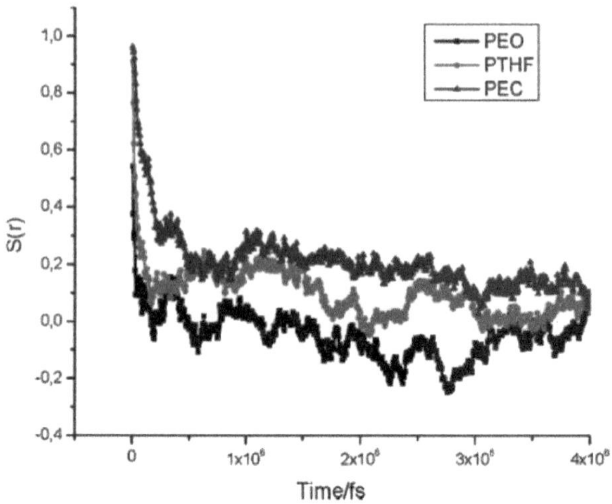

Fig. 5. Order parameter S(r) for precursor polymers from MD simulation.

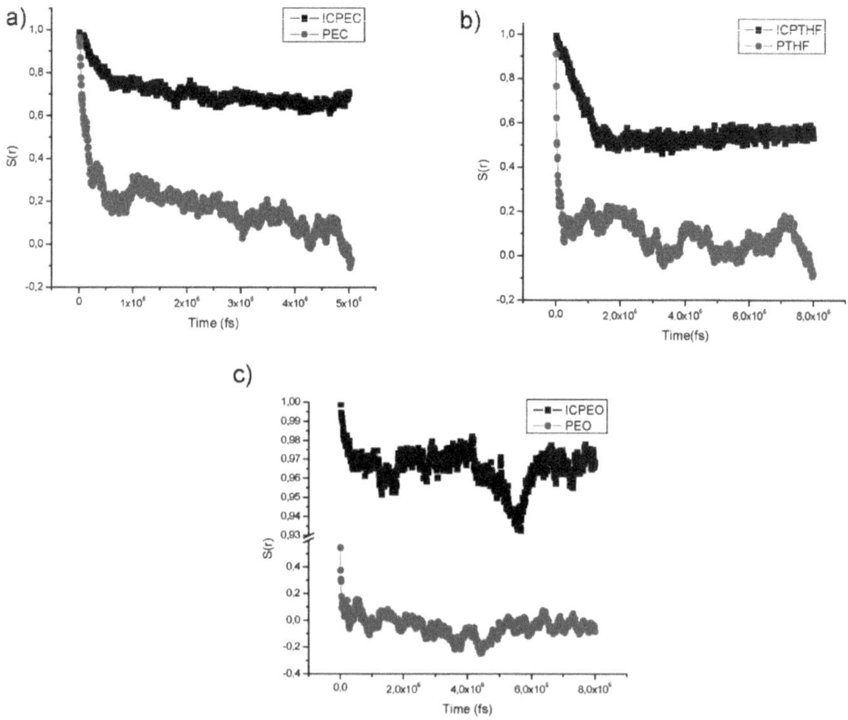

Fig. 6. Order parameter S(r) for a) PEC and IC-CD-PEC, b) PTHF and IC-CD-PTHF, c) PEO and IC-CD-PEO

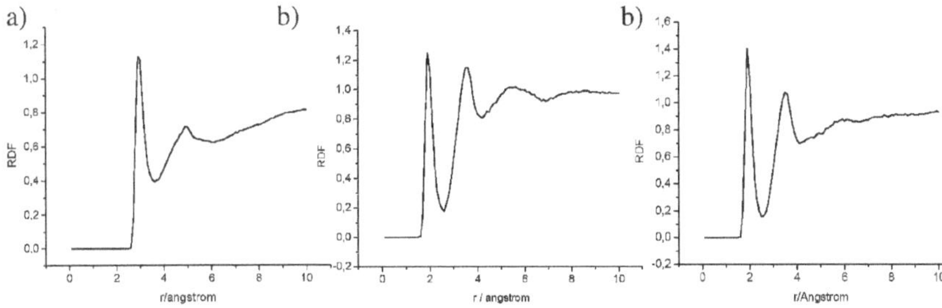

Fig. 7. Radial distribution function (RDF) between oxygen atoms of the precursor polymers and water molecules for: a) PEC, b) PEO and c) PTHF.

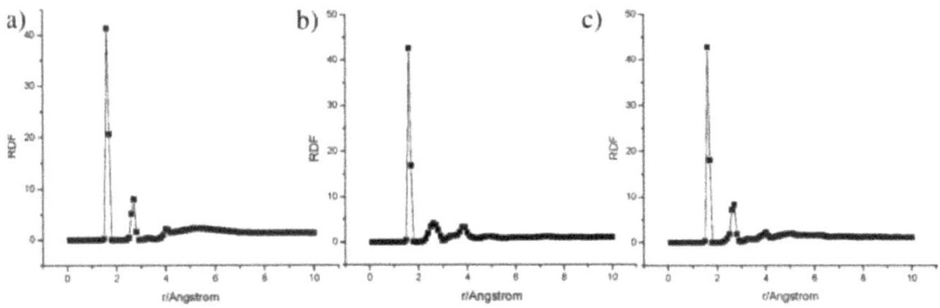

Fig. 8. Radial distribution function between hydrophobic groups of different polymer chains for: a) PEC, b) PEO and c) PTHF.

4. Free energy conformational calculation to the copolymer systems at condensed phase

The di-block copolymers in melt state have been studied by divers researcher groups, due to great interesting from the point view of the dielectric and dynamic-mechanics behavior(Alig et al., 1992; Encinar et al., 2008; Karatasos et al., 1996; Kotaka and Adachi, 1997; Kyritsis et al., 2000; Ma et al., 2003; Moreno and Rubio, 2002; Vogt et al., 1992). Dynamic-mechanical spectroscopy is an important technique to characterize the deformation of polymeric materials applying, for example, a tension to material. The viscoelastic behavior of the polymers can be studied through the determination of properties such as the complex modulus and glass transition temperature. On the other hand, the dielectric relaxation explore the response of a sample submitted an electric field. The permanent dipolar moments of a flexible polymer can be reoriented by an electric field, giving the relaxation mechanics of the fluctuations of dipolar moieties of the sample. For this reason is interesting to know at atomic level the conformational motions that have a polymer or copolymer. The conformational free energy difference (ΔG) for a polymer chain, can be calculated through molecular dynamics simulation at different temperatures, with the aims of obtain a conformational wide sampling. For this purpose, Boltzmann equation(Gilson et al., 1997) was used to obtain ΔG, equation 8.

$$\Delta G = -RTln\frac{f\phi_1}{f\phi_2} \quad (8)$$

Where $f\phi$ is the frequency distribution of the conformational change of the dihedral angle of side chain of PtBa, R is the gas constant and T is the temperature.

The case study was carried out for two systems of diblock copolymer of Polystyrene (PS) and Poly(t-butylacrylate) (PtBa) with different Poly(t-butylacrylate) chain lengths. The two blocks of PS and PtBa are amorphous and immiscible, also these blocks have well separates their glass transitions temperature (Tg). The PS-b-PtBa copolymer is dielectrically active due to PtBa part only. Nevertheless, the PtBa block has no present dipole moment to along the chain, thus the segmental mode is not influenced by other dynamical contributions. Besides, both blocks are active in dynamic-mechanical spectroscopy. The results obtained from the two techniques were compared with the PtBa homopolymer. The experimental results shown that the shape of the relaxation curves has different dependences with the temperature in the copolymers and the PtBa homopolymer. However, the relaxation maps to α-mode were almost equivalent in the copolymer than in the PtBa homopolymer, but the relaxation time to β-mode were different in the copolymer respect to homopolymer. This result was unexpected, because the β-relaxation has a local character due to the PtBa side chains relaxation. The above behavior could be owing to the influence of the PS on the PtBa side chains. For this reason, different environments were simulated in condensed phase and three models were built Figure 9 with the goal of study, if the PS blocks influenced the motion of side chains of PtBa. The parametrization of PtBa and PS monomers were done in the Paratool module of the VMD program. Paratool is a tool that used quantum mechanics methodology to the optimized structures and vibrational frequency to obtain the atomic parameter, both unbounded and bounded generating topology and parameter files, in CHARMM format. The MD simulations were carried out in NAMD program using CHARMM force field.

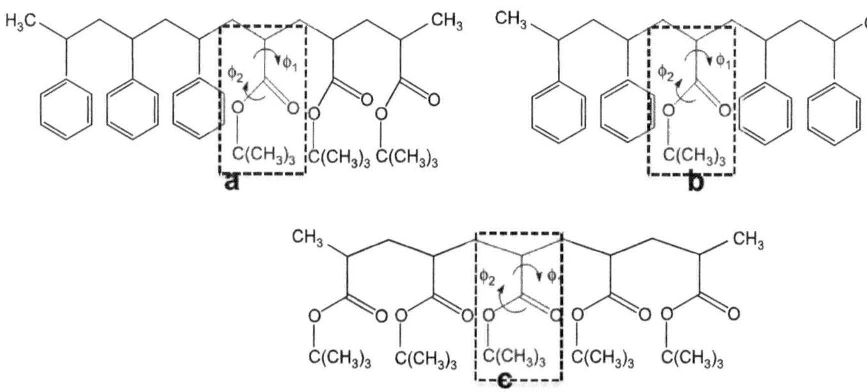

Fig. 9. Different structure models to study the side chain conformational movement of the PtBa containing different environments, frame in dashed line represent the monomer considered in measurements of frequency to the dihedral angle. a) PtBa monomer surrounded by different neighborhood, both PS and PtBa, labeled P(305) b) PtBa monomer with PS neighbors, labeled P(307) and c) PtBa monomer surrounded by PtBa units, labeled PtBa.

The polymeric fragments containing five repeating units of each sequence were placed in a box under periodic boundary condition Figure 10; ten chains were placed in three different boxes for each model. Molecular simulation of 1ns of long time were carried out using Valbond Force Field (CVFF)(Maple et al., 1998) and CHARMM forcé field. The three models were run at different temperatures between 1000K and 1700K with the purpose of obtain a wide conformational sample. A conformational analysis was performed to each model in order to estimate the free energy difference between two conformations for dihedral angles ϕ_1 and ϕ_2, see Figure 9.

The frequency average of the dihedral angles was calculate in the three models at the different temperatures from molecular simulation trajectories Figure 11, and applied to the equation 8. The ΔG values were compared with activation energy E_A from dielectric experiments to the β-relaxation, Table 1.

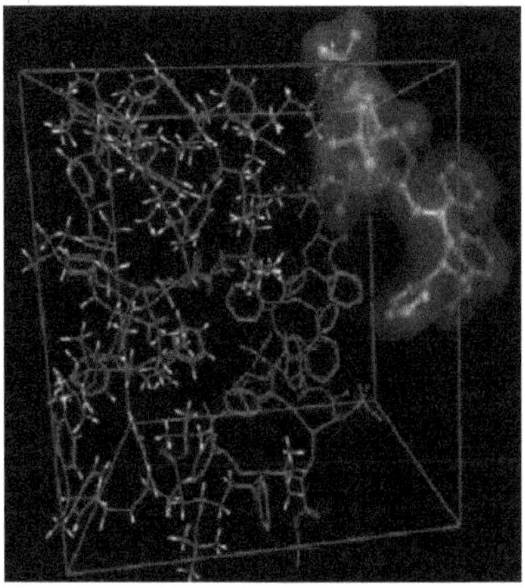

Fig. 10. Box with periodic boundary condition containing ten polymer chains of P(305).

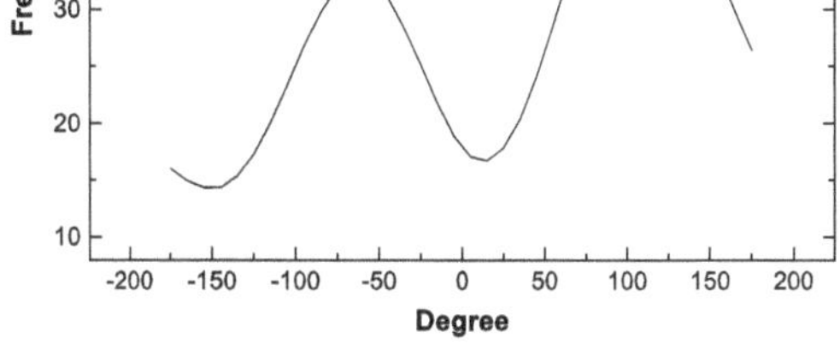

Fig. 11. Frequency distribution versus degree of the dihedral angles, ϕ_1 and ϕ_2 of obtained at 1700K, the numbers in the top peak correspond to area under the curve of these angles. The area values were considered in Boltzmann equation.

Polymer	$10^{-3}\Delta G/R$	$10^{-3}E_A/R$
PtBa	-1.6±0.1	1.5±0.04
P(307)	-1.9±0.1	1.3±0.05
P(305)	-2.2±0.2	1.1±0.1

Table 1. Activation free energy (E_A) to β-relaxation obtained from dielectric experiments and variations.

The experimental data showed that the characteristic time to β-transition is faster in the case of homopolymer compared to the copolymers. This behavior provides evidence that the PS blocks are influencing to the PtBa blocks. From MD simulations was possible to obtain the frequencies distribution to the conformational changes of ϕ_1 and ϕ_2 dihedral angles, Figure 11. Distribution frequencies calculated for P(305), P(307) and PtBa for all temperatures were applied to equation 8 and them the ΔG was obtain from the plots of $\ln(f\phi_1/f\phi_2)$ against 1/T. Conformational free energy had the following trend PtBa>P(305)>P(307). Table 1 shown that the ΔG is unfavorable in the case to PtBa, indeed this mean that there is less mobility of ester group, then the relaxation processes should be more hindered than in the other fragments. These results explain the activation energies values obtain by experimental techniques, where the E_A to PtBa is higher in compare to P(305) and P(307). Therefore, the results obtain by MD simulation of different fragment of the copolymer are in agree with the experimental data. In addition, the model containing the monomer surrounded by PS and PtBa (P305) have a higher mobility, so the activation energy to the relaxation process is less.

We have seen that the molecular simulation methods can be effective to study behavior of the system in solution as well as in condensed phase. The considerations to carried out molecular simulation in polymeric systems are:

1. To know if the process of study occur in solution or condensed phase
2. The time scale that take this process, so with that information is possible decide which methodology can be better to establish an equilibrium between simulation time and accuracy of results. Some process, as the mesoscopic structures, should be studied with coarse-grained methodology.
3. Establish that force field is more appropriate to our polymeric system.
4. In the absence of atomic parameter is necessary get through quantum mechanics and add to the force field, in the case of those of open source, as CHARMM.
5. The atomic parameter can be to get of an easy way through Paratool module available in VMD program or using the web platform of Paramchem.
6. Building models should represent the issue to be solved.
7. Select to appropriate program to carried out the MD simulation
8. Depending of the issue of our system, may be necessary apply to the polymer system constrain. Oftentimes, is required the study of movement of the side chain and not necessary to considered the backbone polymeric movement, in this case can be considerer apply restriction to the movement of the backbone. The restraints to the motion have an important advantage that allows simulation time economy. The important is emphasizing de differences between restraints and constraints. The constraints eliminate the hard degrees of freedom of the systems that correspond to high frequencies of vibration. The constraints can be applied on the hydrogen bond distances and hydrogen bond angles, for this purpose the shake algorithm is used.

Shake algorithm(Andersen, 1983; Ryckaert et al., 1977) to do cyclic computation of all constraints, in an iterative fashion until each constraint be satisfied.

9. Other input parameter to considerer before starting with de MD simulation is the canonical ensemble. Simulation to the polymer at the air-water interface the NPT ensemble is appropriate to take into account.
10. Finally is important the simulation time, it is necessary that the polymer chains to reach equilibrium. For this reason, the molecular simulation energy should be plotted against the time simulation; if the energy is not constant it will require longer simulation.

5. Conclusions

Molecular dynamic simulations can became a powerful tool to study of the behavior of polymer system both in solution and condensed phase. In this chapter, two different works were approached, in both cases the intermolecular interactions were calculated for the purpose of established a correlation between theoretical and experimental results. In the case of inclusion complexes, molecular simulation was carried out to understand the behavior at the air-water interface. The results obtained from molecular simulation trajectories to each system were used to calculate the order parameter and correlated with π-A isotherms from Langmuir experiments. Order parameter established that both PEC and IC-CD-PEC were more ordered systems related to PTHF and IC-CD-PTHF and PEO and IC-CD-PEO. These results were agreement with experimental results. However, radial distribution function was calculated with the purpose of obtain information about orientation of polymeric systems at the interface. From radial distribution function could be established that the interaction between polar groups of the polymers and water molecules took place around 2Å distance, then was possible estimate that there is a diffusion of the polymer to the water. Also, radial distribution function between hydrophobic groups of the different chains was calculated. The distances between polymeric chains were considerably short so the chains adopted a conformation rather widespread; with the purpose of interact among them. In the second case of study, molecular dynamics of Polystyrene-bock-Poly(t-butylacrylate) (PS-b-PtBa) copolymers were carried out at different Polystyrene composition for the purpose of observed the influence of PS in the experimental dielectric relaxation of PtBa. From molecular simulation trajectories the conformational free energy was calculated to the different PS composition in the copolymers. Conformational free energies calculate from Boltzmann equation were correlated with activation free energies calculated from experimental dielectric relaxation. Activation free energy to PtBa was higher compared to P(305) and P(307), this mean that a more hindered change conformational was observed. A less favored change conformational was obtained to PtBa compared to P(305). Therefore, good agree was obtained between the activation free energy results and conformational free energy from molecular simulation. Finally, it is possible conclude that molecular dynamics simulation is a important and challenging tool that can be used to know properties and processes that occurs in polymer systems both dissolution and condensed phase. A important point to take in account is dispose of appropriate atomic parameters, that mean an adequate force field and the other hand, a model that represent the polymeric system.

6. Acknowledgment

This work was supported in part by "Proyecto de Inserción en la Academia CONICYT N°79090038" and InnovaChile CORFO Code FCR-CSB 09CEII-6991.

7. References

Alig, I., Kremer, F., Fytas, G., and Roovers, J., 1992. DIELECTRIC-RELAXATION IN DISORDERED POLY(ISOPRENE STYRENE) DIBLOCK COPOLYMERS NEAR THE MICROPHASE-SEPARATION TRANSITION. Macromolecules 25, 5277-5282.

Almeida, E.W.C., Anconi, C.P.A., Novato, W.T.G., De Oliveira, M.A.L., De Almeida, W.B., and Dos Santos, H.F., 2011. Box-Behnken design for studying inclusion complexes of triglycerides and alpha-cyclodextrin: application to the heating protocol in molecular-dynamics simulations. Journal of Inclusion Phenomena and Macrocyclic Chemistry 71, 103-111.

Andersen, H., 1983. Rattle: A Velocity Version of the Shake Algorithm for Molecular Dynamics Calculations. J. Comput. Phys. 52, 24-34.

Boonyarattanakalin, K.S., Wolschann, P., and Lawtrakul, L., 2011. Molecular dynamics of beta-CD in water/co-solvent mixtures. Journal of Inclusion Phenomena and Macrocyclic Chemistry 70, 279-290.

Darden, T., York, D., and Pedersen, L., 1993. J. Chem. Phys. 98, 10089.

Donald R. P., B., C.B., 2000. Polymer Blends.

Duncan, S.L., Dalal, I.S., and Larson, R.G., 2011. Molecular dynamics simulation of phase transitions in model lung surfactant monolayers. Biochimica Et Biophysica Acta-Biomembranes 1808, 2450-2465.

Encinar, M., Guzman, E., Prolongo, M.G., Rubio, R.G., Sandoval, C., Gonzalez-Nilo, F., Gargallo, L., and Radic, D., 2008. Dielectric and dynamic-mechanical study of the mobility of poly(t-butylacrylate) chains in diblock copolymers: Polystyrene-b-poly(t-butylacrylate). Polymer 49, 5650-5658.

Flory, J.P., 1969. Statistical mechanics of chain molecules, NY.

Flory, J.P., 1995. Principles of polymer chemistry, United state of america.

Gargallo, L., Vargas, D., Becerra, N., Sandoval, C., Saldias, C., Leiva, A., and Radic, D., 2009. Supramolecular Structures, Organization and Surface Behavior at Interfaces. Macromol. Symp. 278, 80-88.

Gargallo, L., Becerra, N., Sandoval, C., Pitsikalis, M., Hadjichristidis, N., Leiva, A., and Radic, D., 2011. Amphiphilic Diblock Copolymers Containing Poly(N-hexylisocyanate): Monolayer Behavior at the Air-Water Interface. Journal of Applied Polymer Science 122, 1395-1404.

Gilson, M.K., Given, J.A., Bush, B.L., and McCammon, J.A., 1997. The statistical-thermodynamic basis for computation of binding affinities: A critical review. Biophysical Journal 72, 1047-1069.

Grubmüller, H., Heller, H., Windemuth, A., and Schulten, K., 1991. Mol. Simul. 6, 121.

Humphrey, W., Dalke, A., and Schulten, K., 1996. VMD: Visual molecular dynamics. Journal of Molecular Graphics 14, 33-&.

Izrailev, S., Stepaniants, S., Balcera, M., Oono, Y., and Schulten, K., 1997. Biophys. J. 72, 1568.

Jorgensen, W., Chandrasekhar, J., Madura, J., Impey, R., and Klein, M., 1983. J. Chem. Phys. 79, 926.

Kalé, L., Skeel, R., Bhadarkar, M., Brunner, R., Gursoy, A., Krawetz, N., and al., e., 1999. J. Comput. Phys. 151, 283.

Karatasos, K., Anastasiadis, S.H., Floudas, G., Fytas, G., Pispas, S., Hadjichristidis, N., and Pakula, T., 1996. Composition fluctuation effects on dielectric normal-mode relaxation in diblock copolymers .2. Disordered state in proximity to the ODT and ordered state. Macromolecules 29, 1326-1336.

Komarov, P.V., Mikhailov, I.V., Alekseev, V.G., Khizhnyak, S.D., and Pakhomov, P.M., 2011. Full-atom molecular dynamics study of structure and stability of filament-like aggregates formed by silver mercaptide molecules. Colloid Journal 73, 482-494.

Korchowiec, B., Korchowiec, J., Hato, M., and Rogalska, E., 2011. Glycolipid-cholesterol monolayers: Towards a better understanding of the interaction between the membrane components. Biochimica Et Biophysica Acta-Biomembranes 1808, 2466-2476.

Kotaka, T., and Adachi, K., 1997. Polymer liquid dynamics studied via dielectric normal mode spectroscopy. Macromolecular Symposia 124, 3-14.

Krishnamoorthy, N., Yacoub, M.H., and Yaliraki, S.N., 2011. A computational modeling approach for enhancing self-assembly and biofunctionalisation of collagen biomimetic peptides. Biomaterials 32, 7275-7285.

Kuttel, M., Brady, J., and Naidoo, K., 2002. J. Comput. Chem. 23, 1236.

Kyritsis, A., Pissis, P., Mai, S.M., and Booth, C., 2000. Comparative dielectric studies of segmental and normal mode dynamics of poly(oxybutylene) and poly(oxyethylene)-poly(oxybutylene) diblock copolymers. Macromolecules 33, 4581-4595.

Lee, H., Baker, J.R., and Larson, R.G., 2006. Molecular Dynamics Studies of the Size, Shape, and Internal Structure of 0% and 90% Acetylated Fifth-Generation Polyamidoamine Dendrimers in Water and Methanol. The Journal of Physical Chemistry B 110, 4014-4019.

Ma, L., He, C.Q., Suzuki, T., Azuma, M., Bin, Y.Z., Kurosu, H., and Matsuo, M., 2003. Molecular mobility of amorphous chain segments of ethylene-methyl methacrylate copolymer films as a function of temperature estimated by positron annihilation, X-ray diffraction, and C-13 NMR. Macromolecules 36, 8056-8065.

Maple, J.R., Hwang, M.J., Jalkanen, K.J., Stockfisch, T.P., and Hagler, A.T., 1998. Derivation of class II force fields: V. Quantum force field for amides, peptides, and related compounds. Journal of Computational Chemistry 19, 430-458.

Moreno, S., and Rubio, R.G., 2002. Dielectric study of the dynamics of poly(oxyethylene) chains in triblock copolymers: Poly(oxyethylene)-b-polystyrene-b-poly (oxyethylene). Macromolecules 35, 5483-5490.

Pan, W.X., Zhang, D.J., and Zhan, J.H., 2011. Theoretical investigation on the inclusion of TCDD with beta-cyclodextrin by performing QM calculations and MD simulations. Journal of Hazardous Materials 192, 1780-1786.

Prosa, T.J., Bauer, B.J., and Amis, E.J., 2001. From stars to spheres: A SAXS analysis of dilute dendrimer solutions. Macromolecules 34, 4897-4906.

Prosa, T.J., Bauer, B.J., Amis, E.J., Tomalia, D.A., and Scherrenberg, R., 1997. A SAXS study of the internal structure of dendritic polymer systems. Journal of Polymer Science Part B-Polymer Physics 35, 2913-2924.

Ryckaert, J.P., Ciccotti, G., and Berendsen, H.J.C., 1977. J. Comput. Phys. 23, 327.

Saldias, C., Gargallo, L., Sandoval, C., Leiva, A., Radic, D., Caballero, J., Saavedra, M., and Gonzalez-Nilo, F.D., 2009. Inclusion complexes containing poly(epsilon-caprolactone)diol and cyclodextrins. Experimental and theoretical studies. Polymer 50, 2926-2932.

Shi, J.H., Hu, Y., and Ding, Z.J., 2011. Theoretical study on chiral recognition mechanism of ethyl-3-hydroxybutyrate with permethylated beta-cyclodextrin. Computational and Theoretical Chemistry 973, 62-68.

Vanommeslaeghe, K., Hatcher, E., Acharya, C., Kundu, S., Zhong, S., Shim, J., Darian, E., Guvench, O., Lopes, P., Vorobyov, I., and MacKerell, A.D., 2010. CHARMM General Force Field: A Force Field for Drug-Like Molecules Compatible with the CHARMM All-Atom Additive Biological Force Fields. Journal of Computational Chemistry 31, 671-690.

Vogt, S., Gerharz, B., Fischer, E.W., and Fytas, G., 1992. LOCAL MOTIONS IN A MICROPHASE-SEPARATED POLY(STYRENE-B-METHYLPHENYLSILOXANE) DIBLOCK COPOLYMER MELT. Macromolecules 25, 5986-5990.

Wick, C.D., and Cummings, O.T., 2011. Understanding the factors that contribute to ion interfacial behavior. Chemical Physics Letters 513, 161-166.

Yamakawa, H., 2001. Modern Theory of Polymer Solutions, NY.

3

Molecular Dynamics Simulation of Permeation in Polymers

Hossein Eslami and Nargess Mehdipour
Persian Gulf University
Iran

1. Introduction

Sorption and diffusion of small molecules through polymers is a topic of broad range of applications in many industrially important phenomena. There is a rapidly growing demand for polymers of specified permeabilities, such as selective memberanes for separation technologies, barrier membranes for packaging applications, foaming, and plasticization. Polymers with a high degree of permeability and permselectivity have been widely used for gas or liquid separation systems based on membranes (Kesting & Fritzsche, 1993), while those with a low degree of permeability have been used in barrier packaging films as containers (Koros, 1990). Therefore, understanding the underlying mechanism of solubility and/or diffusion of peneterants in macromolecular substances is very useful in obtaining a clear picture of the molecular level mechanism of polymer permeability and in design of new membranes.

The first study of gas permeation through a polymer goes back to 1829 by Thomas Graham (Stannett, 1978). According to Graham's postulate, the permeation process involves the dissolution of penetrant followed by transmission of the dissolved species through the membrane. In the late 1870's, Stefan and Exner demonstrated that gas permeation through a membrane was proportional to the product of solubility and diffusion coefficients. Based on the findings of Stefan and Exner, von Wroblewski (1879) constructed a quantitative solution to the Graham's solution-diffusion model. The dissolution of gas was based on Henry's law of solubility, where the concentration of the gas in the membrane was directly proportional to the applied gas pressure. Later Daynes (1920) showed that it was impossible to evaluate both diffusion and solubility coefficients by steady-state permeation experiments. He presented a mathematical solution, using Fick's second law of diffusion, for calculating the diffusion coefficient, which was assumed to be independent of concentration.

Microscopic theories of the diffusion and permeation of penetrants in polymers, on the other hand, were developed during 1950-1970's. In these theories, to explain the mechanism behind the transport of gas molecules through the free volume present in the polymer matrix, the gas-polymer system was defined in terms of liquid molecules traveling through a liquid membrane. The Cohen and Turnbull (1959) theory considered transport through liquid molecules as hard spheres. In this model the gas diffusion in polymers occurs through the redistribution of free volume. Another theory, proposed by Dibenedetto (1963), used the same concept as that of Cohen and Turnbull (1959) regarding free volume distribution, but with a different chain packing theory at the molecular level. An elaborate theory was developed by Pace and Datyner (1979), assuming that the penetrant molecule is moving along the polymer chain bundles and being stopped only by chain entanglements or a crystallite. The penetrnat molecules were further assumed to jump into the adjacent bundles, similar to the Cohen and Turnbull's model. Such a jump event was considered to be the rate controlling step, with the diffusion along the bundle being three times faster than the perpendicular jump of the molecule. None of these theories, however, provides a molecular lever understanding of the permeability in polymers.

Significant advances in the understanding of gas permeation in polymers have been made only in the last few decades with the advent of powerful computers. Monte Carlo (MC) and molecular dynamics (MD) simulations on small-molecule permeation of amorphous polymers have become feasible in recent decades. The simulations now cover a range of different polymers with varieties of chemical complexities ranging from flexible polymers with simple chemical structure, like polyethylene (Boyd & Pant, 1991), to stiff polymers with detailed chemical structure, like polyamides (Eslami & Mehdipour, 2011). In the following we give a detailed discussion of the calculation methods as well as the polymeric samples employed to study the sorption/diffusion mechanism. The main difficulty in the calculation of gas permeabilities stems from the calculation of solubility coefficients. In fact the calculation of gas solubilities necessitates the condition of equilibrium between the permeant (sorbate) in the gas phase and that dissolved in polymer. The equality of the temperature, pressure, and chemical potentials of all species is the necessary condition to establish such an equilibrium situation. The chemical potential is, however, coupled to the number of particles and cannot be easily calculated using molecular simulation methods. Therefore, while there exist excellent reviews on the pearmation of gases in polymers, with the main emphasis on the diffusion process (Müller-Plathe, 1994), in this review we explain the polymer permeation putting more emphasis on the solubility process.

2. Sorption of gases in polymers

Historically, the dual-mode sorption theory (Stannett et al., 1978; Frederickson and Helfand, 1985) describes the concentration C (solubility) of the gas inside a polymer at equilibrium with the gas at a partial pressure p, i.e.,

$$C = k_H p + C_\infty \frac{bp}{1+bp} \tag{1}$$

where k_H is Henry's law solubility coefficient, C_∞ is the saturation concentration of the gas, and b is the affinity coefficient. This model assumes that there are two distinct modes by which a glassy polymer can sorb gas molecules; Henry's law and a Langmuir mechanism which corresponds to the sorption of gases into specific sorption sites in the polymer. Henry's constant has the same physical meaning for glassy polymers as it does for rubbery polymers and liquids, whereas the Langmuir-type term is believed to account for gas sorption into interstitial spaces and microvoids, which are consequences of local heterogeneities and are intimately related to the slow relaxation processes associated with the glassy state of the polymer. Local equilibrium is assumed between the two modes. Figure 1 shows a schematic representation of Henry, Langmuir, and dual-mode sorptions.

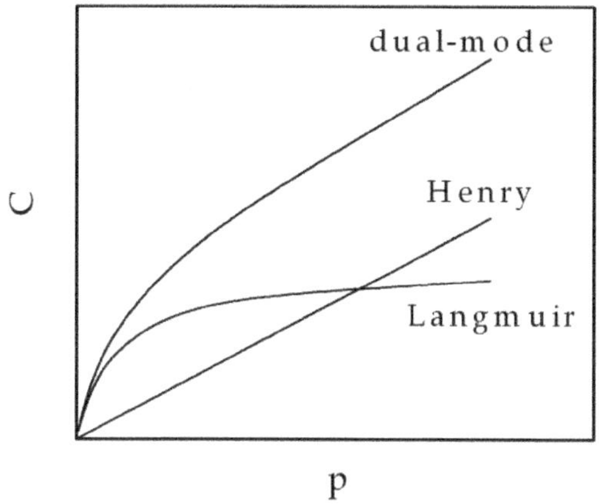

Fig. 1. Schematic representation of Henry, Langmuir, and dual-mode sorptions.

In the low pressure region, Eq. 1 provides the following linear relationship against the pressure:

$$C = (k_H + C_\infty b)P = S_0 P \tag{2}$$

where S_0 is called the apparent solubility coefficient in the zero-pressure limit in glassy polymers.

The determination of the free energy change in sorption of gases in polymer is a problem of primary importance, since the free energy is the main thermodynamic parameter governing the equilibrium between the gas in the pure state and the dissolved one in the polymer. For a system of N particles located at r_1, r_2, ...r_N, the statistical mechanical expression for the Helmholtz free energy, A, reads as (McQuarrie, 1976):

$$A = -k_B T \ln(Q) = \frac{1}{N!\Lambda^{3N}} \int_0^\infty ... \int_0^\infty \exp\left(-U_N\left(r^N\right) / k_B T\right)\, dr^N \tag{3}$$

with

$$\Lambda = \left(\frac{h}{2\pi m k_B T}\right)^{1/2} \tag{4}$$

Here, Q is the canonical partition function, U_N is the potential energy of the system, r^N stands for the whole set of coordinates, r_1, r_2, ...r_N, Λ is the de Broglie wavelength, h is the Planck's constant, m is the molecular mass, k_B is the Boltzmann's constant, and T is the temperature. Assuming pair-wise additiveity of the potential energy between particles, we have:

$$U_N = \sum_{i}^{N} \sum_{j>i}^{N} u_{ij}(r_{ij}) \tag{5}$$

where u_{ij} is the pair potential interacting between particles i and j and r_{ij} is the interparticle distance. The expression for the chemical potential is obtained by taking the logarithm of the ratios of partition functions for a system composed of N particles, with the sorbate (penetrant) density of ρ_s, and a system composed of $N+1$ particles, which is obtained by adding one sorbate particle to the previous N-particle system, i.e,

$$\mu_s = A_{N+1} - A_N = k_B T \ln\left(\rho_s \Lambda^3\right) - k_B T \ln\left(\langle \exp(-\Delta U / k_B T \rangle\right) \tag{6}$$

with

$$\Delta U = U_{N+1} - U_N \tag{7}$$

Here μ_s and ρ_s are the chemical potential and the density of sorbate, respectively, and the brackets indicate the ensemble average. Comparing μ_s with the chemical potential of the ideal gas at the same state, i.e.,

$$\mu_s = \mu_s^{ex} + k_B T \ln\left(\rho_s \Lambda^3\right) \tag{8}$$

various contributions to the chemical potential can be interpreted according to Ben-Naim (1987). In Eq. (8) μ_s^{ex} is called the excess chemical potential of sorbate. The second term on the right hand side of Eq. (8) is the chemical potential of the ideal gas. Now, the chemical potential of the sorbate is split into two terms; a term arising from putting the sorbate molecule at a fixed position in the polymer matrix, μ_s^{ex}, and a term arising from releasing the constraint, i.e., letting the solute molecule move freely, which results in the contribution $k_B T \ln\left(\rho_s \Lambda^3\right)$ to the chemical potential. In fact this term is the contribution of translational motion to the chemical potential.

Looking at the sorption process, we consider the pure sorbate gas s in equilibrium with the dissolved gas in the polymer. Again we connect the thermodynamic states of the system to their corresponding ideal gas states (at the same temperature, and the same density of sorbate s), as it is depicted in Scheme 1.

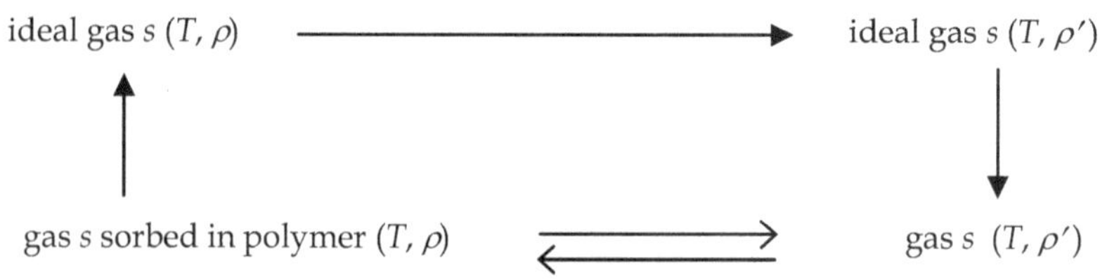

Scheme 1. Description of equilibrium between the gas s in the gaseous and polymeric phases.

Here, ρ and ρ' indicate the densities of gas s in the polymeric and gaseous phases, respectively. The condition of equilibrium for gas s, between the gas phase and the polymer phase is the equality of the chemical potentials, i.e.,

$$\mu_s(T,\rho) = \mu_s'(T,\rho') \tag{9}$$

where μ and μ' are the chemical potentials of the sorbed gas in polymer and in gas phase, respectively. According to the thermodynamic cycle shown in Scheme 1, the equilibrium condition between the pure gas and the gas sorbed in the polymer is formulated as:

$$\mu_s^{ex} = \mu_s'^{ex} - k_B T \ln \frac{\rho}{\rho'} \tag{10}$$

Equation 10 is the main equation governing the equilibrium between the sorbed gas molecules in the polymer and those in the gaseous phase. According to this equation, the main parameter to be determined for the calculation of solubilities is the excess chemical potential. In the following section we study the computational approach, with the main emphasis on the MD simulation methods, for calculation of chemical potentials.

3. Calculation of excess chemical potentials

From the historical point of view, traditional approaches such as the equation-of-state models (Lacombe and Sanchez, 1976 ; Sanchez and Rodgers, 1990 ; von Solms et al., 2005) and the activity coefficient models (Ozkan and Teja, 2005) have been applied to calculate the phase equilibria and sorption of penetrant molecules in polymers. Molecular simulations are the other attractive method for this type of calculation. These methods do not invoke any approximations and predictions are based on well-defined molecular characteristics. In the following sections we describe the application of molecular simulation methods in the case of sorption of gases in polymers. As stated above, the main problem is to calculate the excess chemical potentials, which is given address to in the following sections.

3.1 Widom's test particle insertion method

The Widom's test particle insertion method (Widom, 1963) is an elegant method for calculating the excess chemical potentials. In this method a test particle is momentarily inserted at random points in the simulation box and the interaction energy between the test particle and the host particles are averaged to yield the excess chemical potential. Since the method relies on the statistical accuracy of sampling of configurations that permit the insertion of molecules with low values of the binding energies, its application is more feasible for small-sized solutes. In a dense fluid there is a small probability of finding a cavity to insert a particle in. Therefore, particle insertion becomes a rare event and long simulation times are needed to obtain a good statistics. Similarly, at high densities, the probability of particle deletion is also reduced. The deletion probability depends on the probability of generation of a high density configuration, from which to delete a particle. Due to several limitations, such as poor sampling at high densities or inadequacy of the method in the case of chemical potential calculation for polar systems, the utility of this method to many important applications is precluded. The problem of poor sampling has been solved to some extent by developing such techniques as the umbrella sampling (Shing and K. E. Gubbins, 1981) and map sampling (Deitrick et al., 1989), as the extensions of original Widom's method. The Widom method or its modifications can be used in both MC and MD simulations, but in this method the chemical potential will be calculated at the end of the simulation. This method has widely been applied in calculating the excess chemical potentials of small penetrant molecules in polymers including polypropylene (Müller-Plathe, 1991), polydimethyl siloxane (Sok et al., 1992), polyisobutylene (Müller-Plathe et al., 1993), polystyrene (Eslami & Müller-Plathe, 2007a), and polyimides (Neyertz & Brown, 2004; Pandiyan et al. 2010). However, the method is not applicable in the case of excess chemical potential calculation for polar penetrant in polar polymers. Moreover, the sampling probability gets poor in dense polymeric systems and in the case of big penetrants.

3.2 Thermodynamic integration and fast growth thermodynamic integration methods

Thermodynamic integration method is another useful method for the calculation of the excess chemical potentials. In this method, a parameter λ is used to couple the sorbate with the rest of the system and the excess chemical potential, or coupling work, is obtained by integrating along λ as:

$$\mu_s^{ex} = \int_0^1 \left\langle \left(\frac{\partial U}{\partial \lambda} \right) \right\rangle d\lambda \tag{11}$$

In this method, several simulations are required to obtain a series of points for the integration. To improve the sampling, recently Hess and van der Vegt (2008) have developed a fast-growth thermodynamic integration method. The approach is based on the non-equilibrium work theorem of Jarzynski (1997), which relates the work that is being performed on a system when going from an initial state to a final state along a coordinate λ with the free energy change. Through the applications of this method, polymer microstructure can be considered as a potential landscape as in Figure 2.

In this context, the free energy can be calculated by performing fast growth thermodynamic integration in order to calculate the work required to insert the sorbate in a solvent matrix from sampling the possible coordinates where interaction potentials determine the potential landscape. This method has been applied successfully to calculate the excess chemical potential of relatively large molecules in polymer matrices (Hess & van der Vegt 2008; Hess et al. 2008; Fritz et al., 2009).

3.3 Open system simulations

Since the chemical potential is conjugated with the number of particles, the statistical mechanical ensemble representative for its calculation is the grand canonical ensemble. There are several open-system simulation techniques in the literature, allowing to set the target chemical potential as an independent thermodynamic variable. These methods are preferential to the methods explained above for the calculation of phase equilibria, in that the chemical potential in each phase is set from the beginning as an independent variable, while in closed system simulations the chemical potential is calculated at the end of the run. In the following two subsections we describe the grand canonical ensemble simulations using MC and MD techniques.

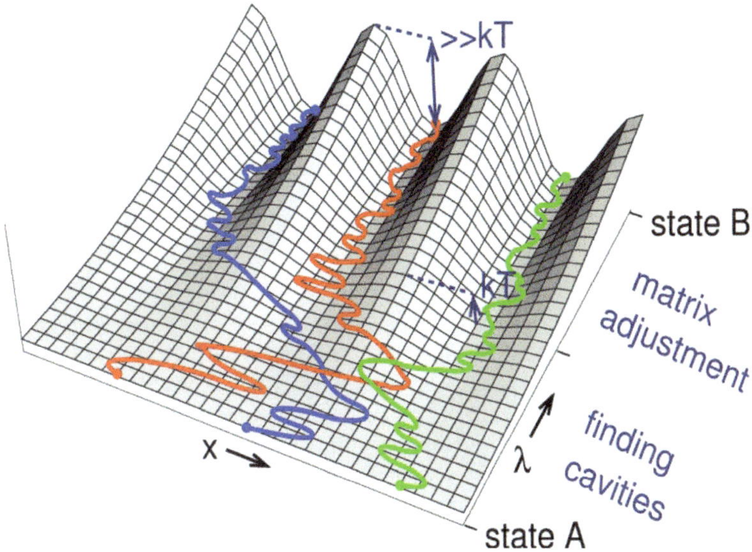

Fig. 2. A schematic one-dimensional representation of a free-energy landscape for an additive molecule (x is a component of its center of mass) in a polymer matrix as a function of the coupling parameter λ. During a fast-growth simulation, a molecule is forced along λ from state A to B, while sampling in x; three example paths are given. With fast-growth thermodynamic integration, it is possible to explore the multiple minima in the free energy landscape by multiple trajectories starting from a flat free energy landscape (a solute not interacting with the matrix). Figure is taken from Hess and van der Vegt (2008) with permission.

3.3.1 Grand canonical ensemble Monte Carlo simulation

The first grand canonical simulation method (Norman & Filinov, 1969) was based on a MC simulation technique. This method was then developed by Adams (1974, 1975). In these methods full particles are inserted into or removed from the simulation box, in a Markov process with a probability, which depends on the target chemical potential. These kinds of movements are compatible with the stochastic nature of moves in MC technique. The main difficulty with the MC simulation methods in the grand canonical ensemble is their inefficiency when applied to the condensed phases. Besides, the insertion or deletion of full particles highly perturbs the simulation box, and it takes some time for the fluid to relax. These methods envisage similar sampling problems, as addressed above in the case of Widom's method, at high densities. However, the sampling problem has been overcome somewhat by searching in the simulation box for suitable cavities and biasing the insertion event (Escobedo & de Pablo, 1995; Shelley & Patey, 1995; Boda et al., 1997). Another problem with the MC simulation methods in the grand canonical ensemble is that these methods do not directly reflect the dynamic information about the system. Therefore, attention has been attracted toward development of the grand canonical ensemble simulations using MD technique.

3.3.2 Grand canonical ensemble molecular dynamics simulation

The discontinuity of the number of particles, introduced through particle insertion/deletion, makes it difficult to do MD simulation in the grand canonical ensemble. Therefore, in some of the existing MD-based simulations in the grand canonical ensemble, the sampling is performed through using the MC-based stochastic procedures (Boinepalli & Attard, 2003). Another alternative, is the introduction of the particle number as a continuous dynamic variable through the MD simulation (Eslami & Müller-Plathe, 2007b; Eslami et al., 2011). This approach is based on the so-called extended system method, in which a Hamiltonian including the kinetic and potential energy terms for the extension variable is constructed and the appropriate equations of motion are solved.

Here we assume that our physical system is composed of N real particles plus one "fractional" particle and is coupled to a heat reservoir and a particle reservoir. The fractional particle is a particle whose potential energy is weighted by a variable λ which varies from zero to one. The inclusion of the fractional particle in the system provides a variable number of particles in a dynamical way. The Hamiltonian of such a combined system; the real N particle system, the fractional particle, and the additional degrees of freedom, i.e., the material reservoir and the heat reservoir, is written as (Eslami & Müller-Plathe, 2007b):

$$H=\sum_{i=1}^{N}\frac{p_{x_i}^2+p_{y_i}^2+p_{z_i}^2}{2m_i s^2}+\sum_{i=1}^{N-1}\sum_{j>i}^{N}U_{ij}+\frac{p_{x_f}^2+p_{y_f}^2+p_{z_f}^2}{2m_f s^2}+\sum_{i=1}^{N}U_{if}+\frac{p_\lambda^2}{2W_\lambda}+U_v+\frac{p_s^2}{2W_s}+U_s \quad (12)$$

where H is the Hamiltonian, subscripts i and j refer to the real particles, subscript f refers to the fractional particle, p_i is the momentum of the i_{th} particle, s is the velocity scaling variable which couples the system to the thermostat, and λ is the particle-number scaling variable which couples the fractional particle to the rest of the system and varies from zero to one. In this equation, the first four terms on the right hand side are the kinetic energy of the real particles, potential energy of interaction between real particles, kinetic energy of the fractional particle, and the potential energy of interaction between real particles and the fractional particle. The number extension variable, λ, and the temperature extension variable, s, have their own kinetic and potential energy terms, shown as the last four terms on the right hand side of Eq. (12). In fact U_λ is the work required to carry a fractional particle from a medium of zero potential to the medium of interest with the potential energy U; i.e., it is the so-called excess Helmholtz free energy.

Adopting one of the real particles (molecules) in the simulation box as the fractional particle, the fractional particle is grown to a full (real) particle or deleted from the system in a dynamical way, by solving the equations of motion. The relevant equation of motion, governing the dynamics of fractional particle (the magnitude of coupling to the ideal gas reservoir) is (Eslami & Müller-Plathe, 2007b):

$$W\ddot{\lambda}=-\sum_{i=1}^{N}\frac{\partial U_{if}}{\partial\lambda}+\mu-\mu^{id} \quad (13)$$

where d $\ddot{\lambda}$ =$d^2\lambda/dt^2$ and μ^{id} is the chemical potential of the ideal gas. In fact, combination of the last two terms on the right hand side of Eq. (13) is the target excess chemical potential. By adding/deleting the fractional particle into/from the simulation box, decision is made either to choose another real particle in the box as the new fractional particle, or to add a new fractional particle to the box. This is done according to the algorithm by Eslami and Müller-Plathe (2007b). On adding the new fractional particle to the box, the coupling parameter, λ, is set close to zero. Similarly, when a real particle is chosen as the new fractional particle, λ is set close to 1. Repeating such a procedure, equilibrium is achieved, in which the density (number) of particles in the system oscillates about an average value, corresponding to the target chemical potential, temperature, and volume.

4. Phase equilibrium calculation

There are several methods for the calculation of phase coexistence points using molecular simulations, such as thermodynamic scaling method (Valleau, 1991; Valleau and Graham, 1990), histogram reweighting method (McDonald and Singer, 1967; Potoff and Panagiotopoulos, 1998), the Gibbs-Duhem integration method (Kofke, 1993a,b), NpT plus test particle method (Boda et al., 1995; 1996), various extensions of it to other ensembles (Moller and Fischer, 1990; Boda et al., 2001), and the Gibbs eensemble MC method (Panagiotopoulos,1987). In the Gibbs eensemble MC method, which is one of the most popular methods for the calculation of phase equilibria, the simulation box is divided into two subsystems and simultaneous simulation of both subsystems are done. To establish the equilibrium (the equality of chemical potentials), particles are exchanged between the two subsystems. This technique has been applied to coexistence properties of simple systems, such as fluids of spherical Lennard-Jones or Yukawa particles (Panagiotopoulos et al., 1998; Rudisill and Cummings, 1989), as well as more complex systems, such as polyatomic hydrocarbons (de Pablo and Prausnitz, 1989; de Pablo et al., 1992) and chain molecules (de Pablo, 1995). There are also reports on the mixed methods in which the molecular simulation approaches have been utilized, to calculate the chemical potentials in the condensed phase and the results from equations-of-state predictions are used to calculate the phase coexistence point (Lim et al., 2003), or to calculate the interaction energy parameters of solvent and polymer, in combination with statistical-mechanical theories for the study of phase equilibria of polymer-solvent mixtures (Jang and Bae, 2002).

Many computational studies of the permeation of small gas molecules through polymers, using afore-cited techniques, have appeared, which were designed to analyze, on an atomic scale, diffusion mechanisms or to calculate the diffusion coefficient and the solubility parameters. Most of these studies have dealt with flexible polymer chains of relatively simple structure such as polyethylene, polypropylene, and poly-(isobutylene) (Müller-Plathe, 1991; Pant & Boyd, 1993; Tamai et al., 1995; Pricl and Fermeglia, 2003; Abu-Shargh, 2004). There are however some reports on polymers consisting of stiff chains. Of these we may address to the works by Mooney and MacElroy (1999) on the diffusion of small molecules in semicrystalline aromatic polymers, by Cuthbert *et al.* (1997) on the calculation of the Henry's law constant for a number of small molecules in polystyrene and the effect of box size on the calculated Henry's law constants, by Lim *et al.* (2003) on the sorption of methane and carbon dioxide in amorphous polyetherimide. In some of these reports the authors have used Gibbs ensemble Monte Carlo method (Vrabec and Fischer, 1995; Panagiotopoulos, 1987) to do the calculations, and in some other cases alternative methods, like the equation-of-state approaches are employed to describe the gas phase. In fact, as explained above, one needs to satisfy the equality of chemical potentials in both phases.

There are, however, some recent methods employing the grand canonical ensemble simulation formalism (to set the target chemical potentials in equilibrating phases), without the necessity to do simultaneous simulations on two boxes (unlike the Gibbs ensemble MC method) to calculate the gas solubilities in polymers. In the following section we explain these methods.

4.1 Grand equilibrium method

In the Gibbs ensemble simulation method one specifies the thermodynamic variables temperature, global composition, and global pressure for the simulation of both phases in separate volumes. Practically, this set of thermodynamic variables is in many cases not convenient and simultaneous simulation of both phases has the disadvantage that fluctuations occurring in one phase influence the other one. Recently a new method, grand equilibrium method, has been developed by Vrabec and Hasse (2002). This method circumvents the afore-cited problems for the study of phase equilibria. The specified thermodynamic variables are temperature and composition and two independent simulations are performed for the two phases without the need to exchange particles in the condensed phase. According to this method for a mixture composed of several components, it is possible to do a simulation in the isothermal-isobaric (*NpT*) ensemble at constant temperature, a constant composition of the condensed phase, and at an arbitrary constant pressure, preferably close to the pressure at the phase coexistence point, to obtain the density of the condensed phase.

In the grand equilibrium method, a simulation of the condensed phase is done to calculate the excess chemical potentials, μ_s^{ex}, and the partial molar volumes, V_s, of all components. One may use the test-particle insertion method (Widom, 1963) to calculate the excess chemical potentials and the partial molar volumes as:

$$\mu_s^{ex} = -k_B T \ln\left\langle \frac{pV}{Nk_B T} \exp(-\Delta U / k_B T) \right\rangle \tag{14}$$

and

$$V_s = \frac{\left\langle V^2 \exp(-\Delta U / k_B T) \right\rangle}{\left\langle V \exp(-\Delta U / k_B T) \right\rangle} - \langle V \rangle \tag{15}$$

Knowing the parameters V_s and μ_s^{ex} from this simulation, the desired excess chemical potentials as functions of pressure are obtained from a first order Taylor series expansion, i.e.,

$$\mu_s^{ex}(p) = \mu_i^{ex}(p^*) + \frac{V_s}{kT}\left(p - p^*\right) \tag{16}$$

where p^* is the target pressure at which the NpT ensemble simulation is done. Once the $\mu_s^{ex}(p)$ is determined from Eq. (16) by one NpT ensemble simulation of the condensed phase, one vapour/gas simulation has to be performed in the pseudo-grand canonical ensemble (pseudo-μVT). In a common grand canonical ensemble (Eslami and Müller-

Plathe, 2007b) the parameters temperature, volume, and the chemical potential of all species are fixed, while in this pseudo-μVT ensemble simulation, the parameters T and V are fixed in the common way, but instead of fixing the chemical potentials, they are set as a function of the pressure in the gas phase. This procedure ensures that equilibrium between the condensed phase and the gas phase is imposed. In a common μVT ensemble simulation (Eslami and Müller-Plathe, 2007b), the chemical potentials are set through insertion and deletion of particles by a comparison between the resulting potential energy change and the desired fixed chemical potential. Here, starting from a low density state point, the gas phase simulation is forced to change its state to the corresponding phase equilibrium state point.

5. Applications

Recently, we have applied the grand equilibrium method for the calculation of solubilities of gases in polymers. The method has been well tested for the case of gas solubilities in polystyrene over a wide range of temperatures and pressures (Eslami and Müller-Plathe, 2007a). To calculate the phase equilibeium points, MD simulations were performed at a specified temperature and concentrations of sorbed gases in polymer. After equilibration, several configurations are extracted at different times from the dynamic simulation and used to insert the test particles, to calculate the excess chemical potentials. Calculating the excess chemical potentials of the sorbed gases in polymer as a function of pressure, a simulation of the gas phase is done in the grand canonical ensemble, to find the phase coexistence point. We have shown in Figure 3 typical results for the zero-pressure limit solubility coefficients of methane in polystyrene (Eslami and Müller-Plathe, 2007a).

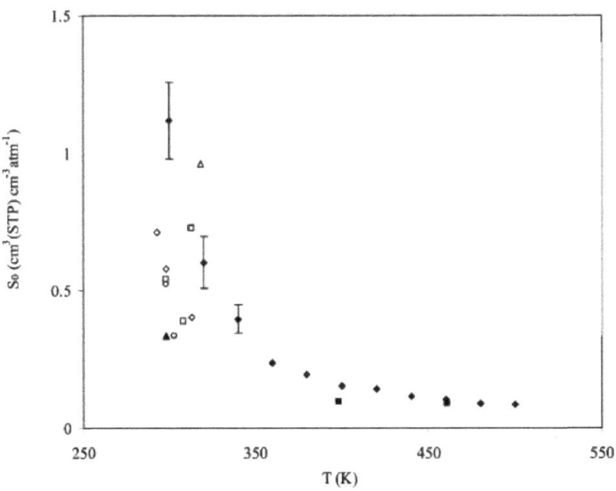

Fig. 3. Comparison of calculated (♦) and experimental (◇ Sada et al. 1987; □ Vieth et al., 1966a; △ Vieth et al., 1966b; ○ Barrie et al. 1980; ▲ Raymond et al. 1990; ■ Lundberg et al. 1969) zero-pressure limit solubility coefficients of methane in polystyrene. Figure is taken from Eslami and Müller-Plathe (2007a).

Another more interesting application of the method is in the case of gas solubility calculation for polar penterants in polar polymers. This is a severe test, as most of the existing methods, like the Widom's test particle insertion method or its modifications, are not applicable in the case of polar fluids. We have recently applied this method to the case of water solubilities of poly(ethylene terephthalate) (Eslami & Müller-Plathe, 2009) and polyamide-6,6 (Eslami & Mehdipour, 2011). Preparing an initially relaxed configuration of the polymer, we have performed a long simulation, in the grand canonical ensemble, of the polymer phase to get the density of water, which corresponds to the target values of μ, V, and T. The number of molecules (density) during the simulation changes until achieving equilibrium, at which the density fluctuates about the average value. We have shown in Figure 4 the variation of number of water molecules in the simulation box at two different temperatures (one above and one below the glass transition temperature).

In Figure 4, the results at T= 280 K belong to $\mu^{ex}{}_s$ = -18.5 kJmole^{-1} in a cubic box with a box length, L, of 3.082 nm. Calculating the partial molar volume of water from the results of this simulation, we expanded the chemical potential of water in the polymer phase as a function of pressure, according to Eq. (16). Then we have performed another simulation of the gas phase in the grand canonical ensemble, as explained above. Here the target chemical potential is varied during the simulation as a function of running-average pressure in the gas phase (see Eq. (16)). The result for time evolution of the density of water in the gas phase (L=50 nm) is also shown in Figure 4. The results in Figure 4 show that the gas phase quickly adjusts its state to the phase coexistence point. Also shown in the same figure are the results of similar calculations at 450 K ($\mu^{ex}{}_s$ = -9.0 kJmole^{-1}). Here the values of L for simulations of polymer and gas phases are 3.128 nm and 7.0 nm, respectively.

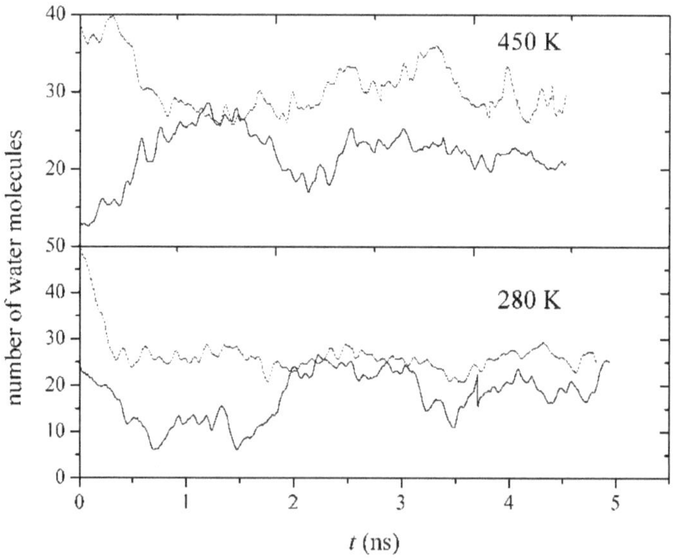

Fig. 4. Time evolution of the number of water molecules in the grand canonical ensemble simulation at 280 K and 450 K. The solid and dotted curves represent the results for water molecules in poly(ethylene terephthalate) and in gaseous phases, respectively.

Similar calculations are done in the case of water solubility of polyamide-6,6 (Eslami & Mehdipour, 2011) over a wide range of temperatures and pressures. This is a particularly interesting example of the applicability of the method, since polyamide-6,6 is a highly polar polymer matrix with the ability of forming a strong hydrogen bond network and can dissolve water to about 10 wt %. We have shown in Figure 5 a typical calculated sorption isotherm for water in polyamide-6,6 at 300 K and its comparison with experiment.

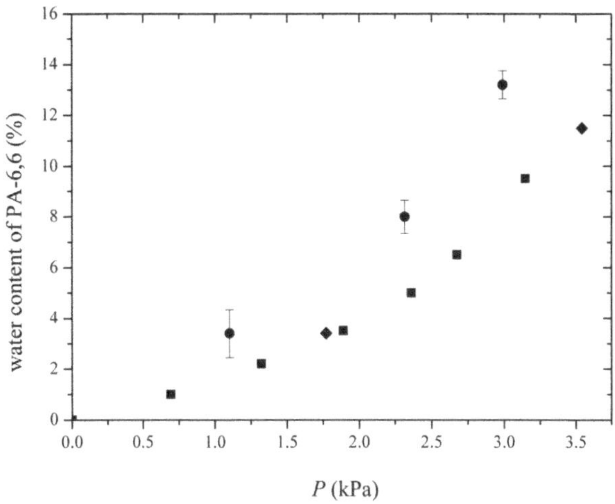

Fig. 5. The sorption isotherm of water in PA-6,6 at 300 K. The symbols represent calculations (●) and experimental data at 298 K (■) (Lim et al., 1999) and 300 K (◆) (Watt, 1980). Figure is taken from (Eslami & Mehdipour, 2011).

6. Diffusion in polymers

Most theories describing the mechanism of diffusion in polymeric materials are based on the free-volume approximation (Vrentas & Duda, 1979). In the free-volume theories, there is a volume which is directly occupied by polymeric molecules, and there is the remainder of the volume, which is called the free volume. A part of the free volume is assumed to be uniformly distributed among the molecules and is identified as the interstitial free volume, which requires a large energy for redistribution and is not affected by random thermal fluctuations. The other part, which is called the hole free volume, is assumed to require negligible energy for its redistribution. Therefore, the hole free volume is being continuously redistributed due to random fluctuations, and is assumed to be occupied by penetrant molecules. This redistribution of hole free volume will move the penetrant molecule with it. According to this model, by movement of segments of the polymer chain, a void will be created adjacent to the penetrant molecule. If the size of this hole is sufficient to host a penetrant molecule and if the penetrant have sufficient energy to jump into the hole, a successful jump of the penetrant molecule is made into the hole. Although the free volume model has been used extensively to describe the mechanism of transport through molten or glassy polymers (Mauritz & Storey, 1990), this model does not show a microscopic view point of penetrant transport in polymers, since it just connects bulk transport properties, like diffusion coefficient, into bulk properties, like molecular volume or thermal expansion coefficient.

The transition-state theory (TST), introduced by Gusev et al. (1993), is another useful method for the calculation of diffusion coefficient of a low-molar-mass substance through the polymer matrix. In the TST it is assumed that the movement of the penetrant from an initial cavity to the saddle point and to a neighboring cavity is a unimolecular rearrangement. For such a transition the reaction trajectory in configuration space is tracked and the transition rate constant is evaluated. In the first studies on the application of TST to study the dynamics of light gases dissolved in rigid microstructures of glassy polycarbonate and rubbery polyisobutylene Gusev et al. (1993), the method was shown to be only capable to study just the dynamics of light gases like He. The method was developed by Gusev and Suter (1993), by Gusev et al. (1994), and by Karayiannis et al. (2004), to calculate the diffusion coefficient of bigger penetrants in glassy polymers.

Computer modeling of molecular systems at a detailed atomistic level has become a standard tool in investigation of sorption and diffusion of small molecules in polymeric media (Müller-Plathe, 1991, 1994, Milano et al., 2002, Mozaffari et al. 2010). MD simulation is a useful tool for exploring the structure and properties of bulk amorphous polymers. The length of the trajectories that can be generated in practice presently is on the order of many nanoseconds. Thus the range of properties that can be studied directly is limited to those that evolve over this time scale. One of the phenomena that appear to be suitable for investigation is the diffusion of small penetrant molecules in an amorphous polymeric matrix. That is, the diffusion coefficients of small penetrants in many rubbery or liquid polymers are such that, at temperatures close to room temperature and above, the average displacement of the diffusant is large enough in a nanosecond interval to be determined via molecular dynamics simulation. Performing such simulations is of practical importance in predicting diffusion coefficients and also in understanding the mechanism of diffusion.

Although there has been significant progress in the use of MD methods in the simulations of diffusion coefficients, early studies were focused on the simulation of gas diffusion in rubbery polymers which could be investigated using full atomistic or united-atom simulations in reasonable computational times (Boyd, 1991; Sok et al., 1992, Pant & Boyd, 1993; Gee & Boyd, 1995). Due to the recent development of improved force fields and the wider availability of sophisticated commercial softwares and high-speed computing facilities, attention is shifting to direct to the more challenging task of simulating the slower diffusional processes occurring in glassy polymers (Han & Boyd, 1996, Milano et al. , 2002, Lime at al., 2003, Mozaffari et al., 2010). In the following section we present the MD simulation results done recently by the authors on the diffusion of gases in polystyrene over a wide range of temperatures and pressures.

6.1 MD simulation of diffusion in polymers (case study)

Recently we have studied the diffusion mechanism of some gases in polystyrene over a wide range of temperatures, ranging from below the glass transition temperature to temperatures well above the glass transition temperature (Mozaffari et al., 2010). Center-of-mass mean-square displacements have been measured during the simulation to calculate the diffusion coefficients. In the limit of long times, which the penetrant molecules perform random walks in the polymer matrix, the mean-square-displacement becomes linear in time, and the diffusion coefficients can be calculated using the Einstein relation:

$$D = \frac{1}{6N} \lim_{t \to \infty} \frac{d}{dt} \left\langle \sum_{i}^{N} [\boldsymbol{r}_i(t) - \boldsymbol{r}_i(0)]^2 \right\rangle \tag{17}$$

The diffusion coefficients of penetrant gases in polystyrene have been calculated over a wide range of temperatures, 300 K-500 K. The motion pattern of penetrant gases in host polymer can be qualitatively studied by monitoring the penetrant's displacement $|r(t)-r(0)|$, from its initial position. We have shown in Figure 6 the displacements of argon and propane in polystyrene at 300 K.

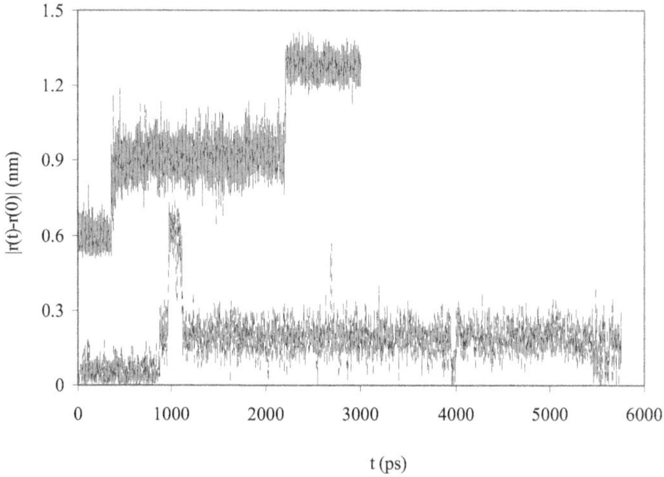

Fig. 6. Displacement of argon (upper curve) and propane (lower curve) molecules from their initial positions at 300 K. In order to avoid overlapping between the curves, the displacements of argon are offset by 0.6 nm. Figure is taken from Mozaffari et al. (2010).

The curves are representative of a common hopping mechanism, showing that for a considerable time interval the penetrants dwell in existing voids in the polymer and occasionally do a jump into the neighboring voids. When dwelling in the voids, the penetrants just perform oscillatory motions around their equilibrium positions, therefore, no net motion of a penetrant molecule occurs with these positional fluctuations. The amplitude of oscillations varies according to the size of voids. From time to time, the penetrants can do a quick jump into their neighboring voids, see Figure 6. The jump frequency depends on penetrant's size, therefore the bigger penetrants (see Figure 6 in the case of propane) can rarely jump between the voids.

Particularly useful information can be obtained by analyzing the trajectories of penetrants in the polymer. The two-dimensional center-of-mass x-y trajectories of nitrogen, as a typical example, in polystyrene at temperatures below (300 K) and above (500 K) the glass transition temperature are indicated in Figure 7.

The results in Figure 7 are indicative of faster movement of penetrants at higher temperatures, as represented by the broadened range of displacements at higher temperatures. This indicates that at higher temperatures the hole free volume redistributes faster and penetrant molecules have higher energy to overcome the activation energy required to jump into new voids.

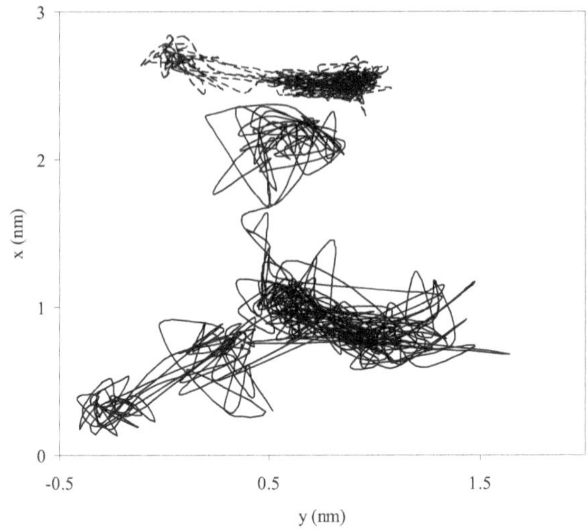

Fig. 7. Typical trajectories of nitrogen molecules in polystyrene at 300 K (upper curve) and 500 K (lower curve). Figure is taken from Mozaffari et al. (2010).

As a measure of diffusion coefficients, the center-of-mass mean-square displacements of carbon dioxide at temperatures below (340 K) and above (480 K) the glass transition temperature, are shown in Figure 8.

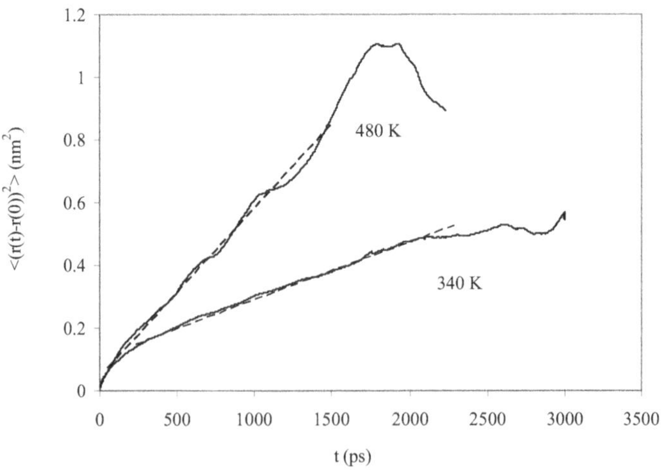

Fig. 8. Center-of-mass mean-square displacement for carbon dioxide at temperatures below and above the glass transition temperature. The curve at 480 K is scaled down by a factor of 15 for the sake of clarity. The dashed lines show the least-squares fits to the linear parts of the curves. Figure is taken from Mozaffari et al. (2010).

The linearity of mean-square displacements versus time, as indicated in Figure 8, confirms Einstein diffusion. In the glassy polymer at times below 500 ps, the penetrant motion is highly anomalous and the diffusion regime begins at longer times. Intercavity jumps rarely occur at this time scale. However, as it can be seen from the results in Figure 8 at high temperatures, the diffusion regime sets in a shorter time. These findings clearly indicate the difference between the diffusion mechanisms in the glassy and melt polymers.

7. Permeation in polymers

For a permeant gas subject to a pressure gradient at two different sides of a membrane, the permeability, P, is defined as

$$P = \frac{(\text{quantity of permeant}) \times (\text{film thickness})}{(\text{area}) \times (\text{time}) \times (\text{pressure drop across film})} \tag{18}$$

The quantity of permeant gas is often expressed as the volume of gas at STP condition. In fact, the pressure difference between up- and down-stream sides of polymer membrane is the driving force of permeation phenomenon. As stated above, penetrant transport through membrane is described by solution-diffusion mechanism. According to this model the permeation occurs in three steps: 1-sorption of gas molecules at upstream side of polymer 2-diffusion of the gas molecules into the polymer matrix 3-desorption of gas molecules at downstream side of polymer. This is indicated schematically in Figure 9.

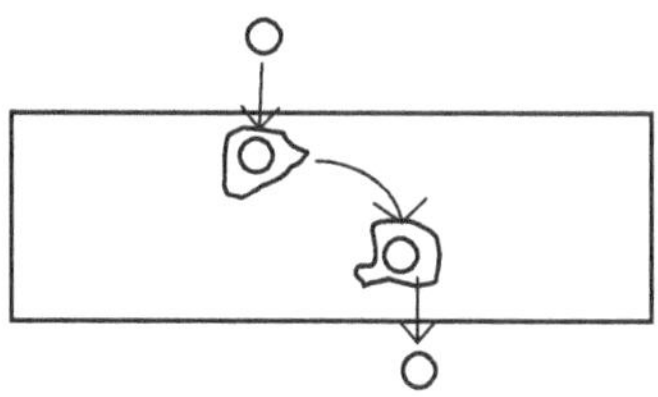

Fig. 9. The schematic of gas permeation in a membrane

The permeability coefficient of gas molecules across a polymer film of thickness L is described by the flux J:

$$J = P\frac{\Delta c}{L} \tag{19}$$

where Δc is the concentration gradient of penetrant molecules through the membrane. Accordingly, the permeation coefficient can be expressed by the product of the diffusion coefficient D, and the solubility coefficient S, as:

$$P = DS \tag{20}$$

Since direct simulation of the permeation process is not possible due to the system size, Eq. (20) is a particularly useful expression connecting the permeability coefficient with the diffusion and solubility coefficients. Therefore, diffusion and solubility coefficients, calculated in MD simulation methods, are used to calculate the permeability coefficients according to Eq. (20).

We have shown in Figure 10 our recently calculated results on the permeability of water in poly(ethylene terephthalate) (Eslami & Müller-Plathe, 2009). The results, indicating the permeability coefficients over a wide range of temperatures, are shown in Figure (10).

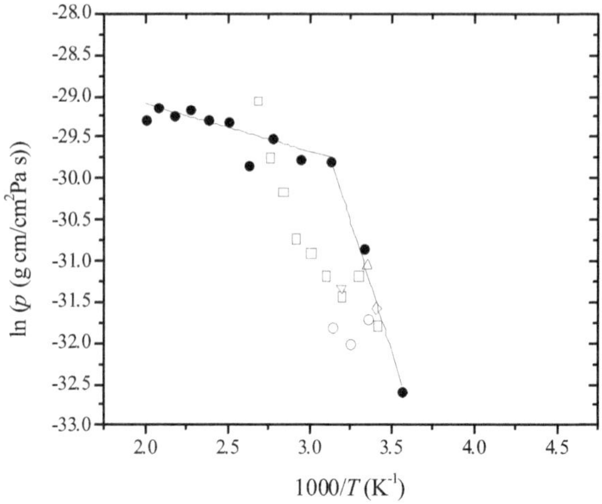

Fig. 10. Comparison of the calculated (●) and experimental permeability coefficients of water in poly(ethylene terephthalate). The experimental data are from (□ Launay et al., 1999); (○ Shigetomi et al., 2000); (◇ Changjing & Junying, 1987); (△ Rueda & Varkalis, 1995); and (▽ Kloppers et al., 1993). Figure is taken from Eslami & Müller-Plathe (2009).

The Arrhenius plot of ln(P) *vs.* $1/T$ shows a break at the glass transition temperature. As a result of calculating a higher solubility coefficient, our calculated permeation coefficients are also higher than the experimental values. Solubility coefficients are higher than the experimental values, especially at lower temperatures. This is trivial for calculations of this type. Similar differences between experimentally and computed solubility coefficients have been observed in previous studies (Cuthbert et al., 1997; Lim et al., 2003, Eslami & Müller-Plathe, 2007a, 2009; Eslami & Mehdipour, 2011). According to Gusev and Suter (1993), errors of the order of (2-4)k_BT are common for the calculation of Helmholtz energies by molecular simulations. It has been speculated that the main contribution to the solubility comes from single holes in the simulated polymer structure, which might not be present in similar proportion in real polymers (Müller-Plathe et al., 1993). Errors of this type become even more serious in cases where the solubility is very low.

We have also compared our calculated permeability coefficients of a number of gases in polystyrene (Mozaffari et al., 2010) with experimental values (Csernica et al., 1987) and with simulation results by Kucukpinar and Doruker (2003). The results in Table 1 indicate that the calculated values are in good agreement with experiment. The results indicate that polystyrene is much permeable to CO_2, compared to other gases studied by Mozaffari et al. (2010), because of the higher solubility coefficient of CO_2 in polystyrene and its relatively higher diffusion coefficient. On the other hand, polystyrene is less permeable to the biggest penetrant molecule studied, propane, because of its very small diffusion coefficient.

An interesting test is to compare the calculated permeability coefficient ratios (selectivities) in the zero pressure limit with the corresponding experimental values (Csernica et al., 1987). This is done in Table 2, in which we have listed the ratios of permeability coefficients at 300 K to that of nitrogen's permeability and compared the results with experimental measurements (Csernica et al., 1987) and with the calculations by Kucukpinar and Doruker (2003). The results in Table 2 show that the calculated ratios are quite close to the experimental ratios. This shows that our calculated permeability coefficients are higher that the experimental values by nearly the same factor. The same conclusion was made in our work on the calculated solubility coefficients (Eslami & Müller-Plathe, 2007a). This indicates that there is an excessive free volume in the polymer sample compared to the experimental polymer samples. Better agreement of the results by Mozaffari et al. (2010) with experiment (Csernica et al., 1987) compared to those of Kucukpinar and Doruker (2003) has been attributed to using a united atom models for both polymer and diffusant gases, in the later case.

penetrant	$P \times 10^{15}$ (cm^3(STP) cm/cm^2 Pa s)		
	(Csernica et al., 1987)	(Kucukpinar & Doruker, 2003)	Mozaffari et al. (2010)
Ar	152	460	284
N_2	45.6	180	109
CO_2	988	2000	2450
CH_4	60.8	130	166
C_3H_8			10.1

Table 1. Comparison of calculated gas permeability coefficients at 300 K (Mozaffari et al., 2010) with experimental values (Csernica et al., 1987) and with simulation results by Kucukpinar and Doruker (2003).

penetrant	P/P_{N2}		
	(Csernica et al., 1987)	Kucukpinar and Doruker (2003)	Mozaffari et al. (2010)
N_2	1.0	1.0	1.0
Ar	3.33	2.55	2.6
CO_2	21.66	11.11	22.47
CH_4	1.33	0.72	1.52
C_3H_8			0.093

Table 2. Comparison of calculated ratios of permeability coefficients at 300 K (Mozaffari et al., 2010) with experimental values (Csernica et al., 1987) and with simulation results by Kucukpinar and Doruker (2003).

8. Summary

In this chapter the polymer permeability is discussed and reviewed from a computational point of view. Although it is impossible to directly simulate the permeability process, the simulation techniques are powerful tools to simulate and to give a molecular level insight to the solubility and diffusion mechanisms of gases in polymers. While there are lots of simulations in the literature, giving address to the diffusion mechanism of gases in glassy and/or rubbery polymers, the computations are less straightforward in the case of gas solubilities. Therefore, the main problem in calculation of gas perameability of polymers, using molecular simulation methods, stems from the calculation of solubility coefficients. In fact the calculation of gas solubilities necessitates the condition of equilibrium between the permeant in the gas/liquid phase and the permeant dissolved in polymer. The equality of the temperature, pressure, and chemical potentials of all species is the necessary condition to establish such an equilibrium situation. The chemical potential is, however, coupled to the number of particles and cannot easily be calculated employing molecular simulation methods.

The applicability and feasibility of different techniques for the calculation of chemical potentials is evaluated and discussed. It is explained that in some straightforward cases, like the thermodynamic integration method, several simulations are needed to calculate the chemical potential. Widom's test particle insertion method (Widom, 1963) is of practical importance in many cases, but the method encounters sampling problems at high densities and in the case of big solutes. The method is also not applicable in the case of chemical potential calculation for polar species. It is shown that recent grand canonical MD simulation methods (Eslami and Müller-Plathe, 2007b; Eslami et al., 2010). are trustable methods for the sake of phase equilibrium calculation. In these methods the chemical potential is set as an independent thermodynamic quantity, and the number of particles in the box is changed in a gradual and dynamical way. The applicability of the method to the challenging cases like gas solubilities in polymers with stiff chemical structures (Eslami and Müller-Plathe, 2007a) and solubility of water (as a polar penetrant) in polar polymers, like polyamide-6,6 (Eslami & Mehdipour, 2011) is reviewed and discussed.

The process of gas diffusion in polymers and different approaches for studying the gas diffusion are addressed to. It is explained that MD simulation well describes the jumping mechanism of diffusions in glassy polymers. According to this mechanism, in a glassy polymer the penetrants dwell in the polymer holes and occasionally do a jump into the neighboring voids (Mozaffari et al., 2010). Therefore, no net motion of a penetrant molecule occurs while the penterant just performs random oscillations inside a hole. From time to time, the penetrants can do a quick jump into their neighboring voids, with a jump frequency depending on penetrant's size, temperature, and chemical structure of the polymer. The permeation coefficients of gases in polymers are, therefore, reviewed in this chapter in terms of gas solubilities and diffusions.

9. References

Abu-Shargh, B. F. (2004) *Polymer* 45, 6383.
Adams, D. J. (1974) *Mol. Phys.* 28, 1241.
Adams, D. J. (1975) *Mol. Phys.* 29, 1307.
Barrie, J. A.; Williams, M. J. L.; Munday, K. (1980) *Polym. Eng. Sci.*, 20, 21.
Ben-Naim, A. (1987) *Solvation thermodynamics*, Plenum, New York.

Boda, D.; Chan, K.-Y.; Szalai, I. (1997) *Mol. Phys.*, 92, 1067.
Boda, D.; Liszi, J.; Szalai, I .(1995) *Chem. Phys. Lett.* 235, 140.
Boda, D.; Winkelmann, J.; Liszi, J.; Szalai, I. (1996) *Mol. Phys.* 87, 601.
Boda, D.; Kristof, T.; Liszi J.; Szalai, I. (2001) *Mol. Phys.* 99, 2011.
Boinepalli S.; Attard, P. (2003) *J. Chem. Phys.* 119, 12769.
Boyd, R. H.; Pant, P. V. K. (1991) *Macromolecules*, 24, 6325.
Changjing, L.; Junying, G. (1987) *Chin. J. Polym. Sci.*, *5*, 261.
Cohn, M. H.; Turnbull, D. (1959) *J. Chem. Phys.*, 31, 1164.
Csernica, J.; Baddour, R. F.; Cohen, R. E. (1987) *Macromolecules*, 20, 2468.
Cuthbert, T. R.; Wagner, N. J.; Paulaitis, M. E. (1997) *Macromolecules* 30, 3058
Daynes, H. A. (1920) *Proc. R. Soc. London Ser. A*, 97, 286.
Deitrick, G. L.; Scriven, L. E.; Davis, H. T. (1989) *J. Chem. Phys.* 90, 2370.
de Pablo, J. J. (1995) *Fluid Phase Equilib.* 104, 195.
de Pablo, J. J.; Prausnitz, J. M. (1989) *Fluid Phase Equilib.* 53, 177.
de Pablo, J. J.; Bonnin, M.; Prausnitz, J. M. (1992) *Fluid Phase Equilib.* 73, 187.
Dibenedetto, A. T. (1963) *J. Polym. Sci. A*, 1, 3477.
Escobedo F. A.; de Pablo, J. J. (1995) *J. Chem. Phys.* 103, 2703.
Eslami, H.; Mehdipour, N. (2011) *Phys. Chem. Chem. Phys.* 13, 669-673.
Eslami, H.; Mojahedi, F.; Moghadasi, J. (2010) *J. Chem. Phys.* 133, 084105.
Eslami, H.; Müller-Plathe, F. (2007a) *Macromolecules* 40, 6413
Eslami, H.; Müller-Plathe, F. (2007b) *J. Comput. Chem.* 28, 1763.
Eslami, H., Müller-Plathe, F. (2009) *J. Chem. Phys.* 131, 234904.
Frederickson, G. H.; Helfand E. (1985) *Macromolecules*, 18, 2201.
Fritz, D.; Herbers, C. R.; Kremer, K.; van der Vegt, N. F. A. (2009) *Soft Matter*, 5, 4556–4563.
Gee, R. H.; Boyd, R. H. (1995) *Polymer* 36, 1435.
Gusev, A. A.; Arizzi, S.; Suter, U. W. (1993) *J. Chem. Phys.* 99, 2221.
Gusev, A. A.; Müller-Plathe, F.; van Gunsteren W. F.; Suter U. W. (1994) *Adv. Polym. Sci.* 116, 207.
Gusev, A. A.; Suter U. W. (1993) *J. Chem. Phys.* 99, 2228.
Han, J.; Boyd, R. H. (1996) *Polymer* 37, 1797.
Hess, B.; Peter, C.; Ozal, T.; van der Vegt, N. F. A. (2008) *Macromolecules*, 41, 2283-2289.
Hess, B.; van der Vegt, N. F. A. (2008) *Macromolecule*, *41*, 7281-7283.
Jang, J. G.; Bae, Y. C. (2002) *J. Chem. Phys.* 116, 3484.
Jarzynski, C. (1997) *Phys. Rev. Lett.*, 78, 2690
Karayiannis, N.; Mavrantzas, V. G.; Theodorou, D. N. (2004) *Macromolecules* 37, 2978.
Kesting R. E.; Fritzsche A. K. (1993) in *Polymeric gas separation membranes*, Wiley, New York.
Kloppers, M. J.; Bellucci, F.; Lantanision, R. M.; Brennen, J. E. (1993) *J. Appl. Polym. Sci.*, *48*, 2197.
Kofke, D. A. (1993a) *Mol. Phys.* 78, 1331.
Kofke, D. A. (1993b) *J. Chem. Phys.* 98, 4149.
Koros, W. J. (1990), *Barrier polymers and structures*, ACS symposium series 423, American Chemical society, Washington DC.
Kucukpinar, E.; Doruker P. (2003) *Polymer*, 44, 3607.
Lacombe, R. H.; Sanchez I. C. (1976) *J. Phys. Chem.* 80, 2568
Launay, A.; Thominetty, F.; Verdu, J. (1999) *J. Appl. Polym. Sci.*, 73, 1131.
Lim, L.; Britt, I. J.; Tung, M. A. (1999) *J. Appl. Polym. Sci.*, 71, 197.
Lim, S. Y.; Tsotsis T. T.; Sahimi M. (2003) *J. Chem. Phys.* 119, 496.

Lundberg, J. L.; Mooney, E. J.; Rogers, C. E. (1969) *J. Polym. Sci., Part A, 7*, 947.
Mauritz, K. A.; Storey, R. F. (1990) *Macromolecules*, 23, 2033.
McDonald, I. R.; Singer, K. (1967) *Discuss Faraday Soc.* 43, 40.
McQuarrie D. A. (1976) *Statistical mechanics*, Harper Collins, New York.
Milano, G.; Guerra G.; Müller-Plathe, F. (2002) *Chem. Mater.* 14, 2977.
Moller, D.; Fischer, J. (1990) Mol. Phys. 69, 463.
Mooney D. A.; MacElroy J. M. D. (1999) *J. Chem. Phys.* 110, 11087
Mozaffari, F.; Eslami, H.; Moghadasi, J. (2010) *Polymer* 51, 300-307.
Müller-Plathe, F. (1991) *Macromolecules*, 24, 6475.
Müller-Plathe, F. (1994) *Acta Polym. Sci.* 45, 259-293.
Müller-Plathe, F.; Rogers, S. C.; van Gunsteren, W. F. (1993) *J. Chem. Phys.* 98, 9895.
Neyertz, S.; Brown, D. (2004) *Macromolecules*, 37, 10109-10122
Norman G. E.; Filinov, V. S. (1969) *High Temp.* (USSR), 7, 216.
Ozkan I. A.; Teja A. S. (2005) *Fluid Phase Equilib.* 487, 228-229.
Pace R. J.; Datyner, A. (1979) *J. Polym. Sci. A: Polym. Phys.*, 17, 437.
Panagiotopoulos, A. Z. (1987) *Mol. Phys.* 61, 813.
Panagiotopoulos, A. Z.; Quirke N.; Stapleton, M.; Tildesley, D. J. (1998) *Mol. Phys.* 63, 527.
Pandiyan, S.; Brown, D., Neyertz, S.; van der Vegt, N. F. A. (2010) *Macromolecules*, 43, 2605–2621.
Pant, K. P. V.; Boyd, R. H. (1993) *Macromolecules* 26, 679.
Potoff, J. J.; Panagiotopoulos, A. Z. (1998) *J. Chem. Phys.* 109, 10914.
Pricl, S., Fermeglia, M. (2003) *Chem. Eng. Comm.* 190, 1267
Raymond, P. C.; Paul, D. R. (1990) *J. Polym. Sci., Part B: Polym. Phys., 28*, 2079-2102.
Rueda, D. R.; Varkalis, A. (1995) *J. Polym. Sci. B: Polym. Phys.*, 33, 2263.
Rudisill, E. N., Cummings, P. T. (1989) *Mol. Phys.* 68, 629.
Sada, E.; Humazawa, H.; Yakushiji, H.; Bamba, Y.; Sakata, K.; Wang, S.-T. (1987) *Ind. Eng. Chem. Res. 26*, 433-438.
Sanchez I. C., Rodgers P. A. (1990) *Pure & Appl. Chem.*, 62, 2107
Shelley, J. C., Patey, G. N. (1995) *J. Chem. Phys.* 102, 7656.
Shigetomi, T.; Tsuzumi, H.; Toi, K.; Ito, T. (2000) *J. Appl. Polym. Sci.*, 76, 67.
Shing, K. S.; Gubbins, K. E. (1981) *Mol. Phys.* 43, 717.
Sok, R. M.; Berendsen, H. J. C.; van Gunsteren, W. F. (1992) *J. Chem. Phys.*, 96, 4699.
Stannett, V. (1978) *J. Membr. Sci., 3*, 97-115.
Stannett, V. T.; Koros W. J., Paul D. R., Lonsdale H. K., Baker R. W. (1978) *Adv. Polym. Sci.*, 32, 69
Tamai, Y.; Tanaka, H.; Nakanishi, K. (1995) *Macromolécules* 28, 2544
Valleau, J. P. (1991) *J. Comput. Phys.* 96, 193.
Valleau, J. P.; Graham, I. S. (1990) *J. Phys. Chem.* 94, 7894.
Vieth, W. R.; Tam, P. M.; Michaels, A. S. (1966a) *J. Colloid Interface Sci., 22*, 360-370.
Vieth, W. R.; Frangoulis, C. S.; Rionda, J. A. (1966b) *J. Colloid Interface Sci., 22*, 454-461.
von Solms N., Michelsen M. L., Kontogeorgis G. M. (2005) *Ind. Eng. Chem. Res.* 44, 3330.
Von Wroblewski, S. (1879) *Ann. Phys.* 8, 29.
Vrabec, J.; Hasse, H. (2002) *Mol. Phys.* 100, 3375.
Vrabec, J.; Fischer, J. (1995) *Mol. Phys.* 85, 781.
Vrentas, J. S.; Duda, J. L. (1979) *AIChE J.* 25, 1.
Watt, I. C. (1980) *J. Macromol. Sci.: A Pure Appl. Chem.*, 14, 245.
Widom B. J. (1963) *J. Chem. Phys.* 39, 2808

4

Backbone Connectivity and Collective Aggregation Phenomena in Polymer Systems

Wen-Jong Ma[1,2] and Chin-Kun Hu[2]
[1]*Graduate Institute of Applied Physics, National Chengchi University, Taipei 11605*
[2]*Institute of Physics, Academia Sinica, Nankang, Taipei 11529*
Taiwan

1. Introduction

A convenient theoretical entry to tackle the protein aggregation problems is to track on the origin of the aggregation phenomena from the viewpoint of polymer physics, enquiring how the underlying factors, such as the geometric packing, the topology and their mutual interplays in presence of solvent, play their roles in the processes. Such a view has been inspired by recent experiments [1, 2] and molecular dynamics simulations [3, 4] which suggest the ubiquitous presence of fibril formation [5] in various natural and laboratory prepared proteins or peptides. While the variety of amino acid sequences interferes with the occurrence of long-range structural ordering, a material-insensitive tendency of aggregation is observed [2, 3, 5]. By coarsening the sequence-sensitive details, the aggregation problem can be formulated in its minimal form as the clustering process of polymer chains [6, 7]. With such simplified models, the approach focuses more on the entropic effect caused by the constraint of chain connectivity [6–8], rather than on the material-dependent characteristics. In this chapter, we summarize our molecular dynamics simulation studies [6, 7] that reveal the relationship between the backbone connectivity of polymer chains and some benchmark features displayed in the aggregation processes of the model polymer chains.

In the model, the backbone connectivity of a polymer chain is realized by assigning a string of monomers with specific monomer-monomer two-body forces, perturbed with three-body and four-body angle dependent interactions [6], with their strengths measured by the parameters, $k_{\rm nn}$, $K_{\rm b}$ and $K_{\rm t}$, for the nearest neighbor (n.n.) interaction, the bending angle and the torsion angle potentials, respectively. In a collection of polymer chains, all the non n.n. pairs along a chain and those pairs on different chains are subject to Lennard-Jones (L-J) pair interactions. The presence of angle potentials, with significantly nonzero $K_{\rm b}$ or $K_{\rm t}$ values, breaks the isotropy surround the backbone. In such a system, the local inter-chain hindrance prevents the chains from clustering into ordered domains. The situation can, however, be reverted by reducing the values of $K_{\rm b}$ and $K_{\rm t}$. We find that the formation of bundle-like domains (Fig. 1) is robust as soon as $K_{\rm b}$ and $K_{\rm t}$ are small enough. The observations are assured even if we introduce a small amount of dispelling background fluid molecules or impose a small fraction of impurity monomer sites.

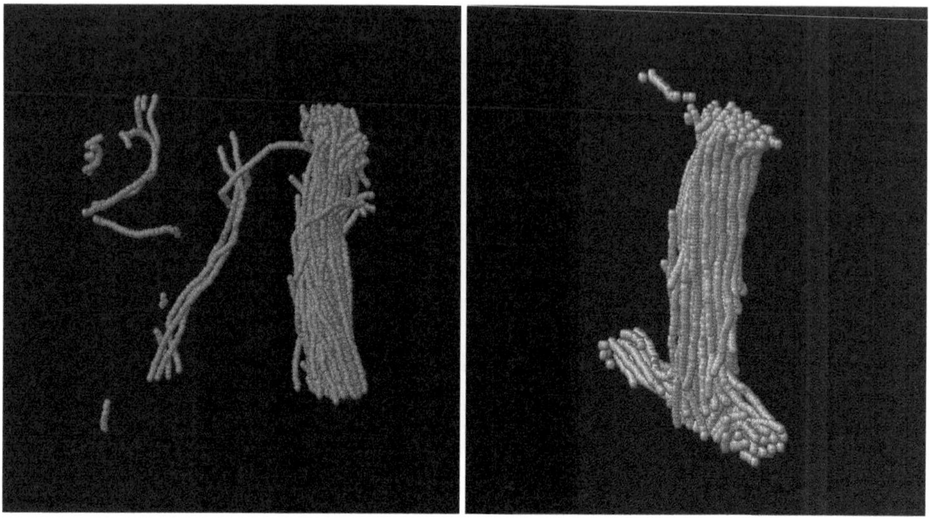

Fig. 1. Bundled domains formed in systems of homopolymer chains, with zero (left plot) or tiny (right plot) values in K_b and K_t. They are snapshots for the systems with rigid bonds, labeled by I-∞ and III, respectively, in Table 1.

In quantifying the overall structural changes over the relaxation processes, we reveal the shared scenario for the cluster forming procedure in a range of systems (see Table 1) . In this scenario, the ordering of the clustered (bundled) domains can occur spontaneously as soon as the energy barriers contributed by the local structural hindrance are overcome. Under such circumstances, the completion of aggregation always occurs at a temperature well above the characteristic temperature determined by the strength ϵ of the site-site L-J pair potential. This is in contrast to the formation of ordered clusters in the two phase coexistence region for a L-J fluid of the same number of monomers. (The system is equivalently obtained by removing those n.n. bonding along the polymer chains.) The clustering can only occur below the critical point temperature, which is about 1.3 times of the characteristic temperature ϵ/k_B (k_B: Boltzmann constant). To have solid-like structured cluster, the temperature has to go below the triple point temperature, which is about $0.6\epsilon/k_B$. The result suggests that the key role played by the chain connectivity is via the lowering of the entropy of the system, so that the formation of ordered domains, becomes feasible at a temperature not necessarily very low for the system to overcome the local structural hindrance.

The bonding strength, parameterized by k_{nn}, of the nearest neighboring monomers along the chains affects the detailed procedures in the formation of clusters during the aggregation processes. More than one stages are present during the processes, with the local alignment of segments in the individual chains followed by the coalescence of these short patches. The second stage is found further to differentiate into several stages with the increased k_{nn}. We allow k_{nn} to have infinity value, which is realized in simulation as a (rigid) bond with fixed length maintained by the constraint force (see Section 2.1). In manipulating the parameters k_{nn}, K_b and K_t in a range of systems, we can obtain more precise information about how the backbone anisotropy prevent the polymer chains from the formation of bundled structure. The scenario unveiled in this study may be useful for the analysis of protein aggregation phenomena in realistic complex biological systems.

system	label	N_P (# chain)	n (# monomer per chain)	N_F (# fluid atom)	$k_{\rm nn}$ (n.n. strength)	$K_{\rm b}$ (bending angle strength)	$K_{\rm t}$ (torsion angle strength)
set I‡	I-1	40	100	0	10	0	0
	I-2	40	100	0	10^2	0	0
	I-3	40	100	0	10^3	0	0
	I-4	40	100	0	10^4	0	0
	I-4	40	100	0	10^4	0	0
	I-∞	40	100	0	∞	0	0
	I-A	80	50	0	∞	0	0
	I-B	160	25	0	∞	0	0
set II§	II-0	40	100	6000	1	10^{-4}	10^{-4}
	II-1	40	100	6000	10	10^{-4}	10^{-4}
	II-2	40	100	6000	10^2	10^{-4}	10^{-4}
	II-3	40	100	6000	10^3	10^{-4}	10^{-4}
	II-4	40	100	6000	10^4	10^{-4}	10^{-4}
	II-∞	40	100	6000	∞	10^{-4}	10^{-4}
set III♮	III	40	100	0	∞	10^{-4}	10^{-4}
set IV†	IV-0	40	100	6000	1	0.1	0.1
	IV-1	40	100	6000	10	0.1	0.1
	IV-2	40	100	6000	10^2	0.1	0.1
	IV-3	40	100	6000	10^3	0.1	0.1
	IV-4	40	100	6000	10^4	0.1	0.1
	IV-∞	40	100	6000	∞	0.1	0.1

Table 1. List of systems
‡ systems of pure isotropic ($K_{\rm b} = K_{\rm t} = 0$) homopolymer chains (see Ref.[6, 7]);
§ systems of chains with tiny backbone anisotropy ($K_{\rm b} = K_{\rm t} = 10^{-4}$) and with ≈ 5% (199 out of nN_P =4000) impurity monomers (see Ref.[6, 7]), mixed with fluid atoms ($N_F > 0$) ;
♮ a pure system of rigidly bonded homopolymer chains with tiny backbone anisotropy and without impurity monomers (see Ref.[7]);
† systems of anisotropic ($K_{\rm b} = K_{\rm t} = 0.1$) chains that fail to aggregate; there being ≈ 5% (199 out of nN_P =4000) impurity monomers (see Ref.[6, 7]) and mixing with fluid atoms ($N_F > 0$).

2. Model and analysis

2.1 Model systems

nearest neighbor bonding

We consider systems containing $N_P = 40$ polymer chains and each chain has $n = 100$ monomers. Two major sets of systems are studied. The first set (set I) are pure ($N_F = 0$) systems of homopolymer chains with no built-in anisotropy with respect to the backbone ($K_{\rm b} = K_{\rm t} = 0$), which can be classified phenomenologically as a bead-spring (if $k_{\rm nn} < \infty$) or a bead-rod (if $k_{\rm nn} = \infty$) model [9, 10]. The nearest neighbors i and j at a distance r_{ij} in our bead-spring chains are connected by the potential

$$V_s(r_{ij}) = \frac{1}{2}k_s(r_{ij} - r_0)^2$$

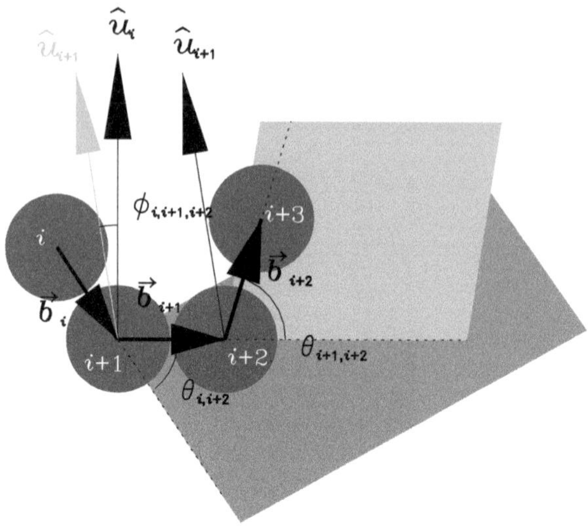

Fig. 2. Bending angle $\theta_{i,j,k}$ and torsion angle ϕ_{ijkl} extended by the consecutive monomers i, $j = i+1, k = i+2$ and $l = i+3$ along a chain.

with its strength constant k_s, the $k_{nn} = 10^s$ multiple of the value k_0 (i.e. $k_s = k_{nn}k_0$) and balance length r_0, for the five systems with $s = 1, 2, 3, 4$ (k_0 and r_0 are given below). They are labelled as I-s (I-1, ..,I-4, etc) systems in Table 1. The bead-rod chains in our model are chains with strict constant nearest neighbor bond lengths and are realized numerically by Lagrange's constraint forces using RATTLE numerical scheme [11]. The system is labelled as system I-∞ in Table 1.

In all these model systems, the pair interactions between the non-neighboring monomers along a chain or between pairs in different molecules, i and j at a separation r_{ij}, are L-J potentials

$$V_{LJ}(r_{ij}) = 4\epsilon((r_{ij}/\sigma)^{-12} - (r_{ij}/\sigma)^{-6}).$$

In terms of the distance parameter σ and the strength parameter ϵ of L-J potential, we choose $k_0 = 1.5552 \times 10^5 \epsilon\sigma^{-2}$ and $r_0 = 0.357\sigma$ [6, 12].

backbone anisotropy along the chains

In the second set (set II) of systems, the degrees of backbone anisotropy along the chains are introduced by perturbing bending potential,

$$\Theta(\theta_{i,j,k}) = K_b c_b (cos\theta_{ijk} - cos\theta_0)^2,$$

determined by the bending angle $\theta_{i,j,k}$ of three consecutive monomers i, j and k along a chain; and the torsion potential [13]

$$\Phi(\phi_{ijkl}) = K_t \sum_{\iota=0}^{3} a_\iota (cos\phi_{i,j,k,l})^\iota$$

as a function of the torsion angle ϕ_{ijkl} extended by four consecutive monomers i, j, k and l (see Fig. 2). The constants c_b, θ_0 [12]; and a_1, a_2 and a_3 [13] are given so that a model polyethelyne

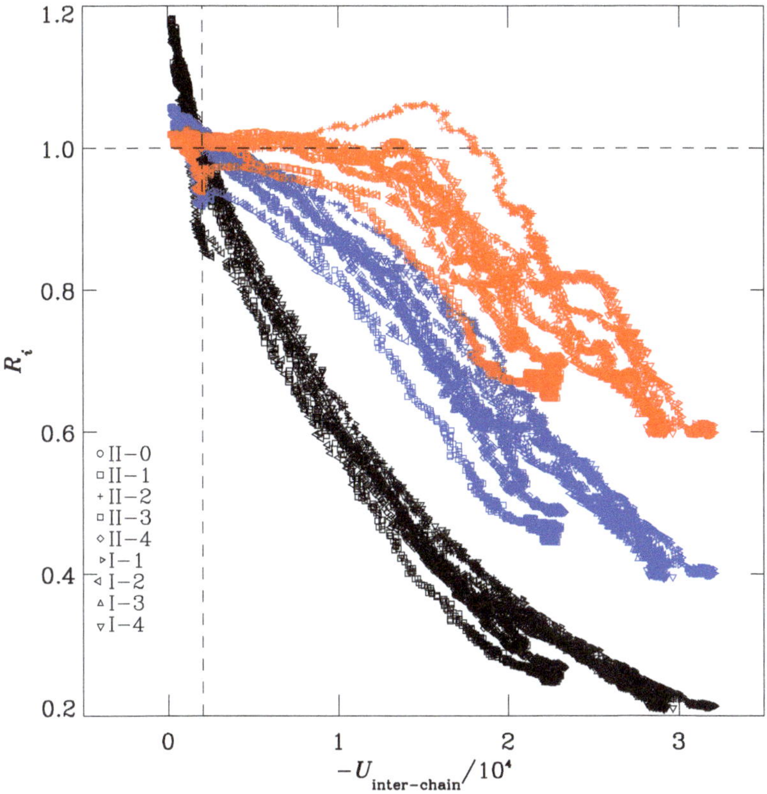

Fig. 3. R_0 (black), R_1 (blue) and R_2 (red) versus $-U_{\text{inter-chain}}$, for systems set I: I-1, I-2, I-3, I-4 and set II: II-0, II-1, II-2, II-3 and II-4, which contain chains with their nearest neighbor connected by spring forces (see Table 1).

chain is obtained by choosing $r_0 = 0.357\sigma$, and let the strength control parameters K_b and K_t be unity [12, 14]. With a range of values for K_b and K_t and a choice of bending angle parameter θ_0, we are able to prepare chains with different degrees of local conformational hindrance [6, 15]. In our simulations, we prepare several systems, each composed of identical chains with $K_b = K_t$, which are allowed to have the value 0, 10^{-4} or 0.1 (see Table 1). The model mimics qualitatively the dispersed local conformational degrees of freedom that is present in those structured monomers, such as the amino-acids in protein molecules.

For systems in set II, the chains is further perturbed by randomly imposing a fraction of 5% monomers to have smaller sizes. In merging with monomer-repelling Lennard-Jones fluid atoms, the heterogeneity is further enhanced by rendering these 5% monomers to interact, in addition to the repulsive force, with an attractive part with the fluid atoms [6, 7]. Under the convention mentioned above, we label those systems as II-s ($s = 0, 1, 2, 3, 4$ or ∞) in Table 1, according to the exponent of the multiple ($k_{nn} = 10^s$ or ∞) of their nearest bonding strength $k_s = k_{nn}k_0$.

Three additional systems of pure ($N_F = 0$) homogeneous rigidly-bonded chains are also studied. Two of them that contain the same rigidly bonded chains, of shorter chain lengths

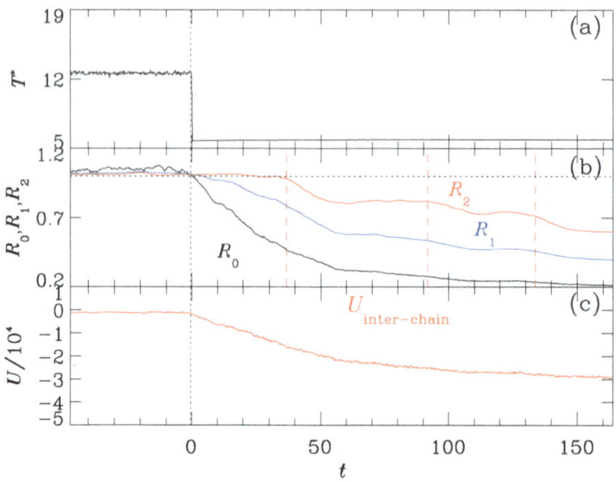

Fig. 4. Time evolutions of (a) instantaneous temperature T^*; (b) the parameters R_0 (solid line), R_1 (dash dot line) and R_2 (dashed line); and (c) the inter-chain potential energy, for system I-4 (adapted from Ref. [6]).

($n = 25$ and $n = 50$, respectively), are included in set I. They are used to underscore the effect of chain length. The other system consists of homopolymer chains to have the same backbone anisotropy ($K_b > 0$ and $K_t > 0$) as those in set II, but in absence of the 5% impurity monomers. We list this system as system III. Table I lists all the systems that are analyzed in this chapter. Each of them contains the same number $nN_p = 4000$ of monomers.

To show that the increased backbone anisotropy indeed hinders the ability to aggregate, we list the systems set IV in Table 1, which are different from their counterparts in set II, only in the 1000 times larger values in K_b and K_t (=0.1).

quenching

According to the scenario of clustering, the polymer chains aggregate when the temperature and the density of the systems fall within the coexistence region of the phase diagram. In the systems of polymer chains with rigid bonds, a quenching can be achieved by the numerical effect for the cases $K_b = K_t \leq 10^{-4}$, that a convergence to satisfy holonomic constraints on bond lengths lead the dynamic system toward the low temperature attractor [7]. The system is, therefore, quenched spontaneously. With the bond constraints replaced by the confining of soft interaction potentials, on the other hand, the incoherence among the springs in different bonds does not favor a global numerical convergence. In order to lower the temperature of the polymer chains in this case, we have to control the simulated system manually by using a thermostat.

Since the distributions of the monomer velocities are known to deviate systematically from the standard Maxwell-Boltzmann with the increased strength of nearest neighbor bonding [16, 17] and are described by the Tsallis q-statistics [18], if the system reaches a "quasi-steady state", it is inappropriate to use the existing thermostating methods that would force the simulated system to converge to a state described by standard Maxwell-Boltzmann statistics. A naive choice without imposing presumed statistics is simply controlling *only* the mean value of the kinetic distributions. In the simplest version of the "velocity scaling" thermostat [19], we

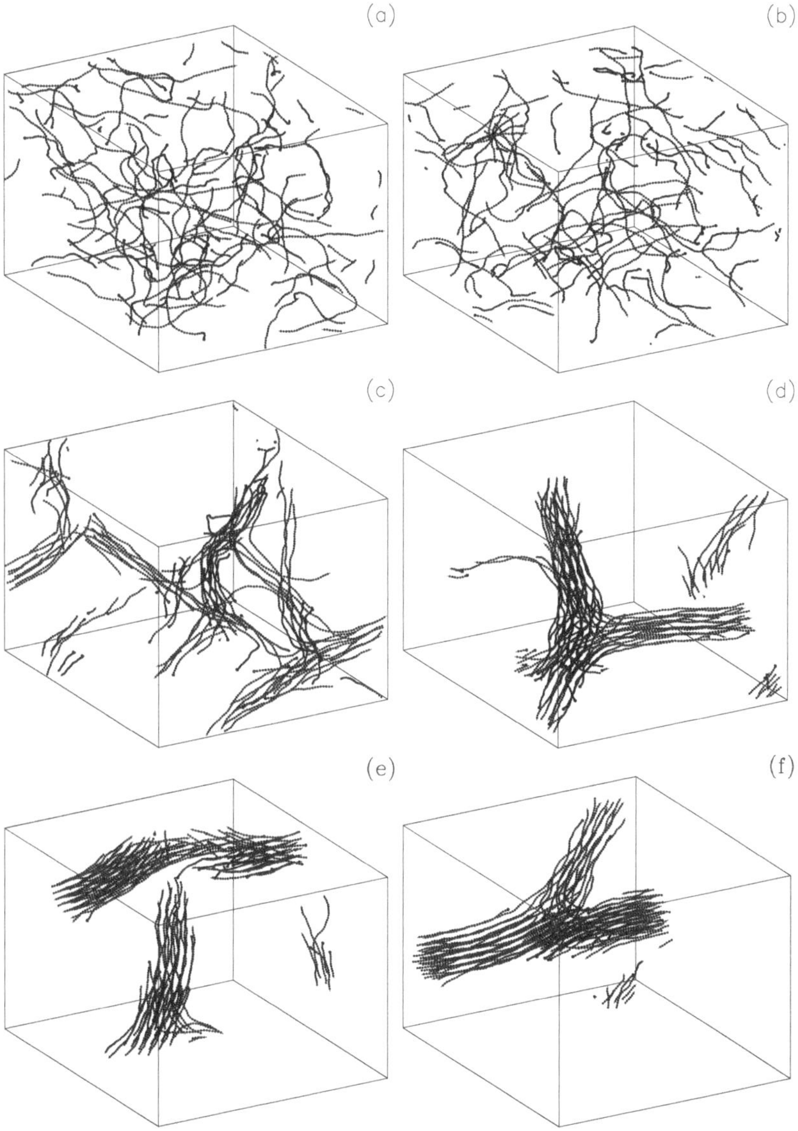

Fig. 5. Snapshots of the configurations of system I-4, for the process described by Fig. 4, in the pure system ($K_b = K_t = 0$) with $k_{spring} = k_4$, (a) before (t=-47, the starting time in Fig. 4) and (b) on the initiation of aggregation ($t = 0$); (c), (d), and (e), respectively, at the onset step of the first, the second and the third fast decaying for R_2 (the time spots of which are marked by vertical lines in Fig. 4(b)); (f) the last step in Fig. 4 (adapted from Ref. [6]).

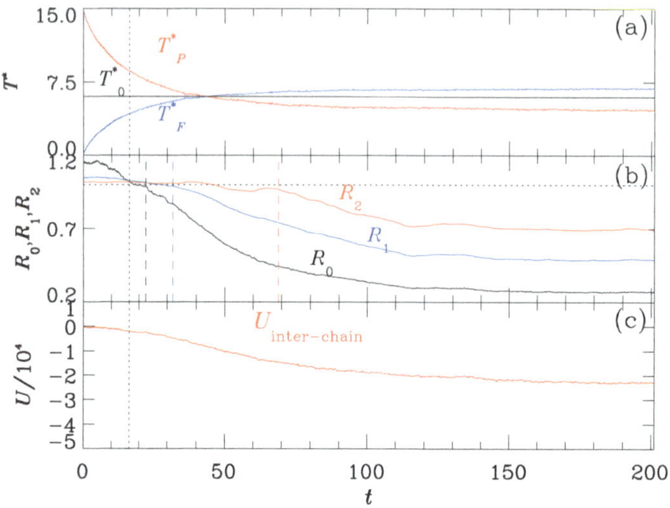

Fig. 6. Same as Fig. 4, for system II-4, which has tiny values for $K_b = K_t = 10^{-4}$. (adapted from Ref. [6]).

re-scale the velocities of the monomers and the L-J molecules at *every* step by a factor forcing the mean kinetic energy for the *whole* system to the value of the target temperature T_0^*.

2.2 Parameters to identify clustering

To quantify the overall structural change of the simulated systems over their non-equilibrium relaxation processes, we define the following parameters. For a system of N_P polymer chains each with n monomers in a simulation cubic box of size $L \times L \times L$ subject to periodic boundary conditions, we use the parameters [7]

$$R_k \equiv \frac{\int_0^{\frac{L}{2}} g_{\text{interchain}}(r) r^k dr}{\int_0^{\frac{L}{2}} \Xi(g_{\text{interchain}}(r)) r^k dr} \tag{1}$$

for $k = 0, 1, 2$, to monitor the degree of clustering, where $\Xi(x) = 1$ if $x \neq 0$ and $\Xi(x) = 0$ if $x = 0$. The parameters are evaluated by finding the interchain pair distribution function $g_{\text{interchain}}(r)$, which is defined as

$$g_{\text{interchain}}(r) 4\pi r^2 dr = \frac{1}{N_m} \sum_{i=1}^{N_m} \{ \frac{V}{N_m - n} \times \sum_{\substack{j(i \text{ and } j \text{ in different chains}) \\ r < r' \leq r+dr}} \delta(|\vec{r}_{ij}| - r') \} \tag{2}$$

for $r < L/2$, where $N_m = nN_P$ is the total number of monomers. R_0, R_1 and R_2 provide information about the conformational features of the clusters at different stages of the growth process. In particular, the condition for uniformly distributed situation, $R_0 = R_1 = R_2 = 1$ can be used as a criteria to locate the time spot of initiation of the aggregation process when such a condition starts to become untrue.

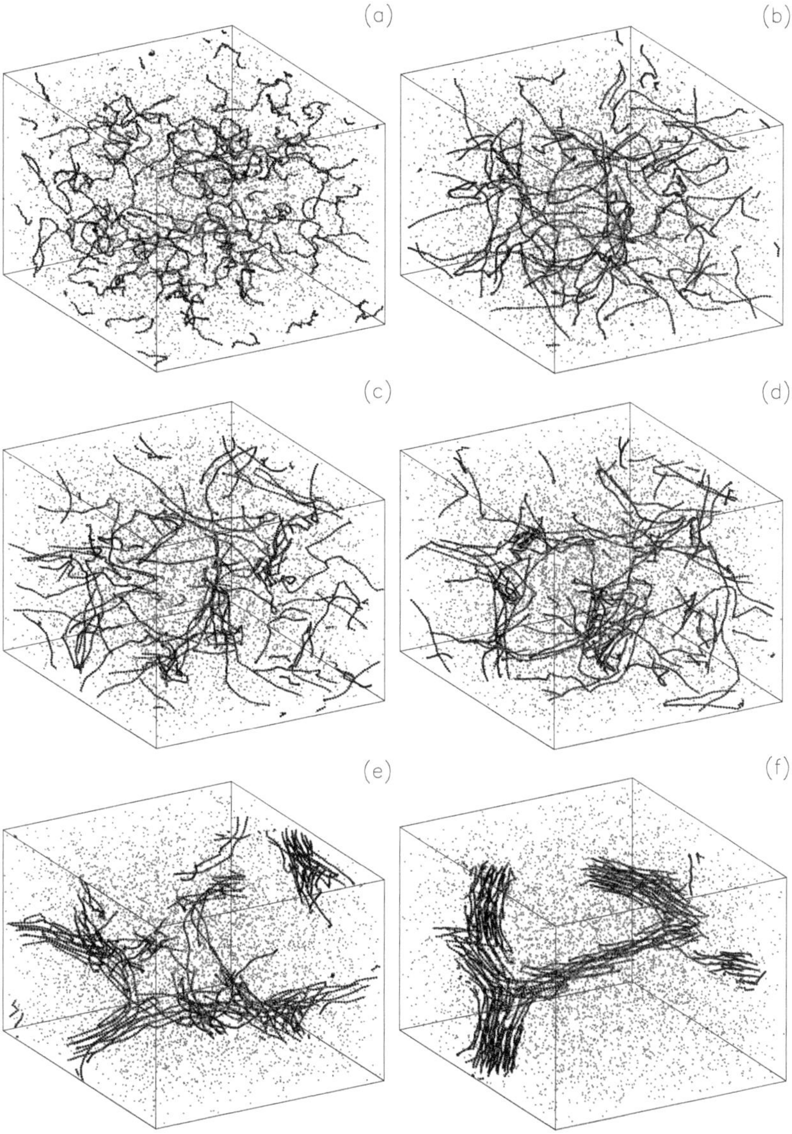

Fig. 7. Snapshots of the configurations for the aggregation process in system II-4, i.e. the mixed system with $K_b = K_t = 1.0 \times 10^{-4}$ and $k_{spring} = k_4$, at (a) t=0; (b) the initiation of aggregation ($R_0 \approx R_1 \approx R_2 \approx 1$); the onset step of fast decaying of (c) R_0, (d) R_1 and (e) R_2; (f) the last step of Fig. 6. The time spots for (c), (d) and (e) are marked by vertical lines in Fig. 6(b) (adapted from Ref. [6]).

The formation of clusters can be identified by the trend to have the lowered total interchain (L-J) potential energy $U_{\text{interchain}}$ and a decreasing virial $v_{\text{interchain}}$ (see Ref. [7]). Since both $v_{\text{interchain}}$ and $U_{\text{interchain}}$ follow the same trend [6, 7] and the latter is subject to much less fluctuations than the former is, we adapt here $U_{\text{interchain}}$ as the major quantity to monitor the process.

3. Aggregation of polymer chains

3.1 Stages of the clustering process

The formation of clusters can be identified by the tendency to have lowered total interchain (L-J) potential energy $U_{\text{interchain}}$. Figure 3 shows the parameters R_0, R_1 and R_2 decrease as the functions of $U_{\text{interchain}}$, over the dynamic relaxation processes for the nine systems with soft n.n. bonding ($k_{\text{nn}} < \infty$), for either pure homopolymer chains (N_F=0 and $K_b = K_t = 0$, i.e. systems I-1, I-2, I-3 and I-4) or mixed heterogeneous chains with backbone anisotropy ($N_F > 0$ and $K_b > 0$ or $K_t > 0$, i.e. systems II-0, II-1, II-2, II-3 and II-4) listed in Table 1. The data are collected in equal time intervals [6, 7]. The increasing of the quantity '$-U_{\text{inter-chain}}$', which is used as the variable for the horizontal axis, corresponds to the direction for the growth of clusters. The parameter R_0 decreases, more or less, smoothly and is concave upward for all systems before reaching the tail regimes with densely accumulated data points, when the size of the clusters virtually saturate under the given temperature and the space constraint by the simulation box. The parameters R_1 and R_2, on the other hand, show the reflected s-shaped decreasing curves before entering the same stalled growth regimes. The reflections in the curves of R_2 are so strong that overall the shapes are close to the letter "z".

The parameter R_0 reflects the increasing contact between the pairs of monomers in different chains during the relaxation processes, over which the line geometry of the chains governs the values. The parameters R_1 and R_2, on the other hand, contain information about the clustering of higher order structures. The three parameters meet at $R_0 = R_1 = R_2 = 1$, corresponding to a spatially uniformly distributed configuration. In simulation, it can be taken as a criteria to find the time spot of initiation of an aggregation process. By the same token, the reflection points in R_1 and R_2 can be identified to divide the processes into stages of growth.

Figure 4 shows the time evolution of a typical process of aggregation in a pure homopolymer system (I-4). The temperature of the system is controlled with a target temperature $T_0^* = 6$, starting from $t = 0$. We can identify the presence of multi-stage plateaus in the subsequent process.

To refine an aggregation process into a number of stages of relaxation, we identify each time spot at which the value of one of the parameters R_0, R_1 or R_2, starts to fall rapidly at the end of each plateau. In general, while the time spot of the triggered rapid decay in R_0 may not be distinguishable from the initiation time (at which R_0=R_1=R_2=1), the onset of the fast falling in the value of R_1 and, especially, that of R_2 can easily be identified. The latter signals the growth transverse to the backbone. The chains in the system I-4 described in Fig. 4 have the stiffest springs among the spring bonded systems in set I (Table 1). The clustering process would become stuck and the system would need to wait until the initiation of the next major growth before there is another rapid fall in the value of R_2. In Fig. 4(b), the fast decaying for both R_0 and R_1 start at $t = 0$. Such time spots for R_2 can be identified (marked by dashed lines) for three sections of plateaus, at t = 36.75, 91.75 and 133.75, respectively. The snapshots at these time spots, together with those at the beginning and the ending of the time span plotted in Fig. 4, are shown in Fig. 5.

3.2 Backbone anisotropy and spring stiffness

The process for a mixed system (system II-4) of polymer chains, with perturbed backbone anisotropy ($K_b = K_t = 10^{-4}$) and impurity monomers, shows similar features (Fig. 6) as those found in system I-4 (Fig. 4). While the chains possess the same strengths in the monomer-monomer bonding springs as those in system I-4, the time evolution in Fig. 6(b) seems to show less tendency to become stuck over the relaxation process, as compared with its counterpart in Fig. 4(b). The corresponding snapshots of those time spots at the onsets of the fast decaying in R_0, R_1 or R_2 are shown in Fig. 7.

A quantitative comparison between the configurations at the onsets of the rapid decaying in R_2 for system I-4 (Fig. 5(c)) and II-4 (Fig. 7(e)) is carried out by plotting the full monomer-monomer pair distribution functions $g(r)$ in Fig. 8. To underscore the effect of the strength of the springs, we also put the data for the configuration at onset of R_2 for system II-0, which is a mixture system having the softest spring bonding in set II listed in Table 1. Each function $g(r)$ shows a power-law like decaying curve in increasing r (see the log-log plot in the inset of Fig. 8), until reaching a minimum, which corresponds to the void regions next to the crowded domains of clusters. The thickness of the segments of the aggregated bundles can be estimated by locating r for such a minimum in $g(r)$. While the two mixed systems (II-0 and II-4, marked by "A" and "B" respectively) have about the same bundle thickness, the packing in each system has its dependence on the strength of the monomer-monomer connecting springs. The plot shows the packing for chains in system II-0 has a crossover (the curve in the inset of Fig. 8), in contrast to the single power-law like behavior for those connected by strong springs (system II-4). The comparison between the data for system I-4 and II-4 indicates that the bundles formed in the pure polymer system (I-4, marked by "C")) is more compact than its counterpart in the mixed system (II-4, marked by "B"). The onset of the (first) fast growing stage requires larger size subunits in the mixed system with backbone anisotropy than those in the isotropic pure system.

Figure 9 shows the R_i parameters versus $U_{\text{interchain}}$ for the three cases with monomers connected by rigid bonds, $k_{nn} = \infty$ (system I-∞, system II-∞ and system III), listed in Table 1. The curves share the same qualitative features as those in Fig. 3. A quantitative comparison of the curves for R_0 shows the larger degree of curving for systems with rigid bonding than the cases with soft bonding, as a result of less flexible patching among packed chains.

3.3 Chain length dependence

For the two pure homopolymer systems ($N_F = 0$) with isotropic ($K_b = K_t = 0$) rigidly bonded ($k_{nn} = \infty$) chains (systems I-A, $n = 50$ and I-B $n = 25$, in Table 1), each containing the same total number of monomers (nN_p=4000), there are two major aggregated clusters in each case. The clusters have bundled structures and are tilted from the axes of simulation boxes, which suggest the independence of the boundary conditions in the formation of bundles. On a close examination of the snapshots of the structures (Fig. 10), we find that the effects of close sphere-packing lead to the tendency to deviate from a parallel bundled structure and lead to the formation of spiral-like local ordering (left plot of Fig. 10). Such orderings cannot, however, extend to the whole chain as the length of the chain becomes larger (middle and right plots of Fig. 10). In Fig. 11, we compare the monomer-monomer pair distributions among the three cases $n = 25$, $n = 50$ and $n = 100$. The humps behind the sharp, delta-function-like peaks, signal the ordering of sphere-packing. They are apparently larger for $n = 25$ and $n = 50$ than those for $n = 100$. The new ordering, other than that imposed by chain connectivity, is stronger for short chains.

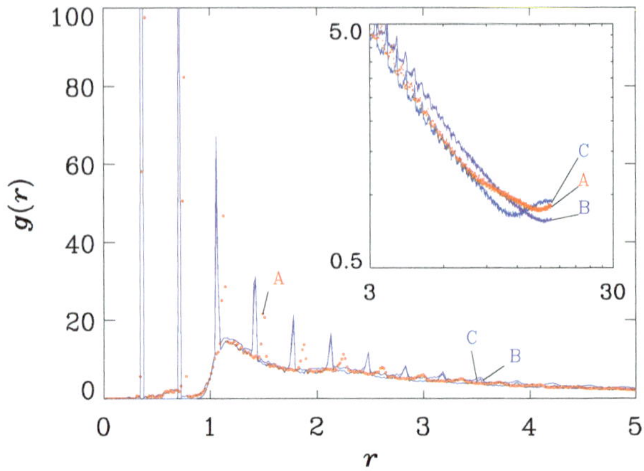

Fig. 8. (Color online) Monomer-monomer pair distribution functions for configurations in the systems II-0 (dots labeled as "A"), II-4 (solid line labeled as "B") and I-4 (solid line labeled as "C") at the onset time for fast falling R_2. The inset shows the data in log-log plot (adapted from Ref. [6]).

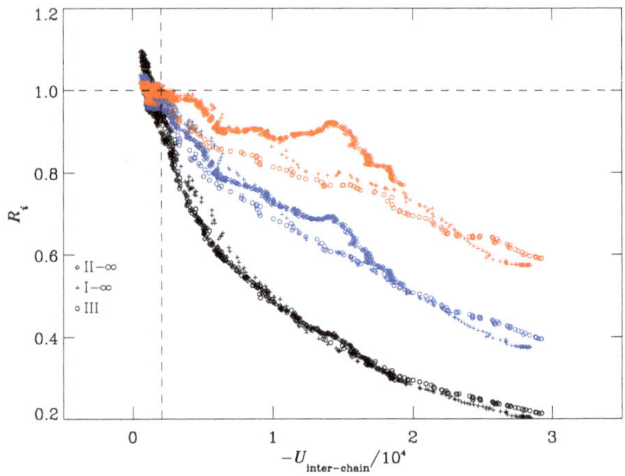

Fig. 9. Same as Fig. 3 for systems of rigidly-bonded chains.

An examination on the time evolutions (Fig 12 and Fig. 13) in these two systems, I-A and I-B, show also corresponding dynamic effects. Since the parameters R_0, R_1 and R_2 feature the line-like, plane-like, and volume-like packed spatial structures, respectively, the returning of R_2 to a larger value in the later stage of evolution indicates the prevailing of three dimensional packing. The latter trend results in the larger values in volume. Figure 14 shows the R_0 and R_1 for $n = 25$ and $n = 50$ follow the decreasing trend in the growth processes, corresponding to reducing $U_{\text{inter-chain}}$ (or increasing $-U_{\text{inter-chain}}$) which is the same as those for systems of longer chains (the cases of $n = 100$ in Figs. 3 and 9), in contrast to the returning of R_2 to a larger value in the later stage. The latter trend is stronger for the shorter chains. The result strongly suggests the relevance of chain length on the structural formation in aggregation phenomena.

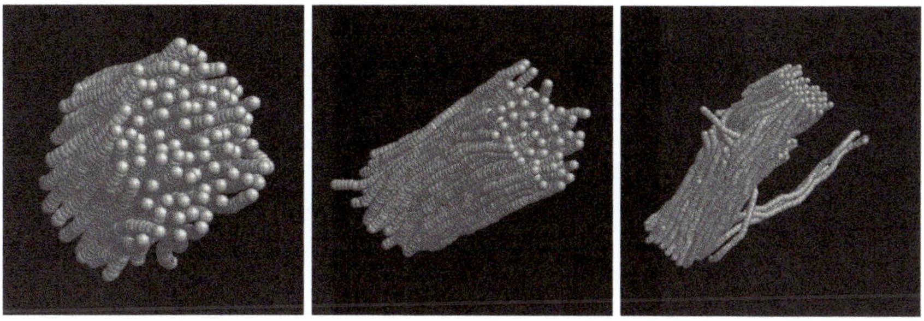

Fig. 10. Snapshots of the largest aggregated clusters for (left plot) $n = 25$, consisting of 81 chains, for (middle plot) $n = 50$), consisting of 51 chains, and for (right plot) $n = 100$, consisting of all 40 chains of the systems (adapted from Ref. [7]).

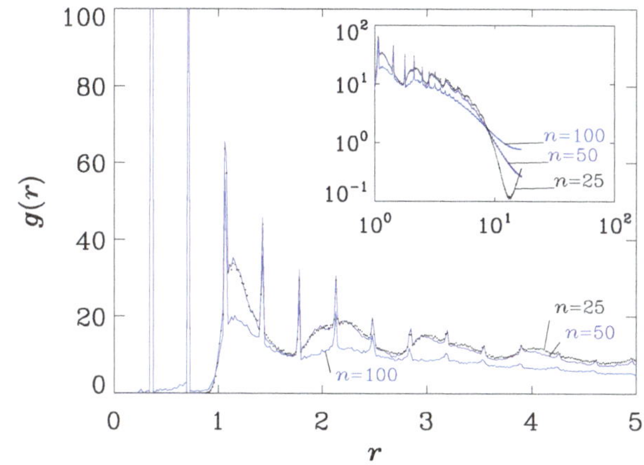

Fig. 11. (Color online) Comparison of monomer-monomer pair distributions of the aggregated configuration for $n = 25$ (blue line), $n = 50$ (red line) and $n = 100$ (black line) (adapted from Ref. [7]).

4. Growth hindered by backbone anisotropy

We have compared the aggregated structures between the systems with zero (system I-4) and with non-zero, but tiny angle potentials (system II-4),. While they both possess bundled clusters (Fig. 1), the non-zero tiny angle potentials break the symmetry to have a branched structure (right plot of Fig. 1). One more sensible question to solve is to ask if the bundled cluster can survive under even larger strengths in angle potentials. The simulations with chains having $K_b = K_t = 0.1$ (set IV in Table 1), which is 1000 times larger than those of the systems in set II have shown no cluster domains formed over time spans equivalent to or larger than those considered for systems in set II [6, 7]. The systems are brought to states at lower temperatures in a stepwise manner, with decreasing target temperature T_0^* at 8, 6 and 3, respectively, in three stages. We do observe the formation of local parallel conformations. But this local ordering can hardly extend to either the larger segment or to the gathering of more chains. The results indicate that the effect of frustration between the straight bundled packing and the local curvature imposed by the angle potentials prevent the aggregation. Such an observation is suggestive in considering more complicated material systems.

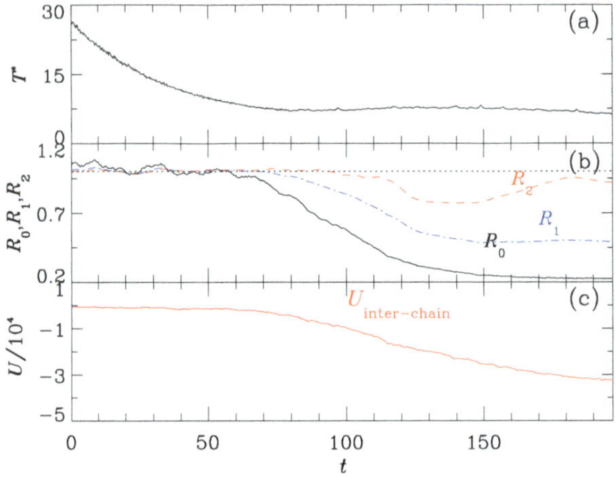

Fig. 12. Same as Fig. 4, for a simulated system with rigidly bonded homogeneous chains ($k_{nn} = \infty$) of length $n = 50$ each (adapted from Ref. [7]).

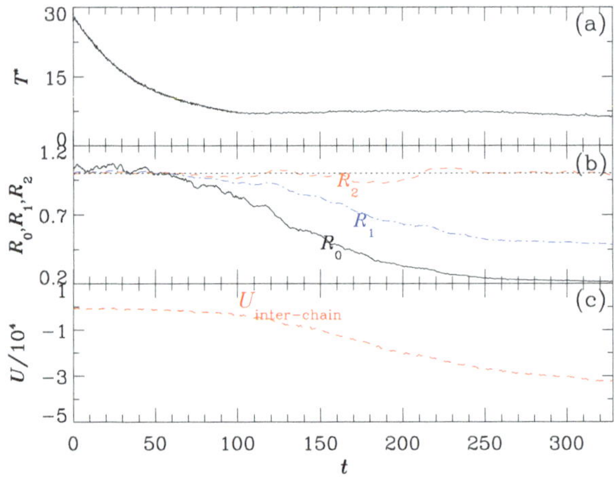

Fig. 13. Same as Fig. 12 for a simulated system with rigidly bonded homogeneous chains ($k_{nn} = \infty$) of length $n = 25$ each (adapted from Ref. [7]).

The typical time evolution of such processes is shown in Fig. 15, for system IV-4. We observe that the values of the parameter R_0 fluctuate in the larger amplitude as compared with those for R_1 and R_2, indicating the occurrence of local alignment between chains fails to survive for triggering the formation of stable structures. In lowering the temperature, the fluctuations change to have the larger amplitude and the longer periods, which seems closer to a stage with monotonic decaying R_0. Entering the regime of $T_P^* \approx 7$, the stronger tendency of alignment between chains signalled by the larger decaying in the value of R_0 still fails to prevent the reverting of the trend, that the enhanced changes in R_0 can only stay with the status as the (larger amplitude) fluctuations. The local mechanical conflicts caused by the angle potentials hinder the pile-up of the packed local segments to form stable higher order structures. We

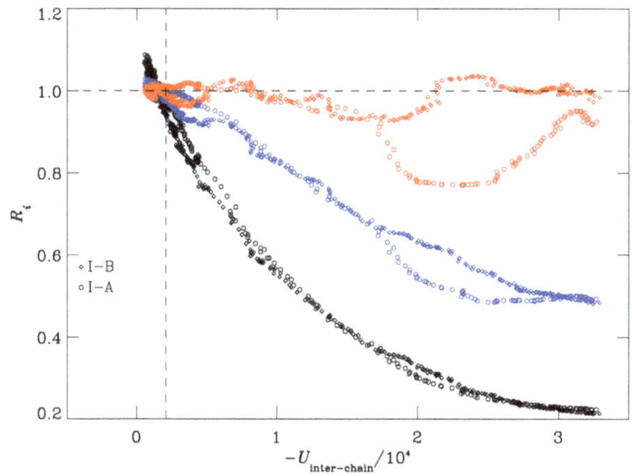

Fig. 14. Same as Fig. 3 for systems with chains of lengths $n = 50$ (circle) and $n = 25$ (diamond), respectively.

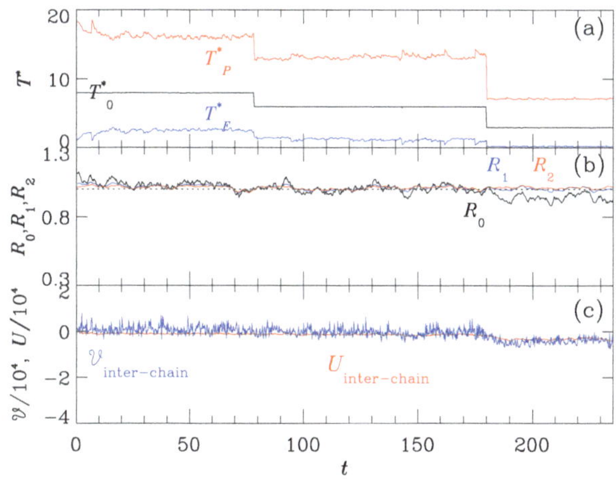

Fig. 15. (Color online) Same as Fig. 4, for the mixed system of polymer and fluid, system IV-4, where $K_b = K_t = 0.1$. (adapted from Ref. [6]).)

conclude that the values of K_b and K_t have to be small for the presence of aggregation process for the polymer chains.

5. Perspectives

The study of aggregations of peptides or proteins and their roles in the occurrence of neurodegenerative diseases, such as Alzheimer's disease [20–22], and the transmissible Creutzfeldt-Jakob disease [23], has become an important cross-disciplined issue for scientists. The important factors for polymer aggregation problems revealed by our analysis, such as the backbone anisotropy and the chain length are suggestive. The suspected origin of Alzheimer's disease at the molecular level is related to the aggregation of proteins $A\beta_{40}$ and $A\beta_{42}$ [20–22] consisting of 40 and 42 residues, respectively. The chain of $A\beta_{40}$ or $A\beta_{42}$ forms two β-strands and such β-strands form β-sheets [21, 22]. A scenario that describes their aggregation is to consider each β-sheet as an effective polymer chain and has a small degree of effective backbone anisotropy. Thus such effective polymer chains has the trend to aggregate into parallel-domain conformation. It will be interesting in the future work to include more refined backbone anisotropy [24, 25] and quantify its effect on the aggregation dynamics.

6. Acknowledgement

This work was supported by Grants NSC100-2112-M-004-001, NSC100-2112-M-001-003-MY2 and NCTS (North).

7. References

[1] T. P. J. Knowles, J. F. Smith, A. Craig, C. M. Dobson and M. E. Welland: *Phys. Rev. Lett.* 96 (2006) 238301.
[2] T. P. J. Knowles, A. W. Fitzpatrick, S. Meehan, H. R. Mott, M. Vendruscolo, C. M. Dobson and M. E. Welland: *Science* 318 (2007)1900.
[3] H. D. Nguyen and C. K. Hall: *Proc. Natl. Acad. Sci.* U. S. A. 101 (2004)16180 .
[4] B. Urbanic, L. Cruz, S. Yun, S. V. Buldyrev, G. Bitan, D. B. Teplow and H. E. Stanley: *Proc. Natl. Acad. Sci.* U. S. A. 101 (2004) 17345.
[5] C. M. Dobson: *Nature* 426 (2003)884.
[6] W. J. Ma and C.-K. Hu: *J. Phys. Soc. Japan* 79 (2010) 100402.
[7] W. J. Ma and C.-K. Hu: *J. Phys. Soc. Japan* 79 (2010) 54001.
[8] R. V. Pappu, X. Wang, A. Vitalis and S. L. Crick: *Arch. Biochem. Biophys.* 469 132 (2008).
[9] M. Doi and S. F. Edwards: *The Theory of Polymer Dynamics* (Oxford University Press, New York, 1986).
[10] R. B. Bird, C. F. Curtiss, R. C. Armstrong and O. Hassager: *Dynamics of Polymer Liquids* Vol.2, 2nd ed. (John Wiley and Sons 1987).
[11] H. C. Andersen: *J. Comput. Phys.* 52, 24 (1983).
[12] J. T. Padding, W. J. Briels: *J. Chem. Phys.* 114 (2001) 8685.
[13] D. Steele: *J. Chem. Soc. Faraday Trans.* 2 81 (1985) 1077.
[14] D. Brown and J. H. R. Clarke: *Comput. Phys. Commun.* 62, (1991) 360.
[15] D. D. Voet and J. G. Voet: *BioChemistry* (John Weily and Sons, 1995) 2nd ed. Chap. 7.
[16] W.-J. Ma and C.-K. Hu: *J. Phys. Soc. Jpn.* 79 (2010) 024005.
[17] W.-J. Ma and C.-K. Hu: *J. Phys. Soc. Jpn.* 79 (2010) 024006.
[18] C. Tsallis: *J. Stat. Phys.* 52 (1988) 479.
[19] L. V. Woodcock: *Chem. Phys. Lett.* 10 (1971) 257.
[20] A. Goate, M. C. Chartierharlin, M. Mullan, J. Brown, F. Crawford, L. Fidani, L. Giuffra, A. Haynes, N. Irving, L. James, R. Mant, P. Newton, K. Rooke, P. Roques, C. Talbot, M. Pericakvance, A. Roses, R. Williamson, M. Rossor, M. Owen and J. Hardy: *Nature* 349 (1991) 704.
[21] A. T. Petkova, Y. Ishii, J. J. Balbach, O. N. Antzutkin, R. D. Leapman, F. Delaglio and R Tycko: *Proc. Natl. Acad. Sci.* USA 99 (2002) 16742.
[22] T. Luhrs, C. Ritter, M. Adrian, D. Riek-Loher, B. Bohrmann, H. Doeli, D. Schubert and R. Riek: *Proc. Natl. Acad. Sci.* USA 102 (2005) 17342.
[23] N. J. Cobb, F. D. Sonnichsen, H. Mchaourab and W. K. Surewicz: *Proc. Natl. Acad. Sci.* USA 104 (2007) 18946.
[24] R. Pellarin and A. Caflisch: *J. Mol. Biol.* 360 (2006) 882.
[25] N. Y. Chen, Z. Y. Su, and C. Y. Mou: *Phys. Rev. Lett.* 96 (2006) 078103.

Part 2

Dynamics of Biomolecules

5

M.Dyna*Mix* Studies of Solvation, Solubility and Permeability

Aatto Laaksonen[1], Alexander Lyubartsev[1] and Francesca Mocci[1,2]
[1]*Stockholm University, Division of Physical Chemistry, Department of Materials and Environmental Chemistry, Arrhenius Laboratory, Stockholm*
[2]*Università di Cagliari, Dipartimento di Scienze Chimiche Cittadella Universitaria di Monserrato, Monserrato, Cagliari*
[1]*Sweden*
[2]*Italy*

1. Introduction

During the last four decades Molecular Dynamics (MD) simulations have developed to a powerful discipline and finally very close to the early vision from early 80's that it would mature and become a computer laboratory to study molecular systems in conditions similar to that valid in experimental studies using instruments giving information about molecular structure, interactions and dynamics in condensed phases and at interfaces between different phases. Today MD simulations are more or less routinely used by many scientists originally educated and trained towards experimental work which later have found simulations (along with Quantum Chemistry and other Computational methods) as a powerful complement to their experimental studies to obtain molecular insight and thereby interpretation of their results.

In this chapter we wish to introduce a powerful methodology to obtain detailed and accurate information about solvation and solubility of different categories of solute molecules and ions in water (and other solvents and phases including mixed solvents) and also about permeability and transport of solutes in different non-aqueous phases. Among the most challenging problems today are still the computations of the free energy and many to it related problems. The methodology used in our studies for free energy calculations is our Expanded Ensemble scheme recently implemented in a general MD simulation package called "M.Dyna*Mix*".

Although several very powerful MD simulation programs exist today we wish to highlight some of the more unique features of M.Dyna*Mix* which is both general purpose package but also designed towards applications which can be directly connected to spectroscopic experiments with solvation state NMR as example. As MD is the only method to obtain dynamical information in realistic (but still limited) time scales we wish to give examples of motional characteristics of molecules in both bulk solvents and in complex biomolecular systems (containing for example DNA, proteins and lipid membranes) in specific non-covalent interactions of varying strength. In addition we will demonstrate the capabilities of

MD simulations in design of orally administrated drugs discussing problems of developing force field models applicable both in physiological water phases and barriers made of biological cell membranes. We will also present theoretical models to couple simulation results with experiments such as NMR relaxation and also powerful schemes to visualize often very complex solvation structures and relate it to intermolecular interactions within the solvation sphere and to the undergoing molecular dynamics behind coordination and solvation. We consider also applications of M.Dyna*Mix* package to simulations of large biomolecular systems, such as DNA molecules in ionic solution and lipid membrane bilayers.

2 The tools: M.Dyna*Mix* software

In this section we describe briefly possibilities of M.Dyna*Mix* package and give a brief manual of its usage, from preparing of molecular structures to analysis of simulation results.

The M.Dyna*Mix* MD simulation package has been developed in our laboratory in 90-ties (Lyubartsev & Laaksonen, 1998a, 2000). This is a scalable parallel general purpose MD software intended for simulations of arbitrary molecular mixtures. The program can employ most of commonly used force fields which are based on the potential energy function including electrostatic and Lennard-Jones interactions, covalent bond and angles, dihedral angles of various types as well as some other optional terms. The long-range electrostatic interactions are treated by the Ewald summation method (Allen & Tildesley, 1987). Various kinds of temperature and pressure control, including separate pressure control in different directions, are included. Two types of MD integrators are implemented: one is based on time-reversible double time step algorithm by Tuckerman et al. (Tuckerman et al., 1992) which is used for simulations of flexible molecular models, and another is based on constraint dynamics with the use of "SHAKE" algorithm (Ryckaert et al., 1977), for simulations of rigid models. The program is written in standard Fortran-77 in a modular manner, the source code is well commented which make it suitable for users modifications. M.Dyna*Mix* version 5.2, released in 2010, includes an option for free energy calculation within the Expanded Ensemble method (Lyubartsev et al., 1992, 1998) which allows automated calculations of free energies and chemical potentials in single simulation run. The program is efficiently parallelized and scales well on up to a hundred processors or cores. The program contains utilities for preparation of molecular description files, as well as analysis suite to perform various calculations from the saved simulation trajectories. The flow chart of the program is schematically shown in **Figure 1**. In this chapter, we give a brief "quick start" introduction to the program, in which we concentrate on the new features appeared after previous publication (Lyubartsev & Laaksonen, 2000). The full reference manual of M.Dyna*Mix* software is available on-line at the distribution site (http://www.mmk.su.se/mdynamix).

2.1 Preparation of molecular description files

Before the start, a user must prepare, for each type of the molecules used in the simulations, a file with extension *.mmol* in which molecular topology and force field parameters are described. This includes list of atoms, their initial coordinates, Lennard-Jones parameters and charges, full list of covalent bonds, angles and torsion angles with the corresponding force field parameters.

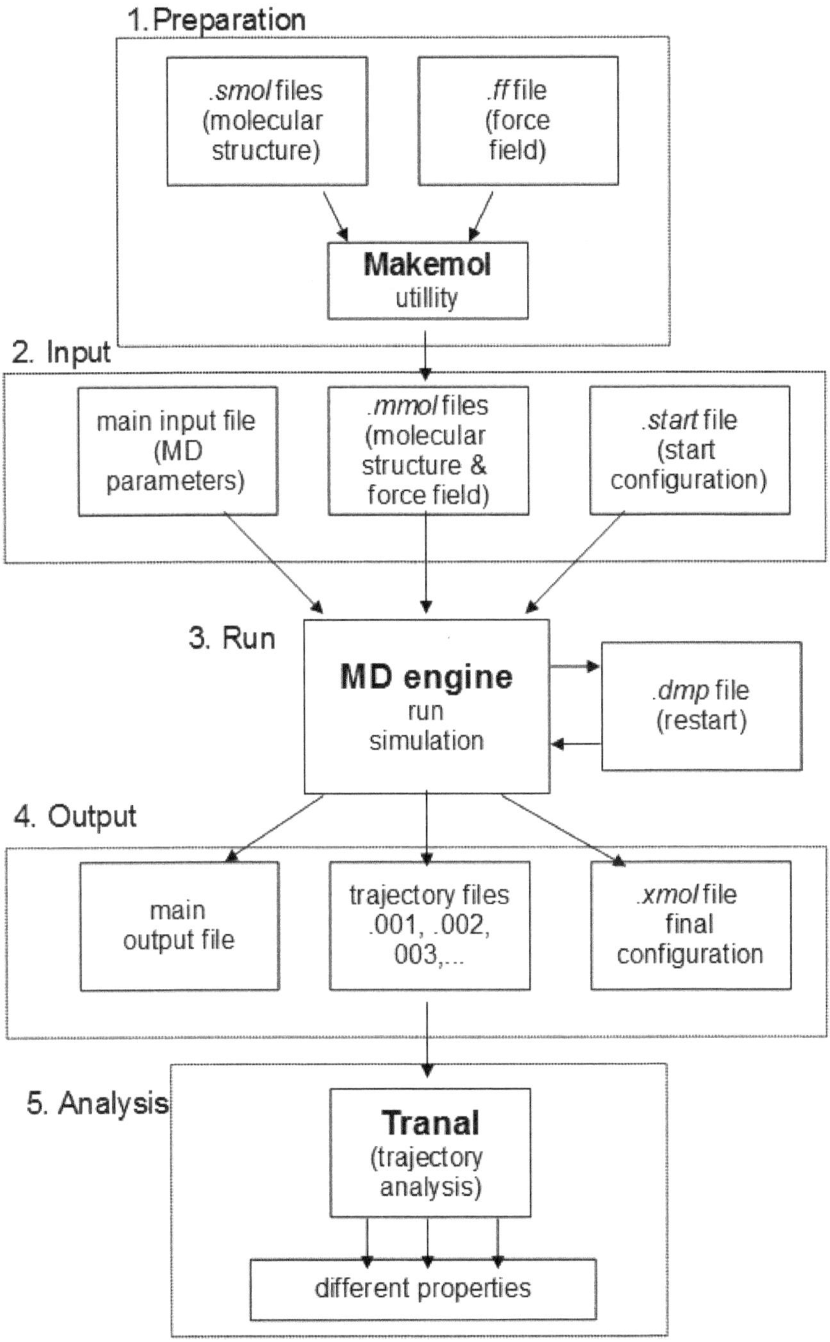

Fig. 1. Flow chart of M.Dyna*Mix* simulation software

Even if *.mmol* files can be created with a simple text editor using available data on molecule structure and force field parameters, their preparation can be significantly simplified by the *makemol* utility. This utility creates a .mmol file from two sources: a "simple molecular structure" (*.smol*) file and "force field" (*.ff*) file containing force field parameters. A *.smol* file contains list of atoms with their initial coordinates, partial charges and force field atom types, and list of bonds. An example of *.smol* file for a methanol molecule is given in Figure **2**. A "force field" file contains force field parameters for a number of atom types (which can be common for different molecules). The *makemol* utility, using atom types specified in the *.smol* file, substitutes corresponding parameters from the force field file and creates the resulting *.mmol* file. It also generates automatically the list of covalent and torsion angles from the list of bonds (present in the *.smol* file) and substitutes parameters for them. The distribution of M.Dyna*Mix* contains examples of Amber94 and CHARMM27 force fields. It also contains utility *char2mdx* to transform general Charmm parameter file to M.Dyna*Mix* force field file.

```
#=======================================================I
#       CH3OH  - Simple molecular file                  I
#       Adopted for Amber94 force field (amber94.ff)    I
#=======================================================I
#      Number of sites
     6
#at        X           Y           Z           Q          type
 O       .0285       .0000       .6948      -.6000        OH
 H      -.8760       .0000      1.0044       .4000        HO
 C       .0285       .0000      -.7322      -.1000        CT
 HC     1.0595       .0000     -1.1022       .1000        H1
 HC     -.4870       .8929     -1.1022       .1000        H1
 HC     -.4870      -.8929     -1.1022       .1000        H1
#    Bonds
      5
# N1     N2
   1      2
   1      3
   3      4
   3      5
   3      6
```

Fig. 2. "*.smol*" file for a methanol molecule

There exists a number of open source molecular editors (for example, kalzium - http://edu.kde.org/kalzium - included into Fedora Linux distribution, or Dundee ProDrg server - http://davapc1.bioch.dundee.ac.uk/prodrg/), that allow to draw a molecule with a mouse and generate a preliminary optimized molecular structure, as well as generate the list of bonds. Next, one needs to assign the atom types for each atom (the notations for atom types are different in different force fields), and specify partial atomic charges, which are usually not included into general force fields, but need to be calculated by a quantum-chemical program for each specific molecule. After the *.smol* file is created, one can use Makemol utility to create the structure-parameter *.mmol* file for the given type of molecules.

2.2 Simulations

Molecular dynamics simulation parameters are described in the main input file. From version 5.0, the format of the main input file was changed significantly compared with the previous publication of the program (Lyubartsev & Laaksonen, 2000). In the new format, all simulation parameters and features are specified in the form "Keyword value(s)", where keyword describes one specific parameter or property of the simulation (for example, time step, temperature, restart control, etc). For missing keywords, default values are used. A complete description of available keywords and their parameters is given in the manual at the M.Dyna*Mix* distribution site (http://www.mmk.su.se/mdynamix).

Finally, in order to start simulation, one needs to specify the start configuration of the simulated system. It is always possible to start from an already prepared (by various means) configuration written in a file in the xmol (xyz) format. This option may be preferable if one wish to simulate a specifically organized system, e.g. lipid bilayer or protein in a folded form. It is also possible to generate the start structure automatically. In this case, molecules are placed with random orientation and their centers of mass on FCC lattice. Clearly, in such case one can expect numerous overlappings of atoms belonging to different molecules. In order to deal with the problem, the following procedure can be used:

1. One starts from a low enough density, and run a short run (of order of 1000 MD steps) with the options "Cut large forces", "Velocity scaling" for temperature control and with pressure barostate switched off. The "Cut large forces" option limits forces acting on atoms by a specified cutoff value, while keeping the direction of forces. As a result, bad atom contacts and overlappings disappear quickly.
2. One proceeds for another few thousands steps specifying expected density of the system and using option "Change volume slowly". The system gradually change volume and increase the density to expected, providing in the end preliminary equilibrated system
3. After the second stage it is possible to switch from "velocity scaling" control of temperature to "Nosé thermostat". It is advisable to continue for some time to run simulation at constant volume, and when the pressure stabilizes, turn on the barostat if constant-pressure simulations are desirable.

2.3 Post-simulation analysis

The main result of a simulation is a set of trajectory files which contain configurations of the system (Cartesian coordinates and optionally velocities of all the atoms in the system) as a function of the simulation time. The trajectory analysis suite Tranal includes a number of utilities to compute a great variety of structural, thermodynamical and dynamical properties of the system, among which are radial and spatial distribution functions, various time correlation functions, diffusion, residence times, dielectric permittivity, order parameters and lateral diffusion for bilayer-like systems, etc. The suite includes the base module which can read trajectories of various formats (including those of other simulation packages), and specialized modules to compute specific properties.

2.4 Expanded ensemble mode

This special feature allows computations of solvation free energies by the expanded ensemble (EE) methodology (Lyubartsev et al., 1992, 1998). In the latest release (v.5.2), the

EE method was implemented together with Wang-Landau algorithm for automatic optimization of weighting factors in expanded ensemble, which allows obtaining accurate free energies in a single MD run. A short description of the algorithm is given below.

The EE method implies a gradual insertion/deletion of the studied solute molecule into/from the solvent. The insertion parameter α is introduced which describes the degree of insertion of the solute, so that $\alpha = 1$ describes the solute molecule fully interacting with the solvent while $\alpha = 0$ represents the solute completely decoupled from the solvent (that is, case $\alpha = 0$ describes pure solvent and the solute molecule in a gas phase of the same volume). The insertion parameter can accept a number of (fixed) values $\{\alpha_i\}$, $i = 0,...,M$ in the range between 0 and 1. During the simulation, which can be carried out by either Monte Carlo (Lyubartsev et al., 1992) or molecular dynamics algorithm (Lyubartsev et al.,1994), attempts are made to change the insertion parameter to another (normally, neighboring) value, with acceptance probability:

$$P(i \to i \pm 1) = \exp\left(-\frac{\left(V^{Ss}(a_{i\pm1}) - V^{Ss}(a_i)\right) + \left(\eta_{i\pm1} - \eta_i\right)}{k_B T} \right) \tag{1}$$

where $V^{Ss}(\alpha)$ is the interaction energy of the solute particle with the solvent corresponding to the given insertion parameter α and η_i are so-called balancing (weighting) factors introduced with the purpose to make distribution over subensembles close enough to the uniform one (Lyubartsev et al., 1992). During the simulation, probabilities of different subensembles ρ_i are defined, and free energy difference between states with fully inserted and fully deleted solute (excess solvation free energy) is computed by:

$$F_M - F_0 = -k_B T \ln \frac{\rho_M}{\rho_0} - \eta_M + \eta_0 \tag{2}$$

In previous works on computation of the solvation free energies (Lyubartsev et al., 1994, 2001), a linear scaling of solute-solvent interaction with α was used, namely $V^{Ss}(\alpha) = \alpha V^{Ss}(1)$. Such a scheme has a shortage that the repulsive core of the Lennard-Jones potential decreases very slowly with decrease of α, which leads to necessity to consider very small α values. Now a new scheme is implemented in which interaction $V^{Ss}(\alpha)$ between the solute and solvent atoms (which is supposed to be a sum of the Lennard-Jones and electrostatic terms) is scaled according to the following:

$$V^{Ss}(a) = a^4 V_{LJ}^{Ss}(1) + a^2 V_{el}^{Ss}(1) \tag{3}$$

where $V^{Ss}{}_{LJ}$ and $V^{Ss}{}_{el}$ are Lennard-Jones and electrostatic parts of the solute-solvent interaction respectively. The rationale between α^4 scaling of the Lennard-Jones interaction is that the effective core radius of the LJ potential scales then as $(\alpha^4)^{1/12} = \alpha^{1/3}$, and thus the effective volume of the core (approximately proportional to the free energy of cavity formation) scales linearly with α. Scaling α^2 for the electrostatic interactions is chosen in order to switch off them faster than the repulsive LJ interaction, to minimize problems with hydrogen atoms of water molecules which do not have LJ potential. With scaling of the solute-solvent interaction described by (3), it is possible to choose α_i points uniformly distributed in the range [0:1] so that a reasonable acceptance probability of the transitions is maintained through the whole range of subensembles.

Another modification of the EE procedure implemented in the new version of M.Dyna*Mix* is the automatic choice of the balancing factors η_i by the Wang-Landau algorithm (Wang & Landau, 2001). In the application to the expanded ensemble technique the Wang-Landau algorithm can be formulated as follows. We start simulations with zero balancing factors. After visiting a subensemble *i*, a small increment $\Delta\eta$ is added to the corresponding value of the balancing factor: $\eta_i(t+dt) = \eta_i(t)+\Delta\eta$, which decreases the probability to go to already visited states and favors to attaining a uniform distribution. After a certain number of steps (a sweep), when the system visited all subensembles and passed several times (default 2) between the end points, the value of the increment $\Delta\eta$ is decreased twice. After a certain number of sweeps (usually 10-12), the value of the increment is becoming very low, and simultaneously the profile of balancing factors is tuned in a way providing uniform walking in the space of subensembles. After that the equilibration stage ends, and production run with fixed balancing factors is made yielding the solvation free energies.

3. Examples of M.Dyna*Mix* applications

In this chapter we concentrate only on some selected types of MD simulations which are just very few of many M.Dyna*Mix* has been used in the past.

3.1 Solubility and permeability

Solubility (or lack of it), in particular in water, but also in other solvents and solvent mixtures is among the most important problems in the development of new drug molecules today, in particular, for poorly water-soluble, orally bioavailable drugs, with a controllable release rate. Computer modelling and simulations (together with combinatorial chemistry, high-throughput screening and genomics/proteomics) have long been key tools in modern drug design but has not been utilized much in solubility studies. After drug molecules are dissolved in the beginning of the administration they start their journey in living organisms to find the target. During this process they are adsorbed and distributed and doing so they need penetrate numerous obstacles, including the blood-brain barrier. To study the molecular details solubility and permeability during this transport, MD simulations together with free energy calculations are highly useful (Lyubartsev et al., 1998c, 2001). Solubility of drug molecules and their permeability across lipid membranes have been studied using M.DynaMix by calculating log*P* partition coefficients from the solvation free energies (Lyubartsev et al. 1999; Åberg et al. 2004).

3.2 Solvation in mixed solvents

Many substances turn out to be readily soluble in solvent mixtures (mixed solvents) but not in the pure components and vice versa (Reichardt, 1988). Selective solvation of ions from salts in mixed solvents is another, early discovered, property. Both IR measurements and computer simulations have confirmed the existence of micro-heterogeneities and conglomerate water structures at higher concentrations of the more organic co-solvent. Many amphiphilic and biomolecular systems are better dissolved in mixtures, as a result of preferential solvation. This Section will highlight the advantage of using MD simulation to study preferential and selective solvation to obtain insight into the complex molecular processes behind them and to give a picture and rules in the delicate balance of interactions between the different molecular species (Bergman & Laaksonen, 1998). We will first use a small disaccharide as a solute in discussing the preferential solvation. However, before discussing the solvation in mixed solvents we highlight a number of key features in hydration and solvation of the disaccharide in several pure solvents.

An important question in studying carbohydrates is the role of the solvent, most often water, and in particular its influence on the structure. Several aspects, such as conformational states, rotation barriers and intra/intermolecular hydrogen bonds, flexibility/rigidity, internal dynamics, hydrophilicity/phobicity, as well as electrostatic interactions are important in studying and describing the full three-dimensional structure of this category of molecules. Even small differences in the molecular geometry or conformation can easily affect the hydration around these inherently flexible molecules. Radial distribution functions (RDF) as well angular distribution functions (ADF) are the most commonly used tools to study solvent structure and conformational populations in computer simulations. Carbohydrates typically have a large number of hydroxyl groups ("half water molecules") which easily engage themselves in the hydrogen bond network of water around them leading to very distinct and sometimes rather spectacular solvation structures.

We have studied a small prototypic carbohydrate molecule *α*-D-Man*p*-(1→3)-*β*-D-Glc*p*-OMe containing both a mannose and a glucose ring connected with a glycosidic bridge, by carrying out molecular dynamics simulations in water, methanol, dimethylsulfoxide (DMSO) and also in the mixture of water and DMSO as the solvents (Vishnyakov et al., 1999, 2000a, 2000b). The mixed solvent allows experimentally low temperature studies at a molar ratio of 3:1 of water and DMSO making the solution more viscous. We use the same concentration in our MD simulations. We will simply call the molecule as "disaccharide" in the text below. Conformations of our disaccharide are well described using the two torsional angles across its glycosidic linkage, namely φ and ψ (**Figure 3**).

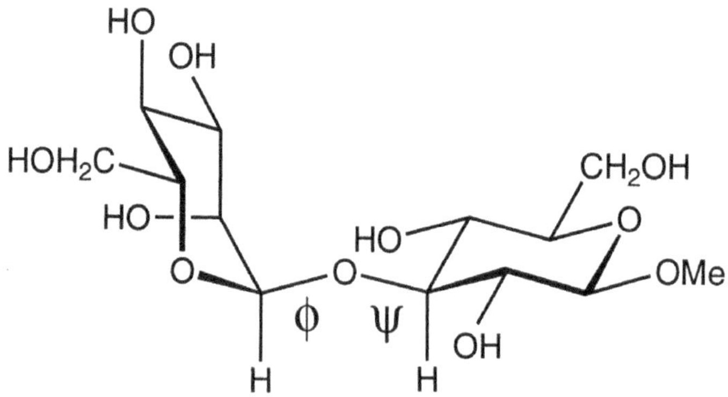

Fig. 3. Molecule *α*-D-Man*p*-(1→3)-*β*-D-Glc*p*-OMe containing both a mannose and a glucose ring connected with a glycosidic bridge.

Our simulations exhibit clearly two main minima (potential wells) on the adiabatic energy surface, one close to *g*- (-60°, we call it "well A") and the other one close to *g*+ (+60°, we call it "well B") with respect to the φ torsional angle. All simulations are carried out using the

M.Dyna*Mix* package (Lyubartsev & Laaksonen, 2000). The solvation structure is presented using spatial distribution functions (SDF) (Kusalik et al., 1999) visualized with the gOpenMol software (Laaksonen 1992; Bergman et al. 1997) http://www.csc.fi/english/pages/g0penMol/tutorials. In all SDFs below the iso-density threshold is three times the bulk density in the simulations except when stated otherwise.

In water solution the disaccharide shows a clear preference for the well A. The hydration is dominated by the first-shell structuring in the solvent. The large number of hydrophilic centers in carbohydrates imposes a strong anisotropic structuring on the surrounding solvent. The hydroxyl groups were found to be extensively hydrogen-bonded to surrounding water molecules (**Figure 4**). For an optimal hydration both the solute and the solvent should have a compatible topology for the hydration requirements of the involved functional groups. The disaccharide molecule is found to be fairly rigid when dissolved in water showing no rotation around the glycosidic linkage.

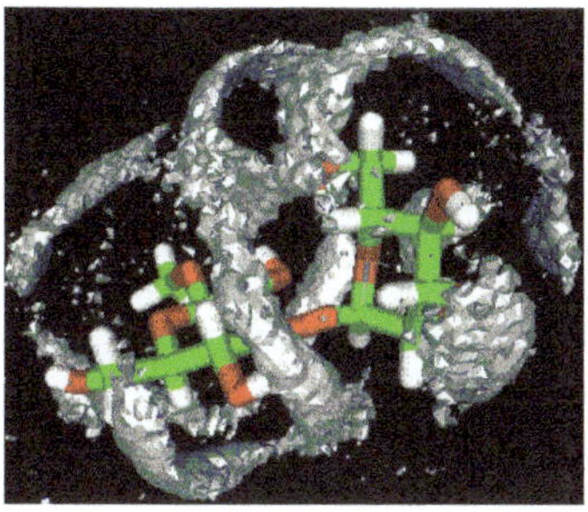

Fig. 4. Hydration structure around *α*-D-Manp-(1→3)-*β*-D-Glcp-OMe showing a belt-like pattern.

In methanol solution (**Figure 5**), the solvation structure, as seen in the SDFs, reflects the ability of methanol to both donate and accept hydrogen bonds. Besides, having in principle the same possibility to form hydrogen bond combinations as water, methanols can even hydrogen bond to the non-hydroxyl oxygens. The methyl groups in methanol restrict the formation of similar solute-solvent hydrogen bonding pattern as in water. However, the hydrogen bonding can be easily seen in the SDFs; e.g. the donation of hydroxyl protons in methanol around O2*m* (upper left), O6*m* (middle right), and O4*g* (lower left). Note, *m* denotes the mannose ring and *g* denotes the glucose ring, while the ordering of the oxygens follows the standard rules. The solvation structure of methanol around the disaccharide is found less extended than in water, but rather more like an average between that of water and that of DMSO presented in the next paragraph.

In DMSO solution the most probable hydrogen-bonding sites for the DMSO oxygens around the disaccharide are depicted in **Figure 6**. Distinct features showing hydrogen bonding to the hydroxyl groups are present. Compared with water, where the corresponding SDFs show distinct belt-like shapes around the entire saccharide, DMSO behaves differently. While four possible combinations of hydrogen bonds between the hydroxyl groups and water are possible, DMSO can only be engaged in one type of H bond as it cannot donate hydrogen. However, distinct regions are observed for HO4*g* (middle left in **Figure 4**), HO2*m* (upper right), and HO2*g* (lower right). Furthermore, the bigger size and much heavier mass slow down the mobility of DMSO and contribute to a more distinct type structure.

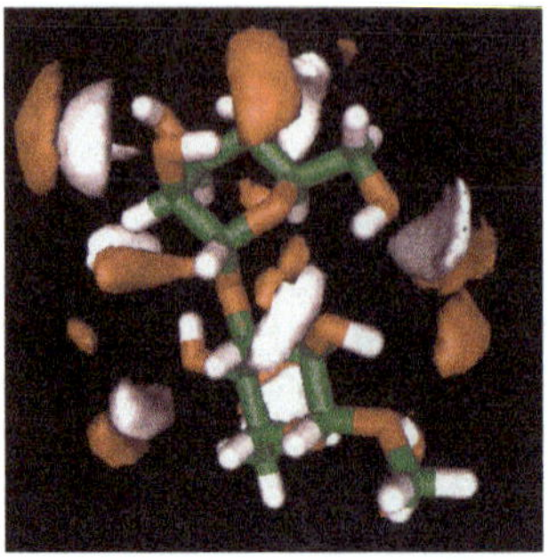

Fig. 5. Solvation structure around α-D-Manp-(1→3)-β-D-Glcp-OMe dissolved in methanol. White - hydrogen, Red - oxygen. For details see the text.

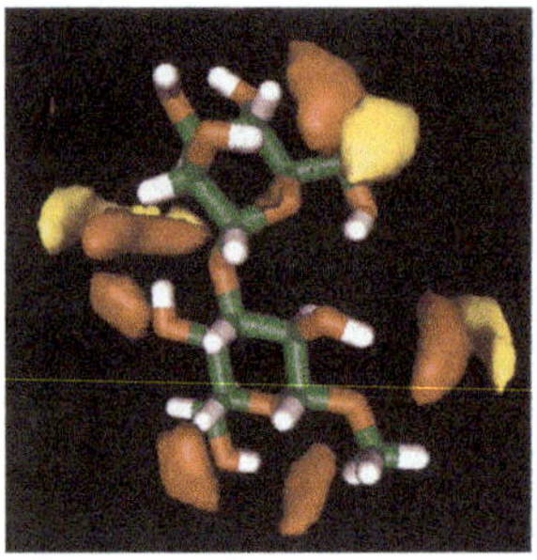

Fig. 6. Solvation structure around α-D-Manp-(1→3)-β-D-Glcp-OMe dissolved in DMSO. Yellow - sulphur, Red - oxygen. For details see the text.

In Water-DMSO mixture (**Figure 7**) we can see that the saccharide is selectively solvated. By combining the information obtained from using the traditional RDFs and the three-dimensional SDFs, we can see the different parts of disaccharide preferentially solvated by either water or DMSO. We can also see that in some parts there is a clear competition between the two components to different hydroxyl groups. For example, DMSO acts as a hydrogen bond acceptor and solvates predominantly the regions close to HO2g and HO2m (see Figure). Water, acting as a hydrogen bond donor, hydrates selectively the regions around O6m and O2g. The SDFs provide a very detailed picture of solvation. Different SDF representations and techniques to analyze solvation graphically are given by Bergman and co-workers. An interesting detail is that when DMSO is acting as a hydrogen bond acceptor to the HO2g hydroxyl group, it can also simultaneously act as an acceptor a specific water molecule hydrogen bonding to O6m. This leads to a hydrogen bonded complex which involves two hydroxyl groups of the disaccharide and one water molecule and DMSO molecule.

We have studied preferential solvation of phenol in equimolar water/acetonitrile and water/ethanol mixtures (Dahlberg & Laaksonen, 2006). In this work we introduce a special "difference spatial distribution function" (ΔSDF) to illustrate the excess densities over the cosolvents in preferential solvation locations around the phenol molecule in these two solvent mixtures. This work was inspired by intermolecular ^{1}H NOESY experiments carried out by Bagno (Bagno, 2002) nicely showing preferential solvation around organic molecules.

Simulations and experiments agree well with each other but reveal also some differences. In fact, this is just another example where NMR and MD simulation techniques can be successfully combined (Odelius & Laaksonen, 1999).

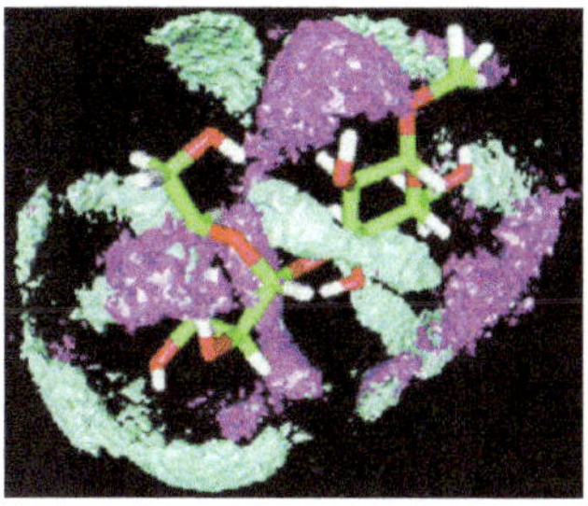

Fig. 7. Preferential solvation structure around α-D-Manp-(1→3)-β-D-Glcp-OMe dissolved in 3:1 water-DMSO mixture. Purple - oxygen in DMSO, Green - oxygen in water. For details see the text.

3.2.1 Mixed solvents

The Water-DMSO system as such is a commonly used mixed solvent. Other examples of such solvents are the binary mixtures of water, alcohols and acetonitrile, including both aqueous and nonaqueous mixtures of them. Mixed solvents are important for example in chemical industry. We have studied the DMSO-water binary system which is known to exhibit strongly non-ideal behavior as a high positive heat of mixing and negative excess mixing volume. It was suggested earlier that two water molecules first hydrogen-bond simultaneously to the sulfoxide oxygen with a third water molecule found hydrogen bonding and bridging the two water molecules, in this way forming a 3:1 water-DMSO complex. It should be straight-forward to study if this is the case using computer simulations. For this purpose we introduced a "multi-particle SDF" in which the local coordinate system was defined based on the three oxygens; the sulfoxide oxygen and the two oxygens in water molecules involved in hydrogen bonding to the sulfoxide oxygen (Vishnyakov et al. 2001). This was possible because of the long enough residence times for these particular hydrogen bonds. It is easy to see from the MP-SDF (**Figure 8**) where the second shell water molecules to the sulfoxide group are found. It becomes clear from the simulations that no third water molecule hydrogen bonding to the first shell water molecules was found.

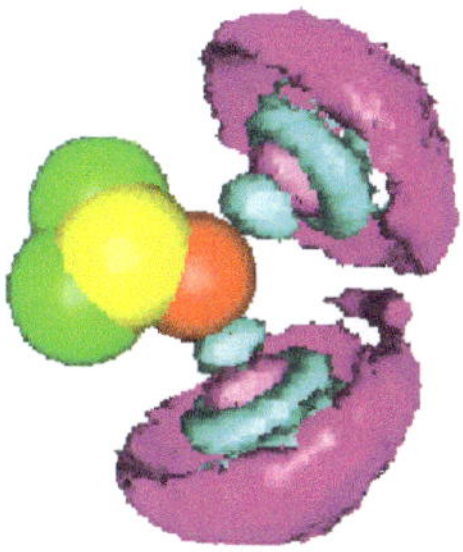

Fig. 8. Multi-particle spatial distribution function showing first and second hydration shell around DMSO. For details see the text.

3.2.2 Hydration of Ni^{2+}

MD simulations of a Ni^{2+} ion in water have been carried out to study the hydration structure around nickel (Egorov et al., 2006). The analysis is extended to the second hydration shell around the divalent ion. The nickel aqua-complex has been treated without any constraints in order to analyze the structure and dynamics, as well as molecular mechanisms of water substitutions and exchange in the first and second shells around the cation. The simulations show that the structure of $[Ni(H_2O)_6]^{2+}$ complex is very stable (**Figure 9**). The main molecular mechanisms contributing to reorientational motion of the whole complex are the "pushes and kicks" given by the second hydration shell water oxygens. The rotations of water molecules themselves in the second shell are only changing the hydrogen positions and do not affect the reorientation of the complex. In spite of frequent exchange of water molecules between the bulk and the second hydration shell the overall dynamical behavior in the first and second shell is very similar.

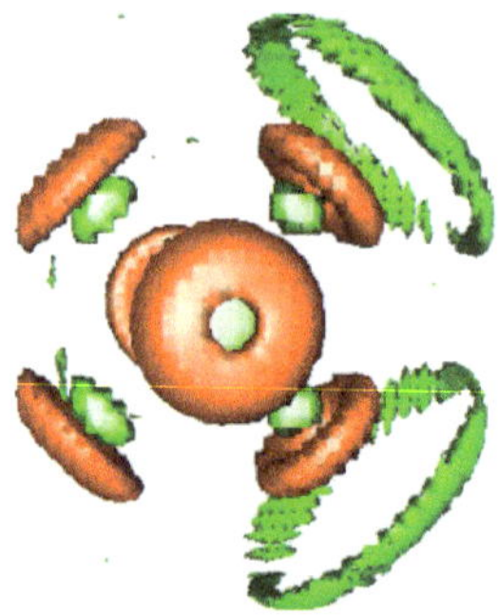

Fig. 9. Structure of $[Ni(H_2O)_6]^{2+}$ complex showing also the second hydration shell around the complex. Green - oxygen in water, Red - hydrogen in water For details see the text.

3.3 Hydration and coordination of counterions around DNA

Nucleic acids are highly negatively charged polyelectrolytes, displaying a considerable sensitivity to their ionic surroundings while undergoing different structural transitions and interacting with other charged species in their surroundings. Since early studies of DNA structure it has been shown that a variation in the counter ion nature or concentration, and in the water content, lead to different conformational properties; these properties depend also on the base pair composition and sequence (Saenger, 1984; Leslie et al., 1980).

Computer simulations and MD in particular are a very powerful technique to obtain detailed information about processes behind hydration and counterion coordination, including the dynamical behaviour of cations around the charged surface of DNA. However the highly charged nature of DNA has lead to a later application of MD simulation compared to that of proteins. In fact, one of the main problems in the simulation of these polyelectrolytes was the treatment of the long range electrostatic interactions, which has been solved with the use of the Ewald summation methods (Allen & Tildesley, 1987; Laaksonen et al., 1989). Proper treatment of these interactions, together with the developments of Force Fields designed for explicit solvent models (Cornell et al., 1995; MacKerell et al., 1995; MacKerell, 2004) and advances in computer power have lead to large use of molecular dynamic simulations in the study of nucleic acids. Here we focus on the modeling, at different levels of sophistication, of nucleic acids in physiological environments with different species of counterion, either mono and multivalent ((Lyubartsev & Laaksonen, 1998b; van Dam et al., 1998; Mocci et al, 2004; Bunta et al., 2007; Korolev et al., 2001, 2002, 2004a, 2004b) and we give some examples of how the MD trajectories can be analyzed and the results checked against experimental data using M.Dyna*Mix* software.

3.3.1 Interactions with monovalent counterions

Until almost the end of the last century it was believed that monovalent counterions did not show any "sign of localization or dehydration due to interactions with double-stranded (ds) nucleic acids" (Anderson et al., 1995) and that the interactions were not affected by the base pair sequence. Since that time, many investigations have lead to a revision of this view, and it is now believed that monovalent cations binds to DNA partially losing the solvation shell, often in a sequence specific way which also depends on the nature of the cation.

MD simulations have been playing a central role in revising this view, and have been of great help in verifying the role of these interactions on the conformational properties of DNA and in understanding the differences in the coordination of counterions around nucleic acids.

As examples of this concept we discuss some of our studies focused on DNA interactions with alkali cations. In order to understand the differences in the behaviour of cations of different size, we performed MD simulations on DNA solutions containing either the physiologically relevant DNA counter ion Na^+, and the smallest and the largest among the alkali ions: Li^+ and Cs^+ (Lyubartsev & Laaksonen, 1998b; van Dam et al., 1998). The simulations were performed mimicking an infinite array of in parallel ordered DNA duplexes, by placing a DNA decamer [$d(ATGCAGTCAG)_2$] along the Z direction of the simulation box and applying periodical boundary conditions.

Important information on ion-DNA interactions can be obtained inspecting the radial and spatial distribution function (RDF and SDF). The RDF gives the probability of finding a counterion type at a distance r from selected DNA atoms, or from DNA axis, relative to the probability expected from a random distribution at the same density. The SDF is the three dimensional distribution of a given atom species in a local coordinate system fixed on some (portion of a) molecule. In the SDF calculated using Tranal utility of M.Dyna*Mix* the value of the function in each point corresponds to the probability of finding there an atom compared to the bulk. Representation of a four dimensional function in two or three dimensions can be done in different ways (Bergman et al., 1998; Kusalik et al. 1999); one of the most used to visualize solvation structure and ion binding modes is the iso-intensity representation, where a surface links the points where the SDF function has a chosen value. RDF calculated between the ions and selected atoms either in the minor or major groove, or in the backbone, are able to show at a glance important differences in the behaviour of the counter ions. For example in **Figure 10** it is shown the RDF between counterions and P atoms.

The smallest ion has a remarkable higher probability to be directly bound to the phosphates compared to the other counterions. On the other hand the RDFs calculated between the counterions and selected atoms in the minor and major grooves, clearly indicate that the smallest ion do not penetrate in the grooves, while the largest has a remarkable preference for the minor groove. Analysis of the SDFs of the water and of the counterions around the binding site is very helpful for understanding these differences in the interactions and to understand the conformational preference of DNA in presence of different counter-ions. In the case of Lithium the SDFs of the water atoms and of the cation around the phosphates groups explained the higher preferences of Li^+ for this site compared to other alkaline ions. In fact the average organization of the water molecules around the phosphates' oxygen atoms contains a hole whose dimensions is perfect for the small Li^+ ions to fit in, keeping a tetrahedral hydration shell, as shown by the SDF in **Figure 11**. The larger alkali ions cannot bind in the same position without largely perturbing this solvation structure.

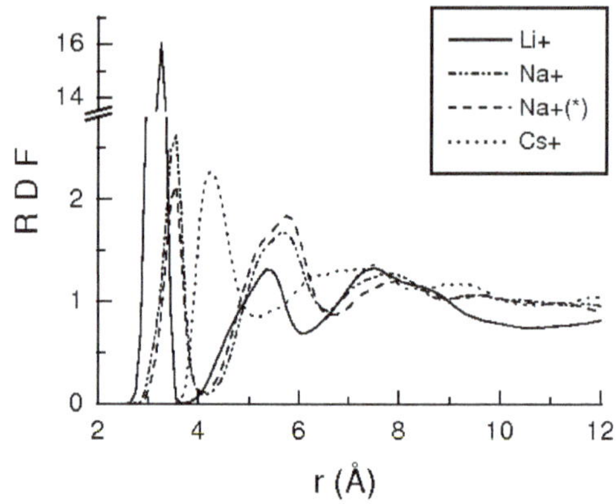

Fig. 10. RDF-s between counter-ions and P atom. Reproduced with permission from J. Biomol. Struct. Dyn., 16(3), 579-592 (1998). Copyright Adenine press (1998)

Fig. 11. SDFs of water's oxygen (violet), and hydrogen (cyan) around DNA around the DNA phosphate group. SDFs are drawn at the level 5. Reproduced with permission from J. Biomol. Struct. Dyn., 16(3), 579-592 (1998). Copyright Adenine press (1998)

The observed difference in the binding modes offered also an important key to the explanation of why Li^+ counterions promote the transitions of DNA from the B conformation to the C-form. In absence of experimental data that could confirm in a direct way the binding preferences observed in the simulations, comparison with experimental data on the diffusion of studied ions in presence of DNA gave an important validation on the used potential model.

In the field of DNA – monovalent counterions, in the last 15 years a large attention has been devoted to the dependence of the binding mode of alkaline counterions on the sequence of DNA base pairs. Among the most studied DNA sequences are the so called A-tracts: uninterrupted sequences of at least 4 A·T base pairs without 5'TpA3' steps. The interest on these sequences is motivated by their peculiar structural features: when inserted in phase with the helix turn A-tracts induce the macroscopic bending of DNA, they are characterized by a narrow minor groove which is known to possess a very ordered hydration structure, called the hydration spine. It has been proposed that sequence specific interactions of monovalent counterions in the minor groove of A-tracts were responsible of these structural features (McFail-Isom et al., 1999). In fact an electrostatic collapse of the A-tract around ions substituting one or more water molecules of the hydration spine in the minor groove of A-tract could induce either the bending, through an electrostatic collapse of the structure around the ion, and the narrowing of minor groove. Unfortunately experimental investigations on the sequence specificity of the binding to macromolecule of ions characterized by high mobility, like Na^+ or K^+, give results whose analysis are not straightforward and leave space to multiple interpretations.

Interesting experimental results had been obtained by NMR dispersion experiments (Denisov & Halle, 2000), measuring the NMR relaxation times of Na-23 in A-T rich sequences solutions at several magnetic fields; therefore we modified the M.Dyna*Mix* code to calculate the NMR relaxation time of this ion. It is important to note that even if the observed relaxation times are in the second time scale, the motion responsible of such decay are in the subpicosecond to the nanosecond time scale, thus perfectly accessible to MD simulations of DNA oligomers in aqueous solution (Mocci et al, 2004; Odelius & Laaksonen, 1999). The coupling of these techniques was used for studying the interactions between Na^+ and a DNA oligomer containing either A-tracts and non A-tract regions, the sequence [d(CTTTTAAAAG)$_2$]. In agreement with previous simulations on A-T rich oligomers (Mocci & Saba, 2003) we observed direct binding with long residence time only in the minor groove of one of the A-tracts. More specifically, a sodium ion, partially losing its hydration waters, intruded in the minor groove substituting one water molecule of the hydration spine (see **Figure 12**), and remained therein stuck for almost the entire simulation.

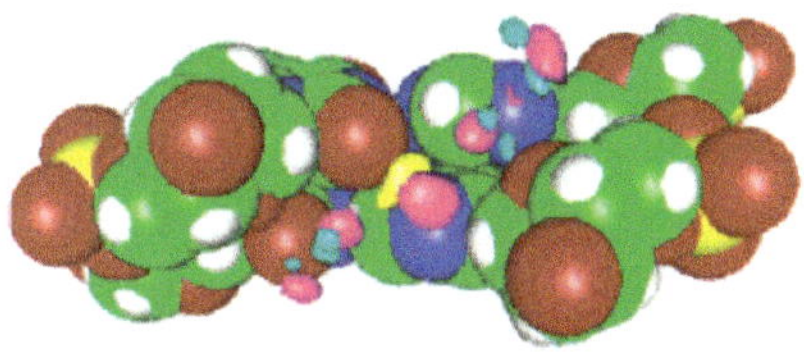

Fig. 12. SDFs of Na+ (yellow), water's oxygen (violet), and hydrogen (cyan) around DNA calculated during the binding to adenine N3 and thymine O2 in the minor groove (ca. 7 ns). The intensity levels are 10 for H and O of water, 1000 for Na^+. Reproduced with permission from J. Phys. Chem. B 108, 16295-16302 (2004). Copyright 2004 American Chemical Society

Combination of the MD results with the quadrupolar relaxation experimental data offered a very good way to verify whether the high occupancy of that binding site, observed in the simulation, was reliable. In fact, the combined approach revealed that the occupancy of the binding sites in the minor groove of uninterrupted adenine sequences should be actually pretty low, indicating that longer samplings would be required for proper evaluation of the occupancy of those binding site.

The MD-NMR combination allowed also obtaining information at the microscopic level on the NMR relaxation processes, showing how the polyion affects the NMR parameters of Na-23. Importantly it indicated that the speed-up of the Na-23 relaxation rate, compared to simple electrolyte solution, is mainly due to the contribution of ions directly bound to the oligomer's surface, while for a long while it has been thought that no sign of dehydration due to the interaction with DNA was exhibited by monovalent counterions (Anderson et al., 1995).

The MD studies discussed above revealed that the preferential binding modes of each cation at the DNA surface are dependent on the hydration structure of DNA and of the cations; such binding modes could have not been investigated without explicit inclusion of the solvent in the simulations. Unfortunately explicit modelling of the solvent is the most computationally demanding part of the calculations and to extend the MD simulations to spatial dimension beyond those of oligomeric DNA fragments it is necessary to use simplified models for the solvent. A method to keep a realistic picture of the DNA-cations specific interactions while simplifying the description of the solvent is based on the use of effective potentials for the interactions between solutes (Lyubartsev & Laaksonen, 1995, 1999). This method is based on reconstruction of effective, solvent-mediated interaction potentials between DNA atoms or group of atoms and the cations from MD simulation with the explicit solvent. More in detail the construction of effective potentials start from the RDF between the solute molecules calculated from a fully atomistic MD simulation with explicit solvent; by means of a reverse Monte Carlo procedure (Lyubartsev & Laaksonen, 1995) the RDF functions are used to construct a set of effective interaction potentials that, when used to run a MD simulation without the solvent should reproduce the same solute-solute RDF. This approach allows obtaining potentials able to maintain the ion specific information also when using an implicit model of the solvent.

3.3.2 Interactions with multivalent counterions

The difficulties in experimentally determining the favourite binding modes with DNA are not limited to monovalent counterions; in facts, also the study of interactions with multivalent molecular cations as polyamines (PA) are affected by the same problem. Polyamines, which are positively charged organic compound having two or more primary amino groups, are present in the living cells and interact with cellular polyelectrolytes like DNA. Despite the fact that PA, especially spermine4+, are largely used as DNA crystallization agents, only a very limited number of crystallographic studies report their presence in the crystallographic cell. Simulations have been extremely useful in understanding why these components are "invisible" in X-ray studies. MD simulations performed mimicking an infinite array of parallel BDNAs (Korolev et al., 2001, 2002, 2004a, 2004b) in presence of either the natural polyamines spermine^{4+} (H_3N^+-$(CH_2)_3$-NH_2^+-$(CH_2)_4$-NH_2^+-$(CH_2)_3$-NH_3^+), spermidine^{3+} (H_3N^+-$(CH_2)_3$-NH_2^+-$(CH_2)_4$-NH_3^+), putrescine^{2+} (NH_3^+-$(CH_2)_4$-NH_3^+) or the synthetic polyamine diaminopropane^{2+} (NH_3^+-$(CH_2)_3$-NH_3^+) revealed that these highly charged cations interact strongly with different groups on DNA. However all of the polyamines adopt disparate binding modes that make their detection pretty difficult with x-ray diffraction. The simulations also showed that PA and Na^+ have a different impact on DNA hydration and structure: while the Sodium cation attracts and organizes the water molecules around DNA, the polyamine pushes water away from the minor groove and induce a significant narrowing of the same. Differences in the binding preferences are observed also among the PA: while a small fraction of divalent polyamines can be found in the major groove, the other two PA are nearly absent. Furthermore differences in the binding modes where observed also between the synthetic and naturally occurring divalent PA, giving some hints on why nature selected only one of the two.

3.3.3 Molecular dynamics of lipid bilayers

In this section we describe application of M.Dyna*Mix* package for simulation of lipid bilayers. Lipid bilayers represent a framework of biological membranes, which are very complex heterogeneous systems consisting of many different types of lipids, sterols, proteins, carbohydrates and various membrane associated molecules which are involved in a variety of cellular processes. Biomembranes surround cells: a membrane separates the interior of a cell from the outside environment. Being selectively permeable, membranes participate in control of the movement of various compounds (substances) into and out of cells. Permeability of drug molecules through membrane is one of the key properties defining efficacy of the drug, since drug molecules have to penetrate numerous membrane barriers in order to reach their targets. Studies of drug solubility in both aqueous and lipophilic environment are thus important for understanding drug transport in living organisms.

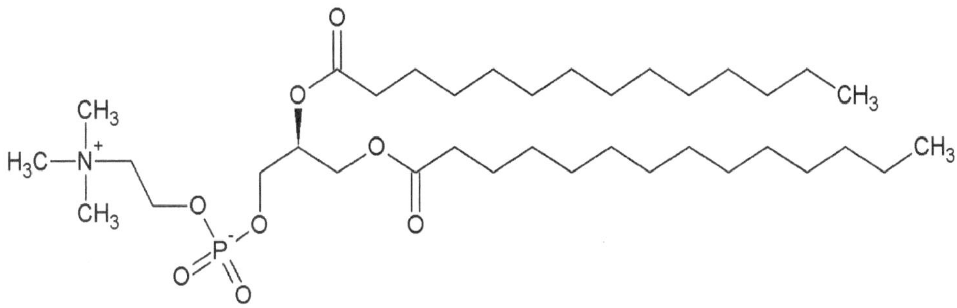

Fig. 13. Dipalmitoylphosphatidylcholine (DMPC) lipid

Lipid molecules constituting biomembranes differ with respect to the type of hydrophilic head-group and occur with a wide variety of hydrophobic hydrocarbon chains. Usually the most abundant phospholipid in animal and plants is phosphatidylcholine which is the key building block of membrane bilayers. An example of such lipid, dipalmitoylphosphatidylcholine (DMPC) is shown in **Figure 13**. It is not surprising that lipid membrane bilayers composed of various phosphatidylcholine lipids have been extensively studied by molecular dynamics as soon as computer hardware allowed to do such simulations. However, obtaining of reliable information on physical-chemical properties of lipid bilayers was, in many studies of the last decades, seriously limited by two factors: 1) necessity to simulate a large (of the order of few tens thousand) number of atoms during long (hundreds of nanosecond) simulation time and 2) not satisfactory reliability of the commonly available force fields to describe behaviour of lipid bilayers consistently, with respect to variation of lipid chemical structure, composition, thermodynamic conditions (Lyubartsev & Rabinovich, 2011). While solution of the first problem is determined by advances in the development of computer hardware, the issue of proper parametrization of the force field requires extensive computational work including numerous test simulation of the bilayer systems and calculations of different experimentally measured properties.

In the last several years we have used M.Dyna*Mix* package in order to improve CHARMM force field (Feller & MacKerell, 2000) which is one of the frequently used in biomolecular simulations. While been fully atomistic, this force field have potential advantages in comparison with the Gromos force field (Berger et al., 1997), which is based on the united atom model. However, recent detailed investigations have shown that the CHARMM force field has not-negligible disagreement with experiment in description of number important properties of lipid bilayers. For example, the CHARMM force field favour to rigid gel-like structures of bilayers composed of saturated lipids, and in order to keep bilayer in a natural liquid crystalline phase one need to apply surface tension (Lyubartsev & Rabinovich, 2011: and references therein). In paper (Högberg et al., 2008) a solution was suggested how to improve the CHARMM force field in order to simulate DMPC lipid bilayer in constant-pressure tensionless simulations. The correction included two changes: 1) scaling of the so-called 1-4 electrostatic interactions (of atoms separated by exact 3 covalent bonds) was introduced, which was tuned to reproduce experimentally measured ration of trans- and gauche conformations in hydrocarbon chains, and 2) partial atom charges in the lipid headgroup were recalculated from high-quality ab-initio calculations. It was demonstrated in paper (Högberg et al., 2008) that these modifications of the CHARMM force field allowed to obtain, in 100 ns constant-pressure simulations, excellent agreement with experimentally measured properties of DMPC bilayer such as area per lipid at zero tension, electron density, structure factor and order parameters. An important feature of these simulations was also that long-range correction to the Lennard-Jones potential was included into pressure calculations. This correction, having physical origin in the dispersion (van-der - Waals) forces, can have an order of 100 - 200 bar in the typical range of cut-off distances, and is known to be important in correct determination of the average bilayer area, as well as affecting phase behaviour of the bilayer (Lyubartsev & Rabinovich, 2011).

In continuation of paper (Högberg et al., 2008), the modified CHARMM force field was used to simulate bilayers consisting of two similar lipids: DSPC (which differ from DMPC lipid displayed in **Figure 13** that it contains 18 carbon atoms in each tail) and DOPC lipid (which exactly as DSPC contains 18 carbon atoms in each tail but have a double bond between 9-th and 10-th carbons in the second tail). Despite very similar chemical structure, bilayers composed of these lipids show different behaviour, and have noticeably different temperatures of the gel - liquid crystalline transition: 24 ^{o}C for DMPC, 53 ^{o}C for DSPC, and 5 ^{o}C for DOPC, that is why in the physiological range of temperatures DMPC and DOPC bilayers exist in a liquid crystalline phase while DSPC is in a gel phase. Simulations of the three bilayers were carried out at 30 ^{o}C, for 128 lipids in the presence of 3840 water molecules (which correspond to fully hydrated state of phosphatydylcholine lipids with 30 water molecules per lipid). In the beginning of simulations, the lipids were arranged in two leaflets, each leaflet being generated by translation of coordinates of one lipid molecule in X and Y direction. The lipid spacing was chosen to correspond to a typical value of area per lipid for liquid-crystalline phase, 64 Å^2. The necessary amount of water molecules was added outside the lipid bilayer, and the system was put into the periodic boundary conditions. The systems were simulated 1 ns under constant volume and then 1 ns under constant pressure and isotropic cell fluctuations. The obtained configurations were considered as starting points for longer simulations with independent cell fluctuations in Z and XY directions. All the systems were simulated after that for 100 ns, with the first 20 ns considered as equilibration.

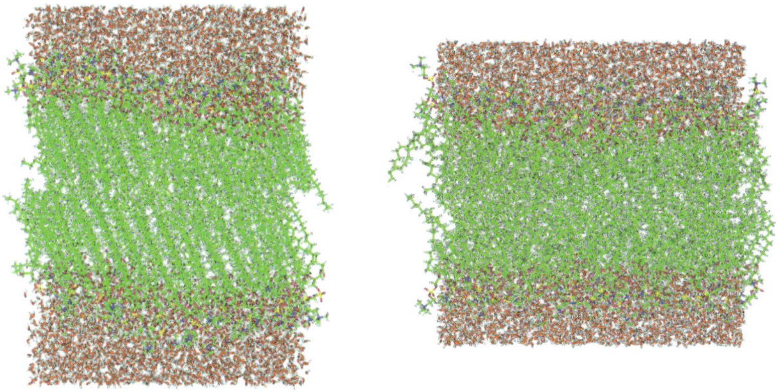

(a) (b)

Fig. 14. DSPC bilayer (a) and DOPC bilayer (b), simulated for 100 ns at 30C and constant pressure (1 bar).

Simulations showed the picture which corresponds well to the behaviour expected from experimental observations: while in DMPC and DOPC bilayers the lipids formed quickly a liquid-crystalline phase, with the area per lipid 59.5 and 62.7 Å^2 respectively, the DSPC lipids became clearly ordered in a tilted structure, with much lower area per lipid (51 Å^2) which is a typical value for a gel phase. Final snapshots of DOPC and DSPC bilayers are shown in **Figure 14**. The structures of two bilayers are strikingly different, taking in mind the fact that the only difference between the two kinds of lipids is presence of one double bond in the middle of the second tails of a DOPC lipid. Nevertheless, this behaviour is what one can expect from experimental observations.

As our exploratory simulations show that we can get reliable, consistent with experiment, behaviour of different bilayer systems, we can use the developed methodologies to address to more challenging problems, related to permeability of different substances across membranes, simulations of membrane proteins, ion channel, effects of other membrane associated molecules (cholesterol, polypeptides, anaesthetics) on membrane properties , and relating observation of these studies with the features essential for biological functioning, thus implementing the idea of a "computer laboratory" for biomembrane research.

4. Acknowledgements

The authors wish to thank SNIC & SNAC for generous allocations of computer time throughout all the years.

5. References

Allen, M. & Tildesley D. (1987). *Computer simulations of Liquids.* Oxford University Press. ISBN 0-19-855375-7, New York

Anderson, C.F. & Record, M.T. Jr (1995). Salt-nucleic acid interactions. *Annual Review of Physical Chemistry,* Vol. 46, pp. 657-700

Bagno, A. (2002). Probing the solvation shell of organic molecules with intermolecular 1H NOESY. *Journal of Physical Organic Chemistry,* Vol. 15, 790–795.

Berger, O.; Edholm O. & Jähnig, F. (1997). Molecular dynamics simulations of a fluid bilayer of dipalmitoylphosphatidylcholine at full hydration, constant pressure and constant temperature. *Biophysical Journal,* Vol. 72, pp. 2002-2013

Bergman, D. & Laaksonen, A. (1998). Topological and spatial structure in the liquid-water–acetonitrile mixture, *Physical Review E: Statistical Physics, Plasmas, Fluids, and Related Interdisciplinary Topics,* Vol. 58, No. 4, pp. 4706-4715

Bergman, D. L.; Laaksonen, L. & Laaksonen, A. (1998). Visualization of solvation structures in liquid mixtures. *Journal of Molecular Graphics and Modelling,* Vol. 15, pp. 301 -306

Bunta, J.; Dahlberg, M.; Erikson, L.; Korolev, N.; Laaksonen, A.; Lohikoski, R.; Lyubartsev, A.; Pinak, M. & Schyman, P. (2007). Solvating, manipulating, damaging, and repairing DNA in a computer. *International Journal of Quantum Chemistry,* Vol. 107, No. 2, pp. 279-291

Cornell, W.; Cieplak, P.; Bayly, C.; Gould, I.; Merz, K. J.; Ferguson, D.; Spellmeyer, D.; Fox, T.; Caldwell, J. & Kollmann, P. J. (1995). Second Generation Force Field for the Simulation of Proteins, Nucleic Acids, and Organic Molecules. *Journal of the American Chemical Society,* Vol. 117, pp. 5179-5197

Dahlberg, M & Laaksonen, A. (2006). Preferential Solvation of Phenol in Binary Solvent Mixtures. A Molecular Dynamics Study. Journal of Physical Chemistry A, Vol. 110, 2253-2258.

Denisov, V. & Halle, B. (2000). Sequence-specific binding of counterions to B-DNA. *Proceedings of the National Academy of Sciences of the United States of America,* Vol. 97, pp. 629-633

Egorov, AV; Komolkin, AV; Lyubartsev, AP & Laaksonen, A. (2006). First and second hydration shell of Ni2+ studied by molecular dynamics simulations Theoretical Chemistry Accounts (2006) Vol. 115, 170–176.

Feller, S.E. & MacKerell, A.D. (2000). An improved empirical potential energy function for molecular simulations of phospholipids. *Journal of Physical Chemistry B,* Vol. 104, pp. 7510-7515

Högberg, C.-J.; Nikitin, A.M. & Lyubartsev, A.P. (2008). Modification of the CHARMM force field for DMPC lipid bilayer. *Journal of Computational Chemistry,* Vol. 29 , No. 14, pp. 2359-2369

Korolev, N.; Lyubartsev, A.P.; Nordenskiöld, L. & Laaksonen, A. (2001). Spermine: An "invisible" component in the crystals of B-DNA: A grand canonical Monte Carlo and Molecular Dynamics simulation study. *Journal of Molecular Biology,* Vol. 308, pp. 907-917

Korolev, N.; Lyubartsev, A.P.; Laaksonen, A. & Nordenskiöld, L. (2002). On the competition between water, sodium ions and spermine in binding to DNA. A molecular dynamics computer simulation study. *Biophysical Journal,* Vol. 82, pp. 2860-2875

Korolev, N.; Lyubartsev, A.P.;. Laaksonen, A. & Nordenskiöld, L. (2004a). A molecular dynamics simulation study of polyamine and sodium DNA. Interplay between polyamine binding and DNA structure. *European Biophysics Journal*, Vol. 33, No. 8, pp. 671 - 682

Korolev, N.; Lyubartsev, A.P.;. Laaksonen, A. & Nordenskiöld, L. (2004b). Molecular dynamics simulation study of oriented polyamine- and Na- DNA: Sequence specific interactions and effects on DNA structure. *Biopolymers*, Vol. 73, No. 5, pp. 542-555

Kusalik, P.G.; Laaksonen, A.: & Svishchev, I.M. (1999). Spatial structure in molecular liquids. In: Molecular dynamics: From classical to quantum methods, Balbuena, P.B., and Seminario, J.M., Eds., Elsevier Science, Amsterdam, Theoretical and Computational Chemistry Vol. 7, pp. 61–97.

Laaksonen, A.; Nilsson, L.G.; Joensson, B. & Teleman, O. (1989). Molecular dynamics simulation of double helix Z-DNA in solution. *Chemical Physics*, Vol. 129, pp. 175-183.

Laaksonen, L. (1992). A graphics program for the analysis and display of molecular dynamics trajectories. Journal of Molecular Graphics Vol. 10, 33-34.

Leslie, A. G. W.; Arnott, S.; Chandrasekaran, R. & Ratliff, R. L. (1980). Polymorphism of DNA double helices. *Journal of Molecular Biology*, Vol. 143, pp. 49-72.

Lyubartsev, A.P.; Martsinovskii, A.A.; Shevkunov, S.V. & Vorontsov-Velyaminov, P.N. (1992). New approach to Monte Carlo calculation of the free energy: Method of expanded ensembles. *Journal of Chemical Physics*, Vol. 96, pp. 1776-1783

Lyubartsev, A.P.; Laaksonen, A. & Vorontsov-Velyaminov, P. N. (1994). Free energy calculations of Lennard-Jones systems and water using the Expanded Ensemble method: A Monte Carlo and molecular dynamics study. *Molecular Physics*, Vol. 82, No. 3, pp 455-471

Lyubartsev, A.P. & Laaksonen, A. (1995). Calculation of effective interaction potentials from radial distribution functions: A reverse Monte Carlo approach. *Physical Review E: Statistical Physics, Plasmas, Fluids, and Related Interdisciplinary Topics*, Vol. 52, pp. 3730-3737

Lyubartsev, A.P. & Laaksonen, A. (1998a). Parallel Molecular Dynamics Simulations of Biomolecular Systems. *Lecture Notes in Computer Science*, Vol. 1541, pp. 296-303

Lyubartsev, A.P. & Laaksonen, A. (1998b). Molecular Dynamics Simulations of DNA in Solution with different counter-ions. *Journal of Biomolecular Structure and Dynamics*, Vol. 16, No. 3, pp 579-592

Lyubartsev, A.P.; Førrisdahl, O. & Laaksonen, A. (1998c). Solvation Free Energies of Methane and Alkali Halide Ion Pairs. An Expanded Ensemble Molecular Dynamics Simulation Study. *Journal of Chemical Physics*, Vol. 108, No. 1, pp. 227-233

Lyubartsev, A.P. & Laaksonen, A. (1999). Effective potentials for ion-DNA interactions. *Journal of Chemical Physics*, Vol. 111, No. 24, pp. 11207-11215

Lyubartsev, A.P. & Laaksonen, A. (2000). M.Dyna*Mix* - a scalable portable parallel MD simulation package for arbitrary molecular mixtures. *Computer Physics Communications*, Vol. 128, No. 3, pp. 565-589

Lyubartsev, A.P.; Jacobsson, S.P.; Sundholm, G. & Laaksonen, A. (2001). Solubility of Organic Compounds in Water/Octanol System. An Expanded Ensemble Molecular Dynamics Simulation Study of log P parameters. *Journal of Physical Chemistry B*, Vol. 105, pp. 7775-7782

Lyubartsev, A.P. & Rabinovich, A.L. (2011). Recent development in computer simulations of lipid bilayers. *Soft Matter*, Vol. 7, pp. 25-39

MacKerell Jr., A. D.; Wiorkiewicz-Kuczera, J. & Karplus, M. (1995). An All-Atom Empirical Energy Function for the Simulation of Nucleic Acids. *Journal of the American Chemical Society*, Vol. 117, No. 48, pp 11946–11975

MacKerell, A. D. (2004), Empirical force fields for biological macromolecules: Overview and issues. *Journal of Computational Chemistry*, Vol. 25, pp. 1584–1604

McFail-Isom, L.; Sines, C. C. & Williams, L. D. (1999). DNA structure: cations in charge?. *Current Opinion in Structural Biology*, Vol. 9, pp. 298-304.

Mocci, F. & Saba, G. (2003). Molecular Dynamics Simulations of AT-Rich Oligomers: Sequence-Specific Binding of Na^+ in the Minor Groove of B-DNA. *Biopolymers*, Vol. 68, pp. 471-485

Mocci, F.; Laaksonen, A.; Lyubartsev, A. & Saba, G. (2004). Molecular Dynamics Investigation of ^{23}Na NMR Relaxation in Oligomeric DNA Aqueous Solution. *Journal of Physical Chemistry B*, Vol. 108, pp. 16295-16302

Odelius, M. & Laaksonen, A. (1999). Combined MD simulation and NMR relaxation studies of molecular motion and intermolecular interactions. In: *Molecular Dynamics. From Classical to Quantum Methods.* Balbuena, P., Seminario, J., Eds.; Theoretical and Computational Chemistry, Vol. 7, pp. 281-324, Elsevier Science, ISBN 10: 0-444-82910-5, ISBN 13: 978-0-444-82910-8 Amsterdam

Reichardt, C. (1988). *Solvents and Solvent Effects in Organic Chemistry*, VCH, Weinheim.

Ryckaert, J.-P.; Ciccotti, G. & Berendsen, H.J.C. (1977). Numerical integration of the cartesian equations of motion of a system with constraints: Molecular dynamics of n-alkanes. Journal of Computational Physics, Vol. 23, pp. 327-341

Saenger, W. (1984) *Principles of Nucleic Acid Structure,* C.R Cantor, Editor, Springer Advanced Text in Chemistry, Springer-Verlag, New York

Tuckerman, M.; Berne, B.J. & Martyna, G.J. (1992). Reversible multiple time scale molecular dynamics. *Journal of Chemical Physics*, Vol. 97, 1990–2001

van Dam, L.; Lyubartsev, A.P.; Laaksonen, A. & Nordenskiold, L. (1998). Self-diffusion and association of Li^+, Cs^+, and H_2O in oriented DNA fibers. An NMR and MD simulation study. *Journal of Physical Chemistry B*, Vol. 102, pp. 10636-10642

Vishnyakov, A.; Widmalm, G.; Kowalewski, J.; & Laaksonen A. (1999). Molecular Dynamics Simulation of the α-D-Manp-(1→3)-β-D-Glcp-OMe Disaccharide in Water and Water/DMSO Solution. Journal of American Chemical Society, Vol. 121(23), 5403-5412.

Vishnyakov, A.; Widmalm, G.; & Laaksonen A. (2000). Carbohydrates exhibit a distinct preferential solvation pattern in binary aqueous solvent mixtures. Angewandte Chemie, International Edition Vol. 39(1), 140-142.

Vishnyakov, A.; Widmalm, G.; & Laaksonen A. (2001). Molecular dynamics simulations of α-D-Manp-(1→3)-β-D-Glcp-OMe in methanol and in dimethylsulfoxide solutions, Journal of Molecular Graphics and Modeling Vol. 19(2), 338-342.

Vishnyakov, A; Lyubartsev, A; and Laaksonen A. (2001). Molecular Dynamics Simulations of Dimethyl Sulfoxide and Dimethyl Sulfoxide-Water Mixture. Journal of Physical Chemistry A, 2001, Vol. 105, 1702-1710.

Wang, F. &. Landau, D.P (2001). Efficient, Multiple-Range Random Walk Algorithm to Calculate the Density of States. *Physical Review Letters,* Vol. 86, pp. 2050-2053.

Åberg, KM,; Lyubartsev, A.P.; Jacobsson, S.P. & Laaksonen, A. (2004). Determination of solvation free energies by adaptive expanded ensemble molecular dynamics. *Journal of Chemical Physics*, Vol. 120, pp. 3770-3776.

6

An Assessment of the Conformational Profile of Neuromedin B Using Different Computational Sampling Procedures

Parul Sharma[1], Parvesh Singh[1], Krishna Bisetty[1] and Juan J Perez[2]
[1]*Department of Chemistry, Durban University of Technology, Steve Biko campus, Durban*
[2]*Department d' Enginyeria Quimica, UPC, ETS d'Enginyers Industrials, Barcelona*
[1]*South Africa*
[2]*Spain*

1. Introduction

Neuromedin B (NMB) (Minamino et al., 1983), a ten residue (GNLWATGHFM-NH_2, Figure 1) neuropeptide, belongs to the ranatensin subfamily of bombesin-like peptides (Erspamer, 1980) which exhibits a wide range of biological responses in the central nervous system and gastrointestinal tract including thermoregulation (Marki et al., 1981), stimulation of the secretion of gastrointestinal hormones (Ghatei et al., 1982), regulation of smooth muscle contraction (Erspamer, 1988), the ability to function as a growth factor in small cell lung cancer cells and murine 3T3 cells (Corps et al., 1985; Cuttitta et al., 1985; Moody et al., 1985). Its mechanism of action involves the initial binding to the three cell surface receptors (Ohki-Hamazaki, 2000) with different pharmacological profile: the neuromedin B receptor (NMB-R or bb1) (Wada et al., 1991), the gastrin-releasing peptide receptor (GRP-R, or bb2) (Corjay et al., 1991), and bombesin receptor subtype 3 (BRS-3, or bb3) (Gorbulev et al., 1992). NMB binds to NMB-R with highest affinity, GRP-R with lower affinity and BRS-3 with lowest affinity (Mantey et al., 1997).

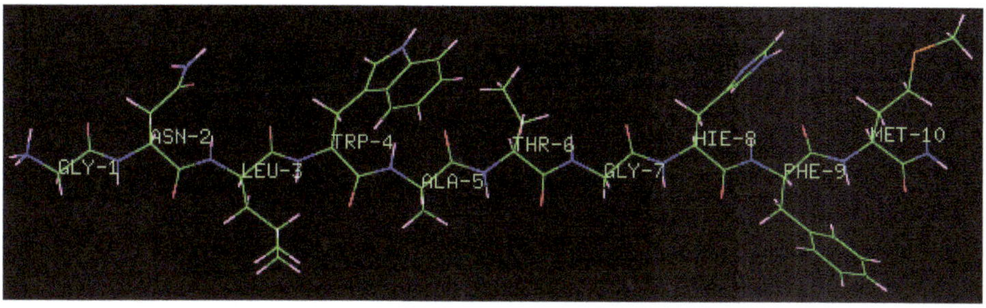

Fig. 1. Extended structure of Neuromedin B showing different residues

A number of spectroscopic studies of NMB including nuclear magnetic resonance (NMR) (Lee & Kim, 1999), Infrared (IR) (Erneand & schwyzer, 1987), Circular Dichroism (CD) and Fluorescence spectroscopy (Polverini et al., 1998) are widely reported in the literature. A recent study of the structure activity relationship (SAR) of bombesin (Glp-Gln-Arg-Leu-Gly-Asn-Gln-Trp-Ala-Val-Gly-His-Leu-Met-NH_2) using alanine scan to determine the contribution of specific residues to a protein's function by mutating the residues to alanine (Horwell et al., 1996), suggested that Trp^4, His^8 and Leu^{12} residues corresponding to Trp^4, His^8 and Phe^9 respectively in NMB, are important for the binding to the NMB receptors (Sainz et al., 1998). It is reported that (Erneand & schwyzer, 1987) in the phospholipids bilayer NMB adopts α-helical conformation in the C-terminal region. Recent studies have demonstrated that small peptides are able to exist in a dynamic equilibrium between folded and unfolded structures, depending on the solvent polarity and their interaction with the membrane phase (Erne et al., 1985). In aqueous solutions small peptides are known to adopt many conformations since the hydrogen bond formation between the polar backbone carbonyl and the amide groups and water solvent effectively competes with an intramolecular hydrogen-bond formation (Erne et al., 1985; Kaiser & Kezdy, 1987; Zhong & Jr. Johnson, 1992). Based on CD, fluorescence and molecular dynamics (MD) studies (Polverini et al., 1998), it has been observed that NMB adopts an α-helical structure in an apolar environment. However, in aqueous solution NMB adopts unordered and very flexible structures. In vacuum 50% of the structures of NMB are helix-like, with a right-handed chirality beginning from the tryptophan residue through to the C terminus and was found to be independent of the initial conformation. Moreover, two-dimensional (2D) NMR studies of NMB suggest that the peptide adopts a relaxed helical conformation from Trp^4 to Met^{10} in a 50% aqueous trifluoroethanol (TFE) solution, and in 150 mM sodium dodecyl sulfate (SDS) micelles. Several reports also suggested that there might be a conformational change to a β-turn type structure upon binding to the receptor (Coy et al., 1988; Rivier & Brown, 1978). Despite being remarkably vital, spectroscopic methods alone cannot provide all the structural details necessary to fully understand the conformational profile of the peptides in solution due to the flexibility of these molecules. Therefore, despite having a great biological and pathological significance, the unique native conformation of NMB has not yet been clearly elucidated on the basis of available spectroscopic results.

Computational studies on the other hand, can provide detailed complementary information about the intrinsic conformational features of the peptide. The methodologies available nowadays to investigate the propensities of a peptide to adopt different conformations are solid enough to provide a reasonable picture of the conformational features of a peptide and the way the solvent affects them. Recently, Generalized Born surface area implicit solvent models (Calimet et al., 2001; Dominy & Brooks, 1999) have been used in bimolecular simulations. This methodology has become popular, especially in molecular dynamics applications due to its relative simplicity and computational efficiency, compared to the more standard numerical solution of the Poisson–Boltzmann (PB) equation. The recent modifications to the standard GB implementations extend its applicability to the entire range from low- to high dielectric environments and thus play an imperative role to reproduce the environment induced by different explicit solvents (Feig & Brooks, 2004; Sigalov et al., 2005).

The present work involves the employment of different computational procedures to explore the configurational space of NMB and to provide an adequate atomic description of the peptide, compatible with the aggregated information provided by different experimental techniques. Specifically, the configurational space of NMB peptide has been explored using standard molecular dynamics (MD), multi-canonical replica exchange molecular dynamics (REMD) and simulated annealing (SA) sampling techniques using the Langevin thermostat. The Onufriev, Bashford, and Case (OBC) implicit water model (Onufriev et al., 2004) has been employed for the current investigations as this solvent model in combination with AMBERff96 is reported to generate a better extent of the helices and β-sheet conformations in peptides (Terada & Shimizu, 2008).

2. Computational methods

2.1 Replica Exchange Molecular Dynamics (REMD)

The leap module of AMBER 9 (Case et al., 2006) was used to generate the extended conformation of NMB with its N-terminal protonated and C-terminal amidated. The extended structure of NMB was energetically minimized untill a convergence criterion of 0.005 kcal mol^{-1} $Å^{-1}$ was achieved. REMD was subsequently performed on the minimized structure using the Generalized Born implicit solvent model (solvent dielectric constant 78.5, surface tension 0.005 cal/$mol^{-1}Å^{2}$) was used to model the effects of solvation (Sitkoff et al., Tsui & Case, 2001). The internal dielectric constant around the peptide was set to 1. The SHAKE algorithm with a relative geometric tolerance of 10^{-5} was used to constrain all bond lengths to their equilibrium distances. Prior to the REMD simulations, standard MD simulations were performed for 5 ns at different temperatures ranging from 200 to 900 K, with a temperature difference of 100 K. In the present study, twelve replicas were used and the temperature of each replica was set to: 277, 300, 326, 354, 385, 419, 457, 498, 544, 595, 651, and 713 K, with a time step of 0.2 fs. The temperature during the MD simulations was regulated by the Langevin thermostat (Wu & Brooks, 2003; Andersen, 1980). Each replica was simulated simultaneously and independently at different replica temperatures. The replica exchange was performed every 2 ps for 50,000 steps during the REMD simulations.

2.2 Molecular Dynamics (MD)

MD trajectory was undertaken using the Generalized Born (GB) approximation at 300 K employing the Langevin coupling algorithm. Internal dielectric constant around the peptide was set to 1, while the external dielectric constant of 78.5 corresponding to water was employed. In order to mimic the physiological conditions a 0.2 M salt concentration was used. SHAKE was used on all bonds involving hydrogen atoms with a time-step of 2 fs.

2.3 Simulated Annealing (SA)

The extended conformation of NMB peptide was energy minimized using the steepest descent method followed by a conjugate gradient method until a convergence of less than 0.001 kcal mol^{-1} $Å^{-1}$ between successive steps was achieved using the SANDER module of AMBER 9 (Case et al., 2006). The SA calculation was performed under implicit solvent conditions using the GB-OBC continuum solvent model (Onufriev et al., 2004). For this purpose, all electrostatic calculations throughout this study were done with the relative permittivity of 80. The minimized starting structure was heated up to 900 K at a rate of 100 K ps^{-1}. This means that the structure was first heated to 200 K, allowed to equilibrate and then reheated to 300 K, and this heating process was repeated until a temperature of 900 K was reached. The high temperature was used to provide the molecules with sufficient kinetic energy to enable them to cross energy barriers between different conformations, as quickly as possible. At this point the structure was slowly cooled from 900 K down to 200 K at a rate of 50 K ps^{-1}. In this technique the system was cooled down at regular time intervals, by decreasing the simulation temperature from 900 K to 200 K in intervals of 50 K. As the temperature approaches 200 K the molecule is trapped in the nearest local minimum conformation. At the end of the annealing cycle, the geometry of the structure was minimized at 200 K, in order to remove the internal strain of the molecule. Information regarding the coordinates and minimized energy data at 200 K is saved separately on a data file, which completes a single cycle of simulated annealing. Subsequently the optimized structure was used as the starting conformation for the next cycle of SA. The 8000 cycles of iterative simulated annealing resulted in a library of 8000 structures accumulated (with each cycle corresponding to a single structure), and ranked according to their energy values. The primary objective of a conformational analysis is in the identification of low energy structures, which forms an important part of understanding the relationship between the structure and the biological activity of a molecule. The biological activity of a drug molecule depends on a single unique conformation hidden amongst all the low energy conformations (Ghose et al., 1989). The search for this so-called bioactive conformation for sets of compounds is one of the major tasks in medicinal chemistry. Only the bioactive conformation can bind to the specific macromolecular environment at the active site of the receptor protein (Jörgensen, 1991). An understanding of the manipulation of the conformational structures of peptides using highly restricted segments ultimately leads to the design of bioactive peptides to fit the three dimensional receptor site requirements. In the identification of low energy structures, the SA strategy employed is widely used in the characterization of low energy conformations (Ghose et al., 1989), and the following protocols were used. Firstly, the structures were rank-ordered by energy every 100 cycles and checked for uniqueness. The efficiency of this process was monitored according to the equation 1:

$$\lambda(N) = \frac{\xi(N)100}{N\xi(100)} \tag{1}$$

The efficiency parameter, λ, was computed every 100 cycles of SA, which is defined as the number of unique conformations, ξ, found after N cycles of SA, ξ (N), divided by N, and adjusted by a coefficient so that the efficiency parameter is unity after the first 100 cycles performed, which completes the criterion of the iterative process (Corcho, 1999). The procedure was terminated in all cases when the calculated efficiency of the process, γ was at least 10% below the starting value. The evolution of this parameter was monitored along the conformational profile, for the peptide.

3. Results and discussion

The sampling efficiency of MD and REMD trajectories was monitored by establishing different conformational patterns attained by NMB during the progress of the simulations. For this purpose, the CLASICO program (Corcho, 2004) was used to compute the pattern profile for every snapshot of MD and REMD trajectories, and is depicted in Figure 2a and Figure 2b, respectively. Accordingly, 105 439 (52.7%) patterns (Figure 2a) were obtained for 200 000 snapshots of MD whereas 68 753 (68.7%) patterns (Figure 2b) were identified for 100 000 snapshots of REMD trajectory. These plots provide a broad estimation of the performance of the different protocols in sampling new patterns. A closer inspection of Figure 2a reveals the appearance of new patterns in a uniform fashion for initial 10 ns trajectory probably due to folding of the peptide. A sharp increase in patterns number was observed for the next 10 ns followed by a slow but regular increase of patterns throughout the trajectory. However, the peptide conformations seem to get trapped (dark areas) in regions of the conformational space at certain intervals clearly suggesting its restrictive nature to explore new patterns. In the case of REMD (Figure 2b), conformations with new patterns were sampled from the start of the simulation and progresses in a uniform fashion during the expansion of the trajectory. Moreover, the presence of less darker regions in the plot (Figure 2b) reveals that new patterns are explored with less restriction, clearly suggesting the better sampling performance of REMD over MD. Moreover, convergence seems to be attained in case of REMD after 100 ns as the appearance of new patterns was almost negligible at the end of trajectory.

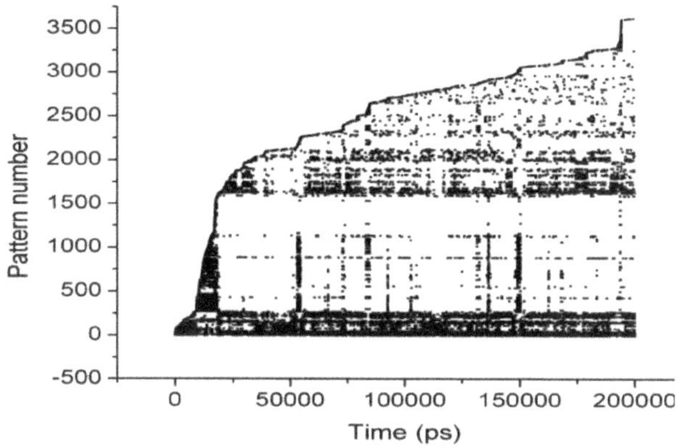

(a)

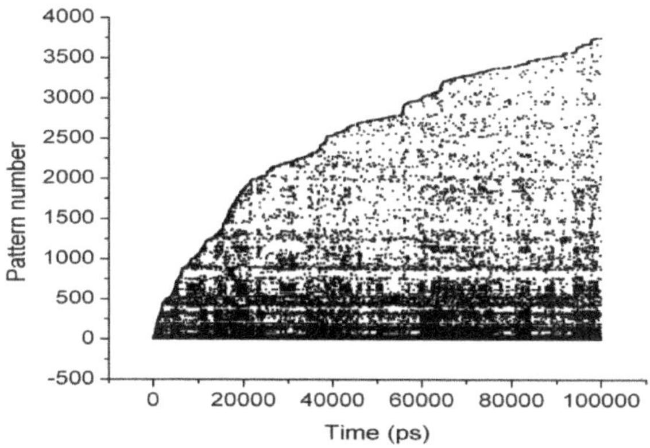

(b)

Fig. 2. Evaluation of new patterns for the NMB in (a) MD and (b) REMD trajectories.

Secondary structure analysis was performed for every snapshot of MD and REMD calculations using the CLASICO program (Corcho, 2004) employing a three-residue window. Figures 3a-3b represents the statistics of the conformational motifs for each residue of the NMB peptide in MD and REMD trajectories, respectively. Figure 3a shows the classification of secondary structures obtained in MD trajectory where the peptide exhibits predominantly β-turns (~25%) between residues 3 to 9 residues with a stronger propensity between residues 5 and 6. Additionally, an α-helical region (4-5%) flanked by residues 3 to 6 was also observed in some of the sampled structures (Figure 3a) with the complete absence of 3_{10}-helical conformations.

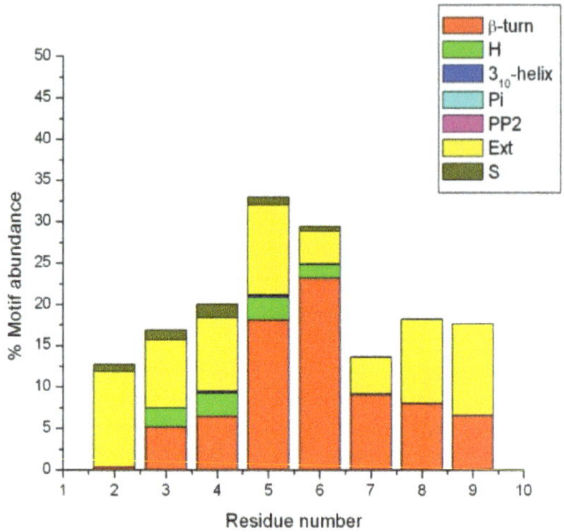

(a)

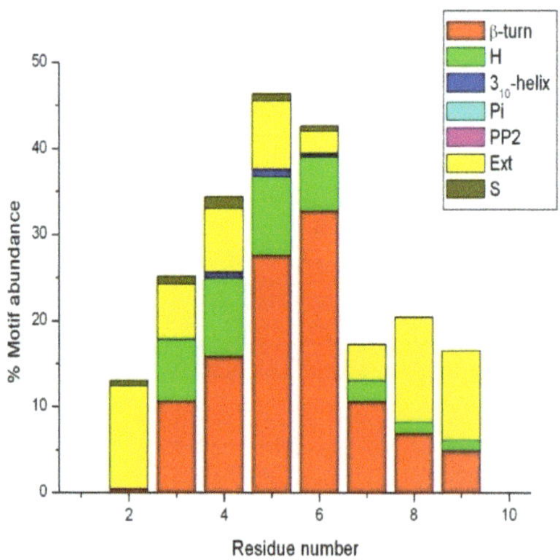

(b)

Fig. 3. Motif abundance for NMB in (a) MD and (b) REMD trajectories. Conformational motifs are labeled: H (α-helix), PI (π-helix), PP2 (polyproline II), Ext (extended), S (β-strand).

To some extent β-strands (2-3%) in a region between residues 2 to 6 were also found in some of the conformations. The REMD protocol (Figure 3b) on other hand, was more efficient at inducing β-turns (~33%) in the sampled conformations flanked by residues 2 to 9 with a strong propensity between residues 5 and 6. The second major conformational motif attained by sampled structures in the REMD trajectory was α-helical region (8-10%) flanked by residues 3 to 9 with a good propensity between residues 3 to 6 whereas a very low propensity between residues 7 to 9. All α-helical conformations were observed to be right-handed, and to a minor extent conformations exhibiting 3_{10}-helical region between residues 4 to 6 were also obtained (Figure 3b). Structures with β-strands between residues 2 to 6 were also observed in REMD calculations.

The □-turn motifs attained by residues of the NMB peptide were further classified into different types using two-residue window of the CLASICO program (Corcho, 2004), and are depicted in the Figures 4a-4b. The motifs obtained in MD trajectory (Figure 4a), shows the predominance of □-turn type I between residues 4 and 9 with a high propensity of type II between residues 6 and 7. To some extent □-turn type III was also observed between residues 3 to 9, with a high propensity between residues 5 and 6 and a low propensity between residues 3 to 4 and 7 to 9. In addition, β-turn type ii (mirror conformation of β-turn type II) was also observed between residues 5 to 8 (Figure 4a). On the other hand, conformations obtained from the REMD trajectory attain preferably □-turn type I between residues 3 to 9 with a strong propensity between residues 3 to 6 (Figure 4b). Structures displaying □-turn type III flanked by residues 3 to 9, with a strong propensity between residues 5 and 6 were also sampled in REMD trajectory. It should be noted that the CLASICO program does not include the first and last residues of the peptide in the secondary structure calculations which accounts for the absence of any of secondary structural features in Figures 3-4.

Since, the computational analysis described above provides an estimation of the average structure of NMB, it was considered worthwhile to compare the results of the different protocols with the reported NMR experiments reported in literature (Lee & Kim, 1999). The average distances corresponding to NMR NOE's were computed independently using the Clasterit algorithm of the CLASICO program (Corcho, 2004). Distances are reported as the average of the distance computed for each snapshot with a tolerance factor of ±1.96 standard deviations, covering a 95% of the variance assuming that they exhibit a normal distribution. In case of long distances (LD), the NOEs considered in the present study include: i) C□-N (bN2) between residues 2-4(m), 4-6 (w), 5-7(w) and 6-8 (w); ii) Cα-N (aN3) between residues 4-7(m), 5-8(w), 6-9(m), and 7-10(m); iii) Cα- N (aN2) between residues 1-3(w), 2-4(m), 4-6(m), 5-7(w), 6-8(m), 7-9(w), 8-10(w), and 9-N(w) [14]. In case of short distances (SD), the NOEs include: i) Cα-N (aN1) between residues 1-2(s), 2-3(s) and 7-8(s); ii) N-N (NN1) between residues 2-3(w), 3-4(s), 6-7(s), 7-8(s), 8-9(s), 9-10(s), 10-N(w). In both cases, m, w and s stand for medium, weak and strong NOEs respectively.

The overlaps between distances obtained from NMR experiments and those computed from the present studies are depicted in Figures 5a-d. These overlapping results compare both long distances (i to i+2 and i to i+ 3 type interactions) and short distances (i to i+1 type interactions) between the atoms. Specifically, Figure 5a reveals that only 12 long distances (LD) in the case of MD are in agreement with the corresponding NMR distances, and clearly suggests the absence of NMR structure in this simulation. Since, only LD accounts for the secondary structures, a good agreement of short distances (SD) in case of MD (Figure 5c) does not make any contribution in the helicity. On the other hand, all computed LD (Figure 5b) and SD (Figure 5d) from REMD calculations corresponds to the NMR distances, clearly revealing the presence of the NMR structure. However, elongation of the computed distances (Figures 5b and 5d) clearly reveals the rapid exchange between NMR and unordered structures in this segment of trajectory. Overall these results reveal that peptide is in a rapid equilibrium between ordered and unordered conformations and suggests low conformational energy barrier between them accounting for the higher flexibility of NMB. Moreover, REMD method is more efficient to induce helicity and β-turns in the peptide and was also successful in sampling the NMR structures which were completely absent in the MD trajectory.

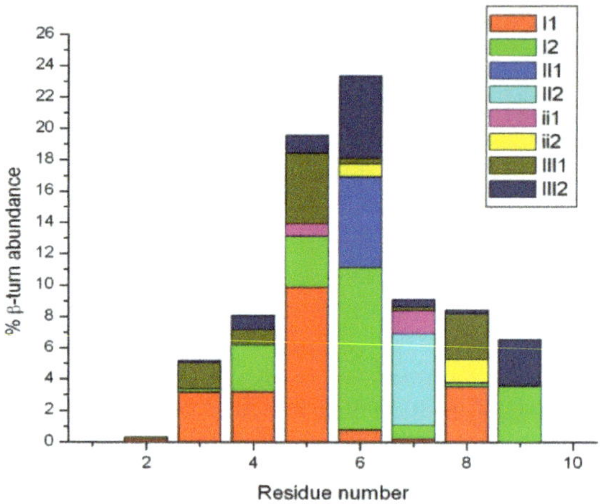

(a)

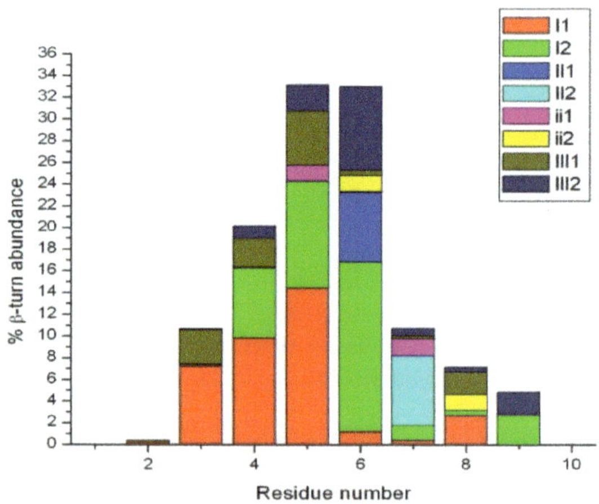

(b)

Fig. 4. Type of β-turns attained by NMB peptide in **(a)** MD and **(b)** REMD trajectories.

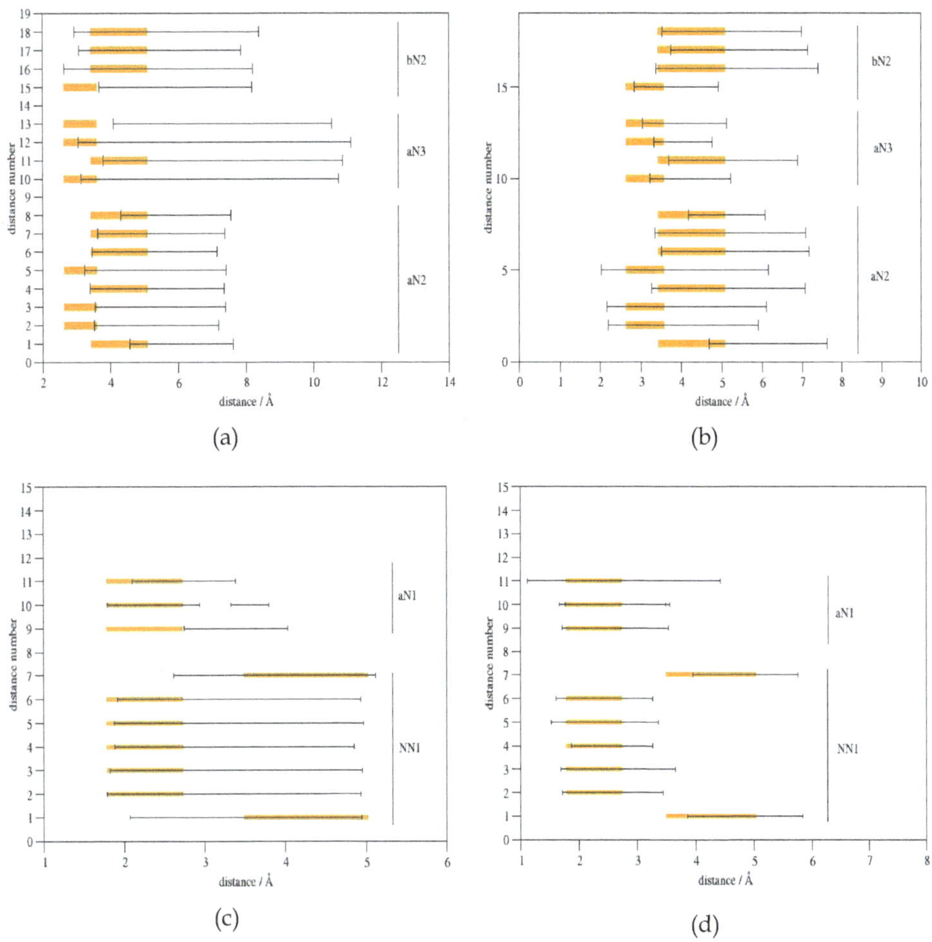

Fig. 5. Comparison of NMR derived long distances (LD) obtained from Lee & Kim, 1999 shown in orange and the computed average distances in a interval containing 95% of the structures for **(a)** MD and **(b)** REMD trajectories. Similar comparison of short distances (SD) for **(c)** MD and **(d)** REMD trajectories.

The conformational searches for NMB presented in this study were performed with the simulated annealing (SA) protocol in an iterative fashion, as a sampling technique. The sampling procedure was stopped after 8000 cycles of an iterative SA process for which the sampling efficiency λ (see Eqn. 1) was found to be less than 10% of the starting value. The evolution of this parameter λ was computed using Eqn. 1 and monitored along the conformational profile, as shown in Figure 6.

The shape of the above figure suggests that the search procedure employed was acceptable both in quality and computer time. Low levels of performance were reached in finding new low energy conformations (λ = 0.1) which is the expected result for most peptide analogues

(Filizola et al., 1998). After completion of 8000 cycles of SA, the resulting 8000 structures were stored in a library. The criterion described in the Methods Section 2.3 was used to compute the total number of unique conformations for the NMB peptide. Of the total 5521 unique conformations, only 205 low energy structures (<5 kcal mol^{-1}) were observed (Table 1).

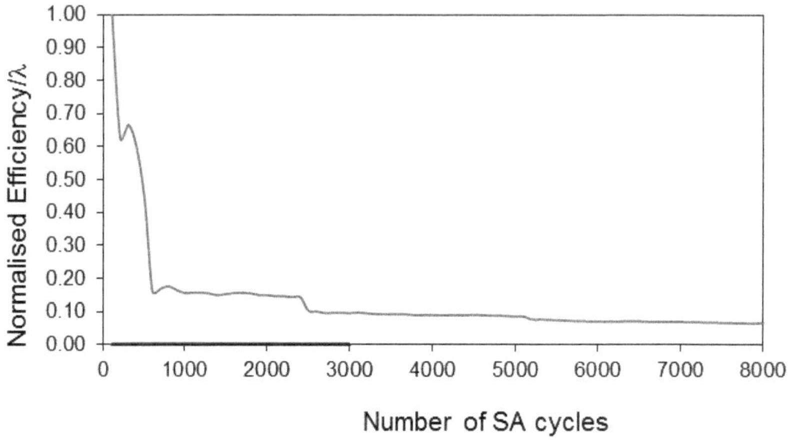

Fig. 6. Normalised efficiency values for the NMB peptide search.

NMB sequence	Total number of unique conformations	Number of classes obtained from cluster analysis	Number of unique conformations within 5 kcal mol^{-1}
H-GNLWATGHFM-NH_2	5521	10	205

Table 1. Summary of conformational analysis obtained from SA

To describe the preferred conformational domains exhibited by the peptide, the low energy structures were clustered into groups according to the values of the root mean square deviation (RMSD) of the distances between the backbone atoms of every structure.

Furthermore the RMSD's for the unique conformations were calculated using the Kleiweg clustering method (Kleiweg et al., 2004) to perform the hierarchical cluster analysis. The clustering can be visually represented by constructing a dendrogram, which indicates the relationship between the items in the data set (i.e. RMSD) and is graphically represented in Figure 7. The dendrogram enables us to identify how many clusters there are at any stage and what the corresponding members of the clusters are. It is a useful tool to show the underlying structure of the data and for suggesting the appropriate number of clusters to choose. A line drawn horizontally across the dendrogram enables one to read off how many clusters there are at any particular distance measured, as shown in Figure 7.

Since the objective of cluster analysis is to determine the representative structures of the conformational space explored, a carefully selected cut-off value is important as this method will avoid choosing subclusters. Accordingly the RMSD value of 0.8 Å was chosen as a cut-off value. Since another goal of this work was to get a better understanding of the structural motifs of the NMB peptide, further conformational analysis for this peptide was carried out on the low energy structures. The 205 unique conformations were thereby classified into ten clusters (D1 to D10), summarized in Table 2. For each of the ten different classes of clusters, a representative of the cluster was chosen on the basis of having the lowest relative energy (designated by E_r min, Table 2).

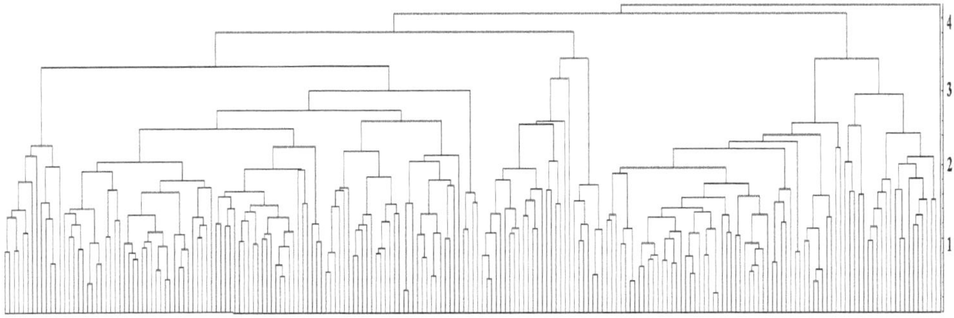

Fig. 7. Dendrogram showing different clusters for NMB classified according to their RMSD (shown along y-axis) using the Kleiweg clustering method.

Cluster name	percentage of structures in the cluster	Structure representative of the cluster	E_r min / Kcal mol^{-1}
D1	0.5	39	4.2
D2	10.2	142	3.9
D3	3.4	140	2.8
D4	41.6	108	4.7
D5	0.5	201	3.8
D6	21.9	87	1.5
D7	6.3	110	2.9
D8	9.2	116	1.9
D9	3.9	193	4.0
D10	2.4	123	4.4

Table 2. Cluster analysis for H-GNLWATGHFM-NH_2.

The five most abundant clusters represented by D2, D4, D6, D7 and D8 in Table 2 corresponds to 10.2%, 41.6%, 21.9%, 6.3% and 9.2% of the total number of structures respectively, clearly suggesting that the bulk of the structures are restricted to small number of clusters. Inspection of Table 2 reveals that 89% of the structures are represented by the most abundant clusters.

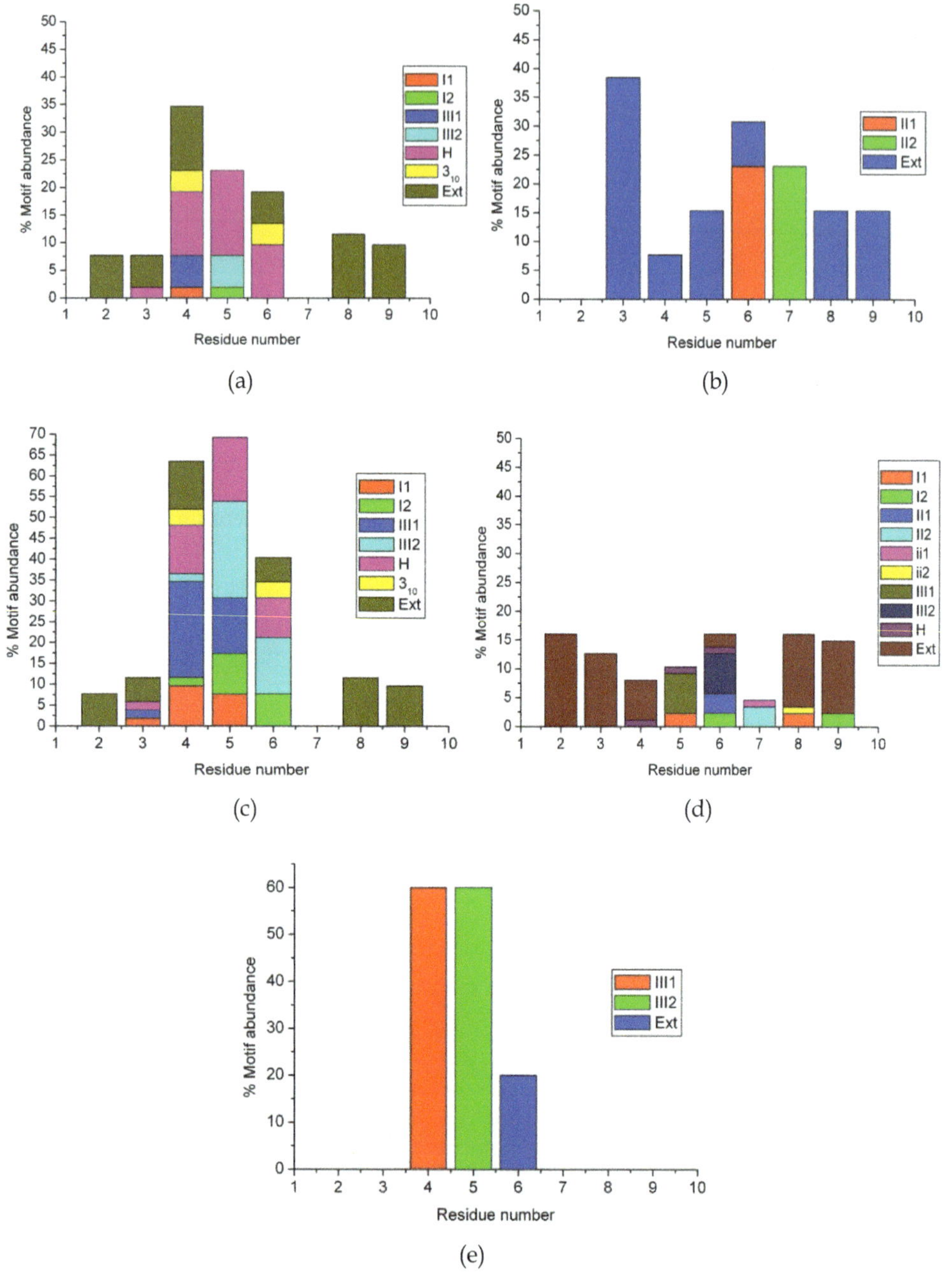

Fig. 8. Conformational motif abundance attained by NMB peptides in (a) cluster D2 (b) cluster D4 **(c)** cluster D6 **(d)** cluster D7 and **(e)** cluster D8. Conformational motifs are labeled: H (α-helix), 3_{10} (3_{10}-helix) and Ext (extended).

All the structures in each of the five most abundant clusters were analyzed to determine the conformational motifs attained by the NMB peptide, using CLASTERIT algorithm of the CLASICO program (Corcho, 2004).

The statistics of all the motifs found in clusters D2, D4, D6, D7 and D8 are depicted in Figures 8a–8e respectively. A closer inspection of Figure 8a reveals that most of the structures in cluster D2 (Table 2) predominantly exhibit an α-helical region between residues 3 to 6, while most of the residues (2-4, 8-9, Figure 8a) prefer to stay in the extended form. The structures of the most populated cluster D4, on other hand, were observed displaying a β-turn type II between residues 6 and 7 while the rest showed only an extended region (Figure 8b). The conformations of the second most abundant cluster D6 (Figure 8c) displayed predominantly an α-helical region between residues 3 to 6 with a stronger propensity between residues 4-5. To some extent the β-turn type III (3_{10}-α helical) and type I between residues 3-6 were also observed in some of the structures. However residue 7 did not show any secondary structural feature as most of the structures in clusters D7 were extended, while some displaying α-helical region between residues 4-6 and β-turns of type II and type I between residues 6-7 and 5-6 respectively, were also part of the cluster (Figure 8d). Almost 60 % of the structures in cluster D5 did not display any ordered conformation except a β-turn type III (3_{10}-α helical) between residues 4 and 5 (Figure 8e).

4. Conclusion

The present results suggest that the peptide adopts different folded and unfolded conformations regardless of the protocols used. REMD under GB conditions sample the new patterns in a uniform fashion and appears to have easily reached the convergence (Figure 6.3b), whereas conformations within the MD simulation seems to get trapped in certain regions of the conformational space making it less efficient. Moreover, the results obtained from REMD and MD computational protocols were compared with the available NMR results of NMB in literature. The comparison indicates that REMD shows good agreement with the reported NMR results. MD results, on the other hand, do not correspond with the reported NMR NOEs, clearly indicating the absence of NMR derived structures in this simulation. Moreover, the results obtained from SA is also in agreement with the corresponding REMD results clearly suggesting the probable existence of both turns and helicity in the NMB peptide, and thus may be responsible for binding of NMB at its receptor site. Hence, the present work provides comprehensive information about the conformational preferences of NMB explored using three different techniques which could be helpful to better understand its native conformation for future investigations.

5. Acknowledgement

Dr P Singh gratefully acknowledges the financial support from the Durban University of Technology, and the National Research Foundation (NRF). KB gratefully acknowledges the experiences and insights gained from the Spanish collaborators through the SA-Spain bilateral agreement. The authors would like to express their acknowledgement to the Centre for High Performance Computing, an initiative supported by the Department of Science and Technology of South Africa.

6. References

Minamino, N.; Kangawa, K. & Matsuo, H. (1983). Neuromedin B: a novel bombesin like peptide identified in porcine spinal cord, *Biochemical and Biophysical Reserach Communications*, Vol.114, pp. 541–548.

Erspamer, V. (1980). Comprehensive Endocrinology (Glass, G.B.J., Ed.). 343-361, Raven Press, New York.

Marki, W.; Brown, M. & Rivier. J. E. (1981). Bombesin analogs: Effects on thermoregulation and glucose metabolism, *Peptides Supplement*, Vol.2, pp. 169-177.

Ghatei, M. A.; Jung, R. T.; Stevenson, J. C.; Hillyard, C. J.; Adrian, T. E.; Lee, Y. C.; Christofides, N. D.; Sarson, D. L.; Mashiter, K.; MacIntyre, I. & Bloom, S.R. (1982). Bombesin - Action on gut hormones and calcium in man, *The Journal of Clinical Endocrinology and Metabolism*, Vol.54, pp. 980-985.

Erspamer, V. (1988). Discovery, isolation and characterization of bombesin-like peptides, *Annals of the New York Academy of Science*, Vol.547, pp. 3-9.

Cuttitta, F.; Carney, D. N.; Mulshine, J.; Moody, T. W.; Fedorko, J.; Fischler, A. & Minna, J. D. (1985). Bombesin-like peptides can function as autocrine growth factors in human small-cell lung cancer, *Nature*, Vol.316, pp. 823-826.

Corps, A. N.; Rees, L. H. & Brown, K. D. (1985). A peptide that inhibits the mitogenic stimulation of Swiss 3T3 cells by bombesin or vasopressin, *Biochemical Journal*, Vol.231, pp. 781-784.

Moody, T. W.; Carney, D. N.; Cuttita, F.; Quattrocchi, K. & Minna, J. D. (1985). High affinity receptors for bombesin/GRP-like peptides on human small cell lung cancer, *Life Sciences*, Vol.37, pp. 105-113.

Ohki-Hamazaki, H. (2000). Neuromedin B, *Progress in Neurobiology*, Vol.62, pp. 297-312.

Wada, E.; Way, J.; Shapira, H.; Kusano, K.; Lebacq-Verheyden, A. M.; Coy, D.; Jensen, R. & Battey, J. (1991). cDNA cloning, characterization, and brain region-specific expression of a neuromedin-B-preferring bombesin receptor, *Neuron*, Vol.6, pp. 421-430.

Corjay, M. H.; Dobrzanski, D. J.; Way, J. M.; Viallet, J.; Shapira, H.; Worland, P.; Sausville, E. A. & Battey, J. F. J. (1991). Two distinct homhcsin receptor subtypes are expressed and functional in human lung carcinoma cells, *Journal of Biological Chemistry*, Vol.266, pp. 18771-18779.

Gorbulev, V.; Akhundova, A.; Biichner, H. & Fahrenholz. F. (1992). *Molecular cloning of a new bombesin receptor subtype expressed in uterus during pregnancy, European Journal of Biochemistry*, Vol.208, pp. 405-410.

Mantey, S.; Weber, C.; Sainz, E.; Akeson, M.; Ryan, R.; Pradhan, T.; Searles, R.; Spindel, E.; Battey, J. F.; Coy, D. H. & Jensen, R. T. (1997). Discovery of a high affinity radioligand for the human orphan receptor, bombesin receptor subtype 3, which demonstrates it has a unique pharmacology compared to other mammalian bombesin receptors, *The Journal of Biological Chemistry*, Vol.272, pp. 26062-26071.

Lee, S. & Kim. Y. (1999). Solution structure of neuromedin B by ^{1}H nuclear magnetic resonance spectroscopy, *FEBS letters*, Vol.460, pp. 263-269.

Erneand, D. & Schwyzer, R. (1987). Membrane structure of bombesin studied by infrared spectroscopy. Prediction of membrane interactions of gastrin-releasing peptide, neuromedin B, and neuromedin C, *Biochemistry*, Vol.26, pp. 6316-6319.

Polverini, E.; Casadio, R.; Neyroz, P. & Masotti, L. (1998). Conformational changes of Neuromedin B and Delta sleep inducing peptide induced by their interaction with lipid membranes as revealed by spectroscopic techniques and molecular dynamics simulations, *Archives of Biochemistry and Biophysics*, Vol.349, pp. 225-235.

Horwell, D.C.; Howson, W.; Naylor, D.; Osborne, S.; Pinnock, R. D.; Ratclii, G. S. & Suman-Chauman, N. (1996). Alanine scan and N-methyl amide derivatives of Ac-bombesin[7-14]. Development of a proposed binding conformation at the neuromedin B (NMB) and gastrin releasing peptide (GRP) receptors, *International Journal of Peptide and Protein Research*, Vol.48, pp. 522-531.

Sainz, E.; Akeson, M.; Mantey, S. A.; Jensen, R. T. & and Battey, J. F. (1998). Four Amino Acid Residues Are Critical for High Affinity Binding of Neuromedin B to the Neuromedin B Receptor, *Journal of Biological Chemistry*, Vol.273, pp. 15927-15932.

Erne, D.; Sargent, D. F. & Schwyzer, R. (1985). Preferred conformation, orientation and accumulation of dynorphin A(1–13), *Biochemistry*, Vol.24, pp. 4261-4263.

Kaiser, E. T. & Kezdy. F. J. (1987). Peptides with affinity for membranes, *Annual Review of Biophysics and Biophysical Chemistry*, Vol.16, pp. 561–581.

Dyson, H. J. & Wright, P. E. (1991). Peptide conformation and protein folding, *Annual Review of Biophysics and Biophysical Chemistry*, Vol.20, pp. 519-538.

Zhong, L. & Jr. Johnson, W. C. (1992). Environment Affects Amino Acid Preference for Secondary Structure, *Proceedings of the National Academy of Sciences USA*, Vol.89, pp. 4462-4465.

Coy, D. H.; Heinz-Erian, P.; Jiang, N.-Y.; Sasaki, Y.; Taylor, J.; Moureau, J.-P.; Wolfrey, W. T.; Gardner, J. D. & Jensen, R. T. (1988). Probing peptide backbone function in bombesin: a reduced peptide bond analogue with potent and specific receptor antagonist activity, *Journal of Biological Chemistry*, Vol.263, pp. 5056-5060.

Rivier, J. E. & Brown, M. R. (1978). Bombesin, bombesin analogs, and related peptides: Effects on thermoregulation, *Biochemistry*, Vol.17, pp.1766-1771.

Schwyzer, R. (1986). Molecular mechanism of opioid receptor selection, *Biochemistry*, Vol.25, pp. 6335-6342.

Dominy, B. & Brooks, C. (1999). Development of a Generalized Born Model Parameterization for Proteins and Nucleic Acids, *Journal of Physical Chemistry B*, Vol.103, pp. 3765-3773.

Calimet, N.; Schaefer, M. & Simonson, T. (2001). Continuum solvent molecular dynamics study of flexibility in interleukin-8, *Journal of Molecular Graphics and Modelling*, Vol.19, pp. 136-145.

Feig, M.; Im, W. & Brooks 3rd, C. L. (2004). Implicit Solvation Based on Generalized Born Theory in Different Dielectric Environment, *Journal of Chemical Physics*, Vol.120, pp. 903-911.

Sigalov, G.; Scheffel, P.; Onufriev, A. (2005). Incorporating variable dielectric environments into the generalized Born model, *Journal of Chemical Physics*, Vol.122, pp. 094511-094515.

Onufriev, A.; Bashford, D. & Case, D. A. (2004). Exploring protein native states and large-scale conformational changes with a modified generalized born model, *Proteins: Structure, Function, and Bioinformatics*, Vol.55, pp. 383-394.

Terada, T. & Shimizu, K. (2008). A comparison of generalized Born methods in folding simulations, *Chemical Physics Letters*, Vol.460, pp. 295–299.

Case, D. A.; Darden, T. A.; Cheatham, T .E. III.; Simmerling, C .L.; Wang, J.; Duke, R. E.; Luo, R.; Merz, K. M.; Pearlman, D. A.; Crowley, M.; Walker, R .C.; Zhang, W.; Wang, B.; Hayik, S.; Roitberg, A.; Seabra, G.; Wong, K .F.; Paesani, F.; Wu, X.; Brozell, S.; Tsui, V.; Gohlke, H.; Yang, L.; Tan, C.; Mongan, J.; Hornak, V.; Cui, G.; Beroza, P.; Mathews, D. H.; Schafmeister, C.; Ross, W. S. & Kollman, P. A. (2006). AMBER 9. University of California, San Francisco.

Sitkoff, D.; Sharp, K. A. & Honig, B. (1994). Accurate Calculation of Hydration Free Energies Using Macroscopic Solvent Models, *Journal of Physical Chemistry*, Vol.98, pp.1978-1988.

Tsui, V. & Case. D. A. (2001). Theory and applications of the generalized Born solvation model in macromolecular simulation, *Biopolymers*, Vol.56, pp.275-291.

Wu, X. & Brooks, B. R. (2003). Self-guided Langevin dynamics simulation method, *Chemical Physics Letters*, Vol.381, pp. 512-518.

Andersen, H. C. (1980). Molecular dynamics simulations at constant pressure and/or temperature, *Journal of Chemical Physics*, Vol.72, pp. 2384-2393.

Ghose, A. K.; Crippen, G. M.; Revankar, G. R.; Smee, D. F.; McKernan, P. A. & Robins, R. K. (1989). Analysis of the in Vitro Antiviral Activity of Certain Ribonucleosides against Parainfluenza Virus Using a Novel Computer Aided Receptor Modeling Procedure, Journal *of Medicinal Chemistry*, Vol.32, pp. 746-756.

Jörgensen, W. L. (1991). Rusting of the lock and key model for protein-ligand binding, *Science*, Vol.254, pp. 954-955.

Corcho, F. J.; Filizola, M. & Perez, J. J. (1999). Assessment of the bioactive conformation of the farnesyltransferase protein binding recognition motif by computational methods, *Journal of Biomolecular Structure & Dynamics*, Vol.5, pp. 1043-1052.

LaFargaCPL: CLASTERIT: Project Info. Available at: https://lafarga.cpl.upc.edu/projects/clusterit [Accessed June 12, 2008].

Corcho, F.; Canto, J. & Perez, J. J. (2004). Comparative analysis of the conformational profile of substance P using simulated annealing and molecular dynamics, *Journal of Computational Chemistry*, Vol.25, pp.1937-1952.

Filizola, M.; Centeno, N. B.; Farina, M. C. & Perez, J. J. (1998). Conformational analysis of the highly potent bradykinin antagonist Hoe-140 by means of two different computational methods, *Journal of Biomolecular Structure & Dynamics*, Vol.15, pp. 639-652.

Kleiweg, P.; Nerbonne, J. & Bosveld, L. (2004). Geographical Projection of cluster composites. In: A. Blackwell, K. Marriott and A. Shimojima (Eds.).

Jain, A. K.; Murty, M. N. & Flynn, P. J. (1999). Data Clustering: A Review. ACM Computing Surveys Vol.31, No.3. pp. 264-323.

7

Practical Estimation of TCR-pMHC Binding Free-Energy Based on the Dielectric Model and the Coarse-Grained Model

Hiromichi Tsurui[1] and Takuya Takahashi[2]
[1]*Juntendo University*
[2]*Ritsumeikan University*
Japan

1. Introduction

To evaluate free energy changes of bio-molecules in a water solution, *ab initio* molecular dynamics (MD) simulations such as Quantum Mechanical Molecular Mechanics (QM/MM) and MD are the most theoretically rigorous methods (Car and Parrinello 1985; Kuhne, Krack et al. 2007), although the calculation cost is far too large for large molecular systems that contain many electrons. Therefore, all-atom MD simulations based on classical mechanics (i.e., Newton's equations) are used for the usual bio-molecular systems. As the conventional free energy perturbation (FEP) method based on all atom MD simulation is a strict method, to elucidate the molecular principles upon which the selectivity of a TCR is based, FEP simulations are used to analyse the binding free energy difference of a particular TCR (A6) for a wild-type peptide (Tax) and a mutant peptide (Tax P6A), both presented in HLA A2. The computed free energy difference is 2.9 kcal mol^{-1} and the agreement with the experimental value is good, although the calculation is very time-consuming and the simulation time is still insufficient for fully sampling the phase space. From this FEP calculation, better solvation of the mutant peptide when bound to the MHC molecule is important to the greater affinity of the TCR for the latter. This suggests that the exact and efficient evaluation of solvation is important for the affinity calculation (Michielin and Karplus 2002). Other FEP calculations of the wild-type and the variant human T cell lymphotropic virus type 1 Tax peptide presented by the MHC to the TCR have been performed using large scale massively parallel molecular dynamics simulations and the computed free energy difference using alchemical mutation-based thermodynamic integration, which agrees well with experimental data semi-quantitatively (Wan, Coveney et al. 2005). However, the conventional FEP is still very time-consuming when searching for so many unknown docking structures because all-atom MD for a large molecular system is a computationally hard task and MD simulations must be done not only in initial and final states but also in many intermediate states.

Recently, the energy representation (ER) method - where only the initial and final states of a molecular system need to be considered and the sampling cost is drastically decreased - is developed for the molecular solvation process by Matubayashi et al. (Matubayasi and Nakahara 2000; Matubayasi and Nakahara 2002) and will be applicable to the calculation of binding free energy. Of course, this ER method can be combined with the approximate models described below. Instead of MD simulations, Monte Carlo simulations are also used for the sampling of the configurations. This type of approach which only considers initial and final states is called an endpoint method.

Most of the calculation cost in all-atom MD involves the sampling of the solvent atom configurations because the number of solvent atoms - such as water and co-solvent ions - is much larger than that of the target bio-molecules, and long-range electrostatic potential is especially time consuming although efficient algorithms such as the Fast Multi Pole method and several Ewald methods are developed for all-atom MD. To decrease the calculation cost of the long-range electrostatic term, a continuum dielectric model - which can calculate the electrostatic free energy term of the system very efficiently - is widely used in many bio-molecular systems and is described in the next section.

In the case of large molecules, the entropy term of solvation change becomes important (Asakura and Oosawa 1954), and the solvent accessible surface area (SA) based calculation method becomes insufficient because the excluded volume effect increases. Therefore, integration equation (IE) theories such as the Ornstein-Zernike equation and the closures - which are developed in molecular liquid theory - promise to evaluate entropy change, including solvation and de-solvation processes (Kinoshita 2006; Kinoshita 2009; Yasuda, Yoshidome et al. 2010). The recent MD software package AMBER also contains such an IE algorithm, 3D-RISM, which is a reference site model employing Cartesian coordinates (Luchko, Gusarov et al. 2010). In particular, the simple morphological theory obtained from this IE approach is now applied to the elucidation of protein folding (Yasuda, Yoshidome et al. 2010) and F1-ATPase mechanisms and has proven to be useful (Yoshidome, Ito et al. 2011).

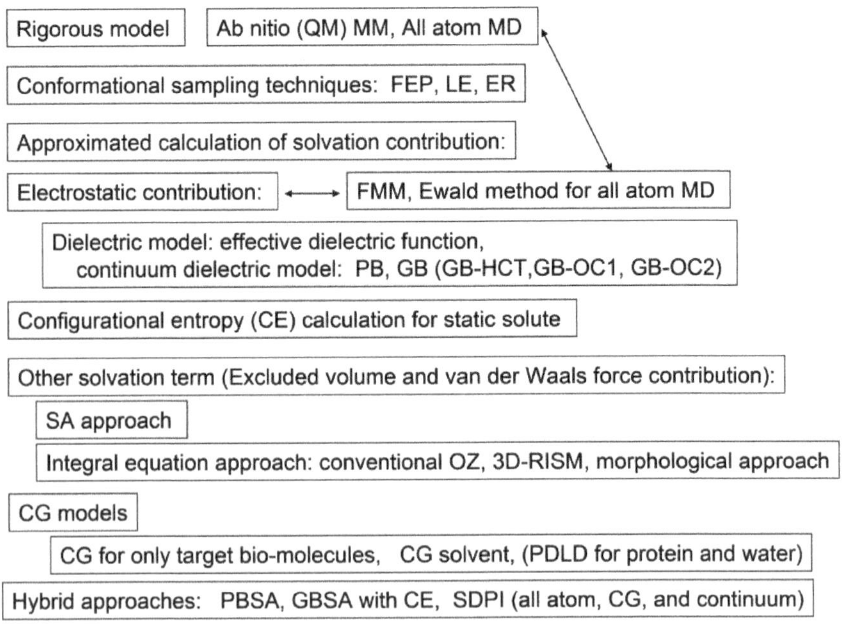

Fig. 1. The relationship among theoretical models and approaches

The other approaches for decreasing the calculation cost of the solvent molecules are coarse-grained (CG) solvent models. The protein-dipole Langevin-dipole (PDLD) model, which can efficiently calculate the electrostatic interaction among permanent dipoles and induced dipoles of proteins and solvent atoms, is one of the coarse-grained solvent models. As the PDLD model is usually used in the outer area of the all-atom region, this is a hybrid approach of CG and all-atom models (Warshel and Levitt 1976; Xu, Wang et al. 1992). Hybrid approaches of all-atom, CG and continuum solvent models are evolving. A smoothly decoupled particle interface (SDPI) model has a switching region that gradually transitions from fully interacting particles to a continuum solvent. The resulting SDPI model allows for the use of an implicit solvent model based on a simple theory that needs only to reproduce the behaviour of bulk solvent rather than the more complex features of local interactions (Wagoner and Pande 2011). Of course, CG models for solute molecules - which are described in the third section - are promising for the understanding of protein folding (Liwo, He et al. 2011) and predictions of the ligand-receptor docking structure, etc.

The relationship among the theoretical models and approaches is summarised in Fig. 1.

2. The dielectric model and the MM-PBSA (GBSA) method

In this section, we describe briefly the principles behind the methods, the differences between PBSA and GBSA and explicit and implicit treatment.

2.1 Principles of the method

A molecule has an atomic polarisability due to its electrons and an orientational polarisability when the molecule is polar and has a permanent electric dipole moment. A high value of the relative dielectric constant (εr=78.4 at 298K) of water is mainly due to its orientational polarisation, where the electric dipole moment is 2.95 Debye. Moreover, the solution in our body contains several co-solvent ions such as Na^+, Cl^-, K^+ and so on for the usual physiological condition. Therefore, electrostatic interactions among bio-molecules are largely decreased by water and solvent ions in a very complicated manner when compared with the *in vacuo* case (Koehl 2006).

To obtain the electrostatic contribution to free energy change, the dielectric model is a good approximation and is widely used to calculate the electrostatic potential of molecular systems in many scientific and technological fields.

First, the use of a simple function of the effective relative dielectric constant is the easiest way to reduce the calculation time required in obtaining electrostatic potentials.

A simple distance-dependent function, 4.5 r, which is proposed by Pickersgill (Pickersgill 1988), can well explain site-directed mutagenesis experiments. Warwicker showed that the simple Debye–Hückel shielding function with a uniform effective relative dielectric constant of 50 was sufficient to explain experimental results when compared with a continuum model (Warwicker 1999). Mehler et al. challenged this problem and proposed a sigmoid function considering the local hydrophobicity and hydrophilicity of protein molecules whose results were also in good agreement with pKa shift measurements (Mehler and Guarnieri 1999). These methods are simple and very fast; however, they all require parameter readjustment for each new system to be studied. Unfortunately, a universal function applicable to all macromolecular systems does not yet exist.

Empirically obtained effective dielectric functions that depend on the inter-atomic distance, r, such as linear functions (εr=*r* or 4*r*) and the sigmoid function is simple, and low calculation-cost method is still used in recent drug design studies for the docking simulations of large molecular systems so as to save on the calculation-cost, although the calculation error is large (Takahashi, Sugiura et al. 2002).

2.1.1 PB approach

On the other hand, the typical dielectric model solves the Poisson equation and treats bio-molecules and water as continuum media which have a specific dielectric constant, although the position-dependent local dielectric constant - which is calculated from the electronic polarisation of atoms and the orientational polarisation of local dipoles - is also possible for a finite difference equation (Nakamura, Sakamoto et al. 1988; Pitera, Falta et al. 2001).

Moreover, the Poisson-Boltzmann (PB) equation, which was first proposed by Gouy in 1910 and was complemented by Chapman in 1913, is widely-used for considering the contribution of solvent ions. The Gouy-Chapman theory, which solves a simple one-dimensional nonlinear PB equation, is often used in a membrane-electrolyte system that has electrical double layers (Forsten, Kozack et al. 1994).

The PB equation is a differential equation and it describes electrostatic interactions between molecules in ionic solutions by using a mean-filed approximation where the correlations among the solvent ions are neglected. The equation in SI units can be written as:

$$\vec{\nabla}\ \left[\varepsilon(\vec{r})\vec{\nabla}\psi(\vec{r})\right] = -\rho(\vec{r}) - \sum_{\mathrm{i}} c_i^{\infty} z_i q \lambda(\vec{r}) \exp\left[\frac{-z_i q \vec{\nabla}\psi(\vec{r})}{k_B T}\right] \quad (1)$$

where $\vec{\nabla}\cdot$ is the divergence operator and $\varepsilon(\vec{r})$ is the position-dependent dielectric, which is set to be constant in the solvent, the bio-molecule and the boundary regions in continuum dielectric models. $\vec{\nabla}\psi(\vec{r})$ is the gradient of the electrostatic potential, $\rho(\vec{r})$ represents the charge density of the solute (i.e., the fixed charges of the bio-molecule), c_i^{∞} represents the concentration of the ion i at a distance of infinity from the solute, z_i is the charge of the solvent ion, q is the charge of a proton, k_B is the Boltzmann constant, T is the temperature and $\lambda(\vec{r})$ is a factor for the position-dependent accessibility of position $\vec{r}$ to the ions in the solution. If the potential is small and the electrostatic energy is negligible compared to the thermal fluctuation, $k_B T$, the equation can be linearised and solved more efficiently.

$$\vec{\nabla}^2\psi(\vec{r}) = \kappa^2\psi(\vec{r}) \quad (2)$$

Here, κ is the Debye shielding parameter, defined as follows:

$$\kappa^2 = \sum_{\mathrm{i}}^{\mathrm{m}} \frac{z_i^2 q^2 c_i^{\infty}}{\varepsilon k_B T} \quad (3)$$

This weak field limit approach is called the Debye–Hückel approximation (Fogolari, Brigo et al. 2002).

To solve the PB equation, there are typically three numerical methods: a finite difference (FD) method is relatively time consuming, but simple and applicable to a complex system which has a position-dependent local dielectric constant. Therefore, the FD method is firstly applied to calculate the electrostatic potential in a protein-solvent system, and the pKa shift of the protein ionisable residues are well-explained (Gilson and Honig 1987) and the effect of the salt concentration on the pKa are also reproduced (Takahashi, Nakamura et al. 1992). The finite element method (FEM) and the boundary element method (BEM) are more powerful and the calculation cost is smaller than the FD method, although only a uniform dielectric constant must be set in each region (Lu, Zhou et al. 2008).

2.1.2 GB approach

One other powerful way to obtain the electrostatic potential based on the dielectric model is the Generalised Born (GB) model, which solves the linearised PB equation by approximating such bio-molecules as proteins and nucleic acids as a set of spheres whose internal dielectric constant differs from the external solvent (Koehl 2006). The functional form of the model is written as:

$$Gs = \frac{1}{8\pi}\left(\frac{1}{\varepsilon_0} - \frac{1}{\varepsilon}\right)\sum_{\mathrm{i,j}}^{\mathrm{N}} \frac{q_i q_j}{f_{ij}} \tag{4}$$

where

$$f_{ij} = \sqrt{r_{ij}^{\ 2} + a_{ij}^{\ 2} e^{-D}} \tag{5}$$

and

$$D = \left(\frac{r_{ij}}{2a_{ij}}\right)^2, a_{ij} = \sqrt{a_i a_j} \tag{6}$$

where ε_0 is the dielectric constant *in vacuo*, ϵ is the dielectric constant of the solvent, q_i is the electrostatic charge on the particle i, r_{ij} is the distance between particles i and j, and a_i is a length defined as the effective Born radius (Still, Tempczyk et al. 1990).

The effective Born radius of an atom represents its degree of burial inside the solute and corresponds to the distance from the atom to the molecular surface. The exact evaluation of the effective Born radii is the central issue for the GB model (Onufriev, Bashford et al. 2004).

To consider the electrostatic shielding effect due to the solvent ions, a simplified function based on the Debye-Hückel approximation is added to the function Gs in the AMBER software package (Case, Cheatham et al. 2005), which is one of the most used packages in the world of bio-molecular simulations, as follows by (Srinivasan, Trevathan et al. 1999):

$$Gs = \frac{1}{8\pi}\left(\frac{1}{\varepsilon_0} - \frac{\exp(-\kappa f_{ij})}{\varepsilon}\right)\sum_{\mathrm{i,j}}^{\mathrm{N}} \frac{q_i q_j}{f_{ij}}. \tag{7}$$

They calculated the solvation free energies, Gs, with this GB model for proteins and nucleic acids, which agreed very well with those of the PB model. The salt-dependence of the electrostatic binding free energy based on the Debye-Hückel approximation is still under investigation (Harris, Bredenberg et al. 2011).

2.1.3 The GBSA (PBSA) approach

GBSA (PBSA) is simply a GB (PB) model with the hydrophobic solvent accessible surface area (SA) term. This is the most commonly used implicit solvent model combination and is widely used in MD simulations for large bio-molecules. This approach is known as MM/GBSA in the context of molecular mechanics. This formulation can well identify the native states of short peptides with a precise stereoscopic structure (Ho and Dill 2006), although the conformational ensembles produced by GBSA models in other studies differ significantly from those produced by an explicit solvent and do not identify the protein's native state (Zhou 2003). In particular, strong charge-charge interaction such as salt bridges are overstabilised due to insufficient electrostatic screening, and the alpha helix population became higher than the native one. These problems are common in PBSA. Variants of the GB model have also been developed to approximate the electrostatic environment of membranes, which have had some success in folding the transmembrane helices of integral membrane proteins (Im, Feig et al. 2003).

There are several kinds of software containing the GB algorithm. For example, the AMBER software package has three types of GBSA models as has as the PBSA model.

The MM-PBSA and GBSA approaches are the endpoint methods and usually only consider the initial unbound state and the final bound state. The binding free energy change, dGbind, is written as:

$$dG\text{bind}=dG\text{gas}+ dG\text{solv} =(dH\text{gas}+dH\text{tr/ro} -TdS)+(dG\text{elsolv}+dG\text{npsolv}). \quad (8)$$

The term dGgas refers to total free energy change and the term dHgas contains the van der Waals and electrostatic interaction energies as well as internal energy variation, such as bond, angle and torsional angle energies *in vacuo* (i.e., gas phase). The terms dHtr/ro denote the energy difference due to translational and rotational degrees of freedom, and becomes 3 RT in the classical limit (i.e., thermal energy is large enough). The term dS refers to the conformational entropy change (Tidor and Karplus 1994; Ben-Tal, Honig et al. 2000). The term dGsolv is the difference between the initial and final solvation free energies and is divided into the electrostatic contribution, dGelsolv, and the nonpolar contribution, dGnpsolv. The term dGnpsolv, which is the sum of a cavity term and a solute-solvent van der Waals term, is calculated from the SA as follows:

$$dG\text{npsolv}=\gamma \text{SA}+b. \quad (9)$$

The surface tension γ and the constant b are 0.00542 kcal mol^{-1} Å^{-2} and 0.92 kcal mol^{-1} respectively, for the MM-PBSA model (Sitkoff, D., K. Sharp and B. Honig. 1994). For GB models, 0.0072 kcal mol^{-1} Å^{-2} and 0 kcal mol^{-1} (Jayaram, Sprous et al. 1998), or else 0.005 kcal mol^{-1} Å^{-2} and 0 kcal mol^{-1} (Gohlke, Kuhn et al. 2004) are used. The SA in AMBER is calculated by using the LCPO algorithm (Weiser, Shenkin et al. 1999; Still, Tempczyk et al. 1990) to compute an analytical approximation to the solvent accessible area of the molecule.

The several types of the GBSA models are not only applied to many protein folding simulations (Zhou 2003), but also to nucleic acid conformational dynamics from massively parallel stochastic simulations, where the ubiquitous helical hairpin conformation is reproduced and folding pathway is investigated (Sorin, Rhee et al. 2003).

2.2 Review of recent work

As mentioned in the previous section 2.1, the dielectric models and the hybrid approaches are widely used in many scientific and technological fields, such as protein folding, molecular docking and drug design, etc. In particular, the binding free energy (BFE) calculation and the prediction of the binding affinity and binding structure between ligands and proteins is the most important aim (Gilson and Zhou 2007) because the major purpose of molecular docking (Zacharias and Fiorucci 2010; Leis and Zacharias 2011) is to predict the experimentally-obtained BFE and the binding site of a receptor to a specific ligand molecule, and drug design is usually supported by suitable molecular docking methods.

For example, the linear interaction energy (Rastelli, Rio et al. 2010) method - which combines two different continuum solvent models - is applied to calculate protein-ligand BFEs for a set of inhibitors against the malarial aspartic protease plasmepsin II, and the explicit solvent LIE calculations and LIE-PB reproduce absolute experimental BFEs with an average unsigned error of 0.5 and 0.7 kcal mol^{-1} respectively (Carlsson, Ander et al. 2006). Moreover, the ligand-water interaction energies - which are calculated from both PB and GB models using snapshots from explicit solvent MD simulations of the ligand and protein-ligand complex - are compared with the explicit solvent MD results. The obtained energy from the explicit water MD agrees well with those from the PB model, although the GB model overestimates the change in solvation energy, which overestimation is caused by consistent underestimation of the effective Born radii in the protein-ligand complex.

Xu and Wang applied the MM-PBSA method to FK506-binding proteins (Xu and Wang 2006) - which are important targets of pharmaceutical interests - and calculated the binding of a set of 12 non-immunosuppressive small-molecule inhibitors to FKBP12 through MD simulations, where each complex is subjected to 1-ns MD simulation conducted in an explicit solvent environment under constant temperature and pressure. The BFE of each complex is then calculated with the MM-PBSA method in the AMBER program and the MM-PBSA computation agrees very well with the experimentally determined BFEs, with a correlation coefficient (R^2) of 0.93 and a standard deviation as low as 0.30 kcal mol^{-1}. The vibrational entropy term given by the normal mode analysis is necessary for achieving this correlation. Moreover, an adjustment to one weight factor in the PBSA model is essential to correct the absolute values of the final binding free energies to a reasonable range, which suggests that the very good correlation is due to the similar properties of ligand molecules and that this artificial weight factor is not universal. A comparison of the MM-PBSA model with a Linear Response Approximation model suggests that the MM-PBSA method seems to be robust in binding affinity prediction for this class of compounds (Lamb, Tirado-Rives et al. 1999).

To systematically evaluate the performance of MM-PBSA and several versions of the MM-GBSA models, extensive calculations of BFEs are done for 59 ligands interacting with six different proteins with the AMBER 9.0 software (Hou, Wang et al. 2011). First, the effects of the length of the MD simulation are explored, ranging from 400 to 4800 ps, and the simulation length has an obvious impact on the predictions. Interestingly, longer MD simulation is not always necessary for achieving better predictions. Second, the effect of a solute dielectric constant (1, 2, or 4) on the BFEs of MM-PBSA is also checked and the predictions are quite sensitive to the solute dielectric constant. Therefore, this parameter should be carefully determined according to the characteristics of the protein/ligand binding interface. Third, conformational entropy often shows large fluctuations in MD trajectories, and a large number of snapshots are necessary to achieve stable predictions. Next, the comparison of the accuracy of the BFEs of three GB models: (1) GB-HCT, the pair wise model by Hawkins et al. (Hawkins, Cramer et al. 1996) parameterised by Tsui and Case (Tsui and Case 2000); (2) GB-OC1 and (Case, Cheatham et al.) GB-OC2, the parameters of which are modified by Onufriev et al. (Onufriev, Bashford et al. 2004) and the GB-OC1 model which gives better results compared to the other two GB models in ranking the binding affinities of the studied inhibitors. This may be explained by the better agreement of GB-OC1 with PBSA. The better performance of MM-PBSA when compared with MM-GBSA in calculating absolute - but not necessarily relative - BFEs is confirmed, which is not surprising because the GBSA is the approximation of PBSA, but it suggests the reliability of the dielectric continuum model itself. Considering its computational efficiency, MM-GBSA gives good relative BFEs and is much faster than MM-PBSA, and can serve as a powerful tool in drug design where the correct ranking of inhibitors is often emphasised and the obtaining of the absolute value of BFEs is not so important.

Interestingly, the successive study of MM-PBSA and MM- GBSA-OC1 using 98 protein-ligand complexes to develop an excellent scoring function by Hou et al. shows that MM-GBSA (success rate 69.4%) outperformed MM-PBSA (45.5%) and many popular scoring functions in identifying the correct binding conformations, and the best prediction of the MM-GBSA model with an internal dielectric constant of 2.0 produced a Spearman correlation coefficient of 0.66, which is better than MM/PBSA (0.49) and almost all the scoring functions used in molecular docking (Hou, Wang et al. 2011). However, the reason why the PBSA underperforms the GBSA is not clear. One possibility is the difference of the SA term and the other possibility is the insufficiency of the conformational sampling of proteins, as the authors are also emphasising the importance of MD calculation time. In any case, MM-GBSA performs well, for both binding pose predictions and binding free-energy estimations and it is efficient at re-scoring the top-hit poses produced by other less-accurate scoring functions.

As AMBER and other software packages - including the PB and GB models - are widely used and drug design is the important issue, many studies concerning ligand-protein docking based on the dielectric model have been done (Rastelli, Rio et al. 2010). The above calculation results of the GB and PB dielectric models are limited to relatively small ligand molecules and receptor proteins, and the size of the complex is not so large compared to so-called super-molecules, such as the immune complex and membrane proteins, etc. To calculate and analyse the BFE of a large, complex T-cell receptor (TCR) and immunogenic peptides (p) presented by class I major histocompatibility complexes (MHC), binding free energy decomposition (BFED) calculations based on the MM–GBSA approach including entropic terms were done on the 2C TCR/SIYR/H-2Kb system and provided a detailed description of the energetics of the interaction (Zoete and Michielin 2007), since this BFED method can detect the important individual side chains for the stability of a protein fold with computational alanine scanning of the insulin monomer (Zoete and Meuwly 2006). A correlation between the decomposition results and experimentally-determined activity differences for alanine mutants of the TCR-pMHC complex is 0.67 when the conformational entropy is neglected, and 0.72 when the entropy is considered. Similarly, a comparison of experimental activities with variations in the BFEs determined by computational alanine scanning yields correlations of 0.72 and 0.74 when the entropy is neglected or taken into account, respectively. In addition, a comparison of the two theoretical approaches for estimating the role of each side chain in the complex formation is given, and a new *ad hoc* approach for decomposing the vibrational entropy term into atomic contributions - the linear decomposition of vibrational entropy (LDVE) - is introduced. The latter allows the rapid calculation of the entropic contribution of interesting side chains to the binding. This approach is justified by the idea that the most important contributions to the vibrational entropy of a molecule originate from residues that contribute most to the vibrational amplitude of the normal modes. The results of the LDVE are very similar to those of the exact but highly computationally demanding method. The BFED approach is also applicable to the design of rational TCR by calculating each amino acid contribution in mutated TCR. As melanoma patients frequently show unusually positive clinical outcomes, it represents an interesting target for adoptive transfer with modified TCR. Sequence modifications of TCR which potentially increase the affinity for this epitope have been proposed and tested *in vitro*. T-cells expressing some of the proposed TCR mutants showed better T-cell functionality, with the improved killing of peptide-loaded T2 cells and better proliferative capacity compared to the wild type TCR expressing cells (Zoete, Irving et al. 2010).

As there are still not many applications for massive simulations with dielectric models to large bio-molecules like the TCR-pMHC complex, more extensive studies are necessary to evaluate the validity of the method and improve its accuracy and performance because the excluded volume effect due to water entropy change in binding will become larger in the larger systems.

2.3 The correlation between calculation-cost and accuracy

It is not easy to state the calculation cost and accuracy exactly because the method is only now developing and the accuracy depends on the system size.

Previous studies have shown a very good correlation between PB and GB results because the GB parameter is modified to achieve better agreement with that of PB (Gohlke, Kuhn et al. 2004; Onufriev, Bashford et al. 2004). Moreover, GB and PB methods also enable the rapid scoring of protein structures when they are combined with physics-based energy functions. The direct comparison of these two approaches on large protein data sets is done with a scoring function based on a GB and PB solvation model and short MD simulations. Against seven publicly available decoy sets, the results of the MM-PBSA approach are comparable to the GB-based scoring function (Lee, Yang et al. 2005).

We also compared the MM-PBSA and MM-GBSA methods. Table 1 shows the comparison of the binding electrostatic free energies of the PB and GB methods for two TCR-pMHC complexes (2gj6 and 3pwp), a complex of A6 and Tax peptide-HLA A2, and A6 with Hud-A2 respectively. Constant regions of TCR were removed (Gregoire, Lin et al. 1996), hydrogen was added and the complexes were neutralised and solvated with TIP3P. The numbers of atoms involved in the systems were 130,545 for 2GJ6 and 127,023 for 3PWP. Calculations were performed with Sander of AMBER 11 for 5 ns. The Gelsolv, which is always largely negative in each case, represents the electrostatic energy contribution due to solvents. The Gnp is the hydrophobic and van der Waals contributions were calculated from the solvent accessible surface area (SA). The difference of the PB and GB results of each case is 3-4%, although the total binding free energy, *d*Gbind, differs by almost 20% because the binding energies *in vacuo*, *d*Ggas, and the contribution of the solvent, *d*Gsolv=*d*Gelsolv+*d*Gnpsolv, have a different sign and cancel each other. We must note that the ratio of the SA contributions between GBSA and PBSA is larger than the Eelsov, although the absolute contribution is 1/10th of the Gelsolv.

	2gj6			3pwp		
	GBSA	PBSA	GB/PB	GBSA	PBSA	GB/PB
Complex						
Gelsolv	-7502.3	-7233.6	1.037	-7367.2	-7112.6	1.036
Gnpsolv	193.3691	152.9235	1.2645	188.3416	151.8496	1.2403
Receptor protein						
Gelsolv	-2512	-2421.1	1.038	-2358.79	-2259.25	1.044
Gnpsolv	73.4152	58.0731	1.2642	71.2015	57.4698	1.2389
Ligand						
Gelsolv	-5340.8	-5151.5	1.037	-5340.08	-5162.4	1.034
Gnpsolv	133.8713	105.4387	1.2697	131.2283	104.4543	1.2563
Difference = Complex-(Receptor + Ligand)						
*d*Gelsolv	350.479	338.916	1.034	331.6509	309.0868	1.073
*d*Gnpsolv	-13.9175	-10.5883	1.314	-14.0882	-10.0745	1.398
*d*Eel_solv+*d*ESA	336.561	328.328	1.025	317.563	299.012	1.062
*d*Ggas	-388.76	-388.76	1	-380.55	-380.55	1
*d*Gsolv	336.561	328.328	1.025	317.563	299.012	1.062
*d*Gbind	-52.199	-60.432	0.864	-62.99	-81.54	0.773

Table 1. A comparison of the binding electrostatic free energies of the PBSA and GB methods for two TCR-pMHC complexes (PDB ID: 2gj6 and 2pwp). The Gelsolv, which is always largely negative in each case, represents the electrostatic energy contribution due to solvents. The Gnpsolv is the hydrophobic and van der Waals contributions are calculated from the solvent accessible surface area (SA). The total binding free energy, *d*Gbind, is the sum of the binding energy *in vacuo*, *d*Ggas, and the contribution of the solvent, *d*Gsolv=*d*Gelsolv+*d*ESA. All energies in the table are given in kcal mol^{-1}.

3. Coarse-grained (CG) simulation

3.1 The limits of all-atom simulations

Even though all-atom simulations provide the most detailed information about the system of interest, its calculation costs are quite high. A system containing a large protein molecule such as several 10^5 to 10^6 Dalton comes up to several 10^5 atoms when solvated in explicit water molecules, and expands to nm^3 in size; hence, the calculation time of less than μsec even using a recent multi-core PC. These figures are too short and still too small to reproduce such biologically interesting phenomena as protein folding, protein-assembly and enzymatic reaction, etc. Therefore, the increase of calculation efficiency is quite an urgent requirement. The calculation cost increases approximately in proportion to the square of the number of atoms, and the time for one step is approximately proportional to the order of the square-root of the mean mass of elements. The number of atoms constituting an amino acid (AA), when polymerized in a peptide, is 7 (Gly) to 24 (Trp), and the mass is between 57 (Gly) and 186 (Trp) - about 5 to 15 times of a C, N or O. When an AA is coarse-grained to 2 to 4 pseudo-atoms, the calculation cost decreased by 2 to 3 orders of magnitude, and the time for a step increases by 2 or 3 orders. In most CG models, the interaction between pseudo-atoms through bonds of less than 5 is described as follows:

Extension potential between two beads

$$U^{bond} = \frac{1}{2}k\left(r - r_0\right)^2 \tag{10}$$

Angle potential between three beads

$$U^{angle} = \frac{1}{2}k\left(\theta - \theta_0\right)^2 \tag{11}$$

Dihedral angle between four beads

$$U^{dihedral} = \frac{1}{2}k\left[1 - \cos\left(n\phi - \phi_0\right)\right] \tag{12}$$

And, the unbound potential between two beads can be expressed as

$$U^{LJ} = 4\varepsilon\left[\left(\frac{\sigma}{r}\right)^{12} - \left(\frac{\sigma}{r}\right)^{6}\right] \tag{13}$$

The whole energy of the system is described as the combination of these elemental potentials. For example, the Head-Gordon et al. model (Brown, Fawzi et al. 2003) is described as:

$$\begin{aligned} H = {} & \sum_{\theta}\frac{1}{2}K_{\theta}\left(\theta - \theta_0\right)^2 + \sum_{\phi}\left[A\left(1 + \cos\phi\right) + B\left(1 - \cos\phi\right) + C\left(1 + \cos 3\phi\right) + D\left(1 + \cos\left[\phi + \frac{\pi}{4}\right]\right)\right] \\ & + \sum_{i,j\geq i+3} 4\varepsilon_H S_1\left[\left(\frac{\sigma}{r_{ij}}\right)^{12} - S_s\left(\frac{\sigma}{r_{ij}}\right)^{6}\right] \end{aligned} \tag{14}$$

where θ, φ, and i, j are summed for all the AAs contained in the peptide. The interaction between non-bonded pseudo-atoms is usually described as the Lennard-Jones potential. The methods for configuration sampling, usually MD (Shih, Arkhipov et al. 2006) and the Monte Carlo simulation (Levy, Karplus et al. 1980; Horejs, Mitra et al. 2011), are the same as those used in all-atom simulations. The equation of motions for MD is principally the same as that used in all-atom simulations, i.e.,

$$m\frac{d^2\vec{r}}{dt^2} = \vec{F} - \Gamma\frac{d\vec{r}}{dt} + \overrightarrow{W} \tag{15}$$

where $\vec{F}$, Γ, and $\overrightarrow{W}$ are external force, friction and thermal noise, respectively. Any modification is made according to the kind of ensemble adopted.

3.2 The difference of CG models between proteins and other molecules

As might be guessed by Fig.2A and B, it is easier to treat a homopolymer by the CG model than to treat a polypeptide or a protein. A homopolymer can be described with rather a few parameters and, under certain circumstances, several components can be coarse-grained as a pseudo-atom (4 styrenes in a dotted circle are treated as one bead). Rheological features such as phase-transition, diffusion coefficient, compressibility, ductility, elasticity and viscosity have been reproduced fairly well (Yaoita, Isaki et al. 2008; Harmandaris and Kremer 2009; Kalra and Joo 2009; Posel, Lísal et al. 2009). On the other hand, peptides and proteins consist of diversified 20 AAs and the particular functions of proteins such as specific binding and enzymatic functions are based on a unique configuration of those characteristic AAs. Therefore, to evaluate the interaction on the CG model is especially difficult due to the effect of averaging specific properties and the anisotropicity of components. Notwithstanding this state of affairs, some CG models have come to predict the docking and binding of proteins fairly well. In this section, representative protein CG models are reviewed and the application of the CG model to the evaluation of TCR-pMHC interaction is foreseen.

3.3 Representative CG models

3.3.1 The one-bead model

Many one-bead models (Taketomi, Ueda et al. 1975; Brown, Fawzi et al. 2003; Jang, Hall et al. 2004) can be deemed as descendants of the Go-model. Go-like models, even though extremely simplified in their format, principally succeeded in reproducing several aspects of protein folding. This is presumably due to the finding that the protein-folding rate and mechanism are largely determined by a protein's topology rather than its inter-atomic interaction (Baker 2000). Those descendant models have equipped their own features, but still have a tendency towards a reference configuration. This might be due to the difficulty of incorporating the geometric and physicochemical aspects of all the AAs in only a few parameters. Recently, the finding that the underlying physicochemical principles of the interaction between the domains in protein folding are similar to those between the binding sites of protein assembly has been accepted (Haliloglu, Keskin et al. 2005; Levy, Cho et al. 2005; Turjanski, Gutkind et al. 2008; Baxter, Jennings et al. 2011). This fact will probably provide another aspect of the application of the CG model to issues of protein-binding.

Miyazawa and Jernigan (MJ) extracted inter-residue potentials from the crystallography of 1168 proteins (Miyazawa and Jernigan 1996). The principle adopted in this method is that the number of residue-residue contacts observed in a large number of protein crystals will represent the actual intrinsic inter-residue interactions. Namely, to regard the effect (contacts in the observed structure) in the same light as the cause (interaction energy) based on "the principle of structural consistency" or "the principle of minimal frustration".

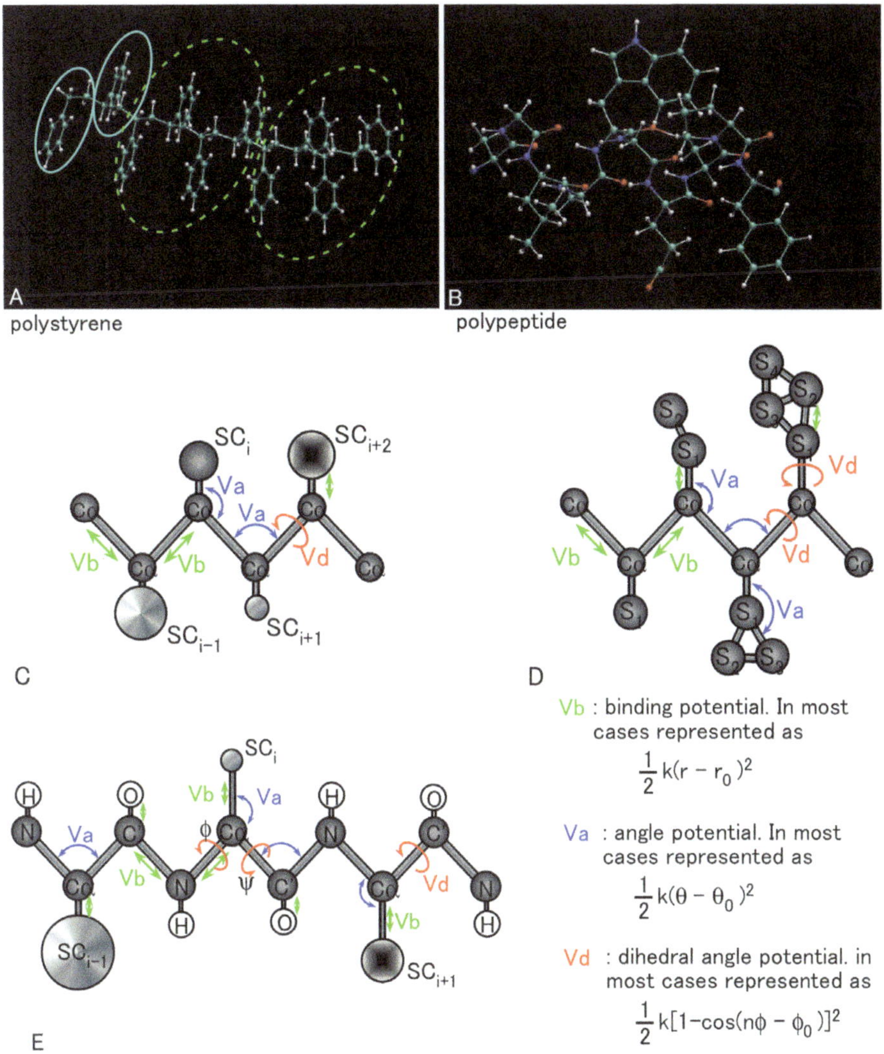

Fig. 2. Coarse-grained models for protein
Homopolymer such as polystytene can be described with rather a few parameters, and in some cases, several units are mapped to one bead (A). Protein consists of heterogenious components, hence more detailed and complicated description (B). Main chain is represented by Cα and each side chain is mapped to one bead, which retains its original geometiric and physico-chemical features (Liwo, Pincus et al. 1993) (C). MARTINI force field maps more beads to a side chain (Marrink, Monticelli et al. 2008), enabled to simulate the release of inner water molecules through stress-sensitive channel enbedded in vesicle membrane (Louhivouri, Lisselada et al. 2008) (D). OPEP model all the atoms of main chain and maps one bead for side chain. As can be guessed, this models is suitable for dealing with the issues where backbone structure such as α-helix and β-sheet play essential roles (Chebaro et al. 2009, Laghaei et al. 2011, Nasica-Labouze et al. 2011) (E).

Adopting this model for the parameters of LJ potentials, Kim and Hummer constructed a one-bead model combined with the Debye-Hückel type potential and performed configure-sampling on replica exchange MC - applied to ubiquitin binding - and obtained good agreement with other experiments (Kim and Hummer 2008). Chakraborty's group applied an MJ matrix to estimate TCR-pMHC and explained the effect of the HLA class I haplotype on TCR repertoire-formation (Kosmrlj, Read et al. 2010). The above mentioned CG models are tabulated in Table 2.

3.3.2 UNRES

Scheraga's group described a CG model which consists of a Cα, side chain centroid (SC) and one dihedral angle (Liwo, Pincus et al. 1993). They searched the conformation space on this model with compactness of the protein as an indicator. The obtained structure was then decoded into an all-atom-backbone with the SC model and then searched further for the lowest-energy structure. Finally, an all-atom model was reconstructed from the obtained structure and searched for the lowest-energy structure on an electrostatically driven Monte Carlo (EDMC) simulation based on the ECEPP/2 potential. They succeeded in predicting *ab initio* the moderate size of proteins (53-235 residues) (Oldziej, Czaplewski et al. 2005). This hybrid method - the sampling of a configuration on a CG model and the estimation of binding energy on an atomistic model - presents quite a reasonable combination of efficiency and accuracy. Their recent accomplishment was a 1 msec simulation of more than 500 AA proteins through massive parallelisation (Scheraga, Maisuradze et al. 2010).

3.3.3 ATTRACT

Zacharias described a docking method of protein-protein or protein-ligand using a reduced protein model and docking algorithm, ATTRACT (Zacharias 2003). An AA is represented with 2 to 3 (Zacharias 2003) or 2 to 4 (Zacharias and Fiorucci 2010) pseudo-atoms and the interactions of specific pseudo-atom pairs, including their size and physicochemical characters, are interweaved into the parameters of the Lennard-Jones potential. ATTRACT assumes that both interacting molecules are rigid, smaller molecules is tried to dock from thousands of sites with 6 degrees of freedom, 3-translational and 3-rotational. Docking includes the minimisation of side chains described as rotamer, hence total minimisation is performed. They applied this CG model and ATTRACT to the Critical Assessment of Prediction of Interest (CAPRI) (Janin 2002) and showed two acceptable bindings out of 6 targets (May and Zacharias 2007) or else obtained better (4 out of 6 targets) prediction by improving the scoring function and docking method (Zacharias and Fiorucci 2010). The estimation of TCR and pMHC binding not only deals with the binding energy of a pre-determined configuration, but also deals with the determination of the binding-configuration, because the TCR-pMHC complex has several binding modes (Wucherpfennig, Call et al. 2009). They showed that it is possible to uncover a binding site by using an electrostatic desolvation profile (Zacharias and Fiorucci 2010) based on ODA method (Fernandez-Recio, Totrov et al. 2005).

3.3.4 The MARTINI force field

The MARTINI force field was originally devised for describing lipids or surfactants, such as dipalmitoylphosphatidylcholine (DPPC), dicapryloyl-PC (DCPC), dodecylphosphocholine

Number of beads	Authors	force field or residual potential	sampling	purpose or reproduced features	year	reference
1	Go and coworkers	Go model	MC	protein folding	1975	1
1	Jang H, Hall CK and Zhou Y	Go like model	MD	protein folding	2004	2
1	Miyazawa & Yernigan	inter residue potential	static model	calculation of MJ-matrix	1996	3
1	Kim and Hummer	MJ matrix	MC	Ubiquitin binding	2008	4
1	Chakraborty and coworkers	MJ matrix	static model	T cell pepertoire formation	2010	5
1	Head-Goadon and coworkers	Go like model	MD	protein folding	2003	6
1	Turjanski et al.	Go like model	MD	protein binding	2008	7
2	Liwo et al.	UNRES and ECEP/2	MC	protein folding	1993	8
2	Bahar and Jernigan	distance dependent statistical potential	static method	peptide-protein binding	1997	9
2	Scheraga and coworkers	UNRES and ECEP/3	MD	structure, folding pathway	2005	10
2	Scheraga and coworkers	UNRES and ECEP/3	REMD by parallelism	millisecond simulation	2010	11
3	Zacharias	reduced protein	ATTRACT	protein-protein docking	2003	12
3	Marrink and coworkers	MARTINI	MD	protein behavior in lipid membrane	2008	13
4	Fiorucci and Zacharias	reduce protein model	modified ATTRACTA	docking, prediction of binding site	2010	14
5	Louhivouri	MARTINI	MD	water release from liposome	2010	15
6	Derreumaux	OPEP	MC	structure and folding of protein	2001	16
6	Derreumaux and Mousseau	OPEP	MD	aggregation mechanism of A $_{16\text{-}22}$	2007	17
6	Derreumaux and coworkers	OPEP	REMD	formation of -amyloid	2009	18
6	Nasica-Labouze et al	OPEP+all atom	REMD	detailed process of amyloid formation	2011	19
6	Laghaei et al	OPEP	HT-REMD	structure and thermodynamics of amylin dimr	2011	20

1 (Taketomi, Ueda et al. 1975); 2 (Jang, Hall et al. 2004);3 (Miyazawa and Jernigan 1996); 4 (Hummer and Kim 2008); 5 (Kosmrlj, Read et al. 2010); 6 (Brown, Fawzi et al. 2003); 7 (Turjanski, Gutkind et al. 2008); 8(Liwo, Pincus et al. 1993); 9 (Bahar and Jernigan 1997); 10 (Liwo, Khalili et al. 2005); 11 (Liwo, Oldziej et al. 2010); 12 (Zacharias 2003); 13 (Marrink, Monticelli et al. 2008); 14 (Fiorucci and Zacharias 2010); 15 (Louhivuori, Risselada et al. 2010); 16 (Derreumaux and Forcellino 2001); 17 (Derreumaux and Mousseau 2007); 18 (Derreumaux, Chebaro et al. 2009); 19 (Nasica-Labouze, Meli et al. 2011); 20 (Laghaei, Mousseau et al. 2011)

Table 2. Representative coarse-grained models for protein and noteworthy results based on them

(DPC) and cholesterol (Marrink, de Vries et al. 2004; Marrink, Risselada et al. 2007). The adoption of a very limited atom type and short range potentials provided very efficient computation, hence the micrometer length in scales and milliseconds in time, and succeeded in the simulation of the spontaneous aggregation of DPPC lipids into a bilayer and the formation of DPC in water. The hydrogen atom is neglected in this model. Heavy four atoms on average are represented as one pseudo-atom (four-to-one mapping) with an exception for ringlike molecules. Ringlike molecules are mapped with higher resolution (up to two-to-one mapping). Interaction sites are classified into 4 types: polar (P), nonpolar (N), apolar (C) and charged (Q). Within a main type, subtypes are distinguished either by a letter denoting the hydrogen-bonding capabilities (d = donor, a = acceptor, da= both, 0 = none) or by a number indicating the degree of polarity (from 1 = lower polarity to 5 = higher polarity). The interaction of each atom-type was parameterised at five levels: attractive (e = 5 kJ/mol), semi-attractive (e = 4.2 kJ/mol), intermediate (e = 3.4 kJ/mol), semi-repulsive (e = 2.6 kJ/mol) and repulsive (e = 1.8 kJ/mol). Non-bonded interactions between the interaction sites i and j are described by the Lennard-Jones potential:

$$U_{LJ}(r) = 4\varepsilon_{ij}\left[\left(\frac{\sigma_{ij}}{r}\right)^{12} - \left(\frac{\sigma_{ij}}{r}\right)^{6}\right] \tag{16}$$

with σ_{ij} representing the effective minimum distance of approach between two particles and ε_{ij} representing the strength of their interaction. This model was extended to deal with proteins (Marrink, Monticelli et al. 2008). The basic parameters are the same as used in the lipid model. Bonded interaction is described by the following set of potential energy functions acting between the bonded sites i, j, k, and l with an equilibrium distance d_b, an angle ϕ_a, and a dihedral angle ψ_i and ψ_{id}:

$$V_b = \frac{1}{2}K_b\left(d_{ij} - d_b\right)^2 \tag{17}$$

$$V_a = \frac{1}{2}K_a\left[\cos\left(\varphi_{ijk}\right) - \cos\left(\varphi_a\right)\right]^2 \tag{18}$$

$$V_d = K_d\left[1 + \cos\left(n\psi_{ijkl} - \psi_d\right)\right] \tag{19}$$

$$V_{id} = K_d\left(\psi_{ijkl} - \psi_{id}\right)^2 \tag{20}$$

where V_b, V_a, V_d and V_{id} represent potential sites for bonding, stiffness, dihedral angle and improper dihedral angle, respectively. The total energy of the system is obtained by summing (17) to (20). The mapping of all AAs is mapped into 4 types of beads or a combination of them. In this mapping, Leu, Pro, Ile, Val, Cys and Met are classified as apolar (C-type), where as Thr, Ser, Asn and Gln are polar (P-type). Glu and Asp are charged (Q-type), and Arg and Lys are modelled by a combination of a Q and an uncharged particle (N-type). The bulky ring-based side chains are modelled by three (His, Phe, and Tyr) or four (Trp) beads. Gly and Ala residues are only represented by the backbone particle. The type of the backbone particle depends on the protein secondary structure; free in solution or in a coil or bend, the backbone has a strong polar character (P-type); as part of α helix or β strand, the interbackbone hydrogen bonds reduce the polar character significantly (N-type). Proline is less polar due to the lack of hydrogen-donor capabilities. More detailed geometrical representation is given in Fig.2 D, illustrating the binding distance, angle, dihedral angle, improper angle and bead configuration. This CG protein model contains directional specificity and heterogeneity in side chains to some extent, hence a feature of a secondary structure (α-helix and β-strand) and the gross physicochemical property, such as being charged, hydrophilic and hydrophobic. They succeeded in the partitioning of AAs in the DOPC bilayer, keeping the AA association (Leu-Leu, Lys-Glu) constant in water, the portioning and orientation of pentapeptides at the border of the water and cyclohexane. The tilt and orientation of hexapeptides in the DOPC bilayer is also reproduced after sub-μ sec to μsec MD simulation on GROMACS software (van Der Spoel, Lindahl et al. 2005). They recently accomplished the simulation of the rapid release of content from a pressurised liposome through a particular mechano-sensitive protein channel, MscL, embedded in the liposomal membrane (Louhivuori, Risselada et al. 2010). The behaviour of this tiny functional organelle, which consists of 5 MscL molecules, 2108 DOPC lipids, 5,444 water beads with an additional 54,649 water beads forming a 4 μm layer around the vesicle, was described in almost atomistic detail. In response to the increase of internal pressure, this vesicle released water molecules by opening the Mscl channel. MD was performed for 40 μs, which corresponds to 160 μs in an all-atom model. This model demonstrated that CG-MD provides for the computer-aided design of super-molecules and organelles of a practically usable size.

3.3.5 The optimised potential for efficient peptide-structure representation (OPEP) model

OPEP is, as shown in Fig.2 E, a CG protein model that uses a detailed representation of all backbone atoms (N, H, Cα, C and O) and reduces each side chain to one single bead with appropriate geometrical parameters and physicochemical properties (Derreumaux and Forcellino 2001). The OPEP energy function, which includes the implicit effects of an aqueous solution - expressed as (21) - is formulated as a sum of local potentials (E_{local}), non-bonded potential ($E_{\text{nonbonded}}$), and hydrogen-bonding potential ($E_{\text{H-bond}}$):

$$E = E_{local} + E_{nonbonded} + E_{H-bond} \tag{21}$$

Local potentials are expressed by:

$$E_{local} = w_b \sum_{bonds} K_b (r - r_{eq})^2 + w_a \sum_{angle} K_\alpha (\alpha - \alpha_{eq})^2 + w_\Omega \sum_{imp-torsions} k_\Omega (\Omega - \Omega_{eq})^2 + w_{\phi,\psi} (\sum_\phi E_\phi + \sum_\psi E_\psi) \tag{22}$$

K_b, K_a, and K_Ω represent force constants associated with changes in bond length, the bond angles of all particles and force constants related to changes in improper torsions of the side chains. The dihedral potentials associated with N-Cα associated are expressed as (23) and Cα-C expressed as as (24), respectively:

$$E_\phi = k_{\phi\psi} (\phi - \phi_0)^2 \tag{23}$$

$$E_{\psi} = k_{\phi\psi}(\psi - \psi_0)^2 \tag{24}$$

The non-bonded functions are expressed by:

$$E_{nonbonded} = w_{1,4}\sum_{1,4} E_{VdW} + w_{C\alpha,C\alpha}\sum_{C\alpha,C\alpha} E_{VdW} + w_{1>4}\sum_{M',M'} E_{VdW} + w_{1>4}\sum_{M',C\alpha} E_{VdW} + w_{1>4}\sum_{M,SC} E_{VdW} + \sum_{Sc,Sc} w_{Sc,Sc}E_{VdW} \tag{25}$$

which includes all the interaction works through more than 3-bonds, and all these functions are expressed as Van der Waals potentials, as shown in (11):

$$E_{VdW} = \varepsilon_{ij}\left(\left(\frac{r_{ij}^0}{r_{ij}}\right)^{12} - 2\left(\frac{r_{ij}^0}{r_{ij}}\right)^{6}\right)H(\varepsilon_{ij}) - \varepsilon_{ij}\left(\frac{r_{ij}^0}{r_{ij}}\right)^{6} H(-\varepsilon_{ij}) \tag{26}$$

Here, the Heavyside function $H(x) = 1$ if $x >= 0$ and 0 of $x < 0$, r_{ij} is the distance between particles i and j, $r_{ij}^0 = (r_i^0 + r_j^0)/2$ with r_i^0 as the Van der Waals radius of particle i.

The hydrogen-bonding potential ($E_{H\text{-bond}}$) consists of two-body and three-body terms (Derreumaux, Maupetit et al. 2007).

This model was originally devised for predicting the structure and folding of proteins (Derreumaux 1999; Derreumaux and Forcellino 2001) and, by combining a Monte Carlo simulation, fairly succeeded in prefiguring basic supersecondary structures. This model, containing all the protein-backbone components, excels in issues where secondary-structure features play an essential role. They combined this potential with MD, which resulted in reproducing the aggregation of Alzheimer's $A\beta_{16\text{-}22}$, (Derreumaux and Mousseau 2007; Wei, Song et al. 2008). In adopting the sampling of Replica Exchange MD (REMD), they obtained an accurate structural description of Alzheimer's Amyloid-β, β-hairpin and Trp-cage peptides (Derreumaux, Chebaro et al. 2009; Derreumaux, Chebaro et al. 2009). A detailed atomic characterisation of oligomer-formation was obtained by combining OPEP, the atomistic model and REMD (Nasica-Labouze, Meli et al. 2011) Their reduced model on REMD enabled the calculation of several tens of μsec in 40 replicas and the full assessment of convergence to the equilibrium ensemble, demonstrating the probability of determining the thermodynamic features of large proteins and assemblies (Laghaei, Mousseau et al. 2011).

As was mentioned above, the main CG models are tabulated in Table 2.

3.4 The trial for the TCR-pMHC and larger systems

At the starting point of the whole immunological synapse (IS) simulation, Wan, Flower and Coveny constructed a ternary complex of TCR-pMHC-CD4 between opposite membranes - which consists of 329,265 atoms - and performed molecular dynamics for 10 ns on 128 processors of SGI Altix (Wan, Flower et al. 2008). It took 23 hours for one ns simulation. This run was not enough to calculate the binding free-energy by MM/PBSA due to the shortness of the simulation time and the lack of entropy evaluation. They intended to simulate a system consisting of four sets of the TCR-pMHC-CD4 complex, made up of about one million atoms. They pointed out the difficulty of the whole IS simulation on the all-atom model due to the too heavy load imposed upon the computer, and pointed out the feasability of adopting the hybrid atomistic/CG simulation for accomplishing the project (Diestler, Zhou et al. 2006).

At present, there have been only very limited trials of evaluated TCR-pMHC binding energy by the CG model. The evaluation of TCR-pMHC binding consists of at least three steps: 1) to determine the binding site, 2) to determine the binding configuration, and 3) to calculate the binding energy. Several works have provided not only the method to determine the binding configuration but also to detect the binding site from the surface nature of its own (Fernandez-Recio, Totrov et al. 2005; Burgoyne and Jackson 2006; Fiorucci and Zacharias 2010). The factors that concern the evaluation of TCR-pMHC binding are: 1) the evaluation of energy from a particular configuration, and 2) the sampling of independent configurations. In most CG models, the calculation of binding energy as the function of the configuration is based on their own parameters (Liwo, Pincus et al. 1993; Miyazawa and Jernigan 1996; Derreumaux 1999; Zacharias 2003; Buchete, Straub et al. 2004; Oldziej, Czaplewski et al. 2005; Zhou, Thorpe et al. 2007; Kim and Hummer 2008; Marrink, Monticelli et al. 2008). The sampling of independent configurations is most time-consuming but critically important process. If the sampling on the CG model reflects the distribution of the atomistic model with reasonable fidelity, it is quite a smart way to sample configurations on a CG-model (Chebaro, Dong et al. 2009), to reconstruct to the atom-scale the structure and then calculate the binding energy on these reconstructed atomistic structures using MM/PBSA. From this point of view, a general method to reconstruct the all-atom from the Cα atom position, RACOGS, was devised and the energy landscapes of both the CG- and the all-atom-model were shown to be quite similar, suggesting the validity of this principle (Heath, Kavraki et al. 2007).

4. Application of GPGPU in molecular dynamics

As mentioned above, all-atom simulation is very expensive, and hence is restricted scope in both time and scale. There have been attempts to breakthrough these circumstances, not only by improving the algorithm but also by devising novel hardware. Special purpose machines for MD have been developed (Susukita, Ebisuzaki et al. 2003; Shaw, Deneroff et al. 2008) and showed fairly good performance (Kikugawa, Apostolov et al. 2009). However, such purpose-specific machines are very expensive and their continuous development is difficult. The recent development of the general purpose graphic processor unit (GPGPU) has had much influence on high performance computing (Giupponi, Harvey et al. 2008). In 2011, three of the top 5 super-computers are constructed mainly on NVIDIA's GPGPU (http://www.top500.org/). Many applications are now being preparing to respond to this momentum, and representative molecular dynamics software such as Amber, CHARM, GROMACS and NAMDA are now being prepared to equip programs working on GPGPUs. Recent representative GPGPUs, such as Tesla C2075, have a performance of 1.03 T Flops on single precision. We calculated the binding energy of two TCR-pMHC complexes, 2GJ6 and 3PWP, on C2075 and compared the results calculated on a Xeon processor. After heating, density-equilibration and equilibration, product runs were performed for 10 runs, corresponding to 5 nsec in total. The results are shown in Table 3. As can be seen from the table, the performance of a Tesla

C2075 is about 40 to 50 cores of a present Xeon CPU. The obtained values are 8% ~ 30% larger by GPGPU than that by CPU. This may be due to the difference of detailed algorithm adopted for the calculations using CPU and GPGPU.

	PDB ID	Product run /run	ΔH
Xeon E5620 (4 core)	2GJ6	35.7 hours	GB -50.3 ± 8.7
			PB -54.1 ± 11.8
	3PWP	34.9 hours	GB -51.3 ± 8.9
			PB -60.1 ± 11.2
Tesla C2075	2GJ6	3.09 hour	GB -53.8 ± 7.64
			PB -63.4 ± 12.5
	3PWP	3.03 hours	GB -61.5± 6.8
			PB -78.0 ± 10.9

Table 3. Comparison of CPU and GPGPU All energies in the table are given in kcal mol^{-1}.

5. Conclusion

Physically meaningful models are rapidly advancing and are being applied to large macro-molecular systems with the rapid evolution of parallel computation and hardware, such as multi-core processors and GPGPUs. Although the exact models become realistic for calculations of large bio-molecules, continuum dielectric models are still useful for the binding free energy calculation and bound complex structure prediction as well as the structure prediction tasks of bio-molecules such as proteins and nucleic acids, etc., because of the high cost performance and fairly good accuracy. In future, hybrid approaches will become promising, where QM model, the all-atom model, the CG model and continuum models are combined with a good conformational sampling technique such as the ER method, and we can choose the optimal hybrid approach according to purpose and the system size.

It has been clear that the calculation of TCR-pMHC binding energy with reasonable efficiency and accuracy is feasible. MMPBSA/GBSA seems quite promising. The sampling method affects both the efficiency and accuracy of the calculation. The combination of sampling on the CG model and energy-calculation on the atomistic model is very reasonable approaches. GPGPUs will be quite important facilities. A combination of those factors will provide for the valid simulation of biologically interesting phenomena for an adequately long time.

6. Acknowledgment

This work was partially supported by JSPS KAKENHI Grant Number 22590194 for HT.

7. References

Asakura, S. & Oosawa, F. (1954). "On interaction between two bodies immersed in a solution of macromolecules." *Journal of Chemical Physics* 22: 1255-1256.
Baker, D. (2000). "A surprising simplicity to protein folding." *Nature* 405(6782): 39-42.

Baxter, E. L.; Jennings, P. A. & Onuchic, J. N. (2011). "Interdomain communication revealed in the diabetes drug target mitoNEET." *Proc Natl Acad Sci U S A* 108(13): 5266-5271.

Ben-Tal, N.; Honig, B.; Bagdassarian, C. K. & Ben-Shaul, A. (2000). "Association entropy in adsorption processes." *Biophys J* 79(3): 1180-1187.

Brown, S.; Fawzi, N. J. & Head-Gordon, T. (2003). "Coarse-grained sequences for protein folding and design." *Proc Natl Acad Sci U S A* 100(19): 10712-10717.

Buchete, N. V.; Straub, J. E. & Thirumalai, D. (2004). "Development of novel statistical potentials for protein fold recognition." *Curr Opin Struct Biol* 14(2): 225-232.

Burgoyne, N. J. & Jackson, R. M. (2006). "Predicting protein interaction sites: binding hot-spots in protein-protein and protein-ligand interfaces." *Bioinformatics* 22(11): 1335-1342.

Car, R. & Parrinello, M. (1985). "Unified approach for molecular dynamics and density-functional theory." *Physical Review Letters* 55(22): 2471-2474.

Carlsson, J.; Ander, M.; Nervall, M. & Aqvist, J. (2006). "Continuum solvation models in the linear interaction energy method." *Journal of Physical Chemistry B* 110(24): 12034-12041.

Case, D. A.; Cheatham, T. E., 3rd; Darden, T.; Gohlke, H.; Luo, R.; Merz, K. M. Jr.; Onufriev, A.; Simmerling, C.; Wang, B. & Woods, R. J. (2005). "The Amber biomolecular simulation programs." *J Comput Chem* 26(16): 1668-1688.

Chebaro, Y.; Dong, X.; Laghaei, R.; Derreumaux, P. & Mousseau, N. (2009). "Replica exchange molecular dynamics simulations of coarse-grained proteins in implicit solvent." *J Phys Chem B* 113(1): 267-274.

Derreumaux, P. (1999). "From polypeptide sequences to structures using Monte Carlo simulations and an optimized potential." *Journal of Chemical Physics* 111(5): 2301-2310.

Derreumaux, P.; Chebaro, Y.; Dong, X.; Laghaei, R. & Mousseau, N. (2009). "Replica Exchange Molecular Dynamics Simulations of Coarse-grained Proteins in Implicit Solvent." *Journal of Physical Chemistry B* 113(1): 267-274.

Derreumaux, P.; Chebaro, Y. & Mousseau, N. (2009). "Structures and Thermodynamics of Alzheimer's Amyloid-beta A beta(16-35) Monomer and Dimer by Replica Exchange Molecular Dynamics Simulations: Implication for Full-Length A beta Fibrillation." *Journal of Physical Chemistry B* 113(21): 7668-7675.

Derreumaux, P. & Forcellino, F. (2001). "Computer simulations aimed at structure prediction of supersecondary motifs in proteins." *Proteins-Structure Function and Genetics* 45(2): 159-166.

Derreumaux, P.; Maupetit, J. & Tuffery, P. (2007). "A coarse-grained protein force field for folding and structure prediction." *Proteins-Structure Function and Bioinformatics* 69(2): 394-408.

Derreumaux, P. & Mousseau, N. (2007). "Coarse-grained protein molecular dynamics simulations." *J Chem Phys* 126(2): 025101.

Diestler, D. J.; Zhou, H.; Feng, R. & Zeng, X. C. (2006). "Hybrid atomistic-coarse-grained treatment of multiscale processes in heterogeneous materials: a self-consistent-field approach." *J Chem Phys* 125(6): 64705.

Fernandez-Recio, J.; Totrov, M.; Skorodumov, C. & Abagyan, R. (2005). "Optimal docking area: a new method for predicting protein-protein interaction sites." *Proteins* 58(1): 134-143.

Fiorucci, S. & Zacharias, M. (2010). "Prediction of protein-protein interaction sites using electrostatic desolvation profiles." *Biophys J* 98(9): 1921-1930.

Fogolari, F.; Brigo, A. & Molinari, H. (2002). "The Poisson-Boltzmann equation for biomolecular electrostatics: a tool for structural biology." *Journal of Molecular Recognition* 15(6): 377-392.

Forsten, K. E.; Kozack, R. E.; Lauffenburger, D. A. & Subramaniam, S. (1994). "Numerical solution of the nonlinear Poisson-Boltzmann equation for a membrane-electrolyte system." *J Phys Chem B* 98(21): 5580-5586.

Gilson, M. K. & Honig, B. H. (1987). "Calculation of electrostatic potentials in an enzyme active site." *Nature* 330(6143): 84-86.

Gilson, M. K. & Zhou, H. X. (2007). "Calculation of protein-ligand binding affinities." *Annual Review of Biophysics and Biomolecular Structure* 36: 21-42.

Giupponi, G.; Harvey, M. J. & De Fabritiis, G. (2008). "The impact of accelerator processors for high-throughput molecular modeling and simulation." *Drug Discov Today* 13(23-24): 1052-1058.

Gohlke, H.; Kuhn, L. A. & Case, D. A. (2004). "Change in protein flexibility upon complex formation: Analysis of Ras]Raf using molecular dynamics and a molecular framework approach." *PROTEINS: Structure, Function, and Bioinformatics* 56(2): 322-337.

Gregoire, C.; Lin, S. Y.; Mazza, G.; Rebai, N.; Luescher, I. F. & Malissen, B. (1996). "Covalent assembly of a soluble T cell receptor-peptide-major histocompatibility class I complex." *Proc Natl Acad Sci U S A* 93(14): 7184-7189.

Haliloglu, T.; Keskin, O.; Ma, B. & Nussinov, R. (2005). "How similar are protein folding and protein binding nuclei? Examination of vibrational motions of energy hot spots and conserved residues." *Biophys J* 88(3): 1552-1559.

Harmandaris, V. A. & Kremer, K. (2009). "Dynamics of Polystyrene Melts through Hierarchical Multiscale Simulations." *Macromolecules* 42(3): 791-802.

Harris, R. C.; Bredenberg, J. H.; Silalahi, A. R.; Boschitsch, A. H. & Fenley, M. O. (2011). "Understanding the physical basis of the salt dependence of the electrostatic binding free energy of mutated charged ligand-nucleic acid complexes." *Biophys Chem* 156(1): 79-87.

Hawkins, G. D.; Cramer, C. J. & Truhlar, D. G. (1996). "Parametrized models of aqueous free energies of solvation based on pairwise descreening of solute atomic charges from a dielectric medium." *J Phys Chem B* 100(51): 19824-19839.

Heath, A. P.; Kavraki, L. E. & Clementi, C. (2007). "From coarse-grain to all-atom: toward multiscale analysis of protein landscapes." *Proteins* 68(3): 646-661.

Ho, B. K. & Dill, K. A. (2006). "Folding very short peptides using molecular dynamics." *PLoS Comput Biol* 2(4): e27.

Horejs, C.; Mitra, M. K.; Pum, D.; Sleytr, U. B. & Muthukumar, M. (2011). "Monte Carlo study of the molecular mechanisms of surface-layer protein self-assembly." *J Chem Phys* 134(12): 125103.

Hou, T.; Wang, J.; Li, Y. & Wang, W. (2011). "Assessing the performance of the MM/PBSA and MM/GBSA methods. 1. The accuracy of binding free energy calculations based on molecular dynamics simulations." *J Chem Inf Model* 51(1): 69-82.

Hou, T.; Wang, J.; Li, Y. & Wang, W. (2011). "Assessing the performance of the molecular mechanics/Poisson Boltzmann surface area and molecular mechanics/generalized Born surface area methods. II. The accuracy of ranking poses generated from docking." *J Comput Chem* 32(5): 866-877.

Im, W.; Feig, M. & Brooks, C. L. (2003). "An implicit membrane generalized born theory for the study of structure, stability, and interactions of membrane proteins." *Biophys J* 85(5): 2900-2918.

Jang, H.; Hall, C. K. & Zhou, Y. (2004). "Assembly and kinetic folding pathways of a tetrameric beta-sheet complex: molecular dynamics simulations on simplified off-lattice protein models." *Biophys J* 86(1 Pt 1): 31-49.

Janin, J. (2002). "Welcome to CAPRI: A Critical Assessment of Predicted Interactions." *Proteins-Structure Function and Genetics* 47(3): 257-257.

Jayaram, B.; Sprous, D. & Beveridge, D. (1998). "Solvation free energy of biomacromolecules: Parameters for a modified generalized Born model consistent with the AMBER force field." *The Journal of Physical Chemistry B* 102(47): 9571-9576.

Kalra, V. & Joo, Y. L. (2009). "Coarse-grained molecular dynamics study of block copolymer/nanoparticle composites under elongational flow." *J Chem Phys* 131(21): 214904.

Kikugawa, G.; Apostolov, R.; Kamiya, N.; Taiji, M.; Himeno, R.; Nakamura, H. & Yonezawa, Y. (2009). "Application of MDGRAPE-3, a special purpose board for molecular dynamics simulations, to periodic biomolecular systems." *J Comput Chem* 30(1): 110-118.

Kim, Y. C. & Hummer, G. (2008). "Coarse-grained models for simulations of multiprotein complexes: application to ubiquitin binding." *J Mol Biol* 375(5): 1416-1433.

Kinoshita, M. (2006). "Roles of entropic excluded-volume effects in colloidal and biological systems: Analyses using the three-dimensional integral equation theory." *Chemical Engineering Science* 61(7): 2150-2160.

Kinoshita, M. (2009). "Importance of Translational Entropy of Water in Biological Self-Assembly Processes like Protein Folding." *International Journal of Molecular Sciences* 10(3): 1064-1080.

Koehl, P. (2006). "Electrostatics calculations: latest methodological advances." *Current opinion in structural biology* 16(2): 142-151.

Kosmrlj, A.; Read, E. L.; Qi, Y.; Allen, T. M.; Altfeld, M.; Deeks, S. G.; Pereyra, F.; Carrington, M.; Walker, B. D. & Chakraborty, A. K. (2010). "Effects of thymic selection of the T-cell repertoire on HLA class I-associated control of HIV infection." *Nature* 465(7296): 350-354.

Kuhne, T. D.; Krack, M.; Mohamed, F. R. & Parrinello, M. (2007). "Efficient and accurate Car-Parrinello-like approach to Born-Oppenheimer molecular dynamics." *Physical Review Letters* 98(6): 066401.

Laghaei, R.; Mousseau, N. & Wei, G. (2011). "Structure and thermodynamics of amylin dimer studied by Hamiltonian-temperature replica exchange molecular dynamics simulations." *J Phys Chem B* 115(12): 3146-3154.

Lamb, M. L.; Tirado-Rives, J. & Jorgensen, W. L. (1999). "Estimation of the binding affinities of FKBP12 inhibitors using a linear response method." *Bioorganic & medicinal chemistry* 7(5): 851-860.

Lee, M. C.; Yang, R. & Duan, Y. (2005). "Comparison between Generalized-Born and Poisson-Boltzmann methods in physics-based scoring functions for protein structure prediction." *Journal of molecular modeling* 12(1): 101-110.

Leis, S. & Zacharias, M. (2011). "Efficient inclusion of receptor flexibility in grid-based protein-ligand docking*." *J Comput Chem* 32(16): 3433-3439.

Levy, R. M.; Karplus, M. & McCammon, J. A. (1980). "Molecular dynamics studies of NMR relaxation in proteins." *Biophys J* 32(1): 628-630.

Levy, Y.; Cho, S. S.; Onuchic, J. N. & Wolynes, P. G. (2005). "A survey of flexible protein binding mechanisms and their transition states using native topology based energy landscapes." *J Mol Biol* 346(4): 1121-1145.

Liwo, A.; He, Y. & Scheraga, H. A. (2011). "Coarse-grained force field: general folding theory." *Phys Chem Chem Phys* 13(38): 16890-16901.

Liwo, A.; Pincus, M. R.; Wawak, R. J.; Rackovsky, S. & Scheraga, H. A. (1993). "Prediction of protein conformation on the basis of a search for compact structures: test on avian pancreatic polypeptide." *Protein Sci* 2(10): 1715-1731.

Louhivuori, M.; Risselada, H. J.; van der Giessen, E. & Marrink, S. J. (2010). "Release of content through mechano-sensitive gates in pressurized liposomes." *Proc Natl Acad Sci U S A* 107(46): 19856-19860.

Lu, B. Z.; Zhou, Y. C.; Holst, M. J. & McCammon, J. A. (2008). "Recent progress in numerical methods for the Poisson-Boltzmann equation in biophysical applications." *Communications in Computational Physics* 3(5): 973-1009.

Luchko, T.; Gusarov, S.; Roe, D. R.; Simmerling, C.; Case, D. A.; Tuszynski, J. & Kovalenko, A. (2010). "Three-dimensional molecular theory of solvation coupled with molecular dynamics in Amber." *Journal of Chemical Theory and Computation* 6(3): 607-624.

Marrink, S. J.; de Vries, A. H. & Mark, A. E. (2004). "Coarse grained model for semiquantitative lipid simulations." *Journal of Physical Chemistry B* 108(2): 750-760.

Marrink, S. J.; Monticelli, L.; Kandasamy, S. K.; Periole, X.; Larson, R. G. & Tieleman, D. P. (2008). "The MARTINI coarse-grained force field: Extension to proteins." *Journal of Chemical Theory and Computation* 4(5): 819-834.

Marrink, S. J.; Risselada, H. J.; Yefimov, S.; Tieleman, D. P. & de Vries, A. H. (2007). "The MARTINI force field: coarse grained model for biomolecular simulations." *J Phys Chem B* 111(27): 7812-7824.

Matubayasi, N. & Nakahara, M. (2000). "Theory of solutions in the energetic representation. I. Formulation." *Journal of Chemical Physics* 113(15): 6070-6081.

Matubayasi, N. & Nakahara, M. (2002). "Theory of solutions in the energy representation. II. Functional for the chemical potential." *Journal of Chemical Physics* 117(8): 3605-3616.

May, A. & Zacharias, M. (2007). "Protein-protein docking in CAPRI using ATTRACT to account for global and local flexibility." *Proteins* 69(4): 774-780.

Mehler, E. L. & Guarnieri, F. (1999). "A self-consistent, microenvironment modulated screened coulomb potential approximation to calculate pH-dependent electrostatic effects in proteins." *Biophys J* 77(1): 3-22.

Michielin, O. & Karplus, M. (2002). "Binding free energy differences in a TCR-peptide-MHC complex induced by a peptide mutation: a simulation analysis." *Journal of Molecular Biology* 324(3): 547-569.

Miyazawa, S. & Jernigan, R. L. (1996). "Residue-residue potentials with a favorable contact pair term and an unfavorable high packing density term, for simulation and threading." *J Mol Biol* 256(3): 623-644.

Nakamura, H.; Sakamoto, T. & Wada, A. (1988). "A theoretical study of the dielectric constant of protein." *Protein engineering* 2(3): 177.

Nasica-Labouze, J.; Meli, M.; Derreumaux, P.; Colombo, G. & Mousseau, N. (2011). "A multiscale approach to characterize the early aggregation steps of the amyloid-forming peptide GNNQQNY from the yeast prion sup-35." *PLoS Comput Biol* 7(5): e1002051.

Oldziej, S.; Czaplewski, C.; Liwo, A.; Chinchio, M.; Nanias, M.; Vila, J. A.; Khalili, M.; Arnautova, Y. A.; Jagielska, A.; Makowski, M.; Schafroth, H. D.; Kazmierkiewicz, R.; Ripoll, D. R.; Pillardy, J.; Saunders, J. A.; Kang, Y. K.; Gibson, K. D. & Scheraga, H. A. (2005). "Physics-based protein-structure prediction using a hierarchical protocol based on the UNRES force field: assessment in two blind tests." *Proc Natl Acad Sci U S A* 102(21): 7547-7552.

Onufriev, A.; Bashford, D. & Case, D. A. (2004). "Exploring protein native states and large-scale conformational changes with a modified generalized born model." *Proteins* 55(2): 383-394.

Pickersgill, R. W. (1988). "A rapid method of calculating charge-charge interaction energies in proteins." *Protein Eng* 2(3): 247-248.

Pitera, J. W.; Falta, M. & Van Gunsteren, W. F. (2001). "Dielectric properties of proteins from simulation: The effects of solvent, ligands, pH, and temperature." *Biophys J* 80(6): 2546-2555.

Posel, Z.; Lísal, M. & Brennan, J. K. (2009). "Interplay between microscopic and macroscopic phase separations in ternary polymer melts: Insight from mesoscale modelling." *Fluid Phase Equilibria* 283(1-2): 38-48.

Rastelli, G.; Rio, A. D.; Degliesposti, G. & Sgobba, M. (2010). "Fast and accurate predictions of binding free energies using MM]PBSA and MM]GBSA." *J Comput Chem* 31(4): 797-810.

Scheraga, H. A.; Maisuradze, G. G.; Senet, P.; Czaplewski, C. & Liwo, A. (2010). "Investigation of Protein Folding by Coarse-Grained Molecular Dynamics with the UNRES Force Field." *Journal of Physical Chemistry A* 114(13): 4471-4485.

Shaw, D. E.; Deneroff, M. M.; Dror, R. O.; Kuskin, J. S.; Larson, R. H.; Salmon, J. K.; Young, C.; Batson, B.; Bowers, K. J.; Chao, J. C.; Eastwood, M. P.; Gagliardo, J.; Grossman, J. P.; Ho, C. R.; Ierardi, D. J.; Kolossvary, I.; Klepeis, J. L.; Layman, T.; Mcleavey, C.; Moraes, M. A.; Mueller, R.; Priest, E. C.; Shan, Y. B.; Spengler, J.; Theobald, M.; Towles, B. & Wang, S. C. (2008). "Anton, a special-purpose machine for molecular dynamics simulation." *Communications of the Acm* 51(7): 91-97.

Shih, A. Y.; Arkhipov, A.; Freddolino, P. L. & Schulten, K. (2006). "Coarse grained protein-lipid model with application to lipoprotein particles." *J Phys Chem B* 110(8): 3674-3684.

Sorin, E. J.; Rhee, Y. M.; Nakatani, B. J. & Pande, V. S. (2003). "Insights into nucleic acid conformational dynamics from massively parallel stochastic simulations." *Biophys J* 85(2): 790-803.

Srinivasan, J.; Trevathan, M. W.; Beroza, P. & Case, D. A. (1999). "Application of a pairwise generalized Born model to proteins and nucleic acids: inclusion of salt effects." *Theoretical Chemistry Accounts: Theory, Computation, and Modeling (Theoretica Chimica Acta)* 101(6): 426-434.

Still, W. C.; Tempczyk, A.; Hawley, R. C. & Hendrickson, T. (1990). "Semianalytical treatment of solvation for molecular mechanics and dynamics." *Journal of the American Chemical Society* 112(16): 6127-6129.

Susukita, R.; Ebisuzaki, T.; Elmegreen, B. G.; Furusawa, H.; Kato, K.; Kawai, A.; Kobayashi, Y.; Koishi, T.; McNiven, G. D.; Narumi, T. & Yasuoka, K. (2003). "Hardware accelerator for molecular dynamics: MDGRAPE-2." *Computer Physics Communications* 155(2): 115-131.

Takahashi, T.; Nakamura, H. & Wada, A. (1992). "Electrostatic forces in two lysozymes: calculations and measurements of histidine pKa values." *Biopolymers* 32(8): 897-909.

Takahashi, T.; Sugiura, J. & Nagayama, K. (2002). "Comparison of all atom, continuum, and linear fitting empirical models for charge screening effect of aqueous medium surrounding a protein molecule." *Journal of Chemical Physics* 116(18): 8232-8237.

Taketomi, H.; Ueda, Y. & Go, N. (1975). "Studies on protein folding, unfolding and fluctuations by computer simulation. I. The effect of specific amino acid sequence represented by specific inter-unit interactions." *Int J Pept Protein Res* 7(6): 445-459.

Tidor, B. & Karplus, M. (1994). "The Contribution of Vibrational Entropy to Molecular Association:: The Dimerization of Insulin." *Journal of Molecular Biology* 238(3): 405-414.

Tsui, V. & Case, D. A. (2000). "Theory and applications of the generalized Born solvation model in macromolecular simulations." *Biopolymers* 56(4): 275-291.

Turjanski, A. G.; Gutkind, J. S.; Best, R. B. & Hummer, G. (2008). "Binding-induced folding of a natively unstructured transcription factor." *PLoS Comput Biol* 4(4): e1000060.

Van Der Spoel, D.; Lindah, E.; Hess, B.; Groenhof, G.; Mark, A.E. & Berendsen, H.J.C. (2005) "GROMACS: fast, flexible and free." J Comput Chem 26(16): 1701-1718

Wagoner, J. A. & Pande, V. S. (2011). "A smoothly decoupled particle interface: New methods for coupling explicit and implicit solvent." *J Chem Phys* 134: 214103.

Wan, S.; Coveney, P. V. & Flower, D. R. (2005). "Molecular basis of peptide recognition by the TCR: affinity differences calculated using large scale computing." *J Immunol* 175(3): 1715-1723.

Wan, S.; Flower, D. R. & Coveney, P. V. (2008). "Toward an atomistic understanding of the immune synapse: large-scale molecular dynamics simulation of a membrane-embedded TCR-pMHC-CD4 complex." *Mol Immunol* 45(5): 1221-1230.

Warshel, A. & Levitt, M. (1976). "Theoretical studies of enzymic reactions: dielectric, electrostatic and steric stabilization of the carbonium ion in the reaction of lysozyme." *Journal of Molecular Biology* 103(2): 227-249.

Warwicker, J. (1999). "Simplified methods for pKa and acid pH-dependent stability estimation in proteins: removing dielectric and counterion boundaries." *Protein science* 8(02): 418-425.

Wei, G. H.; Song, W.; Derreumaux, P. & Mousseau, N. (2008). "Self-assembly of amyloid-forming peptides by molecular dynamics simulations." *Frontiers in Bioscience* 13: 5681-5692.

Weiser, J.; Shenkin, P. S. & Still, W. C. (1999). "Approximate atomic surfaces from linear combinations of pairwise overlaps (LCPO)." *J Comput Chem* 20(2): 217-230.

Wucherpfennig, K. W.; Call, M. J.; Deng, L. & Mariuzza, R. (2009). "Structural alterations in peptide-MHC recognition by self-reactive T cell receptors." *Curr Opin Immunol* 21(6): 590-595.

Xu, Y.; Wang, C. & Shi, Y. (1992). "Improvements on the protein-dipole Langevin-dipole model." *J Comput Chem* 13(9): 1109-1113.

Xu, Y. & Wang, R. (2006). "A computational analysis of the binding affinities of FKBP12 inhibitors using the MM]PB/SA method." *PROTEINS: Structure, Function, and Bioinformatics* 64(4): 1058-1068.

Yaoita, T.; Isaki, T.; Masubuchi, Y.; Watanabe, H.; Ianniruberto, G.; Greco, F. & Marrucci, G. (2008). "Statics, linear, and nonlinear dynamics of entangled polystyrene melts simulated through the primitive chain network model." *J Chem Phys* 128(15): 154901.

Yasuda, S.; Yoshidome, T.; Oshima, H.; Kodama, R.; Harano, Y. & Kinoshita, M. (2010). "Effects of side chain packing on the formation of secondary structures in protein folding." *J Chem Phys* 132(6): 065105(065101-065110).

Yoshidome, T.; Ito, Y.; Ikeguchi, M. & Kinoshita, M. (2011). "Rotation Mechanism of F1-ATPase: Crucial Importance of the Water Entropy Effect." *Journal of the American Chemical Society* 133: 4030-4039.

Zacharias, M. (2003). "Protein-protein docking with a reduced protein model accounting for side chain flexibility." *Protein Sci* 12(6): 1271-1282.

Zacharias, M. & Fiorucci, S. (2010). "Binding site prediction and improved scoring during flexible protein-protein docking with ATTRACT." *Proteins-Structure Function and Bioinformatics* 78(15): 3131-3139.

Zhou, J.; Thorpe, I. F.; Izvekov, S. & Voth, G. A. (2007). "Coarse-grained peptide modeling using a systematic multiscale approach." *Biophys J* 92(12): 4289-4303.

Zhou, R. H. (2003). "Free energy landscape of protein folding in water: Explicit vs. implicit solvent." *Proteins-Structure Function and Genetics* 53(2): 148-161.

Zoete, V.; Irving, M. & Michielin, O. (2010). "MM-GBSA binding free energy decomposition and T cell receptor engineering." *Journal of Molecular Recognition* 23(2): 142-152.

Zoete, V. & Meuwly, M. (2006). "Importance of individual side chains for the stability of a protein fold: computational alanine scanning of the insulin monomer." *J Comput Chem* 27(15): 1843-1857.

Zoete, V. & Michielin, O. (2007). "Comparison between computational alanine scanning and per-residue binding free energy decomposition for protein-protein association using MM-GBSA: application to the TCR-p-MHC complex." *Proteins* 67(4): 1026-1047.

8

Essential Dynamics on Different Biological Systems: Fis Protein, tvMyb1 Transcriptional Factor and BACE1 Enzyme

Lucas J. Gutiérrez[1,2], Ricardo D. Enriz[1,2] and Héctor A. Baldoni[1,3]
[1]Área de Química General e Inorgánica
Universidad Nacional de San Luis (UNSL), San Luis
[2]Instituto Multidisciplinario de Investigaciones Biológicas de San Luis
(IMIBIO-SL, CONICET), San Luis
[3]Instituto de Matemática Aplicada San Luis (IMASL, CONICET), San Luis
Universidad Nacional de San Luis (UNSL)
Argentina

1. Introduction

Proteins and enzymes poses a non-covalent 3D structure and therefore their intrinsic flexibility allows the existence of an ensemble of different conformers which are separated by low-energy barriers. These ranges of available conformers for proteins in solution are due to the relative movements among the different domains. Domain motions are important for a variety of protein functions, including catalysis, regulation of activity, transport of metabolites, formation of protein assemblies, and cellular locomotion.

Considering the importance of these conformational changes it is obvious that the different techniques to evaluate these behaviours are very important in order to understand the biological effects. In the present chapter we report molecular dynamics (MD) trajectories analyzed by essential dynamics method on three different molecular systems of biological interest: i) DNA-bending protein Fis (Factor for Inversion Stimulation), ii) DNA-tvMyb1 (*Trichomonas vaginalis* transcriptional factor) and iii) the BACE1 (β–site amyloid cleaving enzyme 1). Although the general structural characteristics for the above systems are well known, comparatively little information is available about their flexibility and dynamics. This is in part due to difficulties with obtaining such information experimentally. Thus, our primary interest was the comparison between the unligated and the complexed state, because the corresponding conclusions may reveal motions of functional relevance.

2. Methodology

2.1 Molecular dynamics simulations

Twenty-nanoseconds MD simulations were performed for the three systems under study (Fis-protein, DNA- tvMyb1 protein and BACE1 enzyme) in order to relax and investigate the dynamical behaviour of these systems. All the simulations were performed by using the

Amber program (Case et al., 2008). The crystal structure for each system was obtained from the Protein Data Bank (PDB) and such structures were used as the initial model for the different dynamics simulations. The PDB entries were 3IV5, 2KDZ and 1M4H for the DNA-bending protein Fis, DNA-tvMyb1 and BACE1, respectively. An all atom force field was taken from FF99SBilnd (Lindorff-Larsen et al., 2010) for the protein and FF99csc0 (Perez et al., 2007) for the DNA.

Each system was soaked in a truncated octahedral periodic box of TIP3P water molecules. The distance between the edges of the water box and the closest atom of the solutes was at least 10Å. Counterions were added to neutralize the charge of the systems. The entire system was subject to energy minimization in two stages to remove bad contacts between the complex and the solvents molecules. First, the water molecules were minimized by holding the solute fixed with harmonic constraint of strength 100 kcal/molÅ^2. Second, conjugate gradient energy minimizations were performed repeatedly four times using positional restraints to all heavy atoms of the receptor with 15, 10, 5 and 0 kcal/molÅ^2. The system was then heated from 0 K to 300 K in 300 ps and equilibrated at 300 K for another 200 ps. After the minimization and heating, 20 ns dynamics simulations were performed at the NPT assemble (temperature of 300 K and pressure of 1 atm). During the minimization and MD simulations, particle mesh Ewald (PMD) method was employed to treat the long-range electrostatic interactions in a periodic boundary condition. The SHAKE method was used to constrain hydrogen atoms, allowing a time step for all MD is 2 fs. The direct space non bonded cut-off was of 8Å and initial velocities were assigned from a Maxwellian distribution at the initial temperature.

2.2 Essential dynamics

The essential dynamics (ED) method also called Principal Component Analysis (PCA) (Amadei et al., 1993), was used to extract the dimensional subspace in which all biologically relevant motions occur (the so-called essential subspace)(De Groot et al., 1996).

The ED method is based on the diagonalization of the covariance matrix built from atomic fluctuations in a trajectory from which the overall translation and rotations have been removed:

$$C_{ij} = \left\langle \left(X_i - X_{i,0}\right)\left(X_j - X_{j,0}\right)\right\rangle \tag{1}$$

In which X are the separate x, y, and z coordinates of the atoms fluctuating around their average positions X_0. ⟨...⟩ represent the average time over the entire trajectory. Here, to construct the protein covariance matrix we have used C_α atom trajectory. Indeed, it has been shown that the C_α atom contains all the information for a reasonable description of the protein large concerted motions (Amadei et al., 1993). Upon the covariance matrix diagonalization a set of eigenvalues and eigenvectors was obtained. The motions along a single eigenvector correspond to concerted fluctuations of atoms. On the other hand, the eigenvalues represent the total mean square fluctuation of the system along the corresponding eigenvectors.

2.3 Collective movements

To examine domain motions in a protein we calculated the cross-correlation (normalized covariance) matrix, C_{ij}, of the fluctuations of each x, y and z coordinates of the C_α atoms from their average during the last ten nanosecond of the simulation. The displacement vectors Δr_i and Δr_j of atoms i and j, the matrix, C_{ij} is given by (Ichiye & Karplus, 1991):

$$C_{ij} = \frac{\left\langle \Delta r_i \times \Delta r_j \right\rangle}{\left(\left\langle \Delta r_i^2 \right\rangle \left\langle \Delta r_j^2 \right\rangle \right)^{1/2}} \tag{2}$$

Where Δr_i is the displacement from the mean position of the i-th atom and the angle in brackets represent the average time over the entire trajectory. The elements of the cross-correlation matrix take values from -1 to 1. Positive values of C_{ij} represent a motion in the same direction between atoms i and j, and negative C_{ij} values represent a motion in the opposite direction. When C_{ij} is close to zero, the atomic motions are uncorrelated, and their movements are random compared to each other.

PCA was carried out using the PCAZIP software (Meyer et al., 2006). Geometrical analysis was performed using the ptraj module in AmberTools (Case et al., 2008).

2.4 Binding energy decomposition

The MM-GBSA method (Kollman et al., 2000) was applied to the last ten nanosecond of simulation and was used within the one-trajectory approximation. Briefly, the binding affinity for a complex corresponds to the free energy of association written as:

$$\Delta G_{bind} = G_{complex} - \left(G_{receptor} - G_{ligand} \right) \tag{3}$$

In MM-GBSA protocol, the binding affinity in equation (3) is typically calculated using

$$\Delta G = \Delta E_{MM} + \Delta G_{solv} - T\Delta S \tag{4}$$

Where ΔE_{MM} represents the change in molecular mechanics potential energy upon formation of the complex, calculated using all bonded and non bonded interactions. Solvation free energy penalty, ΔG_{solv}, is composed of the electrostatic component (G_{GB}) and a nonpolar component (G_{NP}):

$$\Delta G_{solv} = \Delta G_{GB} + \Delta G_{NP} \tag{5}$$

G_{GB} is the polar solvation contribution calculated by solving the GB equation. Dielectric constants of 1 and 80 were used for the interior and exterior, respectively.

The hydrophobic contribution to the solvation free energy, ΔG_{NP}, is estimated using the equation:

$$\Delta G_{NP} = \alpha \ SASA + \beta \tag{6}$$

Where SASA is the solvent-accessible surface area computed by means of the Linear Combination of Pairwise Overlap (LCPO) method (Onufriev et al., 2000) with a solvent probe radius of 1.4 Å. The surface tension proportionality constant α and the free energy of non polar solvation for a point solute β were set to its standard values, 0.00542 kcal/(mol·Å^2) and 0.92 kcal/mol, respectively (Sitkoff et al., 1994).

3. Results and discussion

3.1 Fis-protein

DNA-binding proteins can broadly be divided into those that recognize their DNA-binding sites with high sequence discrimination and those that bind DNA with little or no obvious sequence preference (Stefano et al., 2010). Examples of the latter class are the nucleoid-associated proteins Fis (Factor for Inversion Stimulation), IHF, UH, and H-NS in eubacteria, chromatin-associated proteins like the HMGB family, and histones in eukaryotes.

Fis protein participates in a wide array of cellular activities such as modulation of DNA topology during growth, (Schneider et al., 1997, 1999) regulation of certain site-specific DNA recombination events, (Betermier et al., 1989; Dorgai et al., 1993; Finkel & Johnson, 1992; Johnson et al., 1986; Kahmann et al., 1985; Weinreich & Reznikoff, 1992) and regulation of the transcription of a large number of genes during different stages of growth, (Kelly et al., 2004; Xu & Johnson, 1995) including ribosomal RNA and tRNA genes and genes involved in virulence and biofilm formation (Bosch et al., 1990; Ross et al., 1990; Falconi et al., 2001; Goldberg et al., 2001; Prosseda et al., 2004; Sheikh et al., 2001; Wilson et al., 2001). In addition, Fis protein can affect various biological processes involved in site-specific DNA recombination, DNA replication, or transcription (Drlica & Rouviere-Yaniv, 1987; Finkel & Johnson, 1992; Schmid, 1990). In some cases, two or more proteins may cooperate in the same process. For example, Fis and HU participate in Hin-mediated DNA recombination (Johnson et al., 1986) and Fis and IHF aid in promoting site-specific recombination of λ DNA (Ball & Johnson, 1991; Johnson et al., 1986; Schneider et al., 1997; Thompson et al., 1987). In other cases, these proteins can play opposing roles, as with Fis and H-NS on transcription of *hns* (Falconi et al., 1996) or IHF and Fis on transcription of *fis* (Pratt et al., 1997).

Regarding the structural aspects, Fis protein is a homodimer composed of two identical 98 amino acid subunits where each Fis subunit contains a β-hairpin (residues 11 to 26) followed by four α-helices (A, B, C, and D) separated by short turns, forming a helix–turn–helix (HTH, residues 74-95) DNA binding motif. Actually it is accepted that Fis protein is joined to non-specific DNA sequences.

The flexion produced by these proteins in the DNA helixes increment the phosphate groups that interacting at the protein flank. The bending of DNA chain is produced because there are two contacts regions between Fis protein and DNA. One is practically fixed during the bending process, while the other slides along the DNA chain. This causes DNA to bend and the twist motions of DNA helix allows the DNA base pair step motions. It appears that there are two different kinds of motions induced by the protein–DNA binding processes. One is the bending or unzipping proceeding along the DNA helical axis, and the other is mainly the base pair opening process which is in the direction orthogonal to the helical axis.

The deviation of the simulated dynamic from the crystal structure was monitored by the temporal evolution of the root-mean-square deviation (RMSD) of C_α atoms. This analysis provides a measure of the structural drift from the initial coordinates as well as the atomic fluctuation over the course of an MD simulation. Most large-scale changes in the overall

RMSD C_α ocurred within 7 ns. From this point, our results indicate that the molecular system is equilibrated and therefore we analysed the last 10 ns of simulation. The value obtained for the RMSD of unligated and complexed forms were 2.40(0.41) and 1.55(0.28) respectively, for the mean value and the standard deviation.

The local protein mobility was analyzed by calculating the average C_α β-factors of unligated and complexed Fis-protein and compared with those values previously reported for the crystal. The β-factors determine the atomic fluctuation in a protein giving information about the flexibility of such structure.

Regarding figure 1a it might be observed that the β-factors obtained for the unligated Fis protein are very similar to those from the experimental data except in determined regions. These regions comprise those amino acids involved in turns, loops and the α_1D y α_2D helixes.

Figure 1b gives the β-factors analysis for DNA bounded Fis-protein. It can be seen that residues 89-91 and 92-102 show a high mobility because they correspond to the terminal residues of the first subunit and the initial residues of the second one, respectively. With respect to the rest of the regions only small differences are observed and these slight differences might be due to the restricted motion in the crystal.

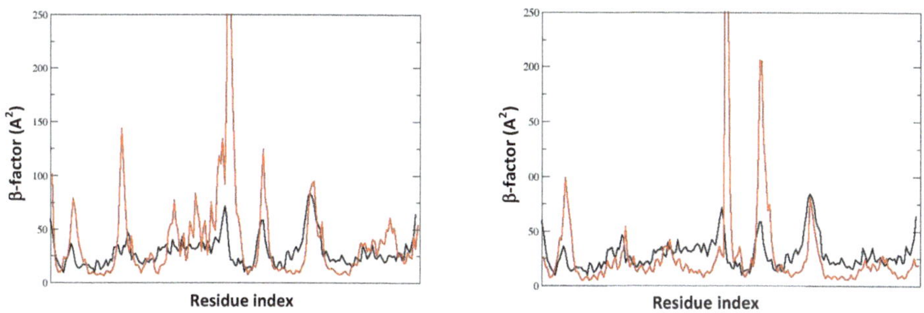

Fig. 1. Thermal factor analysis. MD (red line), experimental (black line).

To analyze the internal motions figure 2 displays the eigenvalues obtained from the diagonalization of the covariance matrix of the atomic fluctuations ranked in a decrement order. The two first eigenvalues represent about 58% and 60% of the collective movement for the unligated and complexed forms, respectively; whereas the last eigenvalues correspond to the small-amplitude vibrations.

In order to evaluate the collective movement, which is determined by the first eigenvector, we projected the last ten ns of the trajectory on the first two eigenvectors. From this resulting trajectory it is possible to calculate a new root-mean-square fluctuation (RMSF) for each residue in order to visualize which residues are the responsible for such movement. Figure 3 shows the RMSF obtained for each residue projected on the first eigenvector for the unligated and complexed forms of Fis protein. The HTH motifs in both subunities displayed a higher mobility at the unligated forms with respect to the complexed form of Fis protein. This is a logical behaviour considering that the HTH motifs are the regions for the binding to DNA. The presences of DNA makes the HTH motifs much more rigid and increase the mobility in the loops 40-49, 51-64 (subunit 1) and 58-70 (subunit 2) as well as in the helixes α_2B.

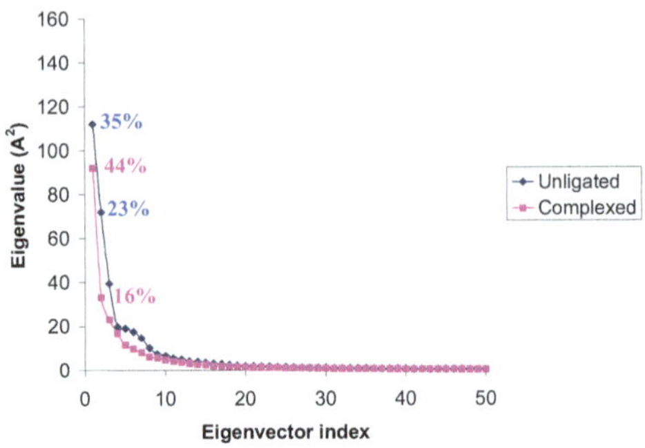

Fig. 2. Comparison between the eigenvalues with the corresponding eigenvector indices obtained from the Ca covariance matrix.

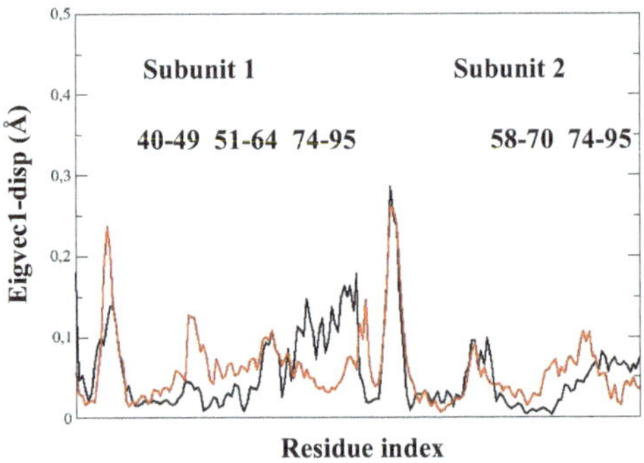

Fig. 3. Residue displacement in the subspace spanned by the first eigenvector. Unligated (black line) and complexed (red line).

Figure 4 gives the movement described the first eigenvector. The red porcupine needles indicate the direction of displacements of motions whereas the size of the needle is proportional to the amount of each displacement. This situation allows us to observe in which direction are moving the previously discussed domains. In short the α_1D and α_2D domain, which are the mainly responsible of the DNA recognition, are moving in opposite directions (negatively correlated). In addition the α_1D domain displayed a higher mobility in comparison to α_2D domain. Thus, we can argue that the DNA is gliding by one of the binding regions (HTH motifs whit high mobility), whereas the other binding region is keeping fixed during the flexion process of DNA (HTH with low mobility).

To study the interdomain motions of the residues that make up the dimmer, we examined the correlated maps for the unligated and complexed forms of Fis-protein. Figures 5 a and b display the Dynamics cross-correlation map (DCCM) among the C_α atoms calculated from the MD simulations for the complexed and unligated Fis-protein, respectively. In both cases, unligated and complexed form, the movements are negatively correlated indicating an expansion and contraction for the binding site located between the HTH motifs of each subunity. On the basis of our results it seems reasonable to argue that this movement is the responsible of DNA flexion.

Figure 6 gives the interaction binding energy decomposition by residue obtained for the complexed Fis protein. In this figure the strength of each interaction might be very well appreciated. Our results are in an almost complete agreement with those reported by Stefano et al. showing that Thr75, Asn73, Gln74, Arg76, Ile83, Asn84, Arg85, Thr87 and Arg89 makes the major contribution for the binding of DNA with Fis protein. It should be noted that only Lys90 is missing in our results in comparison to the experimental data (Stefano et al., 2010).

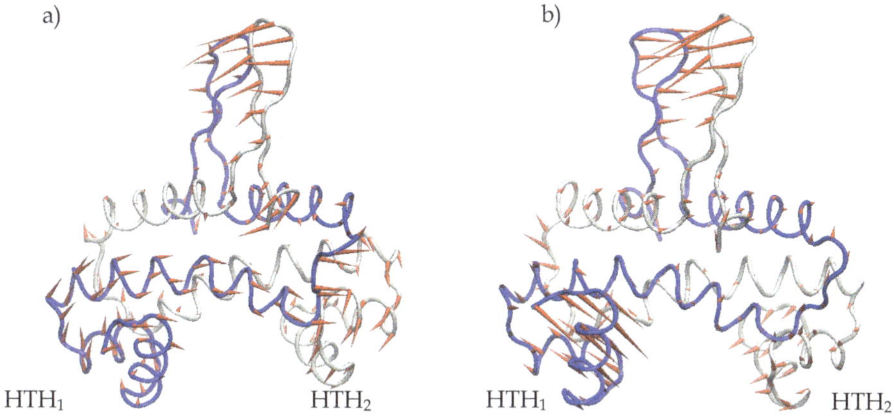

Fig. 4. Porcupine plots obtained for the unligated (a) and complexed (b) forms of Fis protein.

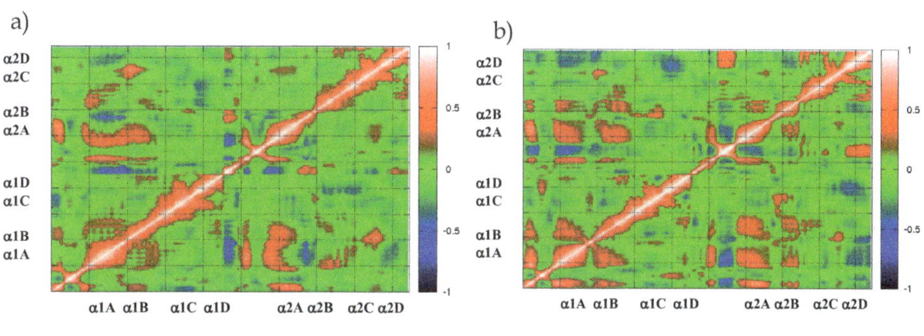

Fig. 5. Cross-correlation matrix obtained for the fluctuations of unligated (a) and complexed (b) forms of Fis protein.

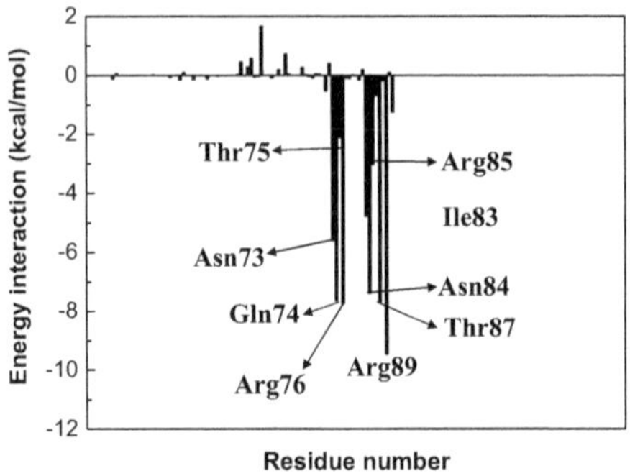

Fig. 6. DNA-Fis protein residue interaction spectrum. The y-axis denotes the interaction energy between the inhibitor and specific residues. The same pattern was observed in both subunities.

3.2 tvMyb1 transcriptional factor

The transcription regulator tvMyb1 was the first Myb family protein identified in Trichomonas vaginalis (*T. vaginalis*), a flagellated protozoan parasite of humans, causative agent of trichomoniasis, the most common non-viral sexually transmitted infection worldwide (WHO, 1995). T. vaginalis infection may cause adverse health consequences, including preterm abort and pelvic inflammatory disease in women, as well as infertility and increased incidence of human immunodeficiency virus transmission in women and men (Cotch et al, 1997; Laga et al., 1993; Martinez-Garcia et al., 1996; Moodley et al., 2002; Sherman et al., 1987; Sorvillo et al., 1998).

T. vaginalis as other pathogens requires iron for its metabolism. This cation is essential for its growth in the human vagina, where the iron concentration is constantly changing through the menstrual cycle. Iron also favours the adherence of the parasite to vaginal epithelial cells, metabolism and multiplication in culture (Gorrell, 1985; Lehker and Alderete, 1992). In spite of the high prevalence of trichomoniasis and the complications associated with the disease, *T. vaginalis* remains one of the most poorly studied parasites with respect to virulence properties and pathogenesis.

Myb proteins contains DNA-binding domains composed of one, two or three repeated motifs (called R1, R2 and R3 respectively) of approximately 50 amino acids, surrounded by three conserved tryptophan residues (Lipsick, 1996). These tryptophans play a role in the folding of the hydrophobic core of the Myb domains, and are generally conserved in all Myb protein. This hydrophobic core generates the helix-turn-helix (HTH) structure of the DNA-binding domain (Sakura et al., 1989). These HTH motifs recognize the major groove of the target DNA sequences.

Recently, Lou et al. reported the structural basis for the $tvMyb_{35-141}$/DNA interaction investigated using nuclear magnetic resonance, chemical shift perturbations, residual dipolar couplings, and DNA specificity (Lou, 2009). In addition, these authors showed that the $tvMyb_{35-141}$ fragment is the minimal DNA-binding domain encompassing two Myb-like DNA-binding motifs designated as R2 and R3 motifs. Both R2 and R3 motifs consists of three helices adopting a HTH conformations. Both motifs are connected by a long loop. The published experimental data indicates that the orientation between R2 and R3 motifs dramatically changes upon DNA binding through a number of key contacts involving residues in helices 3 and 6 to the DNA major groove.

A useful method monitoring the flexibility of a protein is the order parameter S^2 (Peter et al. 2001; Showalter et al. 2007). This normalized autocorrelation function related to the protein N-H bound vector were evaluated from the equilibrium averaged MD trajectories. S^2 gives a measure of the system flexibility, being 1 in a completely rigid system or zero for total flexibility where all the possible conformations are sampled. Figure 7 shows that the calculated S^2, during the last ten ns of simulation, for both complexed and unligated forms of the tvMyb1 protein. This figure is in agreement with those obtained from the NMR measurements (Lou et al., 2009). On the basis of such similitude we can argue that the simulated system follows the same trend to that observed from experimental data.

The S^2 obtained for the R2 (residues 40-80) domain are 0.74 and 0.79 for the complexed and unligated structures, whereas the S^2 obtained for the R3 doma(residues 92-135) in are 0.81 and 0.74 for the complexed and unligated forms, respectively. From these results it might be argue that the R2 of the unligated structure is more rigid than the R2 motif of the complexed form. In turn, the R3 domain of complexed structure is somewhat rigid with respect to the unligated form.

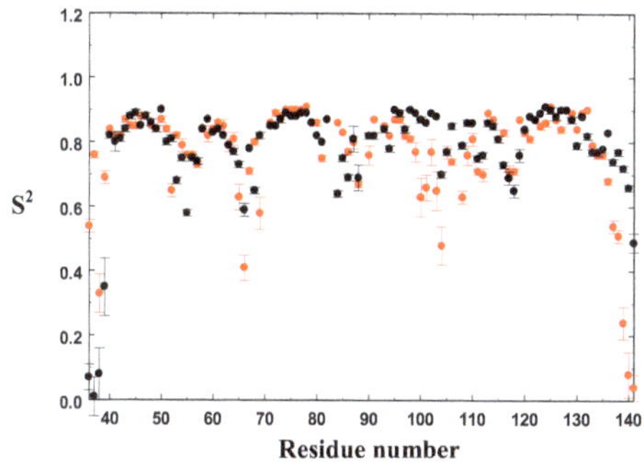

Fig. 7. tvMyb1 backbone N-H bond order parameter profiles from the last 10 ns of trajectory. Complexed form (red circles) and unligated form (black circles).

Figure 8 displays the eigenvalues obtained from the diagonalization of the covariance matrix of the atomic fluctuations ranked in a decrement order. The two first eigenvalues represent about 60% and 65% of the collective movement for the unligated and complexed forms, respectively.

In order to analyse which residues are the responsible of the different movements, we calculate the root-mean-square fluctuation (RMSF) from the equilibrated trajectory. Figure 9a shows the RMSF for each residue projected on the first eigenvector for the unligated and complexed forms of tvMyb1 protein. In this figure it is possible to observe that residues 83-90 on the unligated form show higher fluctuation with respect to that observed for the same residues at the complexed form, the same trend was observed for residues 117-130. In contrast, residues 96-101 display in the complexed form a higher fluctuation than that observed in unligated form.

Figure 9b shows the RMSF for each residue projected on the second eigenvector. In this figure it might be observed that there is not a significant difference in the R2 domain. However residues 79-86 in the unligated form displayed higher flexibility in comparison to the complexed form. On the other hand residue 93-96, 99-104 and 119-121 are more flexible in the complexed form with respect to the unligated form.

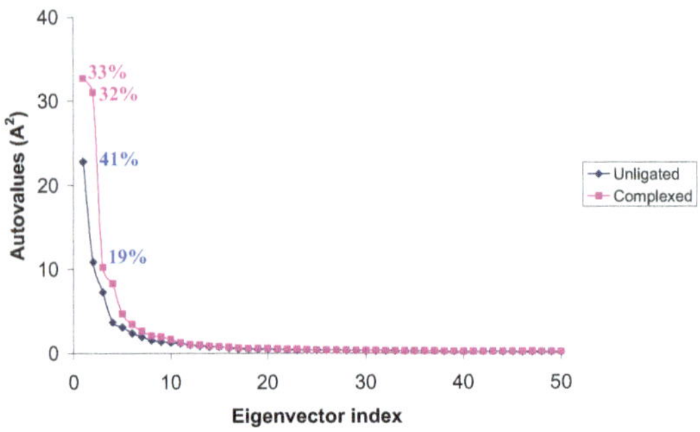

Fig. 8. Eigenvalues plotted against the corresponding eigenvector indices obtained from the Ca covariance matrix.

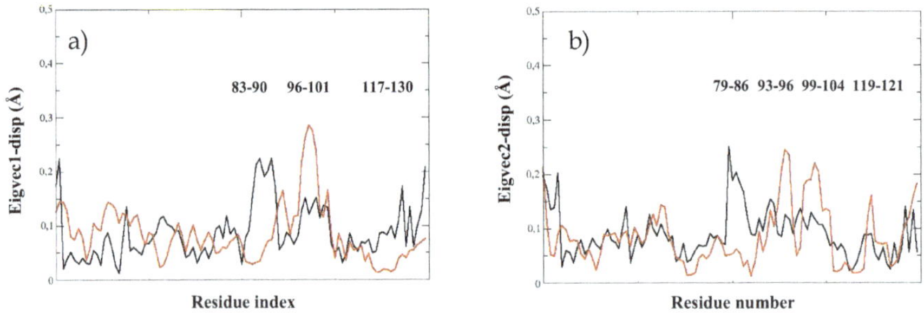

Fig. 9. Residue displacement in the subspace spanned by the first eigenvector (a) and the second eigenvector (b). Unligated and complexed forms are denoted in black and red lines, respectively.

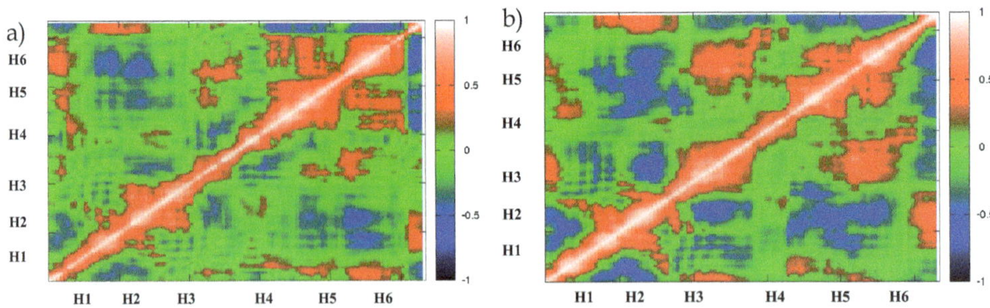

Fig. 10. Cross-correlation matrix obtained for the fluctuations of unligated (a) and complexed (b) forms of tvMyb1 protein.

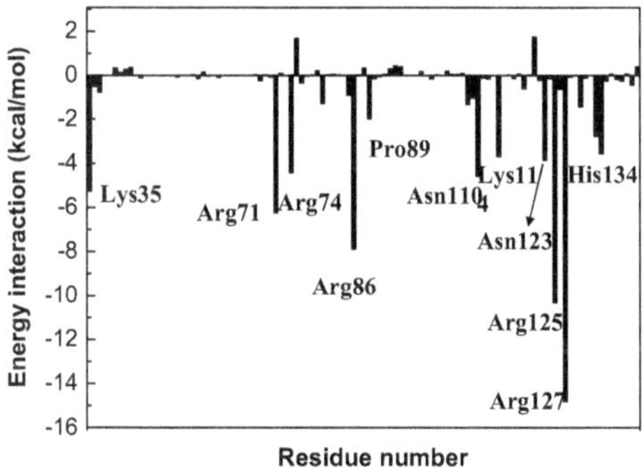

Fig. 11. DNA-tvMyb1 protein residue interaction spectrum. The y-axis denotes the interaction energy between the inhibitor and specific residues.

To better understand the relationship between R2 and R3 domains, we plotted the cross correlation maps for complexed and unligated forms of the tvMyb1 protein (figure 10). Figure 10a gives the cross correlation map obtained for the unligated form. In this figure we can observe that R2 domain (approximately residues 44-72) move in a negatively correlated direction with the R3 domain (approximately residue 97-134). Whereas the H4 and H5 (residues 96-112) move in a positively correlated direction in conjunction with H6 (residues 118-134) in the R3 domain. Figure 10b shows the same behaviour for the complexed form but it should be noted that in this case this movement is somewhat attenuated.

The interaction binding energy decomposition by residue, obtained for the complexed tvMyb1 protein, is show in Figure 11. Our results indicate that residues Lys35, Arg71, Arg74 (R2 domain) , residue Arg86 (loop L1) , and residues Pro89, Asn110, Lys114, Asn123, Arg125, Arg127 and His134 (R3 domain) makes the strongest interactions with the DNA molecule showing that our theoretical results are in agreement with the reported experimental data (Lou, 2009).

3.3 BACE1 enzyme

An estimated of 24 million people Worldwide have dementia, the majority of whom are though to have Alzheimer's disease. The two core pathological hallmarks of Alzheimer's disease are amyloid plaques and neurofibrillary tingles. The amyloid cascade hypothesis suggests that deposition of amyloid β (Aβ) triggers neuronal dysfunction and death in the brain.

$A\beta_{42}$ induces neuronal lipid peroxidation and protein oxidation in vivo and in vitro, possibly by generating radicals (Butterfield, 2001, 2003; Hensley et al., 1995) (78–80). Although the neurotoxicity of $A\beta_{42}$ is related to the generation of H_2O_2 (Behl et al., 1994) (81) the chemistry involved in generating the oxidation products via $A\beta_{42}$ remains unclear. The Aβ is generated from the amyloid precursor protein (APP) rupture by two proteases: β-site amyloid cleaving enzyme and γ-site amyloid cleaving enzyme (γ-secretase). Thus, β-secretace and γ-secretase are attractive targets for the treatment of AD.

β-secretase, also known as BACE1 (Vassar et al., 1999), memapsin-2 (Lin et al., 2000) and Asp 2 is a membrane-anchored member of the aspartyl protease family of hydrolytic enzymes (Lin et al., 2000). In addition to the active site, some proteolytic enzymes contain additional binding pockets, termed exosites, which engage substrates at locations distal to the active site (Krishnaswamy & Betz, 1997; Maun et al., 2003). These binding pockets on some proteolytic enzymes can act as allosteric regulators of the enzyme activity through conformational changes to the active site, in order to cause an augmentation or diminution of the enzyme's catalytic reactivity (Kornacker et al., 2005). While exosites were reported for serine and cysteine proteases, few examples of exosites have been reported for aspartyl proteases (Kornacker et al., 2005). Recently, we have reported the location for the exosite of BACE1 enzyme (Gutierrez et al., 2010). Considering that this exosite is structurally different to the catalytic site, the features of molecules occupying such exosite could be structurally different to the known mimetic peptides. Thus, molecules possessing the ability to bind to this exosite could be an alternative way to design new and selective inhibitors for the BACE1 enzyme. Here we report how is modified the dynamics behaviour of this enzyme by the presence of an inhibitor in the exosite.

The flexibility of the free BACE1 enzyme has already been reported by Caflisch et al. and by Chakraborty et al. and our simulations are in agreement with these previously reported results. Figures 12a and b gives the experimental and calculated β-factors obtained from the last 10 ns of simulation for free and complexed enzyme, respectively. In both Figures it is possible to appreciate the same general trend. However, in the complexed enzyme the mobility is reduced with respect to free enzyme, due to the presence of the inhibitor.

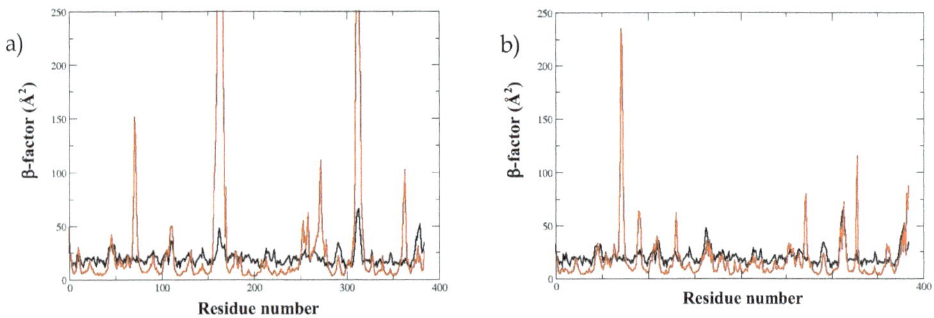

Fig. 12. Thermal factor analysis. MD simulations (red line), experimental data (black line).

In the free and complexed form of BACE1 enzyme it is possible to appreciate different regions with significant mobility. The most representatives are: the 10s loop (residues 9-14), located between two strands at the bottom of the S3 sub pocket; the β-harpin flap (residues 67-77); the A loop (residues 158-167); the F loop (residues 311-318) and D loop (residues 270-273). It is noteworthy to note that the 10s loop, β-harpin flap, A loop loop and F are part of the catalytic cavity.

Figure 13 shows the eigenvalues obtained from the diagonalization of the covariance matrix. Comparing the two models this plot indicates that the presence of the inhibitor in the complexed enzyme decreases the first two eigenvalues drastically with respect to the free form, in accordance with the results shown in Figure 1a) and 1b). The first two eigenvalues account for approximately 59 and 56% of the collective motion of the complexed and free for of the enzyme, respectively and therefore we focussed our study on these two eigenvectors.

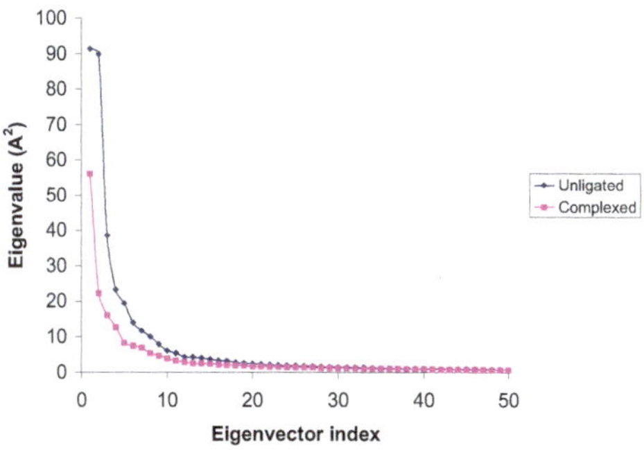

Fig. 13. Eigenvalues plotted against the corresponding eigenvector indices obtained from the Cα covariance matrix.

The motion along any eigenvector can be visualized by projecting all trajectory frames onto a specific eigenvector. Thus, from this new path we calculated the RMSF for the first two eigenvectors (Figure 14). The dynamic behaviour of both forms of the enzyme (i.e. free and complexed) differs in specific areas. In the free form of the enzyme a high mobility was observed for the β-harpin flap and loops A and F, while in the complexed form these regions displayed a similar behaviour. However, loops A and F displayed an attenuated flexibility with respect to the movement on the free enzyme. Another differential behaviour was found in residues 87-93 and 325-330. This behaviour may be relevant because it has been reported that residues 325-330 in the flap are responsible for regulating the ingresses and egress of the BACE1 substrate. (Chakraborty et al., 2011). The collective movement, in for free and complexed form of the enzyme, might be visualized from the porcupine plots in Figures 15 and 16. The collective movement represented by the eigenvectors could be analyzed in two parts a) movement between the N-terminal lobe (residues 1-150) and the C-terminal lobe (residues 151-385), and b) the movement on specific regions that are part of the catalytic cliff.

Figure 14 shows the representation of the first eigenvector for the free and complexed for of the enzyme. Figure 14 a) shows that the C-terminal lobe and N-terminal lobe move in opposite directions. In addition, we see that the β-harpin flap moves toward the catalytic cliff while loops A and F move away, allowing a hinge movement between the lobes. The concerted movements of the above mentioned regions are the ones that dominate the transition between open and closed conformation of BACE1 in agreement with those results reported by Chakraborty. In addition the first eigenvector for the complexed for of the enzyme, Figure 14b displayed an opening between the lobes where the C-terminal lobe moves away from to the N-terminal lobe. It is also possible to appreciate the movement of the β-harpin flap which moves in the opposite direction to the loop 325-330. In the free form of the enzyme, the described movement it is not evident. This finding is very important because these two loops are directly responsible for ingresses and egress of the substrate.

The complexed form of the enzyme (figure 15) displayed a dismissed movement due to the presence of the inhibitor. It is interesting to note that loops A, D and F, which forms part of the exosite, shows a lower movement with respect to the free enzyme due that the inhibitor makes strong interactions with the residues Gln163 (A loop), Trp270 (D loop), Asp311, Thr314 and Asp317 (loop F) in the complexed form of the enzyme as reported by Gutierrez et al. (Gutierrez et al., 2010).

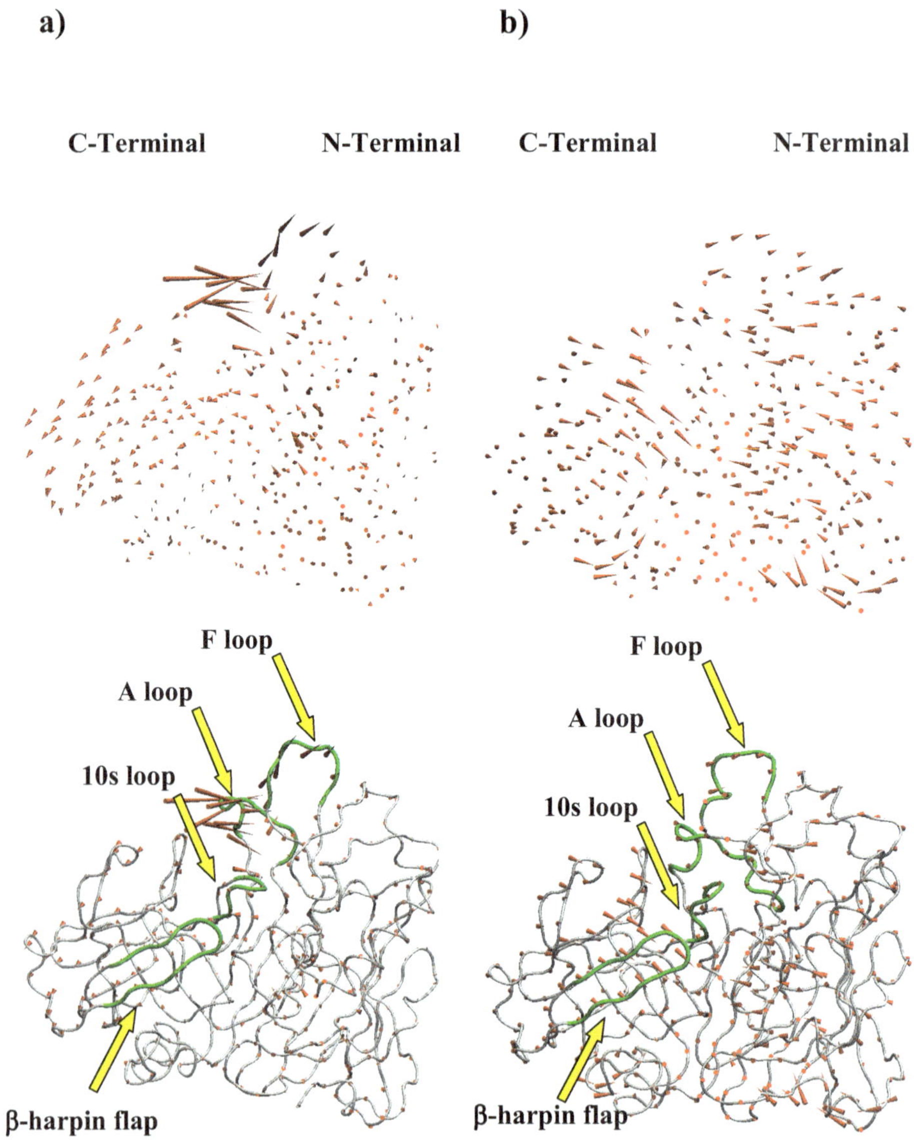

Fig. 14. Porcupine plots obtained for the first a) and second b) eigenvector of the unligated form of BACE1.

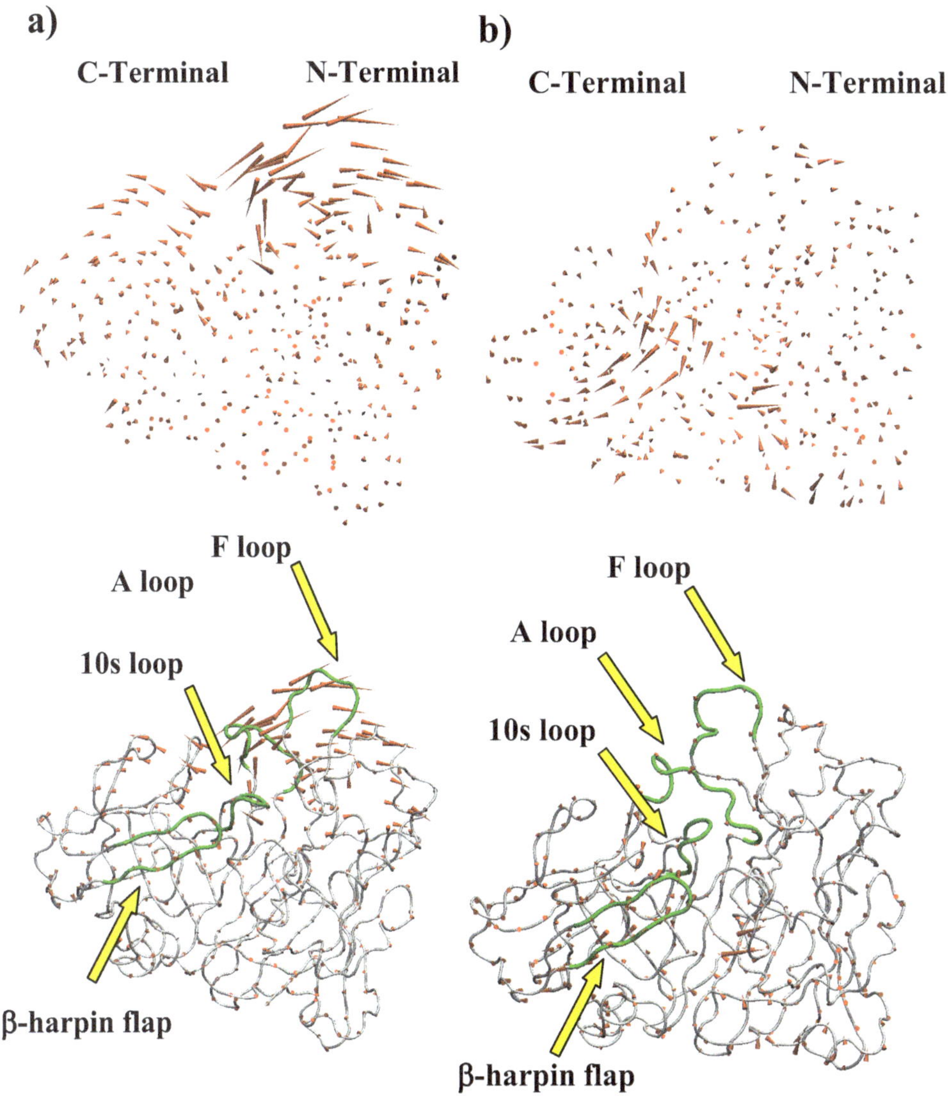

Fig. 15. Porcupine plots obtained for the first a) and second b) eigenvector of the complexed form of BACE1.

4. Conclusion

In the present chapter we reported MD simulations performed on three different molecular systems of biological interest: i) DNA-bending protein Fis (Factor for Inversion Stimulation), ii) DNA-tvMyb1 (*Trichomonas vaginalis* transcriptional factor) and iii) the BACE1 (β-site amyloid cleaving enzyme 1). The model structures proposed accounted for different experimental biological date for these systems indicating that the essential dynamics (ED) method also called Principal Component Analysis (PCA) is a very fruitful tool in order to evaluate the conformational behaviour of molecular systems like reported here. On the other hand, examining the cross-correlation (normalized covariance) matrix, we were able to obtain collective movements which allow examining domain motions for these models. The MM-GBSA methods in turn, allow us to determine binding hot-spots residues as well as the binding energy decomposition giving additional information which in general is very difficult to obtain from experimental data. It is clear that these MD simulations, if it is possible, must be corroborated with accurate experimental data in order to determine their real reaches and limitations. However, it is evident that these theoretical techniques properly applied are very useful and gives additional information to determine the conformational behaviour of complex biological systems.

5. Acknowledgments

This work was supported by Universidad Nacional de San Luis (UNSL), Instituto Multidisciplinario de Investigaciones Biológicas de San Luis (IMIBIO-SL, CONICET), Instituto de Matemática Aplicada San Luis (IMASL–CONICET) and Consejo Nacional de Investigaciones Científicas y Técnicas (CONICET–Argentina). L.J.G. gratefully acknowledges a CONICET fellowship; R.D.E. and H.A.B. are staff members of CONICET.

6. References

Amadei, A., Linssen, A.B.M., & Berendsen, H.J.C. (1993). Essential dynamics of proteins. *Proteins: Structure, Function and Genetics*, Vol.17, (1993), pp.412-425.

Ball, C. A., & Johnson, R. C. (1991). Efficient excision of phage lambda from the *Escherichia coli* chromosome requires the Fis protein. *Journal of bacteriology*, Vol.173, No. 13, (July 1991), pp.4027–4031.

Ball, C. A., & Johnson, R. C. (1991). Multiple effects of Fis on integration and the control of lysogeny in phage l. *Journal of bacteriology*, Vol.173, No.13, (July 1991), pp.4032–4038.

Behl, C., Davis, J. B., Lesley, R., & Schubert, D. (1994). Hydrogen peroxide mediates amyloid beta protein toxicity. Cell, Vol. 77, No. 6, (June 1994), pp. 817–827

Betermier, M., Lefrere, V., Koch, C., Alazard, R., & Chandler, M. (1989). The Escherichia coli protein, Fis: specific binding to the ends of phage Mu DNA and modulation of phage growth. *Molecular Microbiology*, Vol.3, No. 4, (April 1983), pp.459–468, ISSN 0950-382X

Bosch, L., Nilsson, L., Vijgenboom, E., & Verbeek, H. (1990). FIS-dependent trans-activation of tRNA and rRNA operons of Escherichia coli. *Biochimica et Biophysica Acta*, Vol.1050, No. 1, (August 1990), pp.293–301, ISSN:0006-3002

Butterfield, D. A., Drake, J., Pocernich, C., & Castegna, A. (2001). Evidence of oxidative damage in Alzheimer's disease brain: central role for amyloid β-peptide. *Trends in Molecular Medicine*, Vol.7, No. 12, (December 2001), pp.548–554 ISSN 1471-4914

Butterfield, D. A.: Amyloid β-peptide (1–42)-associated free radical-induced oxidative stress and neurodegeneration in Alzheimer's disease brain: mechanisms and consequences. (2003). *Current Medicinal Chemistry*, Vol. 10, No. 24, (December 2003), pp.2651–2659, ISSN 0929-8673

Case, D.A., Darden, T.A., Cheatham, T.E., Simmerling, C.L., Wang, J., Duke, R.E. Luo, R., Crowley, M., Walker, R.C.W., Zhang Merz, K.M., Wang, B., Hayik, S., Roitberg, A., Seabra, G., Kolossváry, I., Wong, K.F., Paesani, F., Vanicek, J., Wu, X., Brozell, S.R., Steinbrecher, T., Gohlke, H., Yang, L., Tan, C., Mongan, J., Hornak, V., Cui, G., Mathews. D.H., Seetin, M.G., Sagui, C., Babin, V., & Kollman, P.A. AMBER, University of California, San Francisco (2008).

Chakraborty, S., Kumar, S. & Basu, S. (2011). Conformational transition in the substrate binding domain of b-secretase exploited by NMA and its implication in inhibitor recognition: BACE1-myricetin a case study. *Neurochemistry International*, Vol. 58, No. 8, (2011), pp. 914-923, ISSN 0197-0186

Cotch, M. F., Pastorek, J. G., Nugent R. P., Hillier, S. L., Gibbs, R. S., Martin, D. H., Eschenbach, D. A., Edelman, R., Carey, J. C., Regan, J. A., Krohn, M. A., Klebanoff, M. A., Rao, A. V., & Rhoads, G. G. (1997). *Trichomonas vaginalis* associated with low birth weight and preterm delivery: the vaginal infections and prematurity study group. *Journal of the American Sexually Transmitted Diseases*, Vol. 24, No. 6, (July 1997), pp.353-60, ISSN 0148-5717

De Groot, B. L., Van Aalten, D.M.F., Amadei, A., & Berendsen, H.J.C. (1996). The Consistency of Large Concerted Motions in Proteins in Molecular Dynamics Simulations. *Biophysical Journal*, Vol.71, No. 4, (October 1996), pp.1707-1713, ISSN 00063495

Dorgai, L., Oberto, J., & Weisberg, R. A. (1993). Xis and Fis proteins prevent site-specific DNA inversion in lysogens of phage HK022. *Journal of Bacteriology*, Vol.175, No. 3, (Febraury 1993), pp.693–700.

Drlica, K., & Rouviere-Yaniv, J. (1987). Histonelike proteins of bacteria. *Microbiology Reviews*, Vol. 51, No. 3, (September 1987), pp.301-319.

Falconi, M., Brandi, A., Teana, A. L., Gualerzi, C. O., & Pon, C. L. (1996). Antagonistic involvement of FIS and H-NS proteins in the transcriptional control of *hns* expression. *Molecular Microbiology*, Vol.19, No. 5, (March 1996), pp.965–975, ISSN 0950-382X

Falconi, M., Prosseda, G., Giangrossi, M., Beghetto, E., & Colonna, B. (2001). Involvement of FIS in the H-NSmediated regulation of virF gene of Shigella and enteroinvasive Escherichia coli. *Molecular Microbiology*, Vol.42, No.2, (October 2001), pp.439–452, ISSN 0950-382X

Finkel, S. E., & Johnson, R. C. (1992). The Fis protein: it's not just for DNA inversion anymore. *Molecular Microbiology*, Vol.6, No. 22, (1992), pp.3257–3265, ISSN 0950-382X

Goldberg, M. D., Johnson, M., Hinton, J. C., & Williams, P. H. (2001). Role of the nucleoid-associated protein Fis in the regulation of virulence properties of enteropathogenic Escherichia coli. *Molecular Microbiology*, Vol.41, No. 3, (August (2001), pp.549–559, ISSN 0950-382X

Gonzalez-Gil, G., Bringmann, P., & Kahmann, R. (1996). FIS is a regulator of metabolism in Escherichia coli. *Molecular Microbiology*, Vol.22, No. 1, (October 1996), pp.21–29, ISSN 0950-382X

Gorrell, T. E. (1985). Effect of culture medium iron content on the biochemical composition and metabolism of Trichomonas vaginalis. *Journal of Bacteriology*, Vol.161, No. 3, (March 1985), pp.1228–1230.

Gutierrez, L. J., Enriz, R. D., & Baldoni, H. A. (2010). Structural and Thermodynamic Characteristics of the Exosite Binding Pocket on the Human BACE1: A Molecular Modeling Approach. *The Journal of Physical Chemistry A*, Vol. 114, (2010), pp. 10261-10269

Hampel, H., Shen, Y., Walsh, D. M., Aisen, P., Shaw, L. M., Zetterberg, H., Trojanowski, J. Q. & Blennow, K. (2009). Biological markers of amyloid β-related mechanisms in Alzheimer's disease. *Experimental Neurology*, Vol. 223, No. 3, (June 2010), pp.334-346, ISSN 0014-4886

Hensley, K., Hall, N., Subramaniam, R., Cole, P., Harris, M., Aksenov, M., Aksenova, M., Gabbita, S. P., Wu, J. F., Carney, J. M., Lovell, M., Markesbery, W. R., & Butterfield, D. A. (1995). Brain regional correspondence between Alzheimer's disease histopathology and biomarkers of protein oxidation. *Journal Neurochemistry*, Vol. 65, No. 5, (November 1995), pp.2146–2156, ISSN 1471-4159

Ichiye, T., & Karplus, M. (1991). Collective motions in proteins: a covariance analysis of atomic fluctuations in molecular dynamics and normal mode simulations. *Proteins: Structure, Function, and Bioinformatics*, Vol.11, No. 3, (November 1991), pp.205-217, ISSN 1097-0134

Johnson, R. C., Bruist, M. F., & Simon, M. I. (1986). Host protein requirements for *in vitro* site-specific DNA inversion. *Cell*, Vol.46, No.4, (1986) pp.531–539, ISSN 0092-8674

Kahmann, R., Rudt, F., Koch, C., & Mertens, G. (1985). G inversion in bacteriophage Mu DNA is stimulated by a site within the invertase gene and a host factor. *Cell*, Vol41, No. 3, (July 1985), pp.771–780.

Kelly, A., Goldberg, M. D., Carroll, R. K., Danino, V., Hinton, J. C., & Dorman, C. J. (2004). A global role for Fis in the transcriptional control of metabolism and type III secretion in Salmonella enterica serovar Typhimurium. *Microbiology*, Vol.150, No.7, (July 2004), pp.2037–2053.

Kollman, P. A., Massova, I. Reyes, C., Kuhn, B., Huo, S., Chong, L., Lee, M., Lee, T., Duan, Y., Wang, W., Oreola Donini, O., Cieplak, P., Srinivasan, J., Case, D. A. & Cheatham, T. E. (2000). Calculating Structures and Free Energies of Complex Molecules: Combining Molecular Mechanics and Continuum Models. *Accounts of Chemical Research*, Vol. 33, (April 2000), No. 12, pp. 889–897, ISSN 0001-4842

Kornacker, M. G., Copeland, R. A., Hendrick, J., Lai, Z., Mapelli, C., Witmer, M. R., Marcinkeviciene, J., Metzler, W., Lee, V., Riexinger, D. J. (2008). Beta secretase exosite binding peptides and methods for identifying beta secretase modulators. *US Patent 7314726*, Available from: http://www.freepatentsonline.com/7314726.html

Kornacker, M. G., Lai, Z., Witmer, M., Ma, J., Hendrick, J., Lee, V. G., Riexinger, D. J., Mapelli, C., Metzler, W., & Copeland, R. A. (2005). An inhibitor binding pocket distinct from the catalytic active site on human β-APP cleaving enzyme. *Biochemistry*, Vol. 44, No. 34, (August 1997), pp.11567-11573.

Krishnaswamy, S., & Betz, A. (1997). Exosites Determine Macromolecular Substrate Recognition by Prothrombinase. *Biochemistry*, Vol. 36, No. 40, (1997), pp.12080-12086

Laga, M., Manoka, A., Kivuvu, M., Malele, B., Tuliza, M., Nzila, N., Goeman, J., Behets, F., Batter, V., & Alary, M. (1993). Nonulcerative sexually transmitted diseases as risk factors for HIV-1 transmission in women: results from a cohort study. AIDS Vol.7, No. 1, (January 1993), pp.95-102, ISSN 0269-9370

Lehker, M. W., & Alderete, J. F. (1992). Iron regulates growth of Trichomonas vaginalis and the expression of immunogenic trichomonad proteins. *Molecular Microbiology*, Vol.6, No. 1, (January 1992), pp.123–132, ISSN 0950-382X

Lin, X., Koelsch, G., Wu, S., Downs, D., Dashti, A., & Tang, J. (2000). Human aspartic protease memapsin 2 cleaves the beta-secretase site of beta-amyloid precursor protein. *Proceedings of the National Academy of Sciences*, Vol. 97, No. 4, (February 2000), pp. 1456-1460, ISSN 1091-6490

Lindorff-Larsen, K., Piana, S., Palmo, K., Maragakis, P., Klepeis, J.L., Dror, R.O., & Shaw, D.E. (2010). Improved side-chain torsion potentials for the Amber ff99SB protein force field. *Proteins*, Vol.78, No.8, (March 2010), pp.1950-1958, ISSN 08873585

Lipsick, J. S. (1996) One billion years of Myb. *Oncogene*, Vol.13, No. 2, (July 1996), pp.223–235. ISSN 0950-9232

Lou, Y. C., Wei, S. Y., Rajasekaran, M., Chou, C. C., Hsu, H. M., Tai, J. H., & Chen, C. (2009) NMR structural analysis of DNA recognition by a novel Myb1 DNA-binding domain in the protozoan parasite Trichomonas vaginalis. *Nucleic Acids Research*, Vol.37, No. 7, (April 2009), pp.2381-2394.

Martinez-Garcia, F., Regadera, J., Mayer, R., Sanchez, S., & Nistal, M. (1996) Protozoan infections in the male genital tract. *The Journal of Urology*, Vol.156, No. 2, (August 1996), pp.340-349, ISSN 0022-5347

Maun, H. R., Eigenbrot, C., Lazarus RA. (2003). Engineering exosite peptides for complete inhibition of factor VIIa using a protease switch with substrate phage. Journal Biological Chemistry, Vol. 13, No. 24, (March 2003), pp. 21823-21830, ISSN 0021-9258

Meyer, T., Ferrer-Costa, C., Perez, A., Rueda, M., Bidon-Chanal, A., Luque, F. J., Laughton, C. A., & Orozco, M. (2006). Essential dynamics: a tool for efficient trajectory compression and management. *Journal of Chemical Theory and Computation*, Vol.2, No. 2, (February 2006), pp.251–258, ISSN 1549-9618

Moodley, P., Wilkinson, D., Connolly, C., Moodley, J., & Sturm, W. (2002). *Trichomonas vaginalis* is associated with pelvic inflammatory disease in women infected with HIV. *Clinical Infectious Diseases*, Vol.34, No. 4, (2002), pp.519-522.

Onufriev, A., Bashford, D., & Case, D. A. (2000). Modification of the generalized Born model suitable for macromolecules. *The Journal of Physical Chemistry B*, Vol. 104, No. 15, (March 2000), pp. 3712–3720.

Pérez, A., Marchán, I., Svozil, D. Sponer, J., Cheatham, T. E., Laughton, C. A. & Orozco, M. (2007). Refinement of the AMBER Force Field for Nucleic Acids: Improving the Description of □/□ Conformers. *Biophysical Journal*, Vol. 92., No. 11, (June 2007), pp. 3817-3829, ISSN 00063495

Peter, C., Daura, X., & van Gunsteren, W. F. (2001). Calculation of NMR-relaxation parameters for flexible molecules from molecular dynamics simulations. Journal of Biomolecular NMR, Vol.20, No. 4, (May 2001), pp.297–310.

Pratt, T. S., Steiner, T., Feldman, L. S., Walker, K. A., & Osuna, R. (1997). Deletion analysis of the *fis* promoter region in *Escherichia coli*: antagonistic effects of integration host factor and Fis. *J. Bacteriol.*, Vol.179, No. 20, (October 1997), pp.6367–6377.

Prosseda, G., Falconi, M., Giangrossi, M., Gualerzi, C. O., Micheli, G., Colonna, B. (2004). The virF promoter in Shigella: more than just a curved DNA stretch. *Molecular Microbiology*, Vol.51, No. 2, (January 2004), pp.523–537, ISSN 0950-382X

Ross, W., Thompson, J. F., Newlands, J. T., & Gourse, R. L. (1990). E. coli Fis protein activates ribosomal RNA transcription in vitro and in vivo. *EMBO Journal*, Vol.9, No. 11, (November 1990), pp.3733–3742.

Sakura, H., Kanei-Ishii, C., Nagase, T., Nakagoshi, H., Gonda, T. J., & Ishii, S. (1989). *Proc. Natl. Acad. Sci.*, Vol.86, pp.5758–5762.

Schmid, M. B. (1990). More than just "histone-like" proteins. *Cell* Vol.63, No. 3, (November 1990), pp.451–453.

Schneider, R., Travers, & A., Muskhelisvili, G. (1997). FIS modulates growth phase-dependent topological transitions of DNA in *Escherichia coli*. *Molecular Microbiology*, Vol.26, No.3, (1997), pp.519–530, ISSN 0950-382X

Schneider, R., Travers, A., & Muskhelishvili, G. (1997). Fis modulates growth phase-dependent topological transitions of DNA in Escherichia coli. *Molecular Microbiology,* Vol.26, No. 3, (October 1997), pp.519–530, ISSN 0950-382X

Schneider, R., Travers, A., Kutateladze, T., & Muskhelishvili, G. (1999). ADNA architectural protein copules cellular physiology and DNA topology in Escherichia coli. *Molecular Microbiology,* Vol.34, No. 5, (December 1999), pp.953–964, ISSN 0950-382X

Sheikh, J., Hicks, S., Dall'Agnol, M., Phillips, A. D., & Nataro, J. P. (2001). Roles for Fis and YafK in biofilm formation by enteroaggregative Escherichia coli. *Molecular Microbiology,* Vol.41, No. 5, (September 2001), pp.983–997, ISSN 0950-382X

Sherman, K. J., Daling, J. R., & Weiss, N. A. (1987). Sexually transmitted diseases and tubal infertility. *Journal of the American Sexually Transmitted Diseases,* Vol.14, No. 1, (1987), pp.12-6, ISSN 0148-5717

Showalter, S.A., & Brüschweiler, R. (2007). Validation of molecular dynamics simulations of biomolecules using NMR spin relaxation as benchmarks: Application to the AMBER99SB force field. Journal of Chemical Theory and Computation. Vol. 3, No 3, (January 2007). pp. 961-975.

Sitkoff, D., Sharp, K., & Honing, B. *J.* (1994). *The Journal of Physical Chemistry,* Vol. 98, No. 7, (February 1994), pp.1978–1988.

Sorvillo F., & Kerndt P. (1998). *Trichomonas vaginalis* and amplification of HIV-1 transmission. *The Lancet,* Vol.351, No. 9097, (January 1998), pp.213-4. ISSN 0140-6736

Stella, S., Cascio, D., & Johnson R. C. (2010). The shape of the Dna minor groove directs binding by the DNA-bending protein Fis. *Genes & Development,* Vol. 24, (August 2010), pp.814-826, ISSN 0890-9369/10.

Thompson, J. F., Moitoso de Vargas, L., Koch, C., Kahmann, R., & Landy, A. (1987). Cellular factors couple recombination with growth phase: characterization of a new component in the lambda site-specific recombination pathway. *Cell,* Vol.50, No. 11, (September 1987), pp.901–908.

Vassar, R., Bennett, B. D., Babu-Khan, S., Kahn, S., Mendiaz, E. A., Denis, P., Teplow, D. B., Ross, S., Amarante, P., Loeloff, R., Luo, Y., Fisher, S., Fuller, J., Edenson, S., Lile, J., Jarosinski, M. A., Biere, A. L., Curran, E., Burgess, T., Louis, J. C., Collins, F., Treanor, J., Rogers, G. & Citron, M. (1999). β-Secretase Cleavage of Alzheimer's Amyloid Precursor Protein by the Transmembrane Aspartic Protease BACE. *Science,* Vol. 286, No. 5440, (October 1999), pp. 735-741, ISSN 0036-8075

Weinreich, M. D., & Reznikoff, W. S. (1992). Fis plays a role in Tn5 and IS50 transposition. *Journal of Bacteriology,* Vol.174, No. 14, (July 1992), pp.4530–4537.

Wilson, R. L., Libby, S. J., Freet, A. M., Boddicker, J. D., Fahlen, T. F., & Jones, B. D. (2001). Fis, a DNA nucleoid-associated protein, is involved in Salmonella typhimurium SPI-1 invasion gene expression. *Molecular Microbiology,* Vol.39, No. 1, (February 2001), pp.79–88, ISSN 0950-382X

World Health Organization. 1995. An Overview of Selected Curable Sexually Transmitted Disease. In: World Health Organization (ed), Global program of AIDS. Geneva, Switzerland, pp. 2–27.

Xu, J., & Johnson, R. C. (1995). Identification of genes negatively regulated by Fis: Fis and RpoS comodulate growth-phase-dependent gene expression in Escherichia coli. *Journal of Bacteriology,* Vol.177, No. 4, (February 1995), pp.938–947.

9

MM-GB(PB)SA Calculations of Protein-Ligand Binding Free Energies

Joseph M. Hayes[1] and Georgios Archontis[2]
[1]*Institute of Organic & Pharmaceutical Chemistry*
National Hellenic Research Foundation, Athens
[2]*Department of Physics, University of Cyprus, Nicosia*
[1]*Greece*
[2]*Cyprus*

1. Introduction

The importance of computational chemistry in modern scientific research is well established. Continuous improvement in software and algorithms for the modeling of chemical interactions has transformed molecular modeling into a powerful tool for many current day research projects. From a medical perspective, one of the ultimate goals in computer-aided drug design (CADD) is the accurate prediction of ligand-binding affinities to a macromolecular target, which can facilitate and speed the routine identification of new candidates in early stage drug discovery projects (Gilson & Zhou, 2007; Hayes & Leonidas, 2010). In particular, structure-based modeling provides an efficient pathway towards exploiting known three-dimensional structural data in the design and proposal of new molecules for experimental evaluation. Docking calculations are now widely used in high-throughput virtual screening of structurally diverse molecules from available compound libraries/databases against specific targets. Once initial "hits" or "lead" molecules are identified (normally low μM inhibitors), modification of their chemical features in the "lead optimization" phase can improve their binding affinities and fine-tune other desirable drug-like properties. However, docking calculations currently have limited success beyond the lead identification stage, where more accurate lower-throughput computational methods are needed. In this regard, the Molecular Mechanics/Generalized Born Surface Area (MM-GBSA) and Molecular Mechanics/Poisson-Boltzmann Surface Area (MM-PBSA) methods calculate binding free energies using molecular mechanics (forcefields) and continuum (implicit) solvation models (Kollman et al., 2000).They have been successfully applied across a range of targets and are implemented in software programs such as Amber (Case et al., 2005), Delphi (Rocchia et al., 2001) and Schrödinger (Du et al., 2011). With a target readership from beginner to expert, the current chapter provides an extensive and critical overview of MM-GB(PB)SA methods and their applications. The theoretical foundation of the MM-GB(PB)SA method is first described. We then discuss key aspects which improve the accuracy of results, and highlight potential caveats due to the approximations inherent in the methods. The chapter concludes with a review of recent representative applications, which illustrate both successes and limitations. The emphasis of this chapter is on structure-based drug design (SBDD) efforts, however, the methods have widespread applicability in other areas such as in supramolecular chemistry.

2. Theoretical background – "Pathway" and "Endpoint" methods

Low-throughput computational approaches for the calculation of ligand binding free energies can be divided into "pathway" and "endpoint" methods (Deng & Roux, 2009; Gilson & Zhou, 2007). In pathway methods, the system is converted from one state (e.g., the complex) to the other (e.g., the unbound protein/ligand). This can be achieved by introducing a set of finite or infinitesimal "alchemical" changes to the energy function (the Hamiltonian) of the system through free-energy perturbation (FEP) or thermodynamic integration (TI), respectively (Kollman, 1993; Straatsma & McCammon, 1992). The fundamentals of FEP and TI methods were introduced many decades ago by John Kirkwood (Kirkwood, 1935) and Robert Zwanzig (Zwanzig, 1954). In recent years, their use in the computation of absolute binding affinities has become feasible due to increases in computational power, the development of more accurate models of atomic interactions (Cornell et al., 1995; MacKerell et al., 1998; van Gunsteren et al., 1996), the clarification of the underlying theoretical framework and the introduction of methodological advances (Boresch et al., 2003; Bowers et al., 2006; Deng and Roux, 2009; Gilson et al., 1997; Gilson & Zhou, 2007; Lee & Olson, 2006; Mobley et al., 2007). Combined with atomistic molecular dynamics (MD) or Monte Carlo (MC) simulations in explicit water solvent models, they are arguably the most accurate methods for calculating absolute or relative ligand binding affinities.

The "alchemical" computation of *differences* in binding affinities (rather than *absolute* affinities) among a set of related ligands for the same target protein is more accurate and technically simpler. A thermodynamic cycle illustrating the basic principles is shown in Figure 1 (Tembe & McCammon, 1984). The horizontal legs describe the experimentally accessible actual binding processes, with free energies $\Delta G_{bind}(L1)$ and $\Delta G_{bind}(L2)$. Since the free energy is a state function, the relative binding free energy $\Delta\Delta G_{bind}$ is exactly equal to the difference of the free energies in the horizontal or vertical legs:

$$\Delta\Delta G_{\text{bind}} = \Delta G_{\text{bind}}(L2) - \Delta G_{\text{bind}}(L1) \qquad (1)$$

$$= \Delta G_{\text{complex}}(L1 \rightarrow L2) - \Delta G_{\text{free}}(L1 \rightarrow L2) \qquad (2)$$

The simulations follow the vertical steps (Eq. (2)) or unphysical processes, by simulations in water solution that gradually change the energy-function of the system from one "endpoint" to the other through a series of intermediate hybrid states.From Figure 1, this involves the stepwise "alchemical" transformation of ligand L1 to L2 both in its 'free' state (unbound) and in the bound complex, through gradual changes in the forcefield parameters describing the ligand interactions. This leads to the free energy changes $\Delta G_{free}(L1 \rightarrow L2)$ and $\Delta G_{complex}(L1 \rightarrow L2)$, respectively. Averaging over both transformation directions is often used to improve the free-energy estimates, although this is not always the case (Lu & Woolf, 2007). These calculations can be accurate, if conducted with the appropriate care. An overview of current state-of-the-art methods for absolute and relative affinity calculations is in (Chodera et al., 2011).

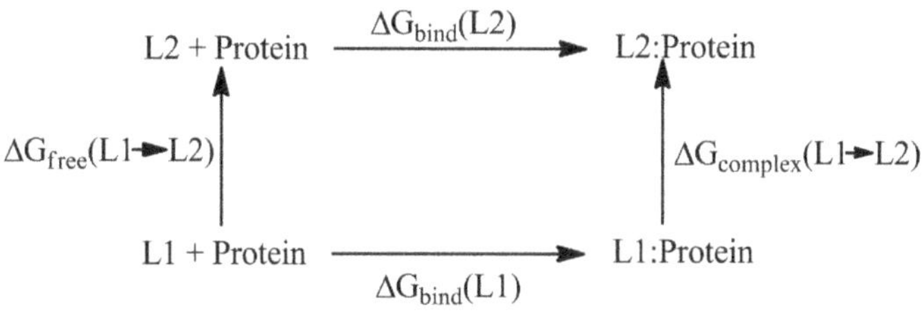

Fig. 1. Thermodynamic cycle linking the binding of two ligands L1 and L2 to a protein in solution.

To conclude, the emerging implementation of biomolecular codes on GPU architectures (Harvey et al., 2009; Stone et al., 2011) and the development of simple free-energy protocols (Boresch & Bruckner, 2011) make atomistic methods of absolute or relative affinities very promising for larger-scale calculations in the near future. Nevertheless, at present they are still relatively time-consuming, and require considerable expertise and planning. They preclude the consideration of more than a few complexes per day on a dedicated CPU cluster with a few tens of nodes. A trade-off between computational expense and accuracy is therefore required when the goal is to investigate and compare the binding strengths of a structurally diverse and/or larger set of ligands via MD simulations. For this purpose, much less computationally demanding „endpoint" methods are often successfully applied, such as the „linear interaction energy" (LIE) (Åqvist et al., 2002) or the molecular mechanics - Poisson Boltzmann (MM-PBSA) (Kollman et al., 2000) and the related molecular mechanics - generalised Born (MM-GBSA) approximation (Gohlke et al., 2003). All these methods compute binding free energies along the horizontal legs of Figure 1, but use only models for the "endpoints" (bound and unbound states). The MM-PB(GB)SA methodology is now described in more detail.

3. The MM-GB(PB)SA methodology

Using MM-GBSA and MM-PBSA methods, relative binding affinities for a set of ligands to a given target can often be reproduced with good accuracy and considerable less computational effort compared to full-scale molecular dynamics FEP/TI simulations. Furthermore, free-energies can be decomposed into insightful interaction and desolvation components (Archontis et al., 2001; Hayes et al., 2011; Polydoridis et al., 2007).

3.1 Thermodynamics & calculation framework

In the MM-GB(PB)SA formulation, the binding free energy of a ligand (L) to a protein (P) to form the complex (PL) is obtained as the difference (Pearlman, 2005):

$$\Delta G_{\text{bind}} = G(PL) - G(P) - G(L) \tag{3}$$

The free energy of each of the three molecular systems P, L, and PL is given by the expression:

$$G(X) = E_{MM}(X) + G_{\text{solv}}(X) - TS(X) \tag{4}$$

In Eq. (4), E_{MM} is the total molecular mechanics energy of molecular system X in the gas phase, G_{solv} is a correction term (solvation free energy) accounting for the fact that X is surrounded by solvent, and S is the entropy of X.

To apply the MM-GB(PB)SA formulation, a representative set of equilibrium conformations for the complex, free protein and free ligand are first obtained by atomistic MD simulations in explicit solvent. In this post-processing phase, the solvent is discarded and replaced by a dielectric continuum. Changes (Δ) in the individual terms (ΔE_{MM}, ΔG_{solv}, $-T\Delta S$) of Eq.(4) between the unbound states and the bound (complex) state are calculated, and contribute to the binding free energies according to Eq.(3). Computation of each of the terms in Eq. (4) is now described in more detail.

E_{MM} is the sum of the bonded (internal), and non-bonded electrostatic and van der Waals energies

$$E_{\text{MM}} = E_{\text{bonded}} + E_{\text{elec}} + E_{\text{vw}} \tag{5}$$

These energy contributions are computed from the atomic coordinates of the protein, ligand and complex using the (gas phase) molecular mechanics energy function (or forcefield). The solvation free energy term G_{solv} contains both polar and non-polar contributions. The polar contributions are accounted for by the generalized Born, Poisson, or Poisson-Boltzmann model, and the non-polar are assumed proportional to the solvent-accessible surface area (SASA), *c.f.* Section 3.2:

$$G_{\text{solv}} = G_{\text{PB(GB)}} + G_{SASA} \tag{6}$$

Finally, the entropy S is decomposed into translational, rotational and vibrational contributions. The first two are computed by standard statistical-mechanical expressions, and the last is typically estimated from a normal-mode (harmonic or quasiharmonic) analysis (Brooks et al., 1995; Karplus & Kushick 1981; Tidor & Karplus 1994). In practice, current software implementations normally determine all three contributions to S as part of a normal-mode analysis.

To improve the accuracy of the computed binding free energies, the various terms of Eq. (4) are averaged over multiple conformations or MD snapshots (typically a few hundred for the E_{MM} and G_{solv} contributions). Depending on the extent of conformational fluctuations in the system under consideration, the convergence into stable values may require relatively long (multi-ns) simulations. The computation of the entropy term, however, requires the extensive minimization of the trajectory conformations for the protein, ligand and complex to local minima on the potential energy surfaces, followed then by normal mode analysis. This procedure is costly and prevents the consideration of a large number of conformations; insufficient sampling can therefore sometimes an issue. To decrease the computational cost, the protein can be truncated beyond a certain cutoff distance and the system minimized using a distance-dependent dielectric, which simulates the deleted surroundings (Kongsted & Ryde, 2009). However, a large variation of the entropy term often results from these 'free' minimizations. Including a fixed buffer region (with water molecules) beyond the cut-off can lead to more stable entropy predictions (Kongsted & Ryde, 2009).

The internal energy terms (E_{bonded}) of the protein and complex can be on the order of a few thousand kcal/mol, and can introduce large uncertainties in the computed binding free energies. This is prevented in the "single-trajectory" approximation (Gohlke & Case, 2004; Page & Bates, 2006), which employs simulations of a single state (the complex) to generate conformations for all three states (complex, protein and ligand). For each MD conformation sampled, the resulting internal energy terms of the protein and ligand are identical in the bound and the unbound states and cancel exactly in Eq. (3). Hence, effectively only the protein-ligand (non-bonded) interaction energies of the E_{MM} term in Eq. (5) contribute to ΔG_{bind}. 'Single-trajectory' simulations significantly reduce computational effort and are generally sufficiently accurate for most applications. The downside of the approximation is that any explicit structural relaxation of the protein and ligand upon binding is ignored. Although charge reorganization can be partly taken into account implicitly by setting the protein/ligand (internal) dielectric constants (Section 3.2) to values larger than ε_{in} = 1-2 (Archontis et al., 2001; Archontis & Simonson, 2005; Schutz & Warshel, 2001; Simonson, 2003), the neglect of explicit structural relaxation may introduce errors depending on the system (Tamamis et al., 2010). Separate MD simulations for the complex, and unbound receptor and ligands may also be performed (the "three-trajectory" approximation) but require greater computational effort, although in theory should yield more accurate results. Indeed, Yang and co-workers have recently shown that including separate simulations for the ligand and accounting for the "ligand reorganization" free energy led to significant improvements in binding affinity predictions for a set of ligands targeting XIAP (Yang et al., 2009). In certain cases, therefore, the added expense of separate simulations may be justified.

3.2 Solvation models – GBSA and PBSA

Proteins function usually inside aqueous solutions or in membrane environments, which are in the vicinity of an aqueous medium. The surrounding solvent can influence protein stability and function, ligand binding and protein-protein association. Since the solvent modifies in a non-trivial manner the intramolecular and intermolecular interactions, an accurate inclusion of solvent effects in biomolecular modeling and simulation is a challenging task.

Currently, the most accurate treatment in molecular simulations is achieved by atomic-detail models that represent explicitly the biomolecule and its surrounding environment. Several water models are used successfully to describe the aqueous environment in atomistic simulations; examples include SPC (Berendsen et al., 1981), SPC/E (Berendsen et al., 1987), TIP3P and TIP4P (Jorgensen et al., 1983), and TIP5P (Mahoney & Jorgensen, 2000). In practice, the explicit inclusion of water leads to a considerable increase in both the size of the simulation system and the computational cost of the simulation itself. Furthermore, the computation of solvation or binding free energies requires an exhaustive sampling of the solvent degrees of freedom. A much less costly approach is to represent the solvent implicitly in the simulation, through the incorporation of additional "potential of mean force" terms (Roux, 2001; Roux & Simonson, 1999) in the gas-phase energy function (e.g., Eq. (7) below). These terms depend only on the atomic coordinates of the solute, and express the solute free energy for a given configuration, after the solvent degrees of freedom have been "integrated out" (Roux, 2001; Roux & Simonson, 1999). Thus, the simulation system has the same number of degrees of freedom as in the gas phase and there is no need for explicit sampling over solvent degrees of freedom. The MM-PB(GB)SA method considered here, combine atomistic simulations in explicit solvent for the generation of representative biomolecular conformations with an implicit-solvent estimation of the binding free energies, in a post-processing step.

Conceptually, most implicit solvent models decompose the solvation process into three sequential steps (Cramer & Truhlar, 1999): i) creation of a cavity in solution to accommodate the biomolecule; ii) switching-on dispersion interactions between the biomolecule and surrounding medium, while all atomic charges are set to zero; and iii) switching-on the biomolecular charges. The solvation free energies of steps i) and ii) are normally assumed to be proportional to the SASA of the biomolecule and represent the *non-polar* contributions (G_{SASA}) to G_{solv} in Eq. (6), although the validity of this approximation has been questioned for step ii) (Levy et al., 2003). With a positive coefficient of proportionality, an increase in the SASA is associated with an unfavorable increase in solvation free energy, which is partly accounted for by the tendency of non-polar residues to be solvent-excluded. The equation typically used is of the form:

$$G_{SASA}(X) = \gamma.SASA + \beta \tag{7}$$

with the γ and β parameter values dependant on the method and solvation model (PBSA or GBSA) used (Rastelli et al., 2010).

Meanwhile, step iii) calculates the contribution to solvation free energy due to the charge /electrostatic interactions of the solute with the surrounding solvent, the *polar* contributions ($G_{PB(GB)}$) to G_{solv} in Eq. (6). In continuum-electrostatics models such as PB and GB, the solute is treated as a low-dielectric cavity embedded in a high dielectric medium. The solute charges are in the simplest and most common approximation centered on the individual atoms. The resulting solvation free energy of a molecule X is expressed as (Simonson, 2003):

$$G_{\mathrm{PB(GB)}}(X) = \frac{1}{2} \sum_{i,j \in X} q_i q_j g_{ij}{}^{PB(GB)} \tag{8}$$

where the summation is over all the atomic charges $\{q_i\}$. The quantity $g_{ij}{}^{PB(GB)}$ is determined using the PB model by numerical solution of the Poisson or Poisson-Boltzmann equation (depending on the existence of salt), or using the GB model by an analytical expression with the functional form (Simonson, 2003; Still et al., 1990):

$$g_{ij}{}^{GB} = \left(\frac{1}{\varepsilon} - 1\right)\left[r_{ij}^{n} + B_{ij}\exp\left(-\frac{r_{ij}^{n}}{AB_{ij}}\right)\right]^{-1/n} \tag{9}$$

The parameters B_{ij} depend on the position (distance from the solute-solvent dielectric boundary) of atoms *i* and *j*, and the shape of the entire biomolecule; ε is the solvent dielectric constant, and r_{ij} is the distance between *i* and *j*. The constants n and A were set to n=2 and A=4 in the original formulation of Still and coworkers (Still et al., 1990).

In the PB model, the solute dielectric constant (ε_{in}) affects the computed functions $g_{ij}{}^{PB}$ and Eq. (8). Meanwhile, in the GB model, the solute dielectric constant drops out from the final expression in Eq.(9), due to the approximations used to arrive at an analytic formula. An ε_{in} value other than 1 can still be used; in this case, the first parenthesis on the right-hand side of Eq. (9) becomes ($1/\varepsilon - 1/\varepsilon_{in}$) and the GB expression yields the free energy of transferring the solute from an infinite reference medium with dielectric constant ε_{in} into solution. A careful discussion of this point is in (Bashford & Case, 2000).

Application of Eq. (8) to a protein:ligand complex (PL) and the dissociated protein (P) and ligand (L) yields the electrostatic (polar) solvation free energy contribution to Eq. (3):

$$\Delta G_{\mathrm{PB(GB)}} = G_{\mathrm{PB(GB)}}(PL) - G_{\mathrm{PB(GB)}}(P) - G_{\mathrm{PB(GB)}}(L) \tag{10}$$

An advantage of these methods is that they facilitate the decomposition of the total solvation free energy into insightful components. Indeed, the summation over atomic charges in Eq. (8) implies that the electrostatic free energy of Eq. (10) can be partitioned into interaction and desolvation components (Archontis et al., 2001; Hendsch & Tidor, 1999):

$$\Delta G_{\mathrm{PB(GB)}} = \sum_{i\in P, j\in L} q_i g_{ij}^{PL} q_j + \left[\frac{1}{2}\sum_{i,j\in P} q_i g_{ij}^{PL} q_j - \frac{1}{2}\sum_{i,j\in P} q_i g_{ij}^{P} q_j\right] + \left[\frac{1}{2}\sum_{i,j\in L} q_i g_{ij}^{PL} q_j - \frac{1}{2}\sum_{i,j\in P} q_i g_{ij}^{L} q_j\right] \tag{11}$$

The notation g_{ij}^X implies that the interaction between charges *i* and *j*, as determined by the function g_{ij}, is in general different in the complex, free protein and free ligand. The first term on the right-hand side of Eq. (11) is the "interaction term" and arises from direct interactions between the protein and ligand charges, which are only present in the complex; the next term [the first brace] is a protein "desolvation" term, arising from the replacement of high-dielectric solvent by the low-dielectric ligand in the protein vicinity, as well as structural relaxation and changes in the charge distribution of the protein. The last term [second brace] corresponds to the desolvation of the ligand.

The "interaction term" can be further decomposed into contributions from specific residues (Archontis et al., 2001; Hendsch & Tidor 1999). For example, the contribution of a protein residue R to this term is given by the following expression

$$\Delta G_{\mathrm{PB(GB)}}^{R} = \sum_{i\in R, j\in L} q_i g_{ij}^{PL} q_j \tag{12}$$

These components provide useful insights on the origin of the binding free energy values. They can help interpret differences in binding affinities for a series of related complexes and guide the design of modified proteins or ligands. For example, in a study of amino acid binding to native and mutant aspartyl-tRNA synthetase, residue decomposition identified the protein residues discriminating between the cognate ligand (aspartic acid) and the analogue asparagine (Archontis et al., 2001). In another study of RNAse A recognition by dinucleotidic inhibitors, a similar decomposition attributed the stronger binding of the most potent inhibitor to interactions with two active site lysines (Polydoridis et al., 2007).

Finally, Hou and co-workers very recently evaluated the performance of MM-GBSA and MM-PBSA for predicting binding free energies based on molecular dynamics simulations (Hou et al., 2011a). Their results showed that MM-PBSA performed better in calculating absolute binding free energies compared to MM-GBSA but not necessarily for the relative binding free energies, sufficient for most applications in computational drug design. Interestingly, in a study of the accuracy of continuum solvation models for drug-like molecules, GB methods typically were more stable and gave more accurate results that the widely used PB methods (Kongsted et al., 2009).

3.3 Selection of MD trajectory conformations/snapshots

A recent study for the binding of seven biotin analogues to avidin suggested that to obtain statistically converged MM-GBSA results, several independent simulations each with sampling times of 20-200 ps (averaging the results) is more effective than a single long simulation (Genheden & Ryde, 2010). 'Single-trajectory' simulations of the complex are generally sufficiently accurate for most applications, and while MD simulation length does have an obvious impact on the accuracy of predictions, longer MD simulations doesn't necessarily mean better predictions (Hou et al., 2011a). For the calculations of the ΔE_{MM} and ΔG_{solv} terms, a large ensemble (e.g. several hundred) conformations are typically extracted in small intervals from the single MD trajectory of the complex. Alternatively, averaging over a select few receptor-ligand binding conformations from the MD trajectory via clustering has proved effective (Section 4.1), as well as more time efficient (Hayes et al., 2011). MM-GB(PS)SA calculations on single (minimized) structures has also recently been proposed and validated (Kuhn et al., 2005; Rastelli et al., 2010), but not necessarily for structures generated from MD simulations (Section 4.2). Meanwhile, for the entropy term calculated using normal mode analysis, fewer snapshots (typically less than a 100) are employed, due to the computational cost involved. As already highlighted (Section 3.1), a larger number of snapshots may be required for more stable and accurate predictions. These calculations, however, are computationally expensive and often not feasible with limited computational resources. Consequently, neglect of the entropy term can in some cases lead to sufficient or more accurate predictions for ranking of ligand binding affinities in certain macromolecular systems (Hayes et al., 2011; Hou et al., 2011a; Rastelli et al., 2010).

3.4 Limitations & caveats of MM-GB(PB)SA calculations

MM-GB(PB)SA methods are widely recognised as valuable tools in CADD applications. However, as with any method they have limitations and caveats, which need to be considered. First, while useful for ranking relative ligand binding affinities, these methods lack the required accuracy for absolute binding free energy predictions (Hou et al., 2011a; Singh & Warshel, 2010).The inclusion of entropic contributions brings the MM-GB(PB)SA values somewhat closer to experimental absolute affinities (Gilson & Zhou, 2007). However, such entropic terms are costly and contain large uncertainties. Force-field inconsistencies may also be an issue: PB and GB results depend strongly on adequate atomic charges and van der Waals radii, which are often optimized for MD simulations. The MM-GB(PB)SA results may be influenced by system-dependent properties, such as the features of the binding site, the extent of protein and ligand conformational relaxation upon association, and the protein and ligand charge distribution (Kuhn et al., 2005; Hou et al., 2011a). Continuum electrostatics models ignore the molecular structure of the solvent; in some cases this might affect the results, particularly when key receptor-ligand interactions are bridged by water molecules, *c.f.* Section 4.1 (Hayes et al., 2011). Furthermore, the value of the protein/ligand dielectric constant is empirically chosen, and takes into account not only the protein and ligand structural relaxation, but also other error-introducing factors such as the ones mentioned above (Archontis & Simonson, 2005; Schutz & Warshel, 2001). Hou and co-workers suggested in a recent MM-PBSA study that the use of ε_{in} = 4 for a highly charged protein-ligand binding interface, ε_{in} = 2 for a moderately charged binding interface and ε_{in} = 1 for a hydrophobic binding interface may improve ligand ranking (Hou et al., 2011a). The lack of a consistent optimum dielectric constant for MM-PBSA calculations has been noted by other workers (see for e.g. (Aleksandrov et al. 2010)), although generally a value ε_{in} =4 often gives satisfactory results (Aleksandrov et al., 2010; Archontis et al., 2001; Thompson et al., 2006). As a final note, MM-GB(PB)SA calculations require some degree of user expertise and planning, from the initial set-up and analysis of the MD simulations through to the binding free energy calculations themselves.

3.5 Variations & extensions of MM-GB(PB)SA

While molecular docking algorithms are computationally efficient methods used to screen a large number of ligands against a given target in reasonable time, generating and post-processing MD ensembles for more than a few receptor-ligand structures in MM-GB(PB)SA calculations is currently impractical. Recently, however, MM-GB(PB)SA methods are receiving plaudits as post-docking methods in virtual screening experiments. Post-processing of single docking poses using MM-GB(PS)SA algorithms can improve correlations between predicted and experimental binding affinities with a number of successes reported, *c.f.* for example (Du et al., 2011; Hou et al., 2011b; Lyne et al.; 2006). This is consistent with the previously mentioned discovery that select single receptor-ligand structures can prove as accurate as sampling over large numbers of MD trajectory snapshots in MM-GB(PB)SA applications (Kuhn et al., 2005; Rastelli et al., 2010). Post-docking MM-GBSA is implemented in Schrödinger software using the program Prime, with options to include receptor and ligand flexibility; the entropy term is neglected by default. Other recent extensions of MM-PBSA exploit quantum mechanics (QM) methods in QM/MM-PBSA calculations (Gräter et al., 2005; Manta et al., 2012; Wang & Wong, 2007). Here, a "hybrid" gas phase energy term ($E_{QM/MM}$) effectively replaces the pure molecular mechanics energy (E_{MM}) term in Eq. (5). Representing the ligand by QM has the advantage, for example, to eliminate a frequently encountered problem of deficient ligand forcefield parameters. However, calculations of this sort are significantly more expensive than MM-GB(PB)SA, and therefore are typically only viable for binding predictions on relatively few ligands.

4. Recent applications of MM-GB(PS)SA

In the present section, we will review recent examples of the use of MM-GB(PS)SA calculations for calculating ligand binding free energies highlighting both successes and limitations.

4.1 Example 1: Phosphorylase kinase ATP-binding site inhibitors

With an aim towards glycogenolysis control in type 2 diabetes, indirubin (IC_{50}> 50 μM), indirubin-3′-oxime (IC_{50} = 144 nM), KT5720 (K_i = 18.4 nM) and staurosporine (K_i = 0.37 nM) were investigated as phosphorylase kinase (PhKγtrnc) ATP-binding site inhibitors (Hayes et al., 2011).

Due to the lack of experimental structural information for binding of these ligands (Figure 2), MD simulations in explicit solvent using Desmond 2.0 (Bowers et al., 2006) were performed for each receptor-inhibitor complex, with 4 ns production runs following an initial equilibration period. A computationally-efficient multiple timestep RESPA integration algorithm was employed with timesteps of 2, 2 and 6 fs for bonded, "near" and "far" non-bonded interactions, respectively. Energy and trajectory atomic coordinate data

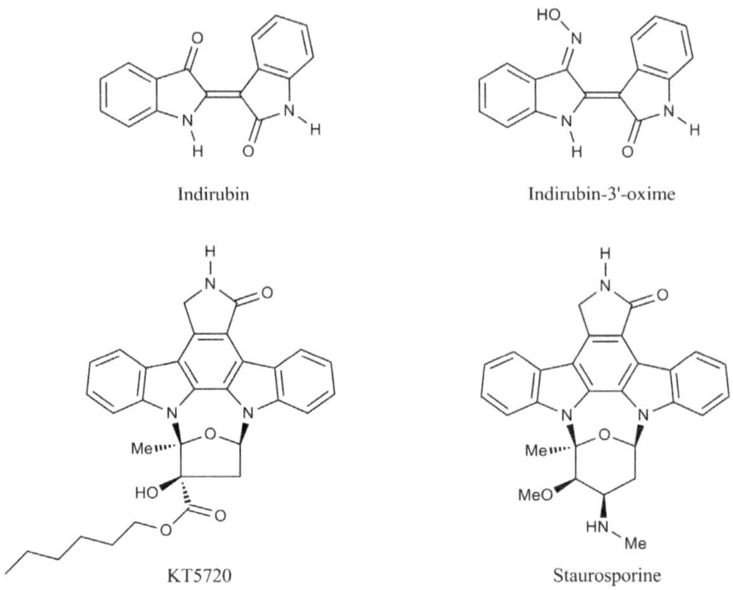

Fig. 2. The ATP-binding site inhibitors of phosphorylase kinase, as studied in Ref. (Hayes et al., 2011).

were recorded every 1.2 and 2.1 ps, respectively. For the trajectory analysis preceding the 'single-trajectory' based MM-GBSA calculations, both VMD (Humphrey et al., 1996) and Desmonds' Maestro simulation analysis tools were employed. The extracted MD trajectory binding site conformations (inhibitors + residues within 7 Å) of each complex were clustered into 10 groups based on atomic root-mean-square-distances (RMSDs). Every second frame (snapshot) of the last 3 ns of the production run (analysis phase) was used in the hierarchial clustering algorithm employed by the Desmond Maestro's Trajectory Clustering module. The representative complex of each of the 10 binding site cluster families (waters and counterions deleted) was then used in MM-GBSA calculations of binding free energies using Eq. (3). Schrödinger software with the MacroModel 9.7 Embrace module was used for the ΔE_{MM} and ΔG_{solv} calculations, while the entropy change, ΔS, was calculated for the minimized representatives using Rigid Rotor Harmonic Oscillator (RRHO) calculations. Using this algorithm, the change in vibrational, rotational and translational (VRT) entropy of the ligands on binding was estimated. Finally, the thermodynamic average ΔG_{bind} values (300 K) were estimated using the corresponding values for the 10 cluster representatives:

$$\Delta G_{bind} = \sum_{i=1}^{10} p_i \Delta G_{bind}(i) \tag{13}$$

The sum i was over the 10 cluster representatives, with p_i defined as the cluster frequency:

$$p_i = \frac{N_i}{N_{total}} \tag{14}$$

where N_i the number of frames in cluster i, and N_{total} the total number of frames.

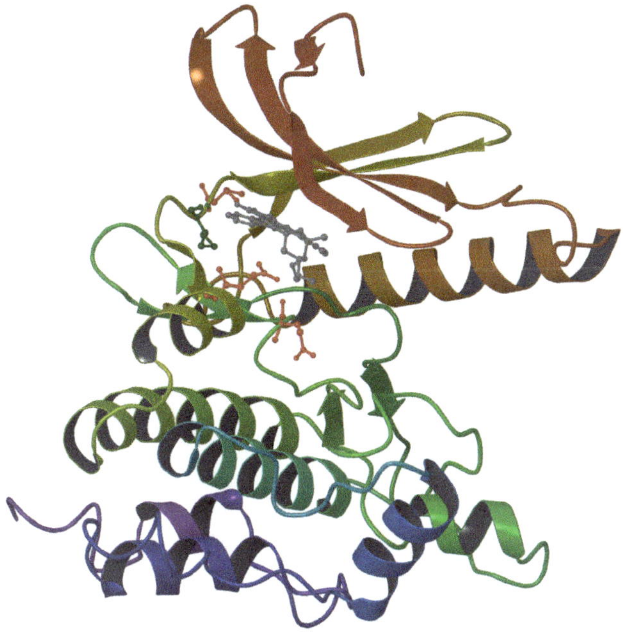

Fig. 3. Predicted binding of staurosporine (gray) at the ATP-binding site of PhKγtrnc. Key interacting residues are colored by type - polar/charged residues shown in red (D104, E110 and E153) and hydrophobic in green (M106). Hydrogen bonds are formed with M106 and D104 backbones (hinge region), E110 and E153.

The relative binding affinities from the final ΔG_{bind} values were generally in agreement with experiment, except the rankings of KT5720 and staurosporine (representative complex shown in Figure 3) were reversed by 0.7 kcal/mol. The discrepancy was accounted for by neglect of certain key contributions to the binding free energies using the MM-GBSA algorithm. Notably, accounting for and estimating the loss of the conformational entropy for the more flexible KT5720 ligand yielded ΔG_{bind} values 1.2 - 2.6 kcal/mol in favour of staurosporine binding and hence in agreement with the experimental rankings. Further, whereas staurosporine had no key receptor-ligand bridging waters, the MD simulations revealed a key role of bridging waters for KT5720 binding. The entropy loss associated with a bound water molecule in a protein-ligand complex is sometimes important (but typically neglected in MM-GB(PB)SA calculations), with an upper bound free energy cost of 2 kcal/mol at 300 K suggested (Dunitz, 1994).

4.2 Example 2: Interpretation of species specificity of compstatin, a peptidic inhibitor of the complement system

The complement system provides the first line of defense against the invasion of foreign pathogens (Mastellos et al., 2003). Its inappropriate activation may cause or aggravate several pathological conditions, including asthma, macular degeneration, rheumatoid arthritis, and rejection of xenotransplantation. The 13-residue compstatin is a promising candidate for the therapeutic treatment of unregulated complement activation (Janssen et al., 2007; Morikis & Lambris, 2005). The conformation of the human C3 - compstatin complex is shown in Figure 4. Compstatin interacts closely with four protein sectors indicated by thick yellow tubes. An important conclusion, drawn from experimental studies with a large number of species, was that compstatin inhibits the key complement component protein C3 from several primate species, but is inactive against the corresponding protein from lower mammals, precluding the testing of compstatin-based analogues in animal models (Sahu et al., 2003). To understand this species specificity, Tamamis and co-workers compared the stabilities of compstatin complexes with human or rat C3 (Tamamis et al., 2010) by atomistic MD simulations and an MM-GBSA analysis. In the simulations of the rat C3 complex, specific protein sectors near the compstatin binding site underwent reproducible localized displacements, which eliminated or weakened critical protein-ligand interactions. In agreement with the simulations and the lack of compstatin activity against rat C3, a MM-GBSA analysis estimated the binding free energy of the human C3 complex to be stronger by -9 kcal/mol relative to the rat C3 complex, in the "single-trajectory" approximation. If protein and ligand relaxation were taken into account by a "three-trajectory" approximation, the relative binding free energy increased (in absolute value) to -19 kcal/mol. Thus, in this system the neglect of relaxation effects introduced a significant error, even though the qualitative conclusion was correct in both cases.

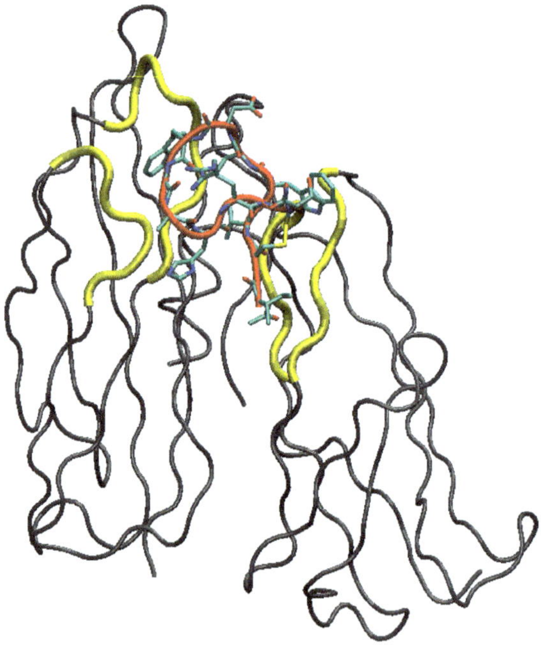

Fig. 4. Representation of the human C3c-compstatin complex. The compstatin main chain is shown in a red tube and licorice representation. The protein C3c is shown as a gray tube; only residues 329-534 and 607-620 are included, corresponding to the simulation system of Ref. (Tamamis et al, 2010). The yellow thick tubes show four protein sectors in proximity to compstatin (488-492, 454-462, 344-349 and 388-393, from left to right). The right-most sector 388-393 moves away from compstatin in the simulations of non-primate C3 complexes (Tamamis et al, 2010; Tamamis et al, 2011).

Guided by the above MM-GBSA study, subsequent simulations investigated the stabilities of compstatin complexes with "transgenic" variants of mouse C3, containing 6-9 substitutions from the human C3 sequence near the compstatin binding site. The MM-GBSA binding affinities of the resulting transgenic complexes were comparable to the one of the human C3 complex, and by -8 to -9 kcal/mol stronger relative to the mouse C3:compstatin complex (Tamamis et al., 2011). More recent simulations have investigated the affinities of a large number of human and mouse or rat C3 complexes with compstatin analogues, producing promising results (Tamamis et al., 2012). Thus, the combined study of a series of related protein-ligand complexes by atomistic simulations and an efficient evaluation of the corresponding affinities, such as in the MM-GB(PB)SA formulation, provides a powerful way to design new proteins or ligands.

4.3 Example 3: Fast predictions of binding free energies using MM-GB(PB)SA

Rastelli and co-workers (Rastelli et al., 2010) explored the reliability of using a single energy minimized receptor–ligand complex in MM-GB(PB)SA calculations to estimate ligand binding affinities for a series of structurally diverse inhibitors (Figure 5) of *Plasmodium falciparum* DHFR (Figure 6) with known binding modes and affinities.

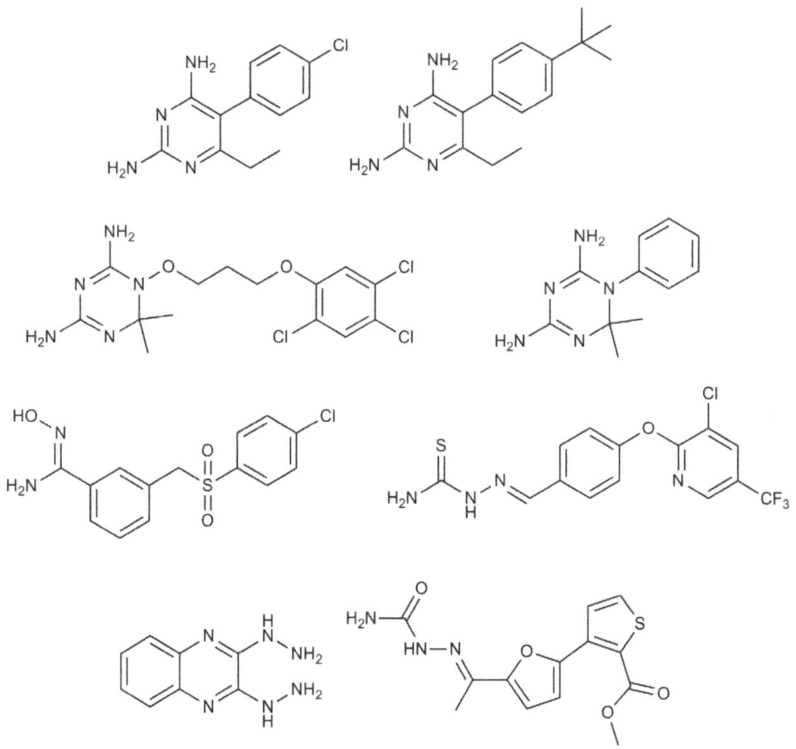

Fig. 5. Sample of the structurally diverse *Pf*DHFR inhibitors studied in Ref. (Rastelli et al., 2010).

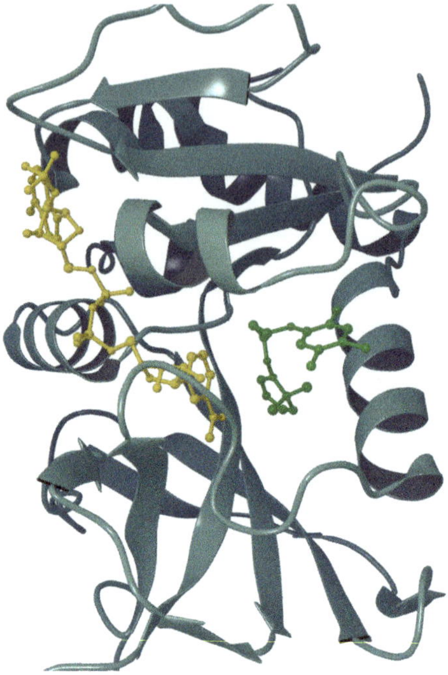

Fig. 6. Wild-type *Pf*DHFR in complex with NADPH (yellow) and the inhibitor WR92210 (green). This protein structure taken from PDB code 1J3I was used for the calculations in Ref. (Rastelli et al., 2010) and as described in Example 3.

They obtained excellent correlations between MM-PBSA or MM-GBSA binding affinities and experimental values, similar to those obtained after averaging over multiple snapshots from periodic boundary MD simulations in explicit water in the traditional sense, but with significant savings on computational time and effort. Different methods were used for generating the structures for the MM-GB(PB)SA calculations from minimizations in implicit and explicit solvent models, to minimization using a distance dependent dielectric function, and finally minimization followed by a short MD simulation and then re-minimization. The approach has been implemented in an automated workflow called BEAR (Binding Estimation After Refinement) which produces both MM-GBSA and MM-PBSA predictions of binding free energies, and is fast enough to be suitable for virtual screening applications (Degliesposti et al., 2011; Rastelli et al., 2009).

5. Conclusion

MM-GBSA and MM-PBSA are computationally efficient, end-point free energy methods that have been widely used to study protein-ligand binding affinities. Even though they lack the sound theoretical foundations of recently developed computationally demanding absolute-affinity free-energy methods (Boresch et al., 2003; Deng & Roux, 2009; Gilson & Zhou, 2007; Lee & Olsen, 2006), their connection with statistical thermodynamics has been established (Swanson et al., 2004). Due to the approximations inherent in MM-GB(PB)SA methods, they are more applicable for ranking ("scoring") of ligand binding affinities rather than to quantitatively predicted absolute binding free energies. They should be regarded as approximate, as they combine a molecular mechanics energy function with a continuum-electrostatics treatment of solvation effects; they include solute conformational entropy effects in an approximate manner (Singh & Warshel, 2010); and ignore the solvent molecular structure. Accurate incorporation of solute entropy (Foloppe & Hubbard, 2006) and solvent effects in binding affinity calculations is challenging, but future extensions and development of MM-GB(PB)SA methods will undoubtedly serve to address these limitations.

6. Acknowledgment

JMH acknowledges funding from the SP4-Capacities Coordination and Support Action, Support Actions, EUROSTRUCT project (CSA-SA_FP7-REGPOT-2008-1 Grant Agreement N° 230146).

7. References

Aleksandrov, A.; Thompson, D. & Simonson, T. (2010).Alchemical free energy simulations for biological complexes: powerful but temperamental. *Journal of Molecular Recognition*, Vol.23, No.2, pp. 117-127, ISSN0952-3499

Åqvist, J.; Luzhkov, V. B. & Brandsdal, B. O. (2002).Ligand binding affinities from MD simulations. *Accounts of Chemical Research*, Vol.35, No.6, pp. 358-365, ISSN 0001-4842

Archontis, G.; Simonson T & Karplus M. (2001). Binding free energies and free energy components from molecular dynamics and Poisson-Boltzmann calculations. Application to amino acid recognition by aspartyl-tRNA synthetase. *Journal of Molecular Biology*, Vol.306, No.2, pp. 307-327, ISNN 0022-2836

Archontis, G. & Simonson, T. (2005). Proton Binding to Proteins: A Free-Energy Component Analysis Using a Dielectric Continuum Model. *Biophysical Journal*, Vol.88, No.6, pp. 3888-3904, ISSN 0006-3495

Bashford, D.; Case, D. A. (2000). Generalized Born models of macromolecular salvation effects. *Annual Review of Physical Chemistry*, 2000, Vol.51, pp. 129-152, ISSN 0066-426X

Berendsen, H. J. C.; Postma, J. P. M.; van Gunsteren, W. F & Hermans, J. (1981). Interaction models of water in relation to protein hydration. In: *Intermolecular forces: Proceedings of the Fourteenth Jerusalem Symposium on Quantum Chemistry and Biochemistry*, B. Pullman, Editor, pp. 331-342, ISBN 902771326X, Reidel Publ. Company, Dordrecht, Holland.

Berendsen, H. J. C.; Grigera, J. A. & Straatsma, T. P. (1987). The missing term in effective pair potentials. *Journal of Physical Chemistry*, Vol.91, No.24, pp. 6269-5271, ISSN 0022-3654

Boresch, S.; Tettinger, F.; Leitgeb, M. & Karplus, M. (2003). Absolute binding free energies: A quantitative approach for their calculation. *Journal of Physical Chemistry B*, Vol.107, No.35, pp. 9535-9551, ISSN 1520-6106

Boresch, S. & Bruckner, S. (2011). Avoiding the van der Waals endpoint problem using serial atomic insertion. *Journal of Computational Chemistry*, Vol.32, No.11, pp. 2449-2458, ISSN 0192-8651

Bowers, K. J.; Chow, E.; Xu, H.; Dror, R. O.; Eastwood, M. P.; Gregersen, B. A.; Klepeis, J. L.; Kolossvary, I.; Moraes, M. A.; Sacerdoti, F. D.; Salmon, J. K.; Shan, Y. & Shaw, D. E. (2006). Scalable algorithms for molecular dynamics simulations on commodity clusters, *Proceedings of the ACM/IEEE Conference on Supercomputing (SC06)*, ISBN 0-7695-2700-0, Tampa, Florida, USA, Nov 11-17, 2006

Brooks, B.R.; Janezic, D. & Karplus M. (1995).Harmonic-analysis of large systems. 1. Methodology. *Journal of Computational Chemistry*, Vol.16, No.12, pp. 1522-1542, ISNN 0192-8651

Case, D. A.; Cheathham, T. E., III; Darden, T.; Gohlke, H.; Luo, R.; Merz, K. M., Jr.; Onufriev, A.; Simmerling, C.; Wang, B. & Woods, R. (2005). Journal of Computational Chemistry, Vol.26, No.16, pp. 1668-1688, ISSN 0192-8651

Chodera, J. D.; Mobley, D. L.; Shirts, M. R.; Dixon, R. W., Branson K. &Pande V.S. (2011). Alchemical free energy methods for drug discovery: progress and challenges. *Current Opinion in Structural Biology*, Vol.21, No.2, pp. 150-160, ISSN 0959-440X

Cornell, W.D.; Cieplak, P.; Bayly C. I.; Gould I. R.; Merz K. M.; Ferguson D. M.; Spellmeyer D. C.; Fox T.; Caldwell J. W. & Kollman P. A. (1995). A second generation forcefield for the simulation of proteins, nucleic acids, and organic molecules. *Journal of the American Chemical Society*, Vol.117, No.19, pp. 5179-5197, ISSN 0002-7863

Cramer, C. J. & Truhlar, D. G. (1999). Implicit solvation models: Equilibria, structure, spectra and dynamics. *Chemical Reviews*, Vol.99, No.8, pp. 2161-2200. ISSN 0009-2665

Degliesposti, G.; Portioli, C.; Parenti, M. D. & Rastelli, G. (2011). BEAR, a novel virtual screening methodology for drug discovery. *Journal of Biomolecular Screening*, Vol.16, No.1, pp. 129-133, ISSN 1087-0571

Deng, Y. & Roux, B. (2009). Computations of standard binding free energies with molecular dynamics simulations. *Journal of Physical Chemistry B*, Vol.113, No.8, pp. 2234-2246, ISSN 1520-6106

Du, J.; Sun. H.; Xi, L.; Li, J.; Yang, Y.; Liu, H. & Yao, X. (2011). Molecular modeling study of Checkpoint Kinase 1 inhibitors by multiple docking strategies and Prime/MM-GBSA. *Journal of Computational Chemistry*, Vol.32, No.13, pp. 2800-2808, ISSN 0192-8651

Dunitz, J. D. (1994). The entropic cost of bound waters in crystals and biomolecules. *Science*, Vol.264, No.5159, pp. 670-670. ISSN 0036-8075

Foloppe, N. & Hubbard, R. (2006). Towards predictive ligand design with free-energy based computational methods. *Current Medicinal Chemistry*, Vol.13, No.29, pp. 3583-3608, ISNN 0929-8673

Genheden, S. & Ryde, U. (2010). How to obtain statistically converged MM/GBSA results. *Journal of Computational Chemistry*, Vol.31, No.4, pp. 837-846, ISSN 0192-8651

Gilson, M. K.; Given, J. A.; Bush, B. L. & McCammon, J. A. (1997). The statistical-thermodynamical basis for computation of binding affinities: A critical review. *Biophysical Journal*, Vol.72, No.3, pp. 1047-1069, ISSN 0006-3495

Gilson, M. K. & Zhou, H-X. (2007). Calculation of protein-ligand binding affinities. *Annual Review of Biophysics and Biomolecular Structure*, Vol.36, pp. 21–42, ISSN 1056-8700

Gohlke, H.; Kiel, C. & Case, D. A. (2003). Insight into protein-protein binding by binding free energy calculation and free energy decomposition for the Ras-Raf and Ras-RalGDS complexes. *Journal of Molecular Biology*, Vol.330, No.4, pp. 891-913, ISSN 0022-2836

Gohlke, H. & Case, D. (2004). Converging Free Energy Estimates: MM-PB(GB)SA Studies on the Protein-Protein Complex Ras-Raf. *Journal of Computational Chemistry* Vol.25, No.2, pp. 238-250, ISSN 0192-8651

Gräter, F.; Schwarzl, S. M.; Dejaegere, A.; Fischer, S. & Smith, J. C. (2005). Protein/ligand binding free energies calculated with quantum mechanics/molecular mechanics. *Journal of Physical Chemistry B*, Vol.109, No.20, pp. 10474-10483, ISSN 1520-6106

Harvey, M. J.; Giupponi, G. & De Fabritiis, G. (2009). ACEMD : Accelerating biomolecularsimulations in the microsecond time scale. *Journal of Chemical Theory & Computation*, Vol.5, No.6, pp. 1632-1639, ISSN 1549-9618.

Hayes, J. M. & Leonidas, D. D. (2010). Computation as a tool for glycogen phosphorylase inhibitor design. *Mini Reviews in Medicinal Chemistry*, Vol.10, No.12, pp. 1156-1174, ISSN 1156-1174

Hayes, J. M.; Skamnaki, V. T.; Archontis, G.; Lamprakis, C.; Sarrou, J.; Bischler, N.; Skaltsounis, A. L.; Zographos, S. E. & Oikonomakos, N. G. (2011). Kinetics, in silico docking, molecular dynamics, and MM-GBSA binding studies on prototype indirubins, KT5720, and staurosporine as phosphorylase kinase ATP-binding site inhibitors: The role of water molecules examined. *Proteins: Structure Function & Bioinformatics*, Vol.79, No.3, pp. 703-719, ISSN 0887-3585

Hendsch, Z. & Tidor, B. (1999). Electrostatic interactions in the GCN4 leucine zipper: substantial contributions arise from intramolecular interactions enhanced on binding. *Protein Science* Vol.8, No.7, pp. 1381-1392, ISSN 0961-8368

Hou, T.; Wang, J.; Li, Y. & Wang, W. (2011a). Assessing the performance of the MM/PBSA and MM/GBSA methods. 1. The accuracy of binding free energy calculations based on molecular dynamics simulations. *Journal of Chemical Information & Modeling*, Vol.51, No.1, pp. 69-82, ISSN 1549-9596

Hou, T.; Wang, J.; Li, Y. & Wang, W. (2011b). Assessing the performance of the Molecular Mechanics/Poisson Boltzmann Surface Area and Molecular Mechanics/ Generalized Born Surface Area methods. II. The accuracy of ranking poses generated from docking. *Journal of Computational Chemistry*, Vol.32, No.5, pp. 866-877, ISSN 0192-8651

Humphrey, W.; Dalke A. & Schulten, K. (1996). VMD: visual molecular dynamics. *Journal of Molecular Graphics*, Vol.14, No.1, pp. 33-38, ISSN 0263-7855

Jannsen, B. J. C.; Hallff, E. F.; Lambris, J. D. & Gros, P. (2007). Structure of compstatin in complex with complement component C3c reveals a new mechanism of complement inhibition. *Journal of Biological Chemistry*, Vol.282, No.40, pp. 29241-29247, ISSN 0021-9258

Jorgensen, W. L.; Chandrasekhar, J.; Madura, J. D.; Impey, R. W. & Klein, M. L. (1983). Comparison of simple potential functions for simulating liquid water.*Journal of Chemical Physics*, Vol.79, No.2, pp. 926-935, ISSN 0021-9606

Karplus, M. & Kushick, J. N. (1981). Method for estimating the configurational entropy of macromolecules. *Macromolecules*, Vol.14, No.2, pp. 325-332, ISSN 0024-9297

Kirkwood, J. G. (1935). Statistical mechanics of fluid mixtures. *Journal of Chemical Physics*, Vol.3, No.5, pp. 300-313, ISSN 0021-9606

Kollman, P. (1993). Free energy calculations: Applications to chemical and biochemical phenomena. *Chemical Reviews*, Vol.93, No.7, pp. 2395-2417, ISSN 0009-2665

Kollman, P.A.; Massova, I.; Reyes, C.; Kuhn, B.; Huo, S.; Chong, L.; Lee, M.; Lee, T.; Duan, Y.; Wang W.; Donini, O.; Cieplak, P.; Srinivasan, J.; Case, D. A. & Cheatham, III, T. E. (2000). Calculating structures and free energies of complex molecules: combining molecular mechanics and continuum models. *Accounts of Chemical Research*, Vol.33, No.12, pp. 889-897, ISSN 0001-4842

Kongsted, J. & Ryde, U. (2009). An improved method to predict the entropy term with the MM/PBSA approach. *Journal of Computer Aided Molecular Design*, Vol.23, No.2, pp. 63-71, ISSN 0920-654X

Kongsted, J.; Söderhjelm, P. & Ryde, U. (2009). How accurate are continuum solvation models for drug-like molecules. *Journal of Computer Aided Molecular Design*, Vol.23, No.7, pp. 395-409, ISSN 0920-654X

Kuhn, B.; Gerber, P.; Schulz-Gasch, T. & Stahl, M. (2005). Validation and use of the MM-PBSA approach for drug discovery. *Journal of Medicinal Chemistry*, Vol.48, No.12, pp. 4040-4048, ISSN 0022-2623

Lee, M. S. & Olson, M. A. (2006). Calculation of absolute protein-ligand binding affinity using path and endpoint approaches. *Biophysical Journal*, Vol.90, No.3, pp. 864-877, ISSN 0006-3495

Levy, R. H., Zhang, L. Y., Gallicchio, E. & Felts, A. K. (2003). On the nonpolar hydration free energy of proteins: Surface area and continuum solvent models for the solute-solvent interaction energy. *Journal of the American Chemical Society* Vol.125, No.31, pp. 9523-9530, ISSN 0002-7863

Lyne, P. D.; Lamb, M. L. & Saeh, J. C. (2006). Accurate prediction of the relative potencies of members of a series of kinase inhibitors using molecular docking and MM-GBSA scoring. *Journal of Medicinal Chemistry*, Vol.49, No.16, pp. 4805-4808, ISSN 0022-2623

Lu, N. & Woolf, T. B. (2007). Chapter 6: Understanding and improving free energy calculations in molecular simulations: Error analysis and reduction methods. In: *Free Energy Calculations: Theory & Applications in Chemistry & Biology (Springer Series in Chemical Physics Vol.86)*, Chipot, C. & Pohorille, A., Springer-Verlag, ISBN 978-3-540-38447-2, Berlin-Heidelberg.

Manta, S.; Xipnitou, A.; Kiritsis, C.; Kantsadi, A. L.; Hayes, J. M.; Skamnaki, V. T.; Lamprakis, C.; Kontou, M.; Zoumpoulakis, P.; Zographos, S. E.; Leonidas D. D. & Komiotis (2012). Contrary to docking and SAR analysis, a 3'-axial CH_2OH substitution on glucopyranose does not increase glycogen phosphorylase inhibitory potency. QM/MM-PBSA calculations suggest why. *Chemical Biology & Drug Design*, in press.

Mahoney, M. W. & Jorgensen, W. L. (2000). A five-site model for liquid water and the reproduction of the density anomaly by rigid, nonpolarizable potential functions. *Journal of Chemical Physics*, Vol.112, No.20, pp.8910-8922, ISSN 0021-9606

MacKerell, A.D.; Bashford, D.; Bellott M.; Dunbrack, Jr., R. L.; Evanseck, J. D.; Field, M. J.; Fischer, S.; Gao, J.; Guo, H.; Ha, S.; Joseph-McCarthy, D.; Kuchnir, L.; Kuczera, K.; Lau, F.T.K.; Mattos, C.; Michnick, S.; Ngo, T.; Nguyen, D. T.; Prodhorn, B.; Reiher, III, W. E.; Roux, B.; Schlenkrich, M.; Smith, J. C.; Stote, R.; Straub, J.; Watanabe M.; Wiórkiewicz-Kuczera, J.; Yin, D. & Karplus, M. (1998). All-atom empirical potential for molecular modeling and dynamics studies of proteins. *Journal of Physical Chemistry B*, Vol.102, No.18, pp. 3586-3616, ISSN 1089-5647

Mastellos, D.; Morikis, D.; Isaacs, S. N.; Holland, M. C.; Strey, C. W. & Lambris, J. D. (2003). Complement – Structure, functions, evolution and viral molecular mimicry. *Immunologic Research*, Vol.27, No.2-3, pp. 367-385, ISSN: 0257-277X.

Mobley, D. L.; Graves, A. P.; Chodera, J. D.; McReynolds, A. C.; Shoichet, B. K. & Dill, K. A. (2007). Predicting absolute binding free energies to a simple model site. *Journal of Molecular Biology*, Vol.371, No.4, pp. 1118-1134, ISSN 0022-2836

Morikis, D. & Lambris, J. D. (2005). Structure, dynamics, activity and function of compstatin and design of more potent analogs. In: *Structural biology of the complement system*,

Morikis D, Lambris J. D.Editors, pp. 317-340, CRC Press/Tayor& Francis Group, ISBN 0824725409, Boca Raton, FL

Page, C. S. & Bates, P. A. (2006). Can MM-PBSA Calculations Predict the Specificities of Protein Kinase Inhibitors? *Journal of Computational Chemistry,* Vol.27, No.16, pp.1990-2007, ISSN 0192-8651

Pearlman, D. A. (2005). Evaluating the molecular mechanics Poisson-Boltzmann surface area free energy method using a congeneric series of ligands to p38 MAP kinase. *Journal of Medicinal Chemistry,* Vol.48, No.24, pp. 7796-7807, ISSN 0022-2623

Polydoridis, S.; Leonidas, D. D.; Oikonomakos, N. G. & Archontis G. (2007). Recognition ofribonuclease A by 3'-5'-pyrophosphate-linked dinucleotide inhibitors: A molecular dynamics/continuum electrostatics analysis. *Biophysical Journal,* Vol.92, No.5, pp. 1659-1672, ISSN 0006-3495

Rastelli, G.; Degliesposti, G. ; Del Rio, A. & Sgobba, A. (2009). Binding estimation after refinement, a new automated procedure for the refinement and scoring of docked ligands in virtual screening. *Chemical Biology & Drug Design,* Vol.73, No.3, pp. 283-286, ISSN 1747-0277

Rastelli, G.; Del Rio, A.; Degliesposti, G. & Sgobba M. (2010).Fast and accurate predictions of binding free energies using MM-PBSA and MM-GBSA. *Journal of Computational Chemistry,* Vol.31, No.4, pp. 797-810, ISSN 0192-8651

Rocchia, W; Alexov, E. &Honig, B. (2001). Extending the applicability of the nonlinear Poisson-Boltzmann equation: Multiple dielectric constants and multivalent ions. *Journal of Physical Chemistry B,* Vol.105, No.28, pp. 6507-6514, ISSN 1089-5647

Roux, B. & Simonson, T. (1999). Implicit solvent models. *Biophysical Chemistry,* Vol.78, No.1-2, pp. 1-20, ISSN 0301-4622

Roux, B. (2001). Implicit solvent models. In: *Computational Biochemistry and Biophysics,* Becker, O.M., Mackerell, A. D. Jr, Roux, B., Watanabe, M. Editors, pp. 133-151, Marcel Dekker Inc., ISBN 0-8247-0455-X, New York

Sahu, A.; Morikis, D. & Lambris, J. D. (2003).Compstatin, a peptide inhibitor of complement, exhibits species-specific binding to complement component C3. *Molecular Immunology,* Vol.39, No.10, pp. 557-556, ISSN 0161-5890.

Schutz, C. N. & Warshel, A. (2001). What are the dielectric constants of proteins and how to validate electrostatic models? *Proteins-Structure, Function and Genetics,* Vol.44, No.4, pp. 400–417, ISSN 0887-3585

Simonson, T. (2003). Electrostatics and dynamics of proteins. *Reports of Progress in Physics,* Vol.66, No.5, pp. 737-787, ISSN 0034-4885

Singh, N. & Warshel, A. (2010). Absolute binding free energy calculations: On the accuracy of computational scoring of protein-ligand interactions. *Proteins: Structure, Function & Bioinformatics,* Vol.78, No.7, pp.1705-1723, ISSN 0887-3585

Still, W.; C.; Tempczyk, A.; Hawley, R. & Hendrickson, T. (1990).Semianalytical treatment of solvation for molecular mechanics and dynamics. *Journal of the American Chemical Society,* Vol.112, No.16, pp. 6127-6129, ISSN 0002-7863

Stone, J. E.; Hardy, D. J.; Isralewitz, B. & Schulten, K. (2011). Chapter 16: GPU algorithms for molecular modeling. In: *Scientific Computing with Multicore & Accelerators,* Dongarra J.; Bader J.A & Kurzak J., pp. (351-371), Chapman & Hall/CRC Press, ISBN 9781439825365

Straatsma, T. P. & McCammon J. A. (1992). Computational alchemy. *Annual Review of Physical Chemistry*, Vol. 43, pp. 407-435, ISSN 0066-426X

Swanson, J. M.; Henchman, R. H. & McCammon, J. A. (2004). Revisiting free energy calculations: a theoretical connection to MM/PBSA and direct calculation of the association free energy. *Biophysics Journal*, Vol.86, No.1, pp. 67-74, ISSN 0006-3495

Tamamis, P.; Morikis, D.; Floudas, C. A. & Archontis, G. (2010). Species specificity of the complement inhibitor compstatin investigated by all-atom molecular dynamics simulations. *Proteins: Structure, Function & Bioinformatics*, Vol.78, No 12, pp. 2655-2667, ISSN 0887-3585

Tamamis, P; Pierou, P.; Mytidou, C.; Floudas, C. A.; Morikis, D. & Archontis, G. (2011). Design of a modified mouse protein with ligand binding properties of its human analog by molecular dynamics simulations: The case of C3 inhibition by compstatin. *Proteins: Structure, Function & Bioinformatics*, Vol. 79, No. 12, pp.3166-3179, ISSN 0887-3585

Tamamis, P; de Victoria, A. L.; Gorham, R. D.; Bellows-Peterson, M. L.; Pierou, P.; Floudas, C. A.; Morikis, D. & Archontis G. (2012). Molecular dynamics in drug design: New generations of Compstatin analogs. *Chemical Biology & Drug Design*, in press, ISSN 1747-0285

Thompson, D.; Plateau, P. & Simonson, T. (2006).Free-energy simulations and experiments reveal long-range electrostatic interactions and substrate-assisted specificity in an aminoacyl-tRNAsynthetase.*Chembiochem.* Vol.7, No.2, pp. 337-344, ISSN 1439-4227

Tembe, B. L. & McCammon, J. A. (1984). Ligand-receptor interactions. *Computers & Chemistry*, Vol.8, No.4, pp. 281-283, ISSN 0097-8485

Tidor, B. & Karplus, M. (1994). The contribution of vibrational entropy to molecular association - the dimerization of insulin. *Journal of Molecular Biology*, Vol.238, No.3, pp. 405-414, ISSN 0022-2836

Van Gunsteren, W. F.; Billeter, S. R.; Eising, A. A.; Hünenberger, P. H.; Krueger, P.; Mark, A. E.; Scott, W. R. P. & Tironi, I. G. (1996). Biomolecular Simulation: The GROMOS96 Manual and User Guide; vdf Hochschulverlag AG an der ETH Zürich and BIOMOS b.v. Zürich, Groningen

Wang, M. & Wong, C. F. (2007). Rank-ordering protein-ligand binding affinity by a quantum mechanics/molecular mechanics/Poisson Boltzmann-surface area model. *Journal of Chemical Physics*, Vol.126, No.2, 026101, ISSN 0021-9606

Yang, C-Y.; Sun, H.; Chen, J.; Nikolovska-Coleska, Z. & Wang, S. (2009). Importance of ligand reorganization free energy in protein-ligand binding affinity prediction. *Journal of the American Chemical Society*, Vol.131, No.38, pp. 13709-13721, ISSN 0002-7863

Zhou, H-X. & Gilson, M. K. (2009).Theory of free energy and entropy in noncovalent binding. *Chemical Reviews*, Vol.109, No.9, pp. 4092-4107, ISSN 0009-2665

Zhou, H-X. & Gilson, M. K. (2009).Theory of free energy and entropy in noncovalent binding. *Chemical Reviews*, Vol.109, No.9, pp. 4092-4107, ISSN 0009-2665

Zwanzig, R. W. (1954). High-temperature equation of state by a perturbation method. I. Nonpolar gases. *Journal of Chemical Physics*, Vol.22, No.8, pp. 1420-1426, ISSN 0021-9606

Part 3

Dynamics of Plasmas

10

Micro-Heterogeneity in Complex Liquids

Aurélien Perera[1], Bernarda Kežić[1,2], Franjo Sokolić[2] and Larisa Zoranić[2]
[1]*Laboratoire de Physique Théorique de la Matière Condensée (UMR CNRS 7600), Université Pierre et Marie Curie, 4 Place Jussieu, F75252, Paris cedex 05, France*
[2]*Department of Physics, Faculty of Sciences, University of Split, Nikole Tesle 12, 21000, Split*
[1]*France*
[2]*Croatia*

1. Introduction

What is micro-heterogeneity and why is it important to the theory of liquids in particular, and for physical chemistry and even Physics in general, to deserve a full chapter devoted to it?

Physics was initially about discovering the sets of laws that would describe motions of single objects, such as planets or more casually a stone under the gravitational field of earth. The focus was a *single object*. Statistical thermodynamics was a revolution in the sense that it could describe matter as a set of multitudes of objects, in par with thermodynamics, which was all about heat and energy. It could describe various states of matter, gases liquids and solids. However, while the statistical description of gases and solids was still about describing single particles under the field created by surrounding particles, liquids stood in a very special place, since it was recognized very early in the 20th century that it was necessary to consider *correlations* between particles. This was thought to be such a serious obstacle that the great soviet Physicist Lev Landau said once that a true "theory of liquids could be neither convincing nor useful" (deGennes, 1977). It was the founders of a true statistical theory of liquids (Frish & Lebowitz, 1974) that showed how the formalism of correlation functions could be advantageously used in understanding various properties of the liquid phase. Now, we know how to calculate various physical properties, that are related to the continuum description of matter in its various disordered phases, from the statistical collection of microscopic objects and their correlations. In other words, we know how to go from single objects to a huge collection of them.

Now, we can ask a new question: can new "objects" emerge from such a statistical description of a discretized continuum? This type of question is partially answered by the high brow theories of fundamental particles, quantum field theories and such, wherein photon and electron can be "emerged" from a peculiarly structured vacuum, by analogy to how sound phonons emerge from structured solids (Wen, 2004). However, all these theories heavily rely on Quantum Mechanics, which is a whole different story. Here, we are particularly interested in the *classical* version of this question. What we intend to show here is that ordinary aqueous mixtures are the theater of the emergence of new objects, that are themselves made of the same constituents, but that are grouped in a particular way. This is what we call here "micro-heterogeneity" (MH). A very good example of micro-heterogeneity is the micelle, which emerges at the critical micellar concentration (CMC), from a disordered assembly of water and surfactant molecules(Floriano et al., 1999; Poland & Scheraga, 1965). This same example also serves to understand the difficulty of the concept of micro-heterogenity: what is the role played by the constituents in the appearance of such a new object, and how does the stability of such an object is affected by various microscopic and macroscopic parameters?

The peculiar nature of aqueous mixtures, as seen from various thermodynamical properties, has been acknowledged from the sixties by Frank et al. (Frank & Yves, 1966) and more recently by many authors such as Desnoyer (de Visser et al., 1977), Davies (Davies, 1993), and more systematically by Koga (Koga, 2007). They have observed that, for binary mixtures, many physical quantities such as the vapour pressure, for example, showed few weak kink-type "anomalies" when the mole fraction of one of the component was varied. This was in contrast with the monotonous behaviour of the same quantities for mixtures of ordinary polar substances, such as benzene-toluene, for example (Rowlinson & Swinton, 1982). For example, McAlister has reported in 1960 the viscosities for benzene-toluene and methanol-toluene(McAlister, 1960). The first has a monotonous almost linear variation in function of mole fraction, while that for the H-bonding substance has an "S" shape with 2 slope changes. These changes in slope were not signs of any known types of phase transitions: no discontinuities in the first or second derivatives of the Gibbs free energy were observed(Koga, 2007) The question was then: what could be the nature of the changes at the molecular level that could produce these changes in slope? Such typical changes in slope are shown in Fig.1 below for very different physical properties. The arrows indicate the position of the slope changes.

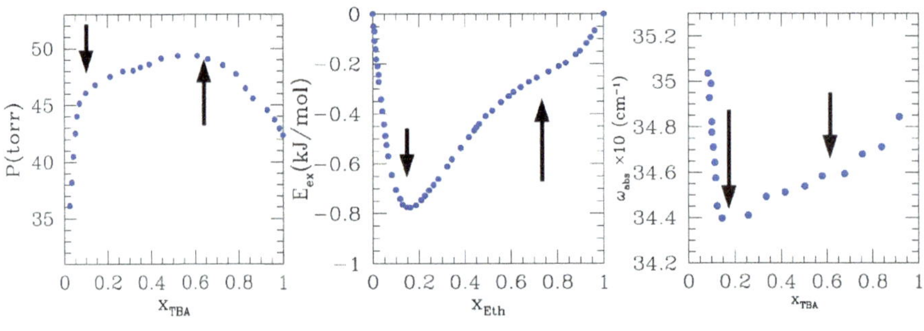

Fig. 1. Left: vapour pressure of aqueous-tbutanol mixture (Koga et al., 1990). Middle: excess enthalpy of aqueous-ethanol mixture (Lama & Lu, 1965) Right: absorbance frequencies of aqueous-Tbutanol mixture (Pradhan et al., 2008), a function of the respective alcohol mole fraction. The arrows indicate slope changes mentioned in the text

Since this is all about physical chemistry, these changes were traced back into the corresponding changes in enthalpy H or entropy S, and it was noticed that rather large compensating changes in these quantities were responsible for small changes in the Gibbs free-energy $G = H - TS$. This is the so-called enthalpy-entropy compensation that is often invoked in the "bio" context(Wiggins, 2008). Subsequent investigations did not lead to any particular clarifications based on the strict thermodynamical arguments, and the research along this direction has stalled.

Computer simulations could be, in principle, the ideal tool to observe the microscopic molecular arrangements that could lead to kink-like variations in macroscopic quantities. However, computer simulations are severely restricted by system size considerations, that are due to the computational handling of the description of the individual motions of millions of particles. The leading paradigmatic dogma of computer simulations is that it is not necessary to study systems of the size of the Avogadro number, and that the much smaller size of thousands of particles is often more than enough to have a reliable statistical estimate of major thermodynamical quantities, such as the enthalpy, for example. However, recent advances in computer studies has confirmed that his paradigm fails for mixtures of complex liquids such as aqueous mixtures (Kezic et al., 2011; Mijakovic et al., 2011; Perera & Sokolic, 2004), as we will show later. The origin of such failure is still under debate, and can be attributed to two sources. The first is the accuracy of the force fields that describe the interaction between molecules, and which can be questioned. Accordingly, the force fields should be modified to account for these specific features under each conditions (Chitra & Smith, 2002; Lee & van der Vegt, 2005; Smith, 2004; Weerasinghe & Smith, 2003; 2005). The second is the existence of intrinsic phenomena, such as the micro-heterogeneity, that cannot be described properly with systems sizes currently in use (about few thousand particles), and that may require computational resources out of proportions in view of the modest scientific interest such systems could *a priori* require (Kezic et al., 2011; Mijakovic et al., 2011; Perera & Sokolic, 2004). These two different explanations differ from a fundamental point of view. The first attitude is to modify the interactions in order to *account* special extended correlations that can be associated with the appearance of large scale structures. However, since correlations are the consequence of interactions, this appear at first as some bootstrap procedure. The second attitude is to modify the computational constraint such that the large scale structures can be seen without artifacts related to periodicity. This second point of view is inspired from the simple idea that, if micro-heterogeneity is associated to a new object, with no precise shape, then current system sizes are simply to small to accomodate enough such "objects" for proper statistical sampling, not counting the fact that periodical boundary conditions may distort these "objects" artifically if the system is too small. From this perspective, the first attitude looks like introducing pseudo-potentials to adapt the system to small scales. If this is the case, is it possible to detect artifacts that they might introduce. We will show below how this discussion is illustrated by the case of aqueous acetone-water mixtures.

For now, it is instructive to compare these two points of view for the case of spontaneous micelle formation of long alcohol chains in water. Although the exact value of the CMC will depend on the various force fields available for water, the proper accouting of the formation of a *single* small micelle, of diameter 50Å, which is in par with experimental values (Tanford, 1974), would need a cubic simulation box of size $0.3\mu m$, which would contain $2048 \times 10^3 \approx 2$ million water molecules. Therefore, simulating realistic micelle formation of say few tens of such micelles is beyond the reach of current computational power. Needless to say, simulating realistic micro-emulsions is totally out of reach, except through simplified models(Floriano et al., 1999). The other point of view would require to modify the various interactions in order to enforce micelle formation within reasonable system sizes. It turns out that such calculations are possible, but often at the cost of eliminating a realistic solvent(Floriano et al., 1999). This example illustrates the dilemna offered by the two point of view.

What is the origin of this whole problematic, as sketched above for the case of micelle formation? Water and most solutes tend to micro-segregate, which means that solutes and water occupy different partitions in space, each partition being of few molecular sizes, but of no specific shape. This was suspected almost a half a century ago from thermodynamical arguments (Frank & Yves, 1966). A recent paper by the group of Soper in *Nature* (Dixit et al., 2002) acknowledged that even simple system as the methanol-water mixture exhibits micro-segregation. Following this re-discovery, many authors, and in particular our group, investigated different aqueous mixtures by computer simulations. It was quickly aknowledged that, unlike simple liquids and their mixtures, such as argon, or carbon dioxide, for example, molecules made of very different subgroups had a strong tendency to self-segregate. Alcohol molecules are a perfect example, which contains both the hydrophilic OH group and the hydrophobic CH_n methyl groups. Liquids made of such molecules exhibit a strong local order, due to the tendency of OH and methyl groups to self-segregate, and which distinguishes them from ordinary Lennard-Jonesium liquids, or even polar liquids such as benzene, for example. The source of the difficulty mentioned in the beginning of this paragraph, is that complex liquids contain *two* scales of description, the original microscopic scale related to molecular size, just like simple liquids, and the newly emerged scale related to the size of the segregated domains, as well as the corresponding time scales. Therefore, computer simulations need some rescaling to accomodate a full statistical description of the larger scale phenomena. This problem hides in fact another fundamental problem, that of the description of disorder.

Liquids are fundamentally thought to be disordered systems (with few exceptions such as liquids crystals). To be more specific about this issue, we need to explain how order differs from disorder at the level of the statistical description of liquids. We will do this in the next section. For now, while order has a very precise statistical microscopic description, disorder is considered as generic. We feel that aqueous mixtures belong to a very special type of disorder, caracterised by the micro-heterogeneity, and where fluctuations in the number of particles in a given volume play an important role. It is this problem that is not well described by finite size simulations, for the very simple reason that such system contain *two* scales of description, the original microscopic scale related to molecular size, and the newly emerged mesoscopic scale related to the segregated domains, and needs special approaches that we will describe later in this chapter. In the next section, we will try to answer the question posed above: how to distinguish between different types of local orders in simple and complex liquids.

2. Statistical description of liquids

There are very good text books about the theory of liquids, among which the Hansen-MacDonald's celebrated *Theory of Simple Liquids* is the reference(Hansen & McDonald, 2006) we follow here to elaborate the theoretical description. If a liquid is made of N molecules in a volume V, each molecule i being described by its position $\vec{r}_i$ and orientation $\mathbf{\Omega}_i$, the latter which is a set of Euler angles, then we use here the shorthand notation $i = (\vec{r}_i, \mathbf{\Omega}_i)$ to denote a generalised "position". The instantaneous microscopic density of such a liquid at position 1 is given by:

$$\rho(1) = \sum_{i=1}^{N} \delta(1 - i) \tag{1}$$

where $\delta()$ is a Dirac symbol. This expression simply states that a given molecule among the N in the sample is found at position 1. As such, this is an instantaneous snapshot of the system, when the position 1 is varied through the sample. In order to have a statistical estimate, one must average this microscopic density over some statistical ensemble, say here the Canonical ensemble when N, V and the temperature T are fixed. One has then the first observable of the system, the microscopic density defined as:

$$\rho^{(1)}(1) = <\rho(1)> \tag{2}$$

where the bracket <> denotes the statistical average defined for any microscopic quantity A as:

$$<A> = \frac{1}{z_N} \int d1...dN\, A\, \exp[-\beta V(N)] \tag{3}$$

where the integral is carried over the generalized positions of the N molecules, and $V(N)$ is the total interaction energy of the N particles, $\beta = 1/k_B T$ is the Boltzmann factor, with T the temperature and k_B the Boltzmann constant. $Z_N = \int d1...dN \exp[-\beta V(N)]$ is the Canonical ensemble partition function.

If $\rho(1)$ is considered as a random variable, then one can construct a whole family of statistical correlations, such as the pair correlation function, which is defined as:

$$\rho^{(2)}(1,2) = <\rho(1)\rho(2)> \tag{4}$$

and more generally, the n-body correlation function

$$\rho^{(n)}(1,...,n) = <\prod_{i=1}^{n} \rho(i)> \tag{5}$$

In practice, only $\rho^{(1)}(1)$ and $\rho^{(2)}(1,2)$ are important quantities. Indeed, most thermodynamical quantities, such as for example the pressure or the enthalpy, can be expressed solely as weighted averages of these two functions. For the matter that concerns us here, a detailed discussion of these 2 functions is at order.

2.1 One-body and two-body functions

The one-body function $\rho^{(1)}(1)$ describes the order of the system. If the system is subject to a symmetry breaking field, such as an electric or magnetic field, or under the presence of a wall, then such a system is spatially and/or orientationally inhomogeneous, and the distribution of each particle needs to be specified with respect to the field. Hence, $\rho^{(1)}(1)$ depends explicitly on the variable 1. For example, if the liquid is next to a wall, then if the z-axis is chosen to be perpendicular to the wall, one has $\rho^{(1)}(1) = \rho^{(1)}(z_1)$, where z_1 is the z-coordinate of particle 1 to the wall. If the symmetry breaking field is a magnetic field $\vec{B}$, then $\rho^{(1)}(1) = \rho^{(1)}(\Omega_1.\vec{B})$ which depends on the orientation of molecule 1 with respect to the field. The mathematical reason for such situation is to be found in Eq.(3): the interaction energy V(N) includes the interaction of each of the particles with the field, hence the average in Eq.(2) also depends on the field's position. In contrast, for a disordered system, the interaction energy depends only on the position of all the particles. Therefore, the integral in Eq.(2) integrates out all coordinates, and one has simply the number density of the system:

$$\rho^{(1)}(1) = \rho = \frac{N}{V} \tag{6}$$

This is a very important result, because it states that the usual statistical description of a liquid cannot differentiate a Lennard-Jonesium liquid from an alcohol at the level of the one-body distribution. In other words, the strong local order of the second type of liquid is not addressed by this function. A *crystal* of any of the two liquids will have a non-trivial $\rho^{(1)}(1)$ which will contain the specificities of the crystallin order. But for both *liquids*, it will be the *same* boring number density $\rho = \frac{N}{V}$.

This has a very strong implication: if we want to differentiate the two liquids according to their local orders, we need to look for one step higher in the correlation functions, namely the two-body correlation function. It is convenient to introduce the distribution function $g(1,2)$ as a measure of the correlation between $\rho^{(2)}(1,2)$ and $\rho^{(1)}(1)\rho^{(1)}(2)$:

$$\rho^{(2)}(1,2) = \rho^{(1)}(1)\rho^{(1)}(2)g(1,2) \tag{7}$$

More generally, one defines the n-body distribution functions as

$$\rho^{(n)}(1,...,n) = \prod_{i=1}^{n} \rho^{(1)}(i)g^{(n)}(1,..,n) \tag{8}$$

For the case of a disordered system, from Eq.(6) one has

$$\rho^{(2)}(1,2) = \rho^2 g(1,2) \tag{9}$$

which indicates that both functions differ only by a scalar ρ^2. Since correlations decouple when particles are infinitely far apart, the relation above contains the very important limiting law. We will see that this law is violated in computer simulation for intrinsic reasons.

$$\lim_{r\to\infty} g(1,2) = 1 \tag{10}$$

Much the same way one can define pair correlation functions between individual sites on each molecules. This is particularly convenient when considering molecules made of sites, which is often the case for realistic molecules. The force fields $v(1,2)$ between 2 molecules that we will consider here are expressed entirely in terms of site-site interactions:

$$v(1,2) = \sum_{i,j} v_{ij}(r_{ij}) \tag{11}$$

where the sum runs over all the sites i on molecule 1 and j on molecule 2, with r_{ij} being the radial distance between 2 such sites. In such case, one defines site-site correlation functions as $g_{ij}(r_{ij})$. In practice, we will consider site-site interactions made of 2 terms, a Lennard-Jones (LJ) term that accounts for repulsive and dispersion interactions, and a Coulomb terms that handles electrostatic interaction between the partial charges q located at the center of each sites:

$$v_{ij}(r_{ij}) = 4k_BT\epsilon_{ij}[(\frac{\sigma_{ij}}{r_{ij}})^{12} - (\frac{\sigma_{ij}}{r_{ij}})^6] + k_BT\frac{q_iq_j}{r_{ij}} \tag{12}$$

The diameters σ_{ij} and energy wells ϵ_{ij} are handled as usual in terms of the individual site diameters σ_i and energy wells ϵ_i by the Lorentz rule $\sigma_{ij} = (\sigma_i + \sigma_j)/2$ and Berthelot rule $\epsilon_{ij} = \sqrt{\epsilon_i \epsilon_j}$.

Many thermodynamical properties can be expressed in terms of these correlation functions. For example, the isothermal compressibility $\kappa_T = (1/\rho)(\partial\rho/\partial P)_T$ can be expressed with any of the correlations between pairs of sites (i,j) (Hansen & McDonald, 2006)

$$\kappa_T = \frac{1 + \rho \int d\vec{r}(g_{ij}(r) - 1)}{\rho k_B T} \tag{13}$$

Similary, the configurational part of the internal energy per particle is given by

$$E/N = \frac{\rho k_B T}{2} \sum_{ij} x_i x_j \int d\vec{(r)} g_{ij}(r) v_{ij}(r) \tag{14}$$

Structure factors can be defined by Fourier transforms of the site-site correlation functions

$$S_{ij}(k) = 1 + \rho\sqrt{x_i x_j} \int d\vec{r}[g_{ij}(r) - 1] \exp(-i\vec{k}.\vec{r}) \tag{15}$$

where it is assumed that x_i stands for the mole fraction of species to which site i belongs, which is a convenient shorthand notation. These quantities can be directly compared with those extracted from the scattering intensity obtained by neutron or X-ray scattering techniques (Hansen & McDonald, 2006), and are a useful alternate description of the microscopic structure of liquids.

2.2 The difference in disorder between a simple and a complex liquid

Now, we have to figure out how to distinguish between a disordered Lennard-Jonesium liquid and a disordered alcohol by using only $g(1,2)$ in each case. In order to do that efficiently, we are going to consider on one hand the methanol molecule as described by an interaction of the type Eq.(11) with OPLS force field parameters (Jorgensen, 1986), and on the other hand the same model but stripped of all the partial charges that account for H-bond interactions, and that are at the origin of the complexity of this liquid. Both liquids are simulated in ambient conditions, and the density is taken to be that of pure methanol. We call this second model the bare model. In Fig.2 below, we have plotted typical site-site correlation functions for each of the two liquids, as well as the corresponding structure factors.

Both models show very similar methyl sites M-M correlations because these sites are uncharged in both models. Such correlations are typical of dense liquids. This is also seen from the correesponding structure factors shown in Fig.3.

However, the correlation between the oxygen atoms show an entirely different packing structure between the two models. The bare model shows weak O-O correlations contrary to the OPLS model. There is a very sharp first neighbour correlations, which is due to the strong hydrogen bonding interaction of the OPLS model, but the correlations further apart look very much featureless, with very weak oscillations that indicate weak packing structure, in opposition to those of the bare model. In fact, these weak correlations are due to the

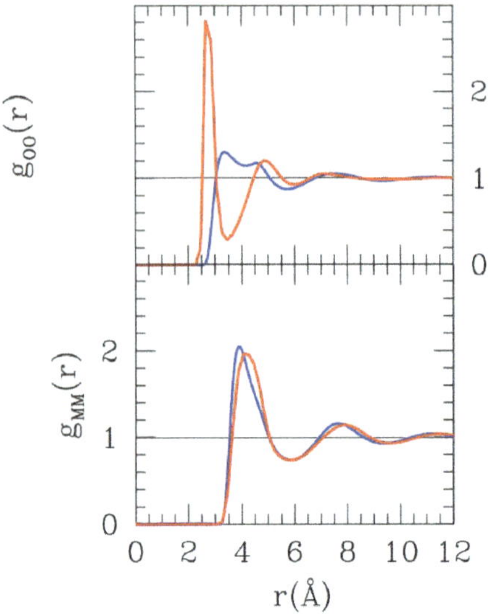

Fig. 2. Site-site correlation functions for OPLS methanol (red) and bare model(blue) (top) Oxygen-oxygen correlations (bottom) methyl-methyl correlations

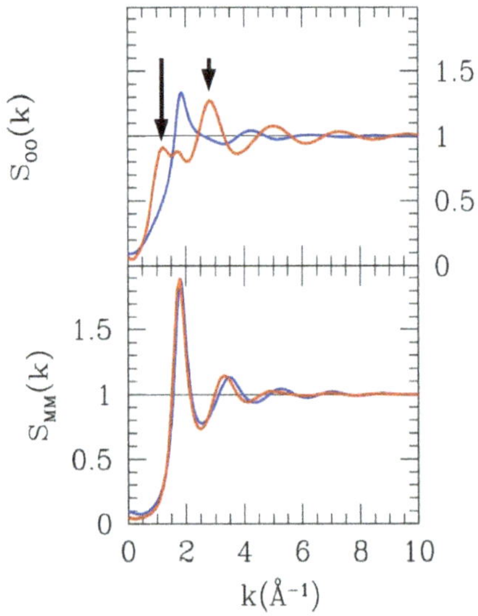

Fig. 3. Structure factors corresponding to the figure above. The arrows in the top panel show the 2 peak structure discussed in the text

fact that methanol molecules form chain-like clusters attached by O-H..O hydrogen bonds, therefore the OO correlations exist only along such chains and are spatially weaker than the packing correlation, although they are very strong along the chains. The periodicity of these oscillations is modulated by the average size of the clusters, say $\sigma_C > \sigma$. Therefore, the main peak of the OO structure factor will be centered at a *smaller* wave number $k_{max} \approx 2\pi/\sigma_C < k_M$. Although both liquids are disordered, methanol exhibits considerable local order which can be tracked through specific features of some site-site correlations. In fact, methanol has *emerged meta-molecules* -the clusters- *within* its disordered structure, and the pre-peak of the structure factor indicates the presence of such objects, just like the main peak of the structure factor of the simple liquid indicates the size of the core of the particles. These newly emerged objects are to be compared with the micelles that we mentioned above. We see that the order in complex liquids is not just near neighbour ordering, such as dipole-dipole alignment for example, but it is all about *the emergence of new larger "particles"*. This is the main message of this whole chapter, that we will develop with other examples. When considering neat liquids, we will characterise its peculiar local structure by the wording *micro-structure*, while in the case of mixtures we will use the wording *micro-heterogeneity*. Indeed, mixtures of alcohol and simple liquids, such as the methanol-acetone mixture for example, exhibit local segregation, which is due to the fact that methanol molecules tend to self-associate through H-bond. Once this difference between ordinary liquids and complex liquids is admitted, we can ask other questions that are central to liquids, namely how density fluctuations affect the local order and the stability of the liquid state.

2.3 Local order, correlation length and stability

The notion of disorder is intimately related to the concept of stability. In fact, when local order develops too much, the system can undergo a mechanical breakdown and loose its stability. This is what happens when a gas-liquid phase transition occurs, for example. If we cool a gas, or increase its density, molecules start to cluster in larger groups, and the local order increases while the gas becomes metastable. This increase of order can be tracked through the intrinsic correlation length ξ, which should be distinguished from the molecular size σ_M and the emergent cluster size σ_C. The clusters of the metastable gas phase are characterised by the correlation length ξ which is a measure of the size of the correlations. This correlation lenght is about σ_M in stable phase, and increases rapidly in the metastable region. One way to define this length is to look at the decay of the center-of-mass correlations, which behave as:

$$\lim_{r\to\infty} g_{CC}(r) = 1 + \frac{A}{r}\exp[-\frac{r}{\xi}] \tag{16}$$

This is an exact relation that is always true in disordered systems. It can be derived through the Ornstein-Zernike equation that is central to the theory of liquids(Hansen & McDonald, 2006). One sees that, if the correlation length diverges, the correlation functions develops a algebraic $1/r$ tail. This occurs exactly at the spinodal and witnesses the loss of the mechanical stability of the system.

Associated complex liquids, such as methanol, also undergo gas-liquid phase separation through the same mechanism described above. It is interesting to ask how the intrinsic clusters affect the universal gas-liquid phase separation. In fact, this problem of the stability is more interesting in the case of mixtures, which can undergo a liquid-liquid phase separation controled by the same physical phenomena of the loss of mechanical stability due to an increase of the correlation length ξ. This is a serious source of problems for mixtures when they are studied by computer simulations. How to distinguish between micro-phase separation and true phase separation within a computer simulation? Micro-phase separation is akin to micelle formation, when the alcohol molecules self-segregates themselves in domains separated from water. If one simulates a systems too small to accomodate a single micelle, then it is quite probable that one would witness a full phase separation, which is a misleading picture. The only way to answer this question properly is to increase the size of the system, which may not be feasible if the number of molecules to handle becomes too large.

2.4 Kirkwood-Buff integrals and computer simulations

The integrals of the correlation functions can be related to the mechanical stability. For each pair of sites (a_i, b_j) with a_i on molecule of species i and b_j on molecule of species j, the corresponding site-site correlation function are noted $g_{ij}(r)$ we defined running integrals (RKBI) $G_{ij}(r)$ as:

$$G_{ij}(r) = 4\pi \int_0^r dt\, t^2 [g_{ij}(t) - 1] \tag{17}$$

In 1951 Kirkwood and Buff(Kirkwood & Buff, 1951) showed how the integrals of the correlation functions $G_{ij} = G_{ij}(r \to \infty)$ were related to physical quantities such as the isothermal compressibility κ_T, the total volume V and the partial molar volumes $\bar{V}_a$ of each species a , as well as density derivatives of the chemical potentials $(\partial\mu_i/\partial\rho_j)_{T,\rho_k}$. These relations can be inverted to express the Kirkwood-Buff integrals (KBI) G_{ij} themselves in terms of the physical quantities. For a binary mixture, these relations can be condensed into the form:

$$G_{ij} = G_{12}\delta_{ij} + (\beta\kappa_T - \frac{\bar{V}_1\bar{V}_2}{D})(1-\delta_{ij}) - \frac{1}{x_i}(\frac{\bar{V}_j}{D} - V)\delta_{ij} \tag{18}$$

where x_i is the mole fraction of species i , δ_{ij} is a Kronecker symbol, and

$$D = \frac{\rho_1}{\rho_2}(\partial\beta\mu_1/\partial\rho_1)_T = \frac{\rho_2}{\rho_1}(\partial\beta\mu_2/\partial\rho_2)_T \tag{19}$$

where $\rho_i = x_i\rho$ is the partial density of species i and the second equality holds because of the Gibbs-Duhem equality $\rho_1\delta\mu_1 + \rho_2\delta\mu_2 = 0$.

This inversion operation was initially thought to provide some insight into the structure of the aqueous mixtures(Ben-Naim, 1977). In a landmark paper(Matteoli E. & Lepori L., 1984), Matteoli and Lepori have provided the behaviour of the experimental KBI for a variety of binary mixtures. The reproduction of this data by computer simulations and its interpretation is still a challenging open problem.

In 1961 Lebowitz and Percus have shown(Lebowitz& Percus, 1961) that the asymptote of $g_{ij}(1,2)$ as obtained in a finite size system, even when periodically extended is not 1, as would be expected from Eq(16) but rather

$$\lim_{r\to\infty} g_{ij}(1,2) = 1 - \frac{1}{N\rho\sqrt{x_i x_j}}(\frac{\partial\rho_i}{\partial\beta\mu_j}) \tag{20}$$

This relation has been largely ignored, mainly because the exact asymptote of g_{ij} is never required in practice, and most thermodynamical properties are computed directly in simulations, instead of using relations such as Eqs.(13,14). We have checked(Zoranic et al., 2007) for many systems that the incorrect asymptote in Eq.(20) is never a serious problem for many quantities. This problem has resurfaced only recently when the computation of the KBI through Eq.(17) requires the proper asymptote. In the absence of any estimate of the correction, we have devised an empirical way of correcting for the asymptote which allows a very good estimate of the KBI in some systems(Mijakovic et al., 2011; Perera et al., 2007). It consists in shifting the incorrect asymptote value a_{ij} to 1 with the help of a switch function $S_{ij}(r)$:

$$g_{ij}^{(corrected)}(r) = g_{ij}(r)[1 + (1 - a)S_{ij}(r)] \tag{21}$$

with $S_{ij}(r) = 0.5(1 + \tanh((r - R_{ij})/\kappa_{ij})$, where we take the switch distance $R_{ij} = \sigma_i + \sigma_j$ and the switch smoothness $\kappa_{ij} = 1$Å. These values guarantee that the RKBI defined in Eq.(17) is unaltered for the first few neighbours. An illustration of the need for such a correction is given in Fig.4 below for the case of liquid methanol (the OPLS model was used). We look at the O-O and M-M site-site correlation functions (top panel) (O=oxygen atom and M=CH_3 methyl group considered as a pseudo atom) and they appear to go to 1 at long range as expected. A closer look, however, (middle panel) shows that they both go to an asymptote slightly lower (the cyan line), as predicted by Eq(20). We therefore use a shifting function (shown in green) to shift the asymptote by avoiding to shift it for the first neighbours. The bottom panel shows the RKBI for shifted and unshifted functions. Only the functions corrected through Eq(21) have the proper horizontal asymptote, in actual very good agreement with the experimental value(Perera et al., 2007) shown as magenta line. It is seen that the incorrect asymptote leads to a typical curved RKBI as can be seen in several works that are unaware of this problem.

2.5 Illustration of the failure of computer simulations: the acetone-water mixtures

When we tackled the acetone-water mixture back in 1998(Perera & Sokolic, 2004), we were not fully aware of the problems we would meet. Indeed, the credo of computer simulations is that near exact results are to be obtained if "proper" conditions are met. The acetone-water system puts this credo down! The first problem was with force fields. There were 3 force fields available for neat acetone, the OPLS(Jorgensen et al., 1990) and the FHMK(Ferrario et al., 1990) that were both produced around the 1970s and checked to give reasonnably accurate thermophysical properties of liquid acetone. The recently introduced WS force field(Weerasinghe & Smith, 2003) that was adjusted to give proper magnitude of the KBI of the acetone-water mixtures and was supposed to avoid the demixion problems encountered with the older force fields. This model is interesting because it shows that altering the force field instead of properly handling the second scale of the system leads to unphysical behaviour of the KBIs. Fig.5 below shows the KBI as obtained from various experiments and from computer simulations using several combinations of force fields for acetone(OPLS,FHMK,WS) and for water(SPC/E, TIP4P).

There are several striking features in this figure. Perhaps the most apparent is that the KBIs obtained by most simulations are 7 fold larger than the experimental values. The physical reason for such large values is due to the demixing behaviour of these simulated mixtures: they tend to overestimate the correlation length ξ in Eq.(16), hence leading to large values of the KBIs. The second feature is that, the KBI obtained by the WS the force field, which has been adjusted in order to lower the values of the KBIs, does indeed lead to the proper magnitude of these latter, but also to a very different physical behaviour: the water-water KBI monotonously increases with acetone mole fraction instead of going through a maximum

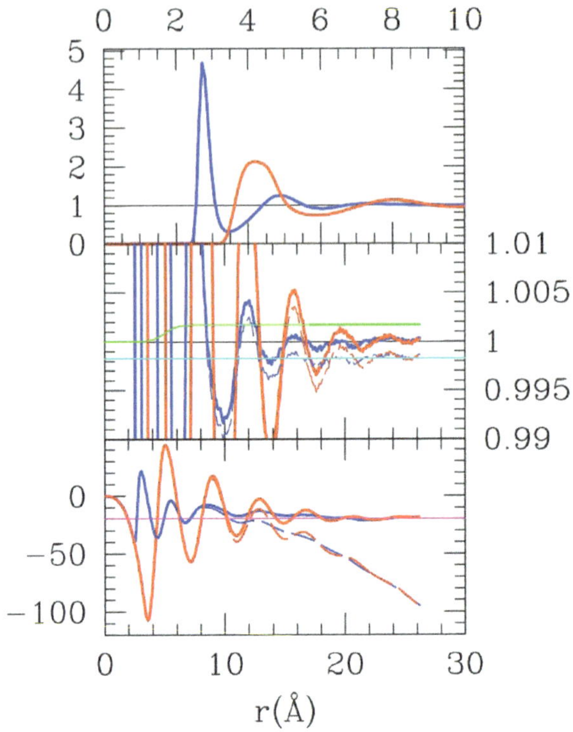

Fig. 4. Illustration of asymptote correction for neat liquid methanol(see Eq(21) and the text for explanations)

around acetone mole fraction 0.6. This latter feature can be interpreted in two different ways: on one hand it could be considered as a small price to pay to avoid demixing and keep the KBI under reasonable scale, on the other hand it could be interpreted as incorrectly describing the structure of the micro-heterogeneity of these systems. This latter interpretation has for now a rather thin theoretical argument, since we don't know how to differentiate between two different topologies of the micro-structure within computer simulations. This is illustrated in Fig.6 below by showing 3 snapshots illustrating the very different scale of the MH in the three different simulations.

It can be seen that the water and acetone microdomains are not of the same size in each of the shots. Snapshot 1 looked like a detail of snapshot 3. However the same simulation performed in a larger system shows a clear demixion (middle snapshot). On the other hand, the model of snapshot 3 shows incorrect KBI behaviour, without maximum, while the two others have the proper KBI behaviour, but are 7 times too large. We believe that proper KBI will be obtained when the size of the micro-domains will be the intermediate between the two extreme shown above. It is particularly difficult to figure out which parts of the acetone force field should be altered in order to arrive at such a result. Since the force fields for both neat systems are relatively good, one may ask if it is the use of the LB rules that creates the difficulties encountered here. It turns out that the Coulomb interaction energies are in general 10 times

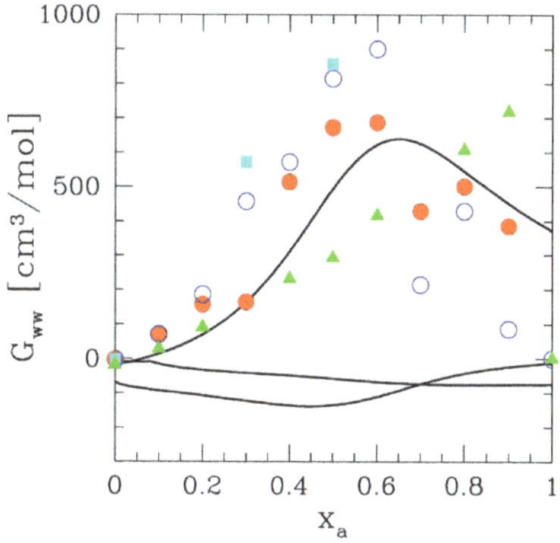

Fig. 5. KBI for acetone-water mixtures. (lines) expt data from (Perera et al., 2006), (green triangles) WS+SPC/E MD data from (Weerasinghe & Smith, 2003), (squares) FHMK+SPC/E, (open circles)OPLS+SPC/E (red circles) OPLS+TIP4P (the last 3 data are shown divided by 7)

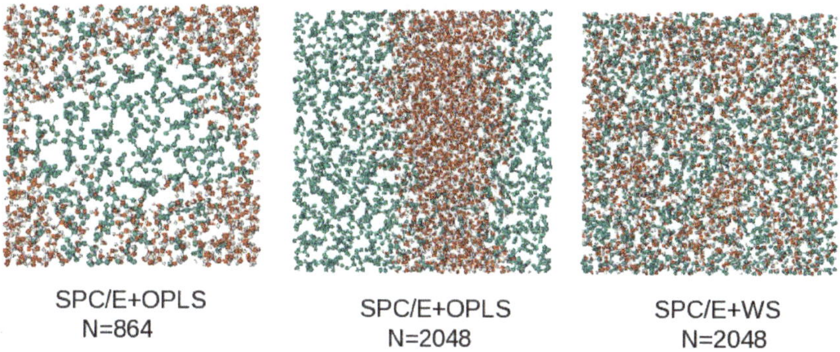

Fig. 6. Snapshots acetone water. (right) OPLS/SPCE N=864, (middle) OPLS/SPCE N=2048, (Left) WS/SPCE N=2048,

larger than the corresponding LJ ones(Zoranic et al., 2007). Hence, it is difficult that a small interaction alone could be responsible for these variation. It seems more appropriate to think that it is a particular combination of changes of both energies that would reproduce the correct MH. Of course, such a change should not alter the neat systems in any undesirable way.

The problem of finding a proper force field for this particular mixture is still an open problem.

3. Micro-structure of complex neat liquids

As stated in the Introduction, we find instructive to differentiate between the local order in neat liquids and mixtures. We will describe the latter in the next section. The local order in simple liquids, such as nitrogen, is simply the dense packing structure found in hard spheres. When considering small polar molecules, such as benzene for example, the preferred orientations of the dipoles tend to create some local order. For the moderate values of the dipole moments, this local order does not alter significatively the hard-sphere liquid packing structure, for example to the point of creating specific clusters. Chain-like stuctures appear only for very large values of dipole moments. The H-bond interaction is very strong, of the order of 200 kJ/mol at the contact distance of 3Åfor SPC/E water model (Zoranic et al., 2007). As a consequence, it creates a strong local order in forms of specific clusters, which can be detected by various experimental techniques(Guo et al., 2003; Ludwig, 2005).

Since clusters are meta-objects, one can re-consider the system as a fluid of clusters floating in the middle of a sea of monomers. This way, one can measure the cluster-cluster interaction and their correlation functions. However, there is a problem of defining the cluster as a well defined entity. In order to achieve this, one needs the cluster distribution function in terms of the monomers. It turns out that clusters in complex liquids of small molecules are not defined as sharply as micelles, and therefore cannot be easily considered as meta-molecules the same way the constituent molecules are themselves. One can rather consider such clustered liquid as a "plasma", in the sense that the meta-particles are "broken" into their constituents. The picture that emerges from this description is that complex liquids are akin to some primordial state of matter. This point of view open new interesting theoretical perspectives from statistical mechanics of liquids.

For now, let us explore the details of self-clustering in some H-bonded liquids. We define the cluster distribution probability in terms of sites. This has the advantage of distinguishing the topology of how the molecules associate into clusters. This probability is defined as:

$$P(n) = \frac{\sum_k s(n,k)}{\sum_m \sum_k s(m,k)} \tag{22}$$

where $s(n,k)$ represents the number of clusters of size n in the configuration k (Perera et al., 2007; Zoranic et al., 2007).

As shown in the Theory section, the presence of specific clusters is detected through the pre-peak of the site-site structure factors that are related to the cluster forming interaction, namely the H-bonding interaction. So, we have two independant tools to detect clusters in neat liquids.

3.1 Acohols

The probability distributions for some of first alcohols in the nomenclatura, namely methanol, ethanol and tert-butanol(TBA) are shown below in Fig.7. Both bonding oxygen O sites and non-bonding methyl M sites are shown. These are obtained by computer simulations of the OPLS models(Jorgensen, 1986), with N=2048 molecules, under ambient conditions.

One can see that all three alcohols have a specific cluster peak about n=5. The pure monomer probability indicates the amount of clustering: a monomer cluster probability higher than that of the specific cluster indicates that most molecules are free monomers. Hence, methanol appears as the least clustered of the three alchohols, while TBA is almost entirely clustered. There may be a topological reason related to geometry of the molecules. Indeed, since the strenght of the H-bonding interactions is the same for all models (the underlying electrostatic forces are the same), the difference in clustering can only come from molecular shapes. Since

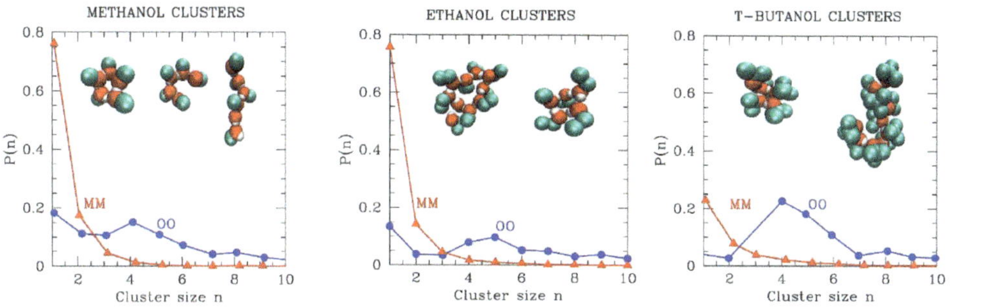

Fig. 7. Cluster distribution probabilities. (Left) methanol, (center) ethanol, (right) tbutanol. Typical cluster shapes are shown in the insets.

methanol molecules form chains through O-O contacts, the methyl groups are disordered and the compactness of the liquid cannot be achieved if all molecules were forming chains. Conversely, TBA molecules are pyramid-like, hence they can form small micelles with the H-bond interactions grouped at the center while the three methyl groups stay outside. This structure can be compactly reproduced through the whole liquid, with defects that do not destroy the compactness of the liquid. These considerations can be confirmed through the analysis of the pair site-site distribution functions shown in Fig.8, together with the corresponding structure factors.

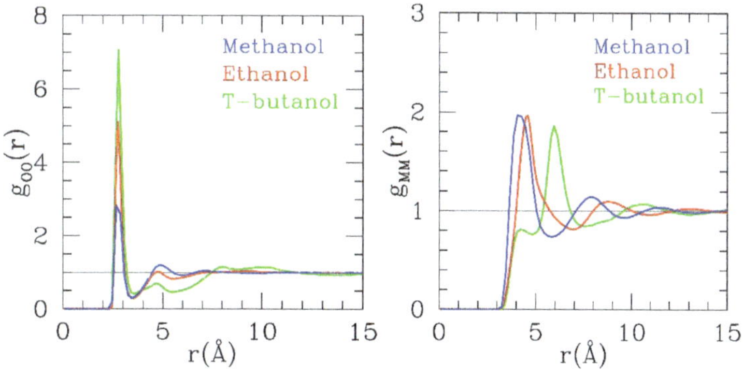

Fig. 8. Site-site distribution functions O-O(left) and M-M(right) for methanol, ethanol and tbutanol.

The O-O correlation functions between the oxygen sites have the same features previously observed for pure methanol in Fig(2): a sharp first peak due to H-bonding, followed by weak and damped oscillatory features. In contrast, the MM correlations look more structured, indicating the packing effects expected in a dense liquid. As stated in paragraph 2.2, the weak oscillations produce a pre-peak in the O-O structure factor shown in the Fig.(9) below, and which witnesses the presence of new emerged entities that are the clusters. It is seen that the pre-peak for TBA is of larger amplitude, which confirms that the corresponding clustering is very strong, as witnessed by the larger specific peak in P(n) for TBA.

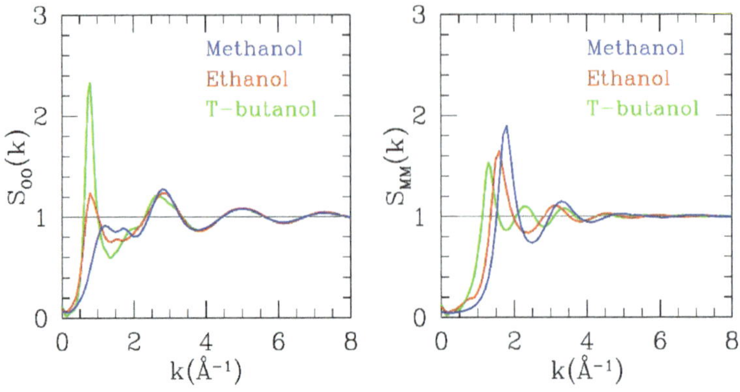

Fig. 9. Structure factors corresponding to site-site correlation functions displayed in 8

3.2 Amides

Amides are an interesting class of small molecules, that are of biological and biochemical interest. Some amides such as formamide are fully H-bonding while others such as N-methyl-formamide (NMF) are weakly H-bonding, and dimethyl-formamide(DMF) is not bonding. We have studied formamide, nmethyl-formamide(NMF) and DMF. For the amides, we have used the Cordeiro force field(Cordeiro et al., 2006), with N=2048 and ambiant conditions as described in Refs.(Zoranic et al., 2007; 2009). The cluster distributions for the three amides are shown in Fig.10 below. It is seen that none of them shows a specific peak as that seen for alcohols. In all three cases, the n-mer probability is always higher than that of the (n+1)-mer, indicating a distribution of cluster similar to that found for Lennard-Jonesium, that is a straight exponential decay.

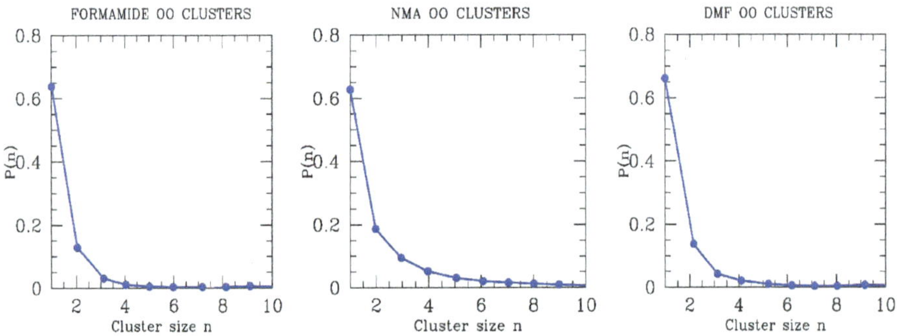

Fig. 10. Cluster distribution probabilities. (Left) formamide, (center) NMF, (right) DMF

The O-N site-site distributions (oxygen-nitrogen) for all three neat amides are shown in Fig.11(left panel). The sharp first peak for formamide and NMF indicate their respective H-bonding tendencies, and this peak is obviously absent for DMF. The structure factor of formamide and NMF (right panel) show a weak-prepeak feature somewhat similar to methanol (Fig.9).

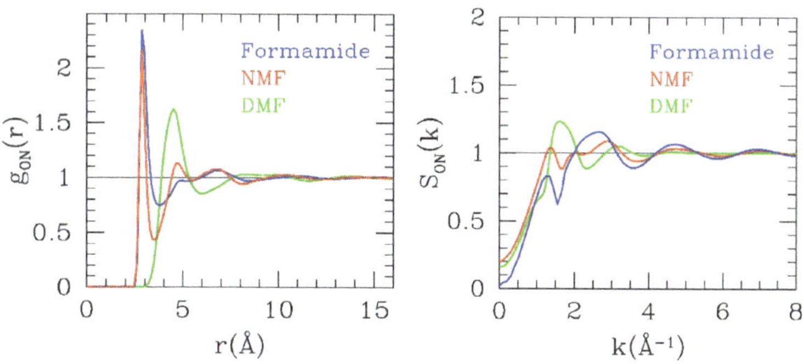

Fig. 11. (Left) O-N (oxygen-nitrogen) site-site distribution functions for FA(blue) NMF(red) and DMF(green); (right) Structure factors corresponding correlations functions in the left panel. (right)

Perhaps the most intriguing feature found here is the absence of specific clustering in formamide which is a H-bonding liquid, in sharp contrast with alcohols. This may be due to the geometry of the molecule, which does not allow dense packing when clusters exits. This argument indicates that strong directional interaction is not enough, and that entropic effects due to packing can frustrate the local ordering. The constraints of packing effects and local ordering are therefore to be tuned in particular ways to give rise to specific clustering. Another typical example of such features is liquid water.

3.3 Water: a very peculiar micro-structure

We have studied clustering in SPC/E (Berendsen et al., 1987) and TIP4P (Jorgensen & Madura, 1985) water, which are very similar, and both widely used in several studies. Fig(12) shows, for SPC/E, that there is no specific cluster peak in the cluster distribution, hence no specific clusters in water. This is equally consistent with similar observations by other authors(Dougan et al., 2004).

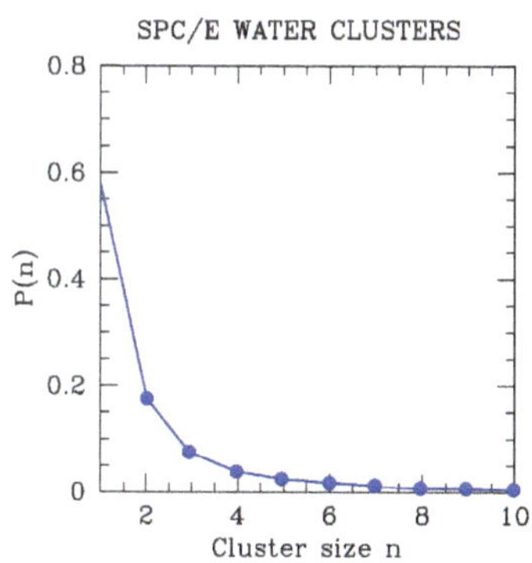

Fig. 12. Cluster distribution probabilities for SPC/E water

A look at the O-O site-site function in Fig(13) does not show any apparent peculiarity in the correlations, aside the strong first peak which witnesses the H-bonding tendency.

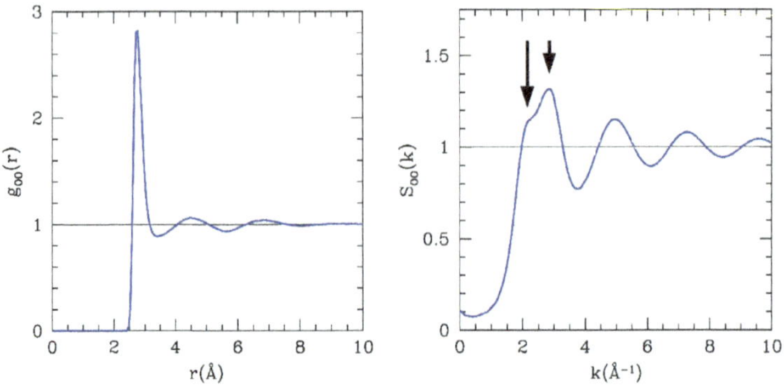

Fig. 13. SPC/E water O-O (oxygen-oxygen) correlations (left) and corresponding structure factor (right) The arrows show the two peaks feature discussed in the text

The O-O structure factor shows a weak shoulder, that is not much of a pre-peak. Hence, pair correlation infomation is consistent with direct cluster calculations: no specific clustering. This is really surprising! Water is often thought to be the best example of a clustering liquid. In fact, a closer look at the long range part of $g_{ww}(r)$ reveals a surprizing *absence* of correlations, a unique feature that have been only pointed out recently by one of us(Perera, 2011). It turns out that water has a very strong and very peculiar local order based on H-bonding, but it does not take the trivial cluster form. This is a much more abstract form of order, based on correlations rather than direct interactions and clusters such as those shown in the inset of Fig.7, for example. This peculiar form of order of liquid water is seen from the fact that the various site-site correlation functions decay abruptly to 1 beyond $R_c \approx 10$Å, and is a very intriguing feature observed in all water models(Perera, 2011). It suggests that every water molecule is the center of specific correlations in the range of R_c, that vanish abruptly beyond. The corresponding entity, we have suggested to name it a "correlon"(Perera, 2011). Since every water molecule is the center of a correlon, it means that correlons are ideal between them and that they do not interact with each other. This is a particularly strange and novel concept, that would require further investigations, namely in how it would serve to explain the various anomalous properties of water. It is noteworthy that a simple two repulsive core model, which has no electrostatic nor H-bonding interactions, is able to reproduce the peculiar structure seen for real water(Perera et al., 2009). This emphasises that the correlon feature may be essentially the consequence of entropic effects related to local competing packing constraints.

4. Micro-heterogeneity: the aqueous-alcohol mixtures

Alcohol molecules are perfect amphiphiles, hence water-alcohol mixtures are akin to micro-emulsions (ME). However, these mixtures are never considered as such, probably because the rich manifestations of ME (micelles of various shapes) are absent from these lesser mixtures. Nevertheless, the basic ingredient found in micro-emulsions, which is the micro-immiscibility is common to both systems. It is only a matter of size of the oil and water domains: in aqueous alcohol mixtures these are of the size of the nanometer, while in ME they are more of the size of the micro-meter. We call the former mixtures **molecular emulsions**, in order to emphasize the analogy by keeping in mind the scale. We will use here this analogy to build a theoretical background of aqueous alcohol mixtures. The major problem is that for many force fields the border between demixing and micro-heterogeneous domain formation is an intrinsic problematic of the simulations. We saw an example of this for the acetone-water mixtures in Section 2.5. The only cure for this problem seems to use much larger system sizes, which is often too expensive to be routinely used. A suitable theoretical procedure that would circumvent this problem would be most welcome.

4.1 The Teubner-Strey theoretical description of micro-emulsions

In a landmark article(Teubner & Strey, 1987), Teubner and Strey described the generic form of the structure factor for micro-emulsions by starting from field theoretic Landau-deGennes Hamiltonian free-energy. Such fre-energy includes local variation of the order parameter, which is in fact the local density of the system. Since molecules sizes are of the order of the Angstrom, and that oil-water domains are of the size of the micro-meter, the five order of magnitude difference is a good justification to consider density variations as being mesoscopic, with molecular details entirely omitted. This description led them to produce the following structure factor

$$S(k) = \frac{A}{c_o + c_2k^2 + a_2k^4} \tag{23}$$

Based on the signs of the three coefficients, they were able to distinguish between 3 regimes. The first regime is governed by pure concentration fluctuations, where the structure factor has the well known Ornstein-Zernike form:

$$S(k) = \frac{A}{c_0 + c_2k^2} \tag{24}$$

where a single correlation length emerges. It is noteworthy that this structure factor is entirely consistent with the asymptotic decay in Eq.(16). This structure factor can only develop a peak at k=0, when the correlation length increases, due to increase of fluctuations, as when nearing a phase transition. The second regime corresponds to well defined oil and water domains, which introduces a second lenght d as the size of the domains. In this regime, the structure facor can have a well defined pre-peak positioned at $k_P \approx 2\pi/d$. The k=0 value of the S(k) is generally smaller than the height of the pre-peak, which indicates that concentration fluctuations do not alter the domains. The last regime is the transition between the previous two, when the domain size is not well distinguished from size of the concentration fluctuations. In this regime, called the Lifshitz regime (Ciach and Godz, 2001), fluctuations form and destroy the domains. In a sense this is akin to a critical point, but where fluctuations can never diverge because domains stabilize their divergences. This seems curiously similar to what we have described in the case of the acetone water mixture. The main problem is that domains in micro-emulsions are 3 orders of magnitude larger than what we have in our aqueous-mixtures. Moreover, the molecular nature of the domains cannot be ignored at such scales. Hence, we cannot use the TS formulation in any straightforward manner by starting from field theoretic free-energy considerations. Nevertheless, one feels that a proper statistical theory of liquid should be able to encompass both micro and molecular emulsions.

4.2 A statistical theory of molecular emulsions

In a recent work on aqueous-TBA mixture (Kezic & Perera, 2011), two of us (Kežić and Perera) have used an extension of the molecular OZ equation to arrive at an expression exactly similar to the TS expression. In order to do that, one starts from the molecular OZ equation, and writes a small-k expansion of it. The molecular OZ equation can be written in matrix form as (Fries & Patey, 1985)

$$\tilde{H}_\chi(k) = \tilde{C}_\chi(k)[I - (-)^\chi \rho \tilde{C}_\chi(k)]^{-1} \tag{25}$$

where ρ is the density, and the matrix indexed by χ contains for elements of the projections on a basis of rotational invariants of the pair and direct correlation functions in Fourier-Hankel space: $A_{mn} = \{\tilde{A}_\chi(k)\}_{mn} = \tilde{a}_{mn\chi}(k)$, where $A = H, C$ and $\tilde{a}_{mn\chi}(k)$ is the Fourier transform of the correlation function $a_{mn\chi}(r)$ with $a = h, c$, the pair and direct correlation functions, respectively. The details of how the projections $a_{mn\chi}$ are related to the full correlation function $a(1,2)$ has been explained in several articles (Fries & Patey, 1985) and is now textbook knowledge(Hansen & McDonald, 2006). In particular, it is important to recall that the $\tilde{a}_{mn\chi}(k)$ are in fact what is called a $\chi-$transform involving projections $\tilde{a}^{mnl}(k)$ each of which are Fourier-Hankel transforms of order l of the functions $a^{mnl}(r)$.

This matricial equation holds also exactly as above for a mixture with the appropriate indexing for handling the correlation functions of the mixture (Kusalik & Patey, 1988). The mathematics of the MOZ that concern the present development are really simple. In short, we wish to examine the behaviour at small-q of any $\tilde{h}_{ij;mn\chi}(k)$. After some algebraic manipulations(Kezic & Perera, 2011) one arrives at the following expression of the desired limit:

$$\tilde{h}_{ij;mn\chi}(k) = \frac{\tilde{t}_{ij;mn\chi}(k)}{1 - \tilde{\gamma}_\chi(k)} \tag{26}$$

where the functions in the numerator and the denominator are related the correlation functions appearing in Eq.(25) The details of these calculations are not relevant to the argument, since the general form above is exactly equivalent to MOZ. Since the function $\gamma_\chi(r)$ involves only direct correlation functions, which are short ranged functions, one can expand this function around q=0, and one has

$$\tilde{h}_{ij;mn\chi}(q \to 0) \approx \frac{\tilde{t}_{ij;mn\chi}(0)}{1 - \tilde{\gamma}_{\chi;0} - q^2\tilde{\gamma}_{\chi;2} - q^4\tilde{\gamma}_{\chi;4} - q^6\tilde{\gamma}_{\chi;6} - ...} \tag{27}$$

where only even powers of q are retained because of the general symmetry of all correlation functions $a(-r) = a(r)$. The small-q expansion involves expanding Fourier-Hankel transforms and it is not so simple to express each of the $\tilde{\gamma}_{\chi;n}$ in terms of integrals of the $\gamma_\chi(r)$ function, as it would have been in case of a simple Fourier transform. But this is a minor detail irrelevant to the generality of the discussion here. The equation above retained to order q^2 alone will lead to the well known discussion in terms of the correlation length(Fisher, 1964). We briefly recall this argument here

$$\tilde{h}_{ij;mn\chi}(q \to 0) \approx \frac{\tilde{t}_{ij;mn\chi}(0)}{1 - \tilde{\gamma}_{\chi;0} - q^2\tilde{\gamma}_{\chi;2}} = \frac{A_{ij;mn\chi}}{\xi^{-2} + q^2} \tag{28}$$

with ξ the correlation lenght expressed as $\xi = \sqrt{-\tilde{\gamma}_{\chi;2}/(1-\tilde{\gamma}_{\chi;0})}$. This function is seen to be a Lorentzian, and its inverse Fourier transform is a Yukawa function:

$$\lim_{r\to\infty} h_{ij;mn\chi}(r) \approx A_{ij;mn\chi}\frac{\exp(-r/\xi)}{r} \tag{29}$$

It is important to note that the correlation length ξ is the same for all projections and it is uniquely defined for the system.

The correlation length is sensitive to density fluctuations in a pure liquid or concentration fluctuations in a mixture. In particular, it is a very useful probe of the approach of any global phase transition regions, since fluctuations are enhanced in their vicinity, hence the correlation length tend to increase and diverge at the limit of stability of the phase -the so-called spinodal. But what happens when the system does not phase separate in global fashion and only micro-segregates, like in all alcohol water mixtures? In order to study such phenomenon, one should retain one more order in the q-expansion, which amounts to explore distances shorter than the domains of critical fluctuations which are of several tens of Angstroms. In principle, there should be no reason to stop the expansion at q^4 or q^6. However, since the expansion at q^4 has been successfully used from micro-emulsions down to aqueous mixtures of relatively short chain alcohol molecules, such as diols, triols and others, there are good reasons to try the TS approximation first.

The molecular level TS equivalent of the MOZ equation reads

$$\tilde{h}^{(TS)}_{ij;mn\chi}(k) = \frac{\tilde{t}_{ij;mn\chi}(0)}{1-\tilde{\gamma}_{\chi;0}-q^2\tilde{\gamma}_{\chi;2}-q^4\tilde{\gamma}_{\chi;4}} = \frac{\tilde{t}_{ij;mn\chi}(0)}{a_2+q^2c_1+q^4c_2} \tag{30}$$

where we have adopted the original TS notation(Teubner & Strey, 1987) in the denominator of the second equality, and where the new coefficients can be redefined in terms of the domain size d and the correlation length ξ as follows(Kezic & Perera, 2011):

$$a_2 = 1-\tilde{\gamma}_{\chi;0} = (\bar{d}^2+\xi^2)^2 \tag{31}$$

$$c_1 = -\tilde{\gamma}_{\chi:2} = 2(\bar{d}\xi)^2(\bar{d}^2-\xi^2) \tag{32}$$

$$c_2 = -\tilde{\gamma}_{\chi:4} = (\bar{d}\xi)^4 \tag{33}$$

where $\bar{d} = d/2\pi$. The definitions in the second equalities have been introduced(Kezic & Perera, 2011) such as the inverse Fourier transform of the TS function reads exactly:

$$h^{(TS)}_{ij;mn\chi}(r) = \int d\vec{q}\,\exp(i\vec{q}.\vec{r})\,\tilde{h}^{(TS)}_{ij;mn\chi}(k) = \frac{\tilde{t}_{ij;mn\chi}(0)}{\pi^2(\bar{d}\xi)^3}\frac{\exp(-r/\xi)}{r}\sin(r/\bar{d}) \tag{34}$$

One sees that the relations between the three TS coefficients(a_2, c_1, c_3) and the combinations of length parameters $(\bar{d}, \xi)$ are stringent, since the moments of the $\gamma_\chi(r)$ functions need to have specific signs to match the algebraic forms and the positivity conditions. Since these functions cannot be obtained unless one calculates the direct correlation function expansion coefficients, this information is unavailable to us. The obtention of such coefficients needs the arsenal of liquid state theory, and reliable output from such theory for realistic molecular system is not well developped at present. Therefore, in our approach to strong microheterogeneous systems, we will test the TS functional form to correlation functions determined by computer simulations.

4.3 Aqueous-tbutanol mixtures: an illustration of the molecular emulsion concept

Tbutanol is a nearly spherical molecule, but it in fact very asymmetrical since it has a bulky group of 4 methylgroups and a strong polar OH head: it is a small amphiphile. Therefore, just like methanol or ethanol, it can be expected to form a molecular-emulsion, in analogy with micro-emulsion of larger alcohols such as 2butoxy-ethanol, for example (Koga, 2007). However, when dealing with computer simulations, aqueous methanol and aqueous ethanol do not have the same problems encountered for aqueous tbutanol. Indeed, the OPLS model of the two former alcohols can be reasonnably simulated with N=2048 system sizes (Kezic et al., 2011; Mijakovic et al., 2011; Perera et al., 2007) and special force fields designed to account for microheterogeneity (Weerasinghe & Smith, 2005) do not alter significantly the results. However, aqueous-tbutanol mixtures with OPLS tbutanol tend to produce very high KBI and consequently it is a good candidate for force field alterations(Lee & van der Vegt, 2005). From experimental point of view, it has also been noticed that tbutanol-water mixture appear as very micro-heterogeneous(Bowron et al., 1998).

Fig.14 below illustrates our considerations on the applicability of the TS structure factor for molecular emulsion, for the case of the water-tbutanol mixture for tbutanol mole fraction $x = 0.2$. This corresponds to the maximum of the experimental Kirkwood-Buff integrals (Perera et al., 2006) and therefore corresponds to a region of high concentration fluctuation where the microheterogeneity should be quite large. Indeed, Fig.15 shows a snapshot from our simulations of the N=2048 particle mixture of the SPC/E-OPLS models mixture. The micro-segregation is quite evident, inducing a clear partitioning of both species.

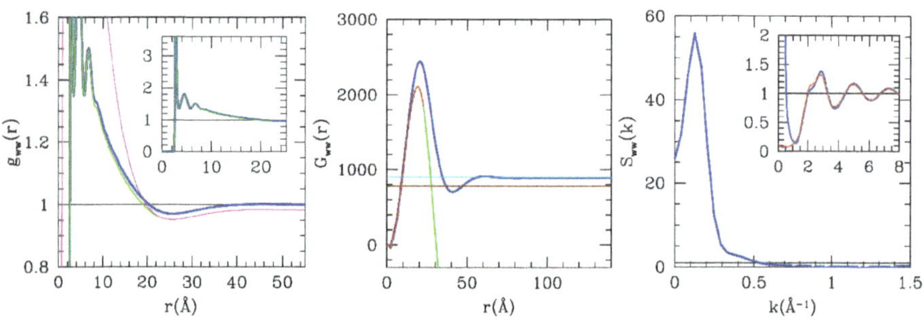

Fig. 14. Water-water correlations of the SPC/E water-OPLS TBA mixture (left) TS treatment (middle) corresponding RKBI and (right) structure factor (right)

The right panel of Fig(14)shows the uncorrected calculated water-water distribution function (in green), together with the version corrected for the TS effect (blue). The fitting function is shown in magenta. Note that a rescaling of the asymptote is required to set the proper asymptotical value to 1, which we do by applying the procedure in Eq(21). The correlation lenght corresponding to the TS fit is found to be $\xi = 8$Å, while the domain size is $\bar{d} = 6.7$Å. It is seen that both quantities are about the same order of magnitude, which correspond to strongly fluctuating domains, as shown in the snapshot of Fig(15). The TS-fit allows to extend the correlations much beyond the initial box size of 50Åas seen in the middle panel. The middle panel shows the RKBI calculated from the various correlation functions shown in the left panel. It is obvious that, if the initial form should have been retained, it would lead to value of the KBI or the order of 2500, which correspond to the peak of the RBKI (as often used in various publications), and too large compared to the various experimental estimates from (Matteoli E. & Lepori L., 1984) shown in brown horizontal line and our own experimental value(Perera et al., 2006) shown in cyan line. The TS extension (blue curve) allows to bring the wrong initial tendency to the value very close to our experimental estimate. The green curve shows the catastroph due to not shifting the incorrect asymptote to 1. Finally, the right panel shows the water-water structure factor. It is obvious that the pre-peak at $k \approx 1.2\text{Å}^{-1}$ is due to the oscillating feature brought by the TS fit. In the inset we plot the structure factor of the pure SPC/E water for comparison (red). It is seen that the major change is just the prepeak feature. In absence of the TS fit, the value at $k = 0$ of S(k) would be around 90, a factor 4 higher than expected from the experiments.

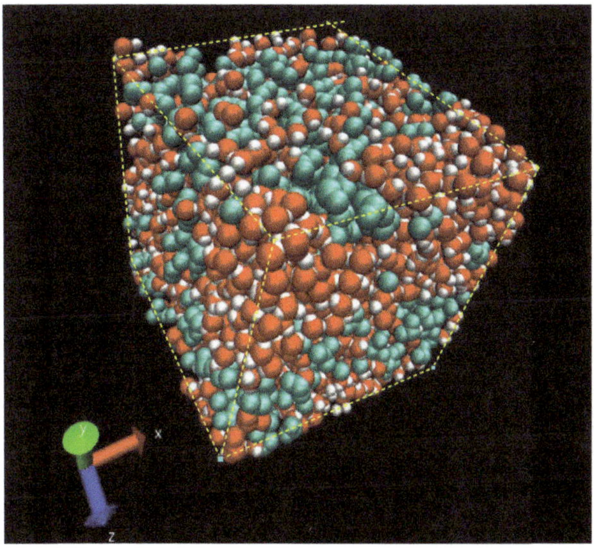

Fig. 15. Snapshot of the TBA/water mixture at TBA mole fraction x=0.2

The fitting form proposed in Eq(34) seems therefore to predict the asymptotical form of the correlations despite small size simulations. This theory predicts the proper effects of fluctuations and domain formation even at molecular scales. The next step is to produce such effects from theoretical considerations on the direct correlation function itself. For this, one needs to develop integral equations techniques beyond the current level of accuracy. Work along these directions have been started(Kezic and Perera, 2011) and is in progress.

5. Conclusion and perspectives

In their well known texbook on liquids and mixtures(Rowlinson & Swinton, 1982), Rowlinson and Swinton open the paragraph on aqueous mixture of non-electrolytes by these words: "No one has yet proposed a quantitative theory of aqueous solutions of non-electrolytes, and such solutions will probably the last to be understood fully". Three decades later, can we say that we have reached an better understanding of these mixtures? The various examples of this chapter indicate that much dark spots remain behind these systems, even when looked from traditional points of view. We still do not know how to clearly explain the various kinks in the properties shown in Fig.1 We have seen that computer simulations are very often unable to give unambiguous interpretations of what exactly are the outcomes of the statistical analysis that arc conducted. The statistical theories of liquids are still in infancy when facing some of the challenges concerning the nature of the correlations in these systems.

In view of such problems, we have proposed to concentrate these various problematics as arising from a single source: the existence of micro-heterogeneity as an intrinsic properties of these mixtures. Form such a viewpoint one can understand this property from two different directions: the fact that it arises from an underlying interacting molecular background, and how macroscopical properties themselves reflect the existence of this feature.

The two most important messages of this chapter are that, first, complex liquids are the seage of a new phenomenon of micro-structure or micro-heterogeneity, and second that computer simulations do not appear as very reliable statistical tools for studying such systems, for reasons intrinsic to the related physical phenomenon. As shown through the chapter, micro-heterogeneity is a physical phenomenon that is originally a spatial and temporal manifestation of simple specific interactions, but that occurs at a much larger scale, and which calls for a novator point of view. The system emerges new "entities" that are the siege of density fluctuations (neat liquids) or concentration fluctuations (mixtures). Instead of undergoing a full phase transition, the system becomes a mixture of these "entities' or "meta-particles", with the emphasis that fluctuations and domains may be interwoven in a way that needs further progresses to figure out how. This is precisely why micelle formation is not a true phase transition: fluctuations are not entirely responsible for the appearance of the micelle like they would if it was a true phase transition. Micellar systems are in fact simpler than the present problem, since each cluster looks like a small spherical particle, while micro-heterogeneity, in general, has no particular shape, as seen from various snapshots. This duality of the fluctuation versus the cluster is not properly handled by simulations, which often smooth out the fluctuations by undergoing a full phase transition for reasons intrinsic to the methodology itself: finite sizes and difficulty to handle stabilizing effects of the fluctuations, to name the two most important of them.

The issue of the competition between fluctuations and intrinsic local organisation is in fact more general than the topic of aqueous mixtures alone: it is about how "particles" could emerge from a discretised background, by the effect of direct interactions and statistical correlations and fluctuations. Our study shows how seemingly methodological problems are in fact hiding fundamental issues about the statistical description of liquids. Beyond this issue, it is the structure of matter itself that could be at stake. As stated in the Introduction, the emergence of "meta-particles" in a composite smaller particles assembly is a new way of looking at an old problem: how to distinguish an "object" from its background. We find it very remarkable that the one most common manifestation of the form of matter, namely liquids, should be the siege of such fundamental issues.

6. References

Allison S. K. ; Fox J. P. ; Hargreaves R. & Bates S. P. (2005). Clustering and microimmiscibility in alcohol-water mixtures: Evidence from molecular-dynamics simulations *Phys. Rev. B*, Vol 71 (2) (January, 2005) 024201-024201-5 DOI: 10.1103/PhysRevB.71.024201

Ben-Naim A. (1977). Inversion of the Kirkwood-Buff theory of solutions: Application to the water-ethanol system, *J Chem. Phys.*, Vol 67 (11) (March, 1977) 4884-4890 ISSN 1089-7690

Berendsen H. J. C. ; Grigera J. R. & Straatsma T. P.(1987). The missing term in effective pair potentials *J. Phys Chem*, Vol 91 (24) (November, 1987) 6269-6271 DOI: 10.1021/j100308a038

Bowron D. T.; Finney J. L. & Soper A. K. (1998). Structural investigation of solute-solute interactions in aqueous solutions of tertiary butanol *J. Phys. Chem.*, Vol. 102 (18) (April, 1998) 3551-3563 DOI: 10.1021/jp972780c

Ciach A. and Godz W. T. (2001) Nonelectrolyte solutions exhibiting structure on the nanoscale *Annu. Rep. Prog. Chem.* Sect. C Vol 97 269-314 DOI: 10.1039/b101392k

Chitra R. & Smith P. E. Molecular association in solution: A Kirkwood-Buff analysis of sodium chloride, ammonium sulfate, guanidinium chloride, urea, and 2,2,2-trifluoroethanol in water *J. Phys. Chem. B* Vol 106(6) 1491-1500 DOI: 10.1021/jp011462h

Cordeiro M. M. Santana W. P.; Cusinato R. & Cordeiro, J.M.M. (2006). Monte Carlo investigations of intermolecular interactions in water-amide mixtures *J. Mol. Struc. Theochem*, Vol 759(1-3) (February, 2006) 159-164 DOI: 10.1016/j.theochem.2005.11.016

Davies, M. I. (1993). Thermodynamic and related studies of amphiphile+water systems. *Chem. Soc. Rev.* Vol 22, 127-134 DOI: 10.1039/CS9932200127

P. G. de Gennes in *Microscopic and Dynamics of Liquids* J. Dupuy and A. J. Dianoux (1977), Plenum Press, New York ISBN 0-306-35733-X

de Visser C., Perron G & Desnoyer J. E. (1977). The heat capacities, volumes, and expansibilities of tert-butyl alcohol-water mixtures from 6 to 65^{0}C *Revue canadienne de chimie* Vol 55, 856-862, DOI: 10.1139/v77-119

Dixit S.; Crain J.; Poon W. C. K.; Finney J. L. & Soper A. K. (2002). Molecular segregation observed in a concentrated alcohol-water solution. *Nature*, Vo416 (6883) (April 2002) 829-32 DOI: 10.1038/416829a

Dougan L., Bates S. P. , R. Hargreaves R., Fox J. P. Crain J. Finney J. L. Reat V. Soper A. K (2004). Methanol-water solutions: A bi-percolating liquid mixtures *J. Chem. Phys.* Vol 121 (13) 6456-6462 DOI:10.1063/1.1789951

Ferrario M., Haughney M., McDonald I. R & Klein M. K.(1990) Molecular Dynamics simulation of aqueous mixtures-methanol, acetone and ammonia *J. Chem.Phys.*, vol 93 5156-5166 DOI:10.1063/1.458652

Floriano M. A, Caponetti E. & A. Z. Panagiotopoulos (1999) Micellization in model surfactant systems *Langmuir* Vol15 3143-3151 DOI: 10.1021/la9810206

Fries P. H. & Patey G. N (1985). The solution of hypernetted-chain approximation for fluids of non-spherical particles - a general method with application to dipolar hard-spheres *J. Chem. Phys.* Vol 82(1) 429-440 DOI: 10.1063/1.448764

Fisher M. E. (1964) Correlation functions+critical region of simple fluids *J. Math. Phys.* Vol 5(7) 944-962 DOI: 10.1063/1.1704197

Frish, H. L. and Lebowitz, J. L (1964), *The Equilibrium Theory of Classical Fluids*, W. A. Benjamin, New York. ASIN: B000PHQPES

Frank F. and Ives D. J. G. (1966). The structural properties of alcohol-water mixtures. *Q. Rev. Chem. Soc.* Vol 20, 1-44. DOI: 10.1039/QR9662000001

Guo J.-H. ; Y. Luo Y. ; A. Auggustsson A. ; Kashtanov S. ; Rubensson J.-E. ; Shuh D. K. ; Agren H.J. & Nordgren J.(2003). Molecular structure of alcohol-water mixtures *Phys. Rev. Lett.*, Vol. 91, No.15 , (October, 2003) 157401-157401-4 DOI: 10.1103/PhysRevLett.91.157401

Hansen J.-P. & McDonald I. R.(2006). *Theory of simple liquids*, Academic Press, Elsevier, ISBN-13:978-0-12-370535-8, Amsterdam, The Netherlands

Jorgensen W. L. (1986). Optimized Intermolecular Potential Functions for Liquid Alcohols. *J Phys Chem*, Vol 90 (7) , (1986) 1276-1284. DOI: 10.1021/j100398a015

Jorgensen W. L. ; Briggs J. M. & Contreras M. L. (1990). Relative Partition Coefficients for Organic Solutes from Fluid Simulations *J Phys Chem*, Vol 94 (4), (1990) 1683-1686 DOI: 10.1021/j100367a084

Jorgensen W. L. & Madura J. D.(1985). Temperature and Size Dependence for Monte Carlo Simulations of TIP4P Water. *Mol Phys*, Vol.56, No. , (1985) 1381-1392 DOI: 10.1080/00268978500103111

Kezic B.; M. Mijakovic M. ; ; Zoranic L. ; Sokolic F. ; Asenbaum A. ; Pruner C.; Wilhelm E. & Perera A. (2011). The microscopic structure of the Ethanol-Water mixtures *J Chem Phys*, (to be published 2011)

Kezic B. and A. Perera (2011) Towards a more accurate RISM integral equation theory of molecular liquids *J. Chem. Phys.* (2011, in prints)

Kezic B. and A. Perera, Aqueous tert-butanol mixtures: a model for molecular-emulsions (in preparation)

Kirkwood J. G.& Buff F. P.(1951). The Statistical Mechanical Theory of Solutions. I *J Chem Phys*, Vol19(6) (1951) 774-777 DOI: 10.1063/1.1748352

Koga Y. *Solution Thermodynamics and its application to aqueous solutions* Elsevier, Amsterdam 2007 ISBN: 978-0-444-53073-8

Koga Y., Siu W. W. Y & Wong T. Y. H (1990). Excess partial molar free energies and entropies in aqueous tert-butyl alcohol solutions at 25.degree.C *J. Phys. Chem* Vol94(19) 7700-7706 DOI: 10.1021/j100382a070

Kusalik P. G and Patey G. N. On the molecular theory of aqueous-electrolytes solutions. 1. the solution of the RHNC approximation for models and finite concentration *J. Chem. Phys.* Vol 88(12) 7715-7738 DOI: 10.1063/1.454286

Lama R. F. & Lu B. C-Y (1965) Excess Thermodynamic Properties of Aqueous Alcohol Solutions. *J. Chem. Eng. Data* Vol 10(3) 216-219 DOI: 10.1021/je60026a003

Lebowitz J. L & Percus J. K,(1961) Long-Range Correlations in a Closed System with Applications to Nonuniform Fluids *Phys. Rev.* Vol122(6) 1675-1691. DOI: 10.1103/PhysRev.122.1675

Lee M. E. & van der Vegt N. F. A. (2005). A new force field for atomistic simulations of aqueous tertiary butanol solutions *J Chem Phys*, Vol. 122, No.11 , (March, 2005) 114509-114509-13

Ludwig R. (2005). Isotopic Quantum Effects in Liquid Methanol *ChemPhysChem* Vol6 (7) 1369-1375 DOI: 10.1002/cphc.200400664

McAllister R. A. (1960) The viscosity of liquid mixtures *A.I.Ch.E. Journal* Vol 6(3) 427-434 DOI: 10.1002/aic.690060316

Matteoli E. & Lepori L. (1984). Solute-solute interactions in water. II. An analysis through the Kirkwood-Buff integrals for 14 organic solutes *J Chem Phys*, Vol 80 (6) (January 1984) 2856-2864 DOI: 10.1063/1.447034

M. Mijakovic M. ; Kezic B.; Zoranic L. ; Sokolic F. ; Asenbaum A. ; Pruner C.; Wilhelm E. & Perera A. (2011). Ethanol-water mixtures: Ultrasonics, Brillouin scattering and molecular dynamics *J Mol Liquids*, Vol 164 (2011) 66-73 doi:10.1016/j.molliq.2011.06.009

Perera A. & Sokolic F. (2004). Modeling nonionic aqueous solutions: The acetone-water mixture *J Chem Phys*, Vol. 121, No.22 , (December, 2004) 11272-11282

Perera A. ; Sokolic F. ; Almasy L. ; Westh P. & Koga Y. (2005). On the evaluation of the Kirkwood-Buff integrals of aqueous acetone mixtures *J. Chem. Phys.*, Vol 123 (2) , (July, 2005) 024503-024503-11 DOI: 10.1063/1.1953535

Perera A. ; Sokolic F. ; Almasy L. & Koga Y. (2006). Kirkwood-Buff integrals of aqueous alcohol binary mixtures *J Chem Phys*, Vo 124 (12) (March, 2006) 124515-124515-9 DOI: 10.1063/1.2178787

Perera A. ; Sokolic F. & Zoranic L. (2007). Microstructure of neat alcohols *Phys Rev E* , Vol. 75, No. 6, (June, 2007) 060502-060506

Perera A. ; Zoranic L.; Sokolic F. & Mazighi R. (2011). A comparative Molecular Dynamics study of water-methanol and acetone-methanol mixtures *J Mol Liquids*, Vol 159 (1) (February, 2011) 52-59 DOI: 10.1016/j.molliq.2010.05.006

Perera A. ; Rispe A. ; Zoranic L. ; Redha M. ; & Sokolic F. (2009). Water-like structure with repulsive double-core interactions *Mol. Phys.*, Vol 107 (13) (2009) 1349-1353 DOI: 10.1080/00268970902877787

Perera A. On the microscopic structure of liquid water *Mol. Phys.* Vol 109 (October 2011) 2433-2441 http://dx.doi.org/10.1080/00268976.2011.617712

Poland D. C & Scheraga H. A. (1965) Hydrophobic Bonding and Micelle Stability *J. Phys. Chem* Vol 69(7) 2431-2442 DOI: 10.1021/j100891a055

Pradhan T., Ghoshal P. & Ranjit R. Excited State Intramolecular Charge Transfer Reaction in Binary Mixtures of Water and Tertiary Butanol: Alcohol Mole Fraction Dependence *J. Phys. Chem. A* Vol112(5) 915-924 DOI: 10.1021/jp0770460

Rowlinson J. S & Swinton F. L. *Liquids and Liquid Mixtures* Butterworth Scientific 1982 (London)

Smith P. E. Cosolvent interactions with biomolecules: Relating computer simulation data to experimental thermodynamic data *J. Phys. Chem. B* Vol108(48) 18716-18724 DOI: 10.1021/jp0474879

Soper A. K.; Dougan L. ; Crain J. & Finney J. L. (2006). Excess entropy in alcohol-water solutions : a simple clustering explanation *J Phys Chem B*, Vol. 110, No. 8 , (March, 2006) 3472-3476

Tanford C. (1974) Thermodynamics of Micelle Formation: Prediction of Micelle Size and Size Distribution *Proc. Natl. Acad. Sci.* Vol 71 (5) 1811-1815 PMCID: PMC388331

Teubner M. & Strey R. (1987). Origin of the scattering peak in microemulsions *J Chem Phys*, Vol 87 (5) (September, 1987) 3195-3201 DOI: 10.1063/1.453006

Weerasinghe S. & Smith P. E. (2003). A Kirkwood-Buff derived force field for acetone and water *J Chem Phys*, Vol. 118 (23) 10663-1067 DOI: 10.1063/1.1574773

Weerasinghe S. & Smith P. E. A Kirkwood-Buff derived force field for methanol and aqueous methanol solutions *J. Phys. Chem. B* Vol 109 (31) 15080-15086 DOI: 10.1021/jp051773i

X. G. Wen *Quantum Field Theory of Many-Body Systems* Oxford University Press 2004, ISBN-10-0198530943

Wiggins P (2008) Life Depends upon Two Kinds of Water. *PLoS ONE* Vol 3(1): e1406. DOI:10.1371/journal.pone.0001406

Zoranic L. ; Sokolic F. & Perera A. (2007). Density and energy distribution in water and organic solvents: A molecular dynamics study *J Mol Liquids*, Vol 136 (3) (December, 2007) 199-205 DOI: 10.1016/j.molliq.2007.08.026

Zoranic L. ; Sokolic F. & Perera A. (2007). Microstructure of neat alcohols: A molecular dynamics study *J Chem Phys*, Vol. 127, No. 2, (July, 2007) 024502-024502-10

Zoranic L. ;Redha M.; Sokolic F. & Perera A. (2007). On the microheterogeneity in neat and aqueous amides: A molecular dynamics study *J Phys Chem C*, Vol 111 (43) (November, 2007) 15586-15595 DOI: 10.1021/jp0736894

Zoranic L. ;Redha M.; Sokolic F. & Perera A. (2009). Concentration fluctuations and microheterogeneity in aqueous amide mixtures *J Chem Phys*, Vol 130 (12) (March, 2009) 1124315-124315-12 DOI: 10.1063/1.3093071

11

Molecular Dynamics Simulations of Complex (Dusty) Plasmas

Céline Durniak and Dmitry Samsonov
*Department of Electrical Engineering and Electronics, University of Liverpool
United Kingdom*

1. Introduction

Complex or dusty plasmas are multi-component plasmas, which, in addition to the usual plasma components *i.e.* ions and electrons, contain micron-sized particles, also called grains or dust. These particles acquire high electric charges, and interact collectively over long distances. Like colloids, complex plasmas form solid- and liquid-like structures with long range correlations and exhibit phase transitions. Unlike colloids, they exhibit a range of dynamic phenomena such as particle-mediated linear and nonlinear waves, shocks, wakes, instabilities, etc. Complex plasma crystallisation has been theoretically predicted (Ikezi, 1986) and subsequently discovered in the early 1990s (Chu & I, 1994; Hayashi & Tachibana, 1994; Melzer et al., 1994; Thomas et al., 1994) giving rise to the interest in the field among the whole physics community.

Particles in complex plasmas can be illuminated by a laser light and easily observed with a video camera yielding their full kinetic information *i.e.* positions and velocities. This makes them useful as model systems for studying various phenomena in solids and liquids at the microscopic level (Thomas & Morfill, 1996). Since it is almost impossible and very expensive to observe the dynamics of real solids and liquids at the kinetic level, model systems are used to study the fundamental properties of phase transitions, diffusion, viscosity and elasticity. These model systems include colloids, granular media and complex plasmas, collectively known as soft matter. Phenomena in complex plasmas have analogues and applications in many different fields of science and technology such as plasma physics, fusion, solid state physics, fluid dynamics, acoustics, optics, material science, nanoscience, nanotechnology, environment protection, space exploration and astrophysics (Fortov et al., 2005a; Merlino & Goree, 2004; Morfill & Ivlev, 2009).

Arguably the most common technique to simulate complex plasmas is molecular dynamics (MD). It solves numerically the equations of motion for each individual particle comprising the system under investigation. It is also applied to simulate biomolecules, polymers, solids, colloids, granular media, atomic nuclei, galaxies, and stellar systems. MD simulations of complex plasmas are used in this chapter to determine their structural and dynamic properties as well as to identify the underlying physical mechanisms of various phenomena.

This chapter is organised as follows. Section 2 describes applications, natural occurrence, scientific significance, and multidisciplinary character of complex plasmas. MD methods and their experimental verification techniques will be detailed in Sections 3 and 4 respectively. Finally, numerical results and their comparison with experiments will be discussed in Section 5 and concluding remarks will be presented in Section 6.

2. Overview of complex (dusty) plasmas

The term "dusty plasmas" originated in astrophysics, where polydisperse (various size and shape) dust grains are exposed to various charged particles, ionised gases, and ionising radiation. Laboratory studies often involve high quality monodisperse (same size) microspheres added to gas discharges to study various complex and collective phenomena that do not occur in natural dusty plasmas. Thus the term "complex plasmas" (Samsonov et al., 2000) is often used instead, by analogy with "complex fluids", where similar complex phenomena are observed in multicomponent fluids.

2.1 Charging and forces

Mixed with a plasma, grains or microparticles collect ions and electrons and typically charge negatively due to higher mobility of electrons (Bronold et al., 2009; Goree, 1994; Melzer et al., 1994). However in the presence of ionising radiation, such as UV light, or thermionic emission, the particle charge may become positive. Typical charges are of the order of $\sim 10^4$ electrons for $\sim 10\ \mu$m diameter particles. The charge value is proportional to the electron temperature and to the particle radius. The time it takes a particle to reach an equilibrium charge, the charging time, is inversely proportional to the particle size and plasma density (Goree, 1994). Its typical value is ~ 100 ns for $\sim 10\ \mu$m grains in typical laboratory conditions. The particle charge is not constant and it fluctuates around an equilibrium. This may cause instabilities and if the charge value changes its sign, it may even result in particle coagulation.

Highly charged particles are affected by electric fields in the discharge and interact with each other electrostatically. Their interaction potential is usually assumed to be of a Yukawa (Debye-Hückel or screened Coulomb) type, if the background plasma is isotropic (Kennedy & Allen, 2003). This approximation has been also shown to be valid for particles levitating in a plasma sheath at the same height (Konopka et al., 2000) as in monolayer complex plasmas. Flowing plasma makes particle-particle interaction anisotropic with regions of negative and positive potentials (ion wake) (Melandsø& Goree, 1995; Vladimirov et al., 2003) and it also affects the charge of downstream particles.

Apart from the electrostatic force, grains are affected by other forces. Gravitational force becomes dominant for particles with a diameter $\gtrsim 1\ \mu$m in the bulk of the discharge. Large particles are pushed by the gravity down into the plasma sheath, where the electric field is strong enough to levitate them. This effect makes it necessary to use microgravity conditions in order to produce large three dimensional (3D) structures. Neutral drag force results from collisions with the gas molecules. It is equivalent to friction and damps particle motion. Streaming ions affect grains via an ion drag. This force is responsible for a void formation (Goedheer et al., 2009; Samsonov & Goree, 1999). Thermophoretic force arises due to a temperature gradient. It can be used for particle levitation (Rothermel et al., 2002). Intense light sources create a light pressure force, which is utilised for grain manipulation (Liu et al., 2003).

Complex plasmas can be characterised by two parameters. The coupling parameter $\Gamma = U/T$ is the average ratio of the electrostatic potential energy to the kinetic energy of particles. The screening parameter $\kappa = a/\lambda_D$ is the ratio of the interparticle distance to the screening (Debye) length. These parameters determine if the complex plasma is in the crystalline or a liquid state (Hamaguchi et al., 1997; Ikezi, 1986; Vaulina et al., 2002). Crystalline plasmas, which have long range correlations, are characterised by large values of the coupling parameter $\Gamma \gtrsim 170$ (Ikezi, 1986). Liquid phase state has smaller values $1 \lesssim \Gamma \lesssim 170$ and short range correlations. Solids and liquids are strongly coupled states. Gaseous state is weakly coupled ($\Gamma < 1$), with uncorrelated particle positions.

2.2 Natural occurrence and significance

Dust is abundant in space, where it is found in planetary rings, comet tails, interstellar clouds, and planetary nebulae. The sources of ionisation are also present, such as charged particles from cosmic rays and stellar winds, gas ionised by the stellar radiation, stellar radiation itself, and various radioactive elements in the dust, which emit charged particles and ionising gamma rays. Dusty plasma effects are believed to be involved in formation of dark spokes in Saturn rings (Hartquist et al., 2003). Charge fluctuations are known to enhance coagulation of particles (Konopka et al., 2005). This effect may influence the models of planet formation. Spacecraft and satellites are often charged by the solar wind (Whipple, 1981). This can cause their malfunction due to electrical breakdown. It also increases the drag force due to enhanced collisions with ions. Lunar and Martian dust can be charged by the solar radiation and levitate above the planet surface (Sternovsky et al., 2002). It poses threat to machinery and spacesuits because of its abrasive properties. Charged ionospheric aerosols affect radio wave propagation (Cho et al., 1996) and often disrupt communications. Ultrafine charged particles in the Earth atmosphere influence cloud formation by providing centres of condensation and thus affect the Earth radiative budget (Boulon et al., 2010) with implications for climate models. Aerosol charging modifies the atmospheric chemistry as well as the formation and transport of pollutants (Aikin & Pesnell, 1998).

2.3 Industrial applications

Particles with designed properties are grown in a plasma environment for various technological applications (Boufendi et al., 2011). These include production of fine powders for ceramics and catalysts, phase separated materials, coatings for solar cells, and nanocoatings for optics. Undesirable dust growth has been observed in ultra-clean etching reactors. As the size of the semiconductor device features has approached 22 nanometers, a single dust particle of a similar size can destroy a whole device, significantly reducing the yield of the manufacturing process. This requires strict measures to prevent dust formation. Fusion devices were found to produce metal dust in significant quantities by evaporating or sputtering their walls. This dust is flammable, radioactive and poses safety hazard. It can contaminate and quench the plasma reducing energy yield.

2.4 Complex plasmas as model systems

One of the most interesting laboratory uses of complex plasmas is to study properties of solids and liquids at the microscopic or kinetic level. Complex plasmas possess a unique combination of properties and share many of them with other model systems such as colloids and granular media. Grains in complex plasmas can be easily observed. They have sizes of a few microns and separation distances of almost a millimetre. Thus complex plasmas remain optically thin over many interparticle distances. Illuminated with a laser, particles can be imaged with a video camera producing images with high contrast. Typical timescales of particle motion are of the order of 10 ms, meaning that a moderately high speed camera is adequate. The damping rate due to the background gas can be below 1 s^{-1}. This makes it possible to observe grain-mediated wave motion. The particle-particle interaction potential is continuous and long range, qualitatively similar to that in real solids and liquids. Complex plasmas exhibit a multitude of dynamic phenomena such as waves (Nunomura et al., 2002; Zhdanov et al., 2003), solitons (Durniak et al., 2009; Samsonov et al., 2002), shock waves (Luo et al., 1999; Samsonov & Morfill, 2008), melting and crystallisation (Knapek et al., 2007), Mach cones (Nosenko et al., 2003; Samsonov et al., 1999), diffusion (Nunomura et al., 2006), heat transport (Nosenko et al., 2008; Nunomura et al., 2005a), and shear flows (Hartmann et al., 2011). Here we will focus on MD simulations of these phenomena in strongly coupled complex plasmas.

3. Numerical models for complex plasmas

Computer simulations are used to describe and predict the behaviour of systems whose complexity makes analytical treatments impossible or very difficult. Numerical models advance our understanding of what basic processes are responsible for different observable phenomena. They can be rerun using exactly the same initial conditions (Durniak & Samsonov, 2010) with altered physical processes (*e.g.* forces) or parameters (*e.g.* damping rate). Examples of numerical simulations of complex plasmas include dynamics of bilayers (Hartmann et al., 2009), self diffusion in two dimensional (2D) liquids (Hou, Piel & Shukla, 2009), phase transition between solid and liquid (Farouki & Hamaguchi, 1992), phonons in a linear chain (Liu & Goree, 2005a), diffusion in 2D liquids (Ott & Bonitz, 2009a), defect dynamics (Durniak & Samsonov, 2010), shear flows (Sanbonmatsu & Murillo, 2001), and nonlinear wave propagation (Durniak et al., 2009). There are two basic types of numerical simulations: stochastic and deterministic. The Monte Carlo method (Sheridan, 2009b) belongs to the first type, while the molecular dynamics (Allen & Tildesley, 1987) and the fluid model (Goedheer et al., 2009) to the second. MD simulations often incorporate stochastic elements in order to simulate Brownian motion and the effects of finite temperature. All these techniques are used to simulate complex plasmas, however here we will only consider the MD method.

3.1 Numerical methods

MD simulations solve numerically the equations of motion for every particle comprising the system. Newtonian equations of motion are used in classical simulations (Allen & Tildesley, 1987). The most common integration techniques employed in complex plasmas include Verlet (Liu et al., 2006), velocity-Verlet (Klumov et al., 2010), Swope (Ott & Bonitz, 2009a), leapfrog (Ma & Bhattacharjee, 2002), predictor-corrector (Farouki & Hamaguchi, 1994), Gear-like (Hou, Piel & Shukla, 2009), Beeman-like (Couëdel et al., 2011; Donkó et al., 2010), Runge-Kutta (Jefferson et al., 2010; Zhdanov et al., 2003), and Runge-Kutta with variable step (Cash Karp) (Durniak et al., 2010). Performance of several integration algorithms for complex plasma problems in presence of a magnetic field has been investigated (Hou, Mišković, Piel & Shukla, 2009).

In order to simulate thermodynamic processes correctly, canonical or microcanonical ensembles are used. This requires conservation of certain thermodynamic quantities (volume, particle number, temperature, energy, etc.) by the algorithm. Numerical errors tend to accumulate and increase the total energy of the system. Energy conserving integrators, such as Verlet and Beeman, are often utilised in order to keep the total energy of the system constant. Since a relatively small number of particles is involved in the simulation, their kinetic energy and thus temperature tend to have significant fluctuations. Constant temperature is maintained by deterministic techniques such as Nosé-Hoover thermostat (Liu et al., 2006) and velocity rescaling (Ohta & Hamaguchi, 2000; Ott & Bonitz, 2009a), or by stochastic methods such as Andersen thermostat (Nelissen et al., 2007) or Langevin dynamics (Schveĭgert et al., 2000). Langevin dynamics is an extension of the MD method, in which a random (Langevin) force and a damping term are added to the equations in order to simulate the effects of liquid or gaseous background on micron-sized particles. This accounts for random kicks by fast moving molecules as well as for the friction force caused by the liquid or gas drag.

Computing pair interactions for all possible pairs of particles in an ensemble is very costly, when the number of particles is large. Since the interaction force decreases with the distance, it is possible to simplify the calculations. Short range interactions allow introduction of a cut-off distance (Vaulina & Dranzhevski, 2006) beyond which the forces are neglected. This method needs to maintain a list of neighbours to keep track of interacting grains in order to increase efficiency. Long range forces can not be easily truncated, however they can be averaged to reduce the computational cost. Several methods are used such as the Ewald summation (Ott et al., 2011), the fast multipole, the tree code, and particle-mesh-based techniques (Frenkel & Smit, 2002; Rapaport, 1995), and an example of the latter the particle-particle particle-mesh method (Donkó et al., 2008). The number of particles in the simulation can be reduced using periodic boundary conditions: a particle exiting a simulation box from one side will re-enter on the opposite side. Particles interact not only with other particles in the simulation box but also with particles in image boxes (Donkó et al., 2010; Hamaguchi, 1999). Free boundaries are also often used in complex plasma simulations.

3.2 Interaction and confinement

Since plasmas contain a mixture of species moving on very different time scales, it is impossible to simulate a meaningful number of them as individual particles even using a supercomputer. Thus complex plasmas are typically simulated as microparticles interacting via an effective potential. The ion-electron plasma is not explicitly included and enters only as a screening parameter in the interaction potential. The neutral gaseous background is simulated as a friction force. This approach is valid in most cases, as comparisons with experiments show. However there were some notable attempts to include all plasma components (Ikkurthi et al., 2009; Joyce et al., 2001). It is also frequently assumed that the particles have a constant charge, which is often a very good approximation for large and highly charged grains.

The most frequently used particle-particle interaction potential in complex plasma simulations is Yukawa, however Coulomb (Lai & I, 1999) as well as a repulsive-attractive potentials have been used (Chen , Yu & Luo, 2005). Other forces can be added *e.g.* due the influence of magnetic fields (Hou, Shukla, Piel & Mišković, 2009), ion drag (Ikkurthi et al., 2009), or external excitations (Durniak et al., 2009). External excitations applied to an equilibrated structure allow to investigate non-equilibrium dynamics (Donkó et al., 2010), and various dynamic phenomena (Jefferson et al., 2010).

Since Yukawa potential is purely repulsive, an additional confinement is needed, if free boundary conditions are used. The most common confining potential is parabolic (Durniak et al., 2009; Ma & Bhattacharjee, 2002; Nelissen et al., 2007), but other shapes have been used such as 4th order (Lai & I, 1999), 10th order (Durniak et al., 2010), soft-wall (Ott et al., 2008), and hard-wall (Klumov et al., 2009). Special shapes of confinement are used in order to obtain particular particle arrangements, *e.g.* a 2D annular potential to obtain rings (Sheridan, 2009a) and anisotropic potentials for elliptical clusters (Cândido et al., 1998), linear chains (Liu & Goree, 2005a), or flat disks (Durniak et al., 2010).

The Langevin dynamics method includes a stochastic force and a dynamic damping force into the equations of motion. The stochastic force does not depend on the particle momentum. It has a zero mean value and a Gaussian probability distribution with a correlation function:

$$\langle L_i(t)\rangle = 0, \quad \langle L_i(t)L_j(t')\rangle = 2\nu m k_B T\delta(t-t')\delta_{ij} \tag{1}$$

where ν is the damping coefficient, k_B is the Boltzmann constant, T the temperature of the system and the indices i and j are linked to particles. The damping force is often chosen to be equal to the gas drag force, which depends on the gas pressure, the kind of gas and the momentum exchange between the gas molecules and the grain surface (Epstein, 1924).

3.3 Our simulation code

The code used to simulate the complex plasmas presented in this chapter is based on an objects-oriented multi-threaded programming. It is assumed that the microparticles comprising a complex plasma interact with each other via a Yukawa potential, their motion is damped by a neutral drag, and that they are confined in an external potential well with free boundaries. A Langevin force is used to study effects of finite temperature. The ions and electrons are not explicitly included in the model. We take into account the interaction of every microparticle with every other one (particle-particle code), thus there is no cut-off for the potential. The equations of grain motion are solved using a fifth order Runge Kutta method with the Cash Karp adaptive step size control (Press et al., 1992). This makes the code precise, simple and stable at the expense of computational efficiency. It is used to simulate a wide range of dynamic phenomena in a system of several thousand particles.

The model is based on the Newtonian equations of motion written for each microparticle:

$$\begin{aligned}
m\ddot{\mathbf{r}}_s &= \mathbf{f}_s^{int} + \mathbf{f}_s^{fr} + \mathbf{f}_s^{conf} + \mathbf{L}_s(t) + \mathbf{f}_s^{ext}, \quad \mathbf{f}_s^{fr} = -m\nu\dot{\mathbf{r}}_s \\
\mathbf{f}_s^{int} &= -\boldsymbol{\nabla}\sum_{j\neq s} U_0, \quad U_0(r_{sj}) = Q^2(4\pi\epsilon_0 r_{sj})^{-1}\exp(-r_{sj}/\lambda_D) \\
\mathbf{f}_s^{conf} &= -m[\Omega_h^2(\mathbf{x}_s + \mathbf{y}_s) + \Omega_v^2\mathbf{z}_s]
\end{aligned} \tag{2}$$

where m is the particle mass, $\mathbf{r} = \mathbf{x} + \mathbf{y} + \mathbf{z}$ is the particle coordinate ($\mathbf{z} = 0$ for 2D) with the subscripts s and j denoting different particles, $\mathbf{f}_s^{fr}$ is the friction force due to collisions with neutrals, $\mathbf{f}_s^{int}$ is the grain-grain interaction force, $\mathbf{f}_s^{ext}$ is the external excitation force, $\mathbf{L}_s(t)$ is the Langevin force (1), ν is the damping rate, Ω_h and Ω_v are respectively the horizontal and vertical confinement parameters of the well ($\Omega_v = \infty$ in 2D case), U_0 is the Yukawa interaction potential, λ_D is the Debye screening length, Q is the particle charge and $r_{sj} = |\mathbf{r}_s - \mathbf{r}_j|$ is the intergrain distance. The overdots denote time derivatives. Q and λ_D are kept constant during the simulation. Dimensionless units are used: the lengths are normalised in terms of the screening length λ_D and the time in terms of $t^\star = \sqrt{4\pi\epsilon_0 m\lambda_D^3/Q^2}$. The units are converted into the dimensional units after the simulation is completed, to facilitate comparison with the experiments. The code records the position, velocity, and potential energy of each particle at specified time steps.

After seeding the grains randomly, the code is run with the external and Langevin forces switched off until the equilibrium is reached and a monolayer crystal lattice or a solid 3D cluster is formed. The structural properties of the resulting crystals are characterised before they are utilised as inputs for simulations of dynamic phenomena. These simulations are performed by applying various excitation forces. A random (Langevin) force is used to simulate the thermal Brownian motions of particles and to obtain phonon spectra. Pulsed excitations are applied to investigate nonlinear waves and structural properties of complex plasmas.

We use the following parameters $m = 5 \times 10^{-13}$ kg, $\lambda_D = 1$ mm, $Q = -16000e$ (where e is the electron charge) in all our simulation runs. Other parameters are listed in Table 1.

3.4 Data and structural analysis

The results of complex plasma simulations are analysed in order to determine their basic microscopic and macroscopic parameters, structural and dynamical properties using standard methods, which are also used for analysing the experimentally obtained particle tracks.

The local orientation of 2D crystalline cells is characterised using the local bond orientational order parameter for each lattice cell: $\psi_6 = \frac{1}{N}\sum_{j=1}^{N} \mathrm{e}^{6i\,\theta_j} = |\psi_6|\mathrm{e}^{i\,\theta_6}$, $\theta_6 = \arctan[\mathrm{Im}(\psi_6)/\mathrm{Re}(\psi_6)]$, where N is the number of nearest neighbours, θ_j is the angle between the x-axis and the bond connecting the central particle with its neighbour j. The average bond orientation angle θ_6 is used to highlight crystal grains separated by strings of defects as well as lattice deformations. The value of $|\psi_6|$ gives the local order parameter, which is equal to one for an ideal crystal.

Delauney triangulation of the lattice is used in order to find the nearest neighbours of each particle and determine their numbers. An ideal hexagonal lattice would have particles with 6 nearest neighbours. Lattice defects are defined as lattice cells that have other numbers of neighbours, such as 5 or 7. Dislocations are pairs of 5- and 7- fold cells, they are also called penta-hepta defects and are characterised by their Burgers vectors, perpendicular to the axis formed by the two defective cells. Defects and dislocations play an important role in melting and plastic deformations (Durniak & Samsonov, 2011; Knapek et al., 2007).

Case	Effect	D	N	Ω_h (Hz)	ν (Hz)	κ	$\varnothing$ (mm)	F_{ex0} (arb.u.)	x_0 (mm)	w (mm)	τ (s)	Excitation space	Excitation time (t)
1.	phonon spectrum	2D	3000	2	1	0.725	50.2	0.01	-	-	-	stochastic	stochastic
2.	soliton collision	2D	3000	0.5	1	1.325	93	3	17	2	0.185	2×Gaussian	parabola
3.	tsunami	2D	3000	2	1	0.725	50.2	1	12	3	0.185	half-Gaussian	parabola
4.	shock	2D	3000	2	1	0.725	50.2	4.5	12	2	0.185	half-Gaussian	parabola
5.	defect dynamics	2D	3000	2	1	0.575	50.2	1	12	2	0.185	half-Gaussian	parabola
6.	plasticity	2D	3000	2	1	0.725	50.2	0.75	17	2	9.2	2×half-Gaussian	half-parabola
7.	clusters	2D	3-150	2	2	0.725	50.2	-	-	-	-	-	-
8.	clusters	3D	150	2	2	0.725	50.2	-	-	-	-	-	-
9.	Mach cone	2D	3000	2	1	0.725	50.2	0.005	12	0.5× 1	-	moving anisotropic Gaussian	$v = 50$ mm/s
10.	melting	2D	3000	3	2	0.575	41.5	8	17	2	0.185	half-Gaussian	parabola

Table 1. Parameter values used in our simulations. D is the dimensionality of the simulation (2D or 3D), N is the number of particles, Ω_h is the horizontal confinement parameter [$\Omega_z = \Omega_h$ for 3D simulation (case 8)], ν is the damping rate, κ is the screening parameter, $\varnothing$ is the initial crystal diameter, F_{ex0} is the amplitude of the excitation force F_{ex}, x_0 is the force offset with respect to the centre of the lattice, w is the width and τ is the duration of the force. Continuous randomly changing excitation force is used to simulate phonon spectra (case 1). Pulsed excitation force is applied on one (cases 3-5,10) or both (cases 2,6) sides of the lattice. Its spatial profile is either Gaussian $F_{ex} \propto \exp\left[-\frac{(x+x_0)^2}{w^2}\right]$ or half-Gaussian, that is Gaussian for $x \leq |x_0|$ and $F_{ex} = const$ for $x > |x_0|$. The temporal profile (cases 2-5,10) is a parabola (inverted and truncated at negative values) $F_{ex} \propto 1 - \left(1 - \frac{t}{\tau}\right)^2$ for $t \leq 2\tau$ and $F_{ex} = 0$ otherwise. It is modified in case 6 to keep a constant value after the maximum is reached ($F_{ex} = const$ for $t \geq \tau$). For cases 1-8 and 10, the excitation force is independent of y. The Mach cones are simulated using an anisotropic Gaussian excitation force $F_{ex} \propto \exp\left[-\frac{(x-x_0-vt)^2}{w_x^2} - \frac{y^2}{w_y^2}\right]$, with $w_x = 0.5$ mm, and $w_y = 1$ mm, moving with a speed $v = 50$ mm/s. The amplitude of the excitation force F_{ex0} is expressed in terms of the parameter $4\pi\epsilon_0\lambda_D^2/Q^2$.

The number density of the lattice is computed using the Voronoi analysis (Voronoi, 1908). A Voronoi cell is defined as a set of points for which a given particle is the nearest. The local number density is proportional to the inverse area in 2D (volume in 3D) of a Voronoi cell. The compression factor of a lattice is calculated as the ratio of the number density n of a strained lattice to its unperturbed number density n_0.

The kinetic temperature of a complex plasma is determined from the velocities of individual particles. The lattice is split into bins and the average bin velocity $\langle \mathbf{v} \rangle$ is calculated. It is then subtracted from the speeds of all particles. The average kinetic energy E in the bin is determined using the mean square random velocity $E = \frac{m}{2}\langle(\mathbf{v} - \langle \mathbf{v} \rangle)^2\rangle$, where m is the particle mass. The kinetic temperature T is found from the relation $E = \frac{d}{2}k_B T$, where k_B is the Boltzmann constant and d is the number of the degrees of freedom.

A correlation analysis is performed in order to assess the structure of complex plasmas. The pair correlation function $g(r)$ gives the probability of finding a specific distance between two particles in the system relative to the probability of finding that distance in a completely random particle distribution of the same density (Crocker & Grier, 1996; Quinn et al., 1996). It measures the translational order of the lattice. For a perfect crystal at zero temperature $g(r)$ is a series of δ-functions. At non-zero temperature peaks have finite widths and decaying amplitude. The position of the first peak of the pair correlation function is determined by the interparticle spacing and is often used to measure the average interparticle distance.

4. Experimental verification

Like any other simulations, numerical models of complex plasmas have to be verified by comparison with the experiments. The code should reproduce the same phenomena as observed in the experiment with similar quantitative characteristics.

4.1 Laboratory complex plasmas

The most frequently used method to obtain high quality complex plasmas in the laboratory is to add premanufactured monodisperse plastic microspheres to the plasma (Thomas et al., 1994). Since the crystalline state is most often the subject of interest, relatively large ($\sim 10\ \mu$m in diameter) particles are used. The reason for that is the lower boundary of the coupling parameter ($\Gamma \propto Q^2/T \gtrsim 170$) required for crystallisation (Sect. 2.1). The value of the kinetic temperature can not be lower than the temperature of the neutral gas, which is close to the room temperature. Thus the coupling parameter can be only increased by increasing the potential energy of the grains or their charge Q, which is a function of the grain size. The dependence of the charge on the particle size makes it also necessary to use monodisperse particles in order to make their charges, potential energies, and levitation heights identical.

Most ground-based experiments utilise a capacitively coupled radio-frequency (rf) gas discharge (Donkó et al., 2010; Durniak & Samsonov, 2011; Käding et al., 2008). The particles are suspended in the plasma sheath, where a strong electric field counteracts the gravity. They typically form between one (Nosenko & Zhdanov, 2009) and a few layers (Quinn & Goree, 2001) creating 2D and quasi-2D structures. In order to obtain 3D structures, gravity has to be compensated (Sect. 2.1). This can be done in a strong electric field such as in striations of a direct current glow discharge (Fortov et al., 2005b), using a thermophoretic force (Arp et al., 2004), or under microgravity conditions on parabolic flights or on board International Space Station (Konopka et al., 2005; Seurig et al., 2007). A Q-machine (Luo et al., 1999) has also been used to create weakly coupled complex plasmas using polydisperse particles.

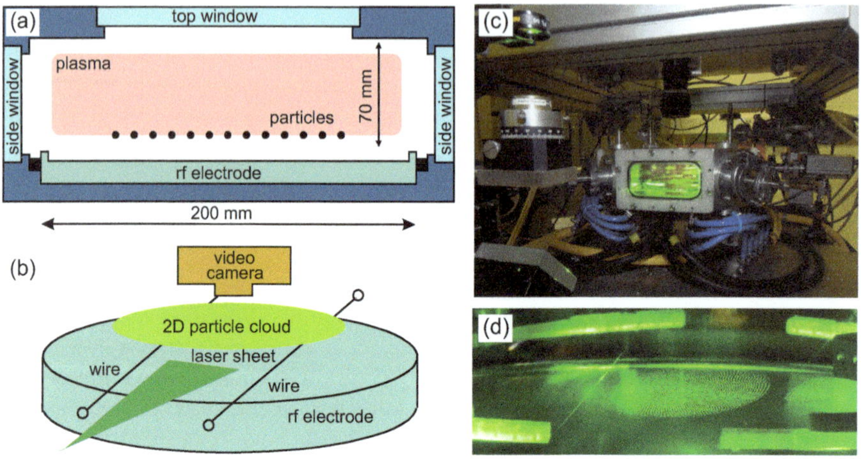

Fig. 1. Experimental setup for ground-based complex plasma experiments. (a) Side view. A discharge is formed in a vacuum chamber. Micron-sized plastic spheres levitate in the plasma sheath above the rf electrode. (b) Oblique view. The particles are illuminated by a laser sheet and imaged with a top view video camera. They are excited electrostatically using the wires stretched across the electrode. (c) Experimental rig. The chamber (shown in the centre) has optical access from 5 sides. The optical system for illumination is shown on the left and the video camera on the top. (d) Close-up view of the experiment. The wires are positioned on both sides of the monolayer particle cloud. This setup has been used to verify complex plasma simulations.

4.2 The experimental setup

Our experiments are performed in a capacitively coupled rf discharge vacuum chamber as shown in Fig. 1. An argon flow (a few sccm) maintains a constant working gas pressure (1-2 Pa) in the chamber. An rf power is applied to the lower disc electrode, which is 20 cm in diameter. The chamber itself is the other grounded electrode. Due to different area of the electrodes and different mobility of ions and electrons, the powered electrode has a negative self-bias voltage, which helps to suspend the particles in the plasma sheath against the gravity. The particles used are monodisperse plastic microspheres 8.9 or 9.19 μm in diameter. They are injected into the plasma through a particle dispenser, levitated in the plasma sheath, and confined radially by a rim on the outer edge of the electrode, forming a monolayer hexagonal lattice of approximately 6 cm in diameter. The particles are illuminated by a horizontal thin (0.2-0.3 mm) sheet of laser light and imaged by a top-view digital camera. Two parallel horizontal tungsten wires, both 0.1 mm in diameter are placed below the particles. Negative pulses applied to one or both wires excite compressional disturbances and deformations.

The experimental results are analysed by identifying the particle positions in all video frames using the intensity weighted moment method (Feng et al., 2007; Ivanov & Melzer, 2007). The particle velocities are calculated by tracking their positions from one frame to the next.

5. Results and discussion

5.1 Structure of complex plasmas

Complex plasmas are formed into different shapes by the confinement potential (see Sect. 3.2). Parameters of the confinement should be selected to ensure that stable structures are formed. The structures of interest include linear 1D chains, monolayer 2D lattices, and 3D balls. Different parameter values result in formation of zigzag lines instead of chains (Sheridan, 2009a), or multiple layers instead of monolayers. The only structure stable against shear perturbations in crystalline monolayers is hexagonal (Durniak et al., 2010). Several different structures exist in bilayer lattices (Donkó & Kalman, 2001) and transition between them is controlled by the layer separation. As the separation increases, the structure changes from hexagonal to staggered square and then to staggered rhombic.

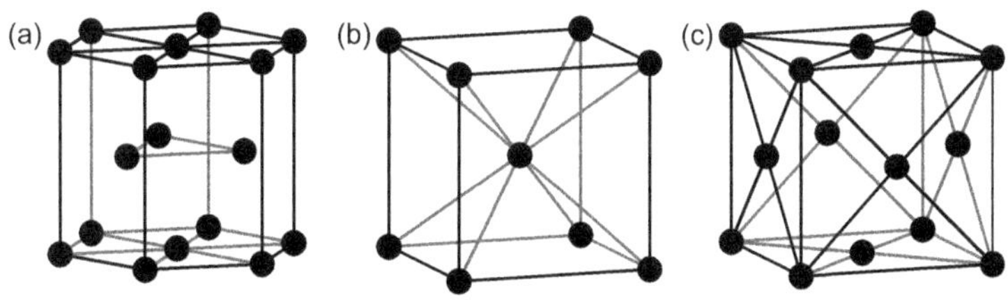

Fig. 2. Crystalline structures commonly observed in 3D complex plasmas: (a) hexagonal close packed (hcp), (b) body centred cubic (bcc), and (c) face centred cubic (fcc).

More crystal structures are observed in multilayer and 3D strongly coupled systems. Face centred cubic (fcc), body centred cubic (bcc), and hexagonal close packed (hcp), illustrated in Fig. 2, have been identified in (Hamaguchi et al., 1997; Klumov et al., 2009). Phase diagrams of the phase transitions between liquid, bcc and fcc phases have been simulated by (Hamaguchi, 1999).

5.2 Linear and nonlinear waves

Complex plasmas sustain waves, which are analogous to those in ordinary solids and liquids. Small amplitude linear waves as well as nonlinear waves, solitons and shocks are observed. Most wave experiments in crystalline complex plasmas are performed in monolayers, since they are easy to obtain at low gas pressure. This results in very low damping and therefore underdamped wave motion can be studied in great details.

5.2.1 Small amplitude waves

Complex plasmas can be in the solid or liquid state (Sect. 2.1). Crystalline monolayer complex plasmas sustain acoustic compressional and shear wave modes, as shown experimentally and numerically in (Donkó et al., 2008; Zhdanov et al., 2003). The phonon spectra are isotropic for long wavelength phonons, but strongly anisotropic in the short wavelength case. The compressional mode has a higher frequency than the shear mode. As the complex plasma crystal melts, the shear mode disappears, while a dust thermal mode appears (Nunomura et al., 2005). Brownian simulation has been used to calculate the cut-off wavenumber for the shear mode in a liquid state and to compare the wave spectra in 2D complex plasma solids and liquids with those from different analytical theories (Hou, Mišković, Piel & Murillo, 2009). It confirmed the existence of the dust thermal mode first observed by (Nunomura et al., 2005). Lattices formed of particles with two different sizes exhibit optical compressional and shear modes. Simulation of bilayers revealed the existence of the optical modes and verified the existence of the $k = 0$ energy (frequency) gap (Hartmann et al., 2009).

The method to compute phonon spectra from either experimental or simulated data relies on the Fourier transform of particle velocities $\mathbf{v}(x,t)$ both in time t and in the x-direction (Nunomura et al., 2002; Zhdanov et al., 2003):

$$\mathbf{V}(k,\omega) = \frac{1}{L_o T_o} \sum_{m=0}^{M} \sum_{n=0}^{N} \mathbf{v}(\mathbf{x}_{mn}, t_{mn}) \exp[i(\mathbf{k}\mathbf{x}_{mn} + \omega t_{mn})]$$

where k and ω are the wave number and the frequency, N and M are the numbers of data points in space and time respectively, L_o is the length of the field of view and T_o is the recording period. The longitudinal and transverse modes are resolved by using v_x and v_y components of the velocity respectively. The resulting phonon spectra are shown in Fig. 3. They are obtained using a MD simulation (Table 1, case 1) with a stochastic force. The theoretical dispersion relations (Donkó et al., 2008; Hou, Mišković, Piel & Murillo, 2009) are calculated for a perfect hexagonal lattice by solving the eigenvalue problem for the dynamical matrix $D_{\mu\nu}$:

$$||\omega^2(k,\varphi) - D_{\mu\nu}(\mathbf{k})|| = 0, \quad D_{\mu\nu}(\mathbf{k}) = \frac{1}{m} \sum_j \frac{\partial^2 U_0(r_j)}{\partial_\mu \partial_\nu} [1 - \cos(\mathbf{k} \cdot \mathbf{r}_j)],$$

where the summation is performed over all particles j with coordinates $\mathbf{r}_j$ and mass m; $U_0(r_j)$ is the Yukawa potential (Eq. 2), and coordinates μ, ν take the values $\{x, y\}$. The spectra depend on the propagation angle or the direction of $\mathbf{k}$. The longitudinal L and transverse T modes are given by:

$$\omega_{L,T}^2(k,\varphi) = \frac{1}{2}\left[D_{xx} + D_{yy} \pm \sqrt{(D_{xx} - D_{yy})^2 + D_{xy}^2}\right]. \tag{3}$$

In the limit of long wavelengths the dispersion relations (Eq. 3) become $\omega_{L,T} = c_{L,T}k$, where $c_{L,T}$ are the longitudinal and transverse wave speeds, which depend on the particle charge and the screening parameter. This can be used to determine the values of Q and κ in experiments.

5.2.2 Nonlinear pulses and solitons

Pulsed excitation applied to a lattice with a laser or a biased wire results in a localised propagating disturbance. This disturbance can be compressional (Nosenko et al., 2002; Samsonov et al., 2002) or shear (Nunomura et al., 2000). If it complies with the Korteveg-de Vries (KdV) equation it is called a KdV soliton. The properties of these solitons include conservation of the soliton parameter ($AL^2 = const$, where A is the soliton's amplitude and L

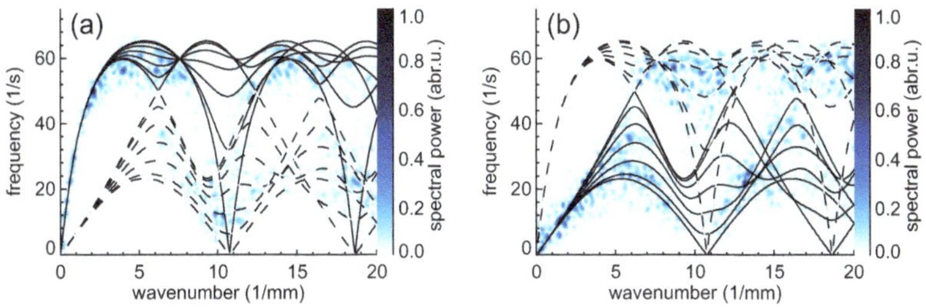

Fig. 3. Phonon spectra in a hexagonal crystalline complex plasma monolayer. (a) Longitudinal and (b) transverse wave modes. The colour scale indicates the normalised spectral power $||V(k,\omega)||$. The solid lines correspond to the theoretical dispersion relations of longitudinal (a) and transverse (b) modes for different propagation directions varying from 0° to 30° in increments of 5°. The dashed lines show complementary transverse (a) and longitudinal (b) modes.

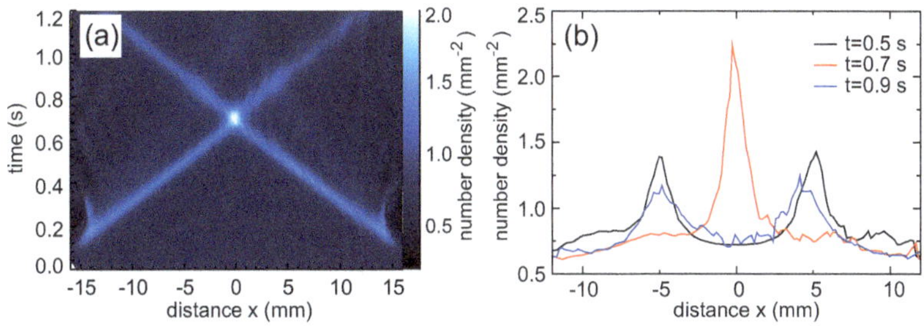

Fig. 4. Collision of counter-propagating solitons. (a) Number density of interacting solitons *vs* distance and time. The collision point corresponds to the peak of number density. (b) Number density versus distance at different times t=0.5 s (before the collision), t=0.7 s (at the collision point), and t=0.9 s (after the collision).

is its width) and a relation between the propagation speed and the amplitude (faster solitons have larger amplitudes). This has been observed in a 2D experiment and a linear chain simulation (Samsonov et al., 2002). Interactions between nonlinear waves is another subject of interest. Experimental and numerical investigation of collisions of counter-propagating solitons in complex plasma monolayers find that solitons with larger amplitude experience larger delays and that the amplitude at the collision point is different from the sum of the initial soliton amplitudes (Harvey et al., 2010). Figure 4 shows a head-on collision of two solitons simulated with the parameters of Table 1, case 2. The amplitudes of the pulses slightly decrease due to the neutral damping as they propagate. The amplitude of the overlapping solitons is lower than the sum of the initial amplitudes.

Waves can gain amplitude due to nonlinear effects even in the presence of damping. A soliton propagating in a lattice with decreasing number density gains amplitude as observed in simulation (Table 1, case 3) and experiment (Fig. 5). It is found that the measured amplitude gain is higher than that predicted by the KdV equation with damping included (Durniak et al., 2009).

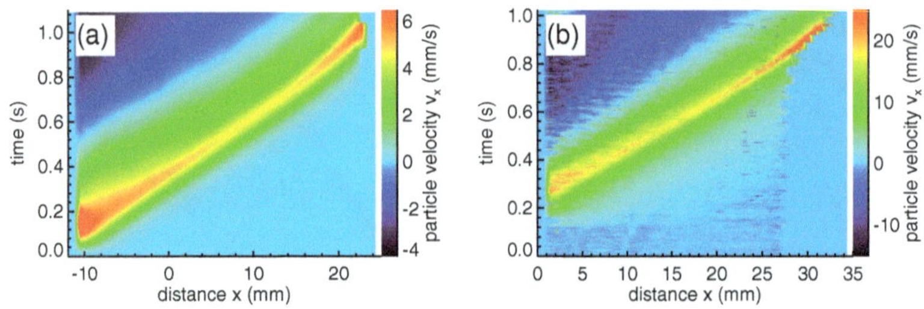

Fig. 5. "Tsunami" effect observed in an inhomogeneous complex plasma lattice. The particle velocity v_x along the wave propagation direction is shown as a function of time and distance in (a) simulation and (b) experiment. The amplitude of the nonlinear wave increases as the number density decreases at later times even in the presence of a damping force. The wave trajectories curve upward as their speeds decrease.

5.3 Shock waves

Shock waves are propagating discontinuities arising from large amplitude perturbations, therefore they cannot be treated as small amplitude waves. Shocks can cause phase transitions and present a challenge for simulations since they are likely to cause numerical instabilities. Experimental observations of shock waves in complex plasmas were reported in (Fortov et al., 2005b; Samsonov et al., 2003; 2004). The structure of a simulated shock (Table 1, case 4) is shown in Figure 6. The shock front has a thickness of a few interparticle distances and an oscillatory structure. It propagates from left to right at a velocity decreasing from 57 to 46 mm/s or a Mach number varying from 1.9 to 1.6. The lattice, initially in the solid state, is melted after the shock. There is a discontinuous jump at the shock front in compression factor, number density, kinetic temperature, and defect fraction (Durniak et al., 2010; Samsonov & Morfill, 2008).

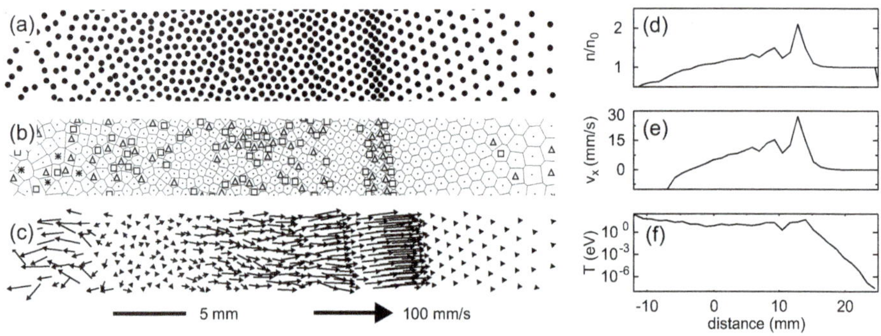

Fig. 6. Simulated shock wave in a monolayer complex plasma visualised as (a) particle positions, (b) Voronoi map with 5-fold defects marked by $\triangle$ and 7-fold by $\square$, and (c) velocity vector map. The oblique propagation is due to the alignment of the crystal lattice. The discontinuity at the shock front is illustrated by (d) compression factor, (e) flow velocity, and (f) kinetic temperature plots computed for a cropped lattice ($-12 \leq x \leq 25$ mm and $|y| \leq 2.5$ mm) at time $t = 0.54$ s. The oscillatory shock structure is clearly visible.

5.4 Defect dynamics and plastic deformation

Lattice defects and dislocations determine mechanical properties of crystals and may be responsible for material fatigue and catastrophic failure. The effect of temperature on defects in 2D Coulomb clusters has been studied numerically by (Nelissen et al., 2007). It was found that the defect mobility strongly depends on the neighbouring defects, that the geometrical defects have different dynamics than the topological defects, and that a fast cooling rate favours formation of a non-equilibrium glass-like state with many defects. Dislocations have been observed to propagate in crystalline complex plasmas supersonically (Nosenko et al., 2007). They interact with nonlinear waves (Durniak & Samsonov, 2010) and they are generated during shear slips in plastic deformations (Durniak & Samsonov, 2011).

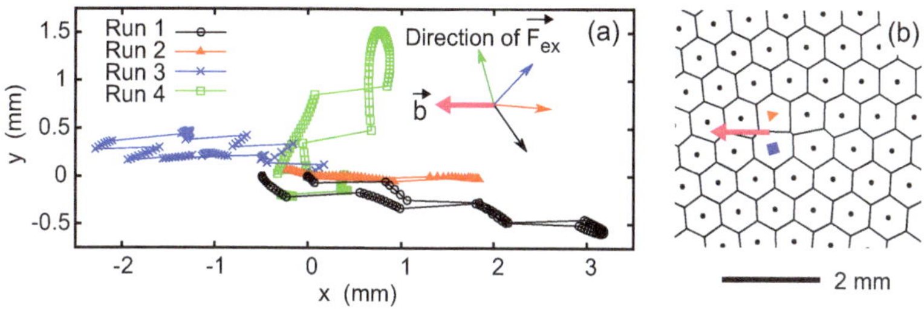

Fig. 7. Interaction of defects with large amplitude waves. (a) Simulated trajectories of isolated dislocations were placed with their initial Burgers vectors **b** pointing left (in $-\mathbf{e}_x$ direction). The direction of the wave excitation force $\mathbf{F}_{ex}$ (and thus the wave propagation direction) is shown with the same colour as the corresponding wave trajectories. The dislocations move either in small increments due to elastic deformations of the lattice, or in large jumps due to lattice structural changes. One can see that the defect jumps occur in directions almost parallel to the Burgers vector regardless of the wave propagation direction. The crystal lattice (b) is initially characterised by a screening parameter $\kappa = 0.725$. The defects are marked by ■ (7-fold defect) and ▲ (5-fold defect) and the Burgers vector by the pink arrow.

The interaction of a dislocation with a wave is simulated using parameters listed in Table 1, case 5. Figure 7a shows the trajectories of an isolated dislocation as the wave passes by from different directions (runs 1-4). The excitation wave has an amplitude between 4.8 and 6.1 mm/s in all runs and propagates at about 38 mm/s. The dislocation either stays at the same lattice site, or jumps to a neighbouring pair of particles. In the former case it displaces roughly in the direction of the wave, while in the latter case it moves almost parallel to its Burgers vector. This result agrees with the experiment (Durniak & Samsonov, 2010).

Plastic deformation under a uniaxial compression is numerically modelled using the conditions of Table 1, case 6. As the strain increases and exceeds the elastic limit, shear slips occur causing stress relaxation. This happens because a uniaxial compression is a superposition of a uniform compression and shear. Complex plasmas are very compressible and do not change their structure under a uniform compression. However their shear strength is not very high, thus their structural failure results in a shear slip. Shear slips are initialised by generation of a pair of dislocations, which move in opposite directions (Durniak & Samsonov, 2011).

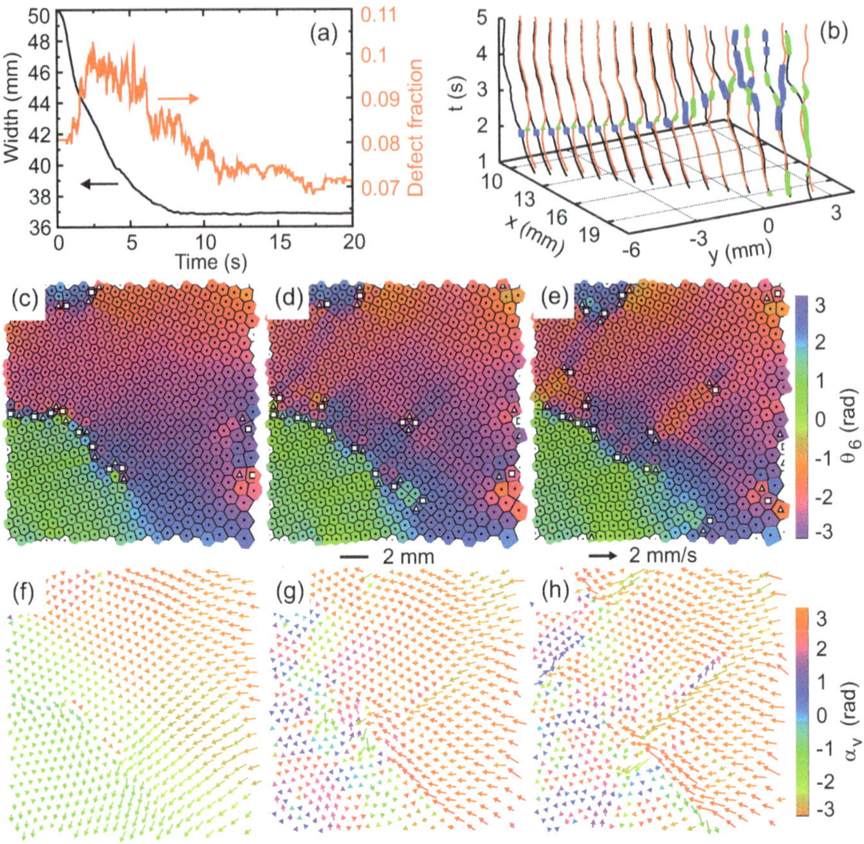

Fig. 8. Plastic deformation of a crystal lattice under a slow uniaxial compression. (a) Time evolution of the crystal width and of the fraction of defective lattice cells (those with other than a 6-fold symmetry). (b) Time evolution of the particle trajectories shows that the local stress is relaxed by a shear slip (trajectories twist at time 2 s). The crystal defects are marked in blue (7-fold defect) and green (5-fold defect). Only the particles along the slip line are shown. Voronoi maps (c-e) with defects marked by △ (5-fold) and □ (7-fold) visualise the lattice structure at (c) t=1.5 s, (d) t=2.04 s, and (e) t=2.3 s. The colour scale shows the bond orientation angle θ_6. A pair of dislocations is generated at t=2.04 s, they separate and move in opposite directions as the slip progresses. Velocity vector maps show the particle velocities at (f) t=1.5 s, (g) t=2.04 s, and (h) t=2.3 s. The colour scale indicates the angle between the velocity vector and the horizontal axis. The slip is shown as particle rows moving in opposite directions.

5.5 Coulomb (Yukawa) clusters

Coulomb (or Yukawa) clusters are systems of up to a few hundred charged particles confined in a 2D or 3D well and interacting via a Coulomb (or Yukawa) potential. They occur in systems of trapped electrons, ions, colloids, and complex plasmas. Since they comprise a small number of particles, they are easy to model numerically (Nelissen et al., 2007). Properties of clusters are different from those of the bulk material and they depend on the cluster size (Yurtsever et al., 2005), the interaction and confinement potentials. The emergence of bulk behaviour in a strongly coupled Yukawa cluster has been studied in (Sheridan, 2007). Clusters are often simulated using the Monte Carlo method, however we will focus here on the MD technique.

Figure 9 shows the structure of 2D and 3D clusters simulated using the parameters of Table 1, cases 7 (2D) and 8 (3D). The particles in these clusters interact via a Yukawa potential and are confined in an isotropic parabolic potential. The structure of the clusters results from the interplay between the particle-particle interaction and the global confinement. The interaction potential favours hexagonal order whereas the confinement induces a circular symmetry. Thus large clusters have a hexagonal inner core and circular outer shells, while small ones tend to contain only shells (Lai & I, 1999). The same effect is observed in 3D clusters (Arp et al., 2004; Totsuji et al., 2002). Clusters assume particularly stable configurations at certain "magic" numbers of particles (Tsuruta & Ichimaru, 1993). The effect of the screening length on the structure of Coulomb balls has been studied in (Bonitz et al., 2006; Käding et al., 2008): particles moved from the outer shells to the inner ones as the screening length increased.

Metastable states of 3D Yukawa clusters can occur with a significantly higher probability than the ground state. The results strongly depend on the screening parameter and the damping coefficient. Slow cooling favours the ground state over the metastable ones (Kählert et al., 2008). The effects of anisotropic confinement and interaction potentials have been studied by (Killer et al., 2011). The structure of spherical clusters was found to be unaffected by a weak ion focus unlike the structure of elongated clusters.

Since clusters have a finite size, they have a finite number of normal modes of oscillations. Knowing the normal modes of a system allows to determine its response to external excitations, *i.e.* its dynamics. Clusters of N particles in a harmonic potential well have $2N$ normal modes. Oscillations of clusters are not purely compressional or shear, however they can be described as compression-like or shear-like (Melzer, 2003). It was shown that in asymmetric potentials both the rotational and breathing modes of elliptical clusters were robust (Sheridan et al., 2007). Melting and defect excitation in Coulomb clusters have been also investigated (Lai & I, 2001; Nelissen et al., 2007).

5.6 Mach cones

Mach cones are propagating V-shaped disturbances. Their existence in a dusty plasma of Saturn rings was first predicted by (Havnes et al., 1995). They were then observed in a monolayer complex plasma (Samsonov et al., 1999), where they were generated by fast particles moving parallel to the main layer. They had a multiple V-shaped structure formed by compressional waves. The properties of Mach cones were confirmed using laser excitation (Melzer et al., 2000). The opening angle of Mach cone obeys the relation $\mu = \sin^{-1}(1/M)$, $M = v/c$ being the Mach number of an object moving at speed v through a medium with an acoustic speed c. Measurements of this angle can therefore be used as a diagnostic tool for complex plasmas to detect inhomogeneity (Zhdanov et al., 2004). Since there are two acoustic wave modes in plasma crystals, compressional and shear (see Section 5.2.1), two types of Mach cones exist. They are distinguished by the particle motion in the cone front. The particles move

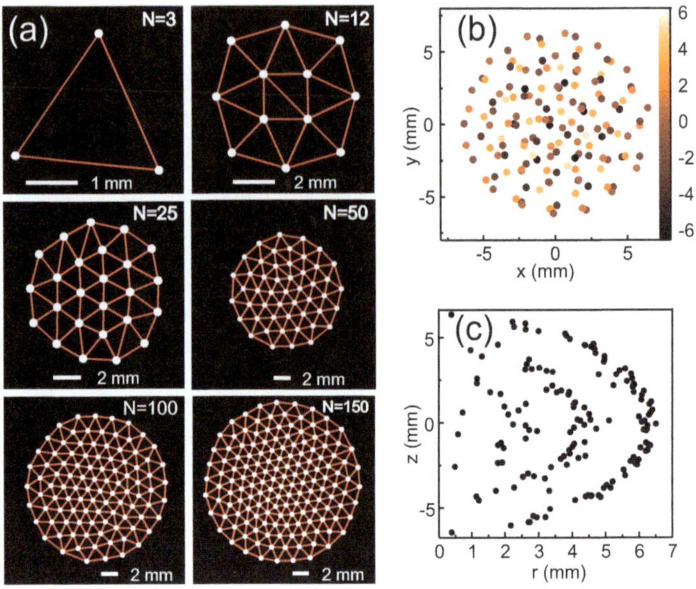

Fig. 9. Simulated Coulomb clusters in (a) 2D and (b,c) 3D. The red lines correspond to the particle bonds calculated by a Delauney triangulation, which reveals the static structure of the clusters. The numbers of particles N in the 2D clusters are shown in (a). The 3D cluster (b) has been been generated using 150 particles. Its shell structure is visualised in (c) by plotting the particle positions in cylindrical coordinates $(r = \sqrt{x^2 + y^2}, z)$. In (b) the colour scale corresponds to the particle's vertical position z.

perpendicular to the cone front in compressional Mach cones and parallel to the front in shear Mach cones (Ma & Bhattacharjee, 2002; Nosenko et al., 2003).

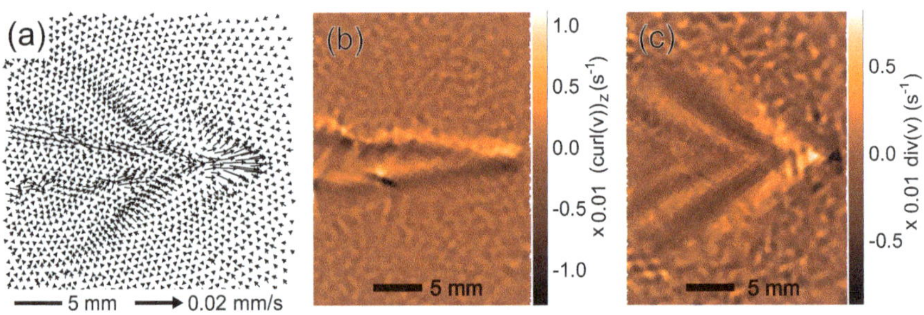

Fig. 10. Simulated Mach cone (wake). The velocity vector map (a) shows particle positions and speeds at $t = 0.55$ s. The Mach cone consists of a shear and a compressional cone. The shear cone is generated by the shear lattice wave and the compressional cone by the longitudinal wave. Their structures are revealed by (b) the vorticity $\nabla \times \mathbf{v}$ and (c) the divergence $\nabla \cdot \mathbf{v}$ maps respectively.

Figure 10 shows a numerically generated Mach cone with parameters listed in Table 1 case 9. The excitation force has been chosen similar to that reported by (Ma & Bhattacharjee, 2002;

Nosenko et al., 2003). We visualise the cones by plotting the velocity vector map (Fig. 10a) and also the vorticity and divergence maps of simulated particle positions, which highlight the shear (Fig. 10b) and compressional (Fig. 10c) cones respectively. The compressional cone has a multiple structure while the shear cone is single. This is due to the fact that the shear wave is slower and thus it can propagate a shorter distance than the compressional before being damped. The angles of the inner and outer compressional cones might be slightly different due to nonlinear effects (Samsonov et al., 2000).

5.7 Phase transitions

MD simulations of charged grains in plasmas have been used to study phase transitions before this became possible experimentally (Farouki & Hamaguchi, 1992). Solid and liquid phases have been predicted as well as a hysteresis at the transition between them. This hysteresis corresponds to a superheated solid and supercooled liquid. Solid superheating was later observed experimentally (Feng et al., 2008). Phase diagrams of Yukawa systems have been computed in (Robbins et al., 1988) predicting liquid as well as solid fcc and bcc structures. Simulations show that strongly screened Yukawa systems have a triple point or a point on the phase diagram where liquid, bcc, and fcc phases coexist, whereas weakly screened Yukawa systems do not have a triple point (Hamaguchi et al., 1997). In experiments, phase transitions can be induced by stochastic laser heating, shear flows (Nosenko & Zhdanov, 2009), shock waves (Knapek et al., 2007) or by changing the discharge power (Rubin-Zuzic et al., 2006). In the latter case a propagating crystallisation front has been observed. It was shown that this process is fundamentally 3D (Klumov et al., 2006). Complex plasma recrystallisation has been simulated by (Hartmann et al., 2010). It was found that the sizes of crystal domains have a power-law time dependence.

Figure 11 shows melting and recrystallisation of a complex plasma lattice excited by a shock wave. The simulation parameters are given in Table 1 case 10. As the shock propagates, the kinetic temperature increases and defects are generated (fig. 11a)). The pair correlation function calculated at different times indicates that the order in the system decreases during melting and then increases reaching almost the initial level at 6 s as the lattice recrystallises.

5.8 Transport phenomena

Transport phenomena are irreversible statistical processes which are responsible for transfer of mass, momentum, or energy in matter. These processes use similar mathematical formalisms and are described by similar equations. The three most commonly considered transport phenomena are diffusion (mass transfer), heat conduction (energy transfer), and viscosity (momentum transfer). Their fundamental nature makes them very important for understanding basic properties of matter. Complex plasmas offer a possibility to study these processes at the level of individual particles and compare experiments directly to MD simulations.

Two simulation techniques are used to study transport phenomena: the equilibrium and nonequilibrium methods (Donkó & Hartmann, 2008). The first method calculates the particle trajectories of a system in a state of statistical equilibrium (Vaulina et al., 2008). The second method applies a perturbation to an equilibrated system and measures the changes it causes

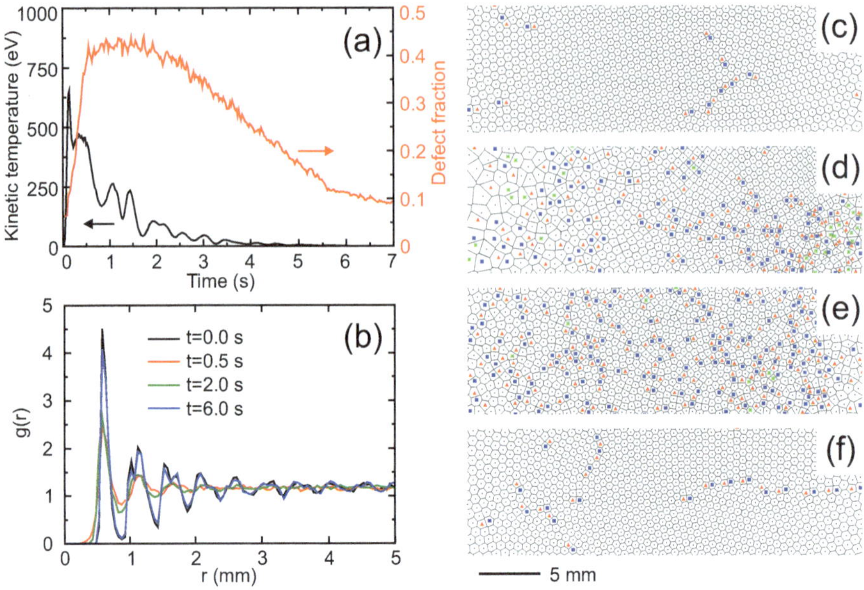

Fig. 11. Crystal melting and recrystallisation caused by a shock wave applied to the lattice. (a) Time evolution of the kinetic temperature and defect fraction. As the lattice crystallises their values return to the initial ones. (b) Pair correlation function at different times. The lattice structure changes from solid (t=0 s), to liquid (t=0.5 and 2 s), and back to solid (t=6 s). Voronoi maps visualise the lattice structure at (c) t=0 s, (d) t=0.5 s, (e) t=2 s, and (f) t=6 s. The lattice defects are marked by ■ (7-fold), ▲ (5-fold), and ∗ (other).

(Sanbonmatsu & Murillo, 2001). If the investigated system exhibits small deviations from the statistical equilibrium, which is the case for equilibrium simulations, the transport coefficients are found from the Green-Kubo relations (Donkó et al., 2009). Nonequilibrium simulations calculate the transport coefficients directly, by computing the diffusion coefficient from the mean square displacement (Ohta & Hamaguchi, 2000), viscosity coefficient from the velocity profile (Sanbonmatsu & Murillo, 2001), or the thermal conductivity from the temperature gradient (Hou & Piel, 2009). Both equilibrium and nonequilibrium methods might produce artifacts due to the finite size of the system or insufficient recording time, thus care should be taken. Another possible problem is the existence of the transport coefficients in particular systems. It was found that in a 2D Yukawa liquids the diffusion coefficient exists at high temperature and the viscosity coefficient at low temperature but not in the opposite limits (Donkó et al., 2009). The thermal conductivity did not appear to exist at high temperature and it could not be evaluated at low temperature due to computational limitations.

The diffusion is governed by the Fick's law: $\mathbf{J} = D\nabla C$, where $\mathbf{J}$ is the diffusion flux, D is the diffusion coefficient, C is the molecular concentration. The diffusion coefficient determines the time dependence of the mean square displacement of all particles $\langle|\mathbf{r}(t) - \mathbf{r}(0)|^2\rangle = 4Dt^{\alpha}$. If $\alpha = 1$ the diffusion is normal, otherwise ($\alpha \neq 1$) it is anomalous: $\alpha > 1$ corresponds to superdiffusion and $\alpha < 1$ to subdiffusion. The temperature dependence of the diffusion coefficient for 3D Yukawa systems has been studied by (Ohta & Hamaguchi, 2000). It was found to be independent of the screening parameter. Different methods of computing the diffusion coefficient at short observation times have been compared by (Vaulina et al., 2008) and the required minimum observation time has been estimated. Anomalous diffusion has been reported in 2D Yukawa systems in some simulations (Hou, Piel & Shukla, 2009; Liu & Goree, 2007) as well as experiments (Juan & I, 1998), which contradicts some other numerical (Vaulina & Dranzhevski, 2006) and experimental results (Nunomura et al., 2006). As shown by (Ott & Bonitz, 2009b) the diffusion exponent depends critically on the neutral gas friction, which makes anomalous diffusion a transient effect in simulations. This certainly does not rule out superdiffusion in experiments, however it is possible that limited fields of view and insufficiently long observation times might obscure anomalous effects and the role of nonequilibrium states in the experimental dissipative-driven systems. The effect of coherent transport on the diffusion exponent, *e.g.* waves and flows is also difficult to rule out.

Shear viscosity in strongly coupled 3D Yukawa system has been studied by (Sanbonmatsu & Murillo, 2001). It was found that the viscosity coefficient has a minimum and that it is nonlocal, with the scale length consistent with the correlation length. The decay of the velocity profile deviates from the one predicted by the Navier-Stokes equation. The minimum of the viscosity coefficient is also observed in 2D Yukawa liquids (Liu & Goree, 2005b). The wavenumber-dependent viscosity, which characterises the viscous effects at different length scales has been computed by (Feng et al., 2011). They have also verified the accuracy of the Green-Kubo relation for static viscosity in the presence of damping. Dynamic shear viscosity has been studied experimentally and numerically by (Hartmann et al., 2011) and shown to exhibit strong frequency dependence. A shear-thinning effect has been demonstrated under static shear.

Heat conductivity in a 2D strongly coupled system has been simulated using nonequilibrium methods by (Hou & Piel, 2009). The results show that the heat conductivity coefficient depends on the damping rate and indicate that it might be temperature-dependent. The experimental results, however, show no temperature dependence within the experimental uncertainty (Nosenko et al., 2008; Nunomura et al., 2005a).

Since the Newton's law of viscosity, the Fourier's law of heat transfer, and the Fick's law of molecular diffusion are very similar, relations between the transport coefficients can be found. The Stokes-Einstein formula relates the diffusion and viscosity coefficients. It has been thoroughly tested in 3D liquids even at the molecular level. In 3D strongly coupled complex plasmas the Stokes-Einstein relation has been verified for a wide range of temperatures down to the solidification point (Donkó & Hartmann, 2008). However this relation is violated in 2D complex plasmas near the disordering transition, remaining valid at higher temperatures (Liu et al., 2006).

6. Conclusion

Methods of molecular dynamics simulations of complex plasmas and their results have been reviewed in this chapter and illustrated by examples of such simulations. Complex plasmas belong to the soft matter class of materials; they are formed by mesoscopic and highly charged particles immersed in a plasma. They are underdamped due to their gaseous background and thus ideally suitable for studying dynamic effects at the particle level. Complex plasmas resemble colloids and some effects such as phase transitions, diffusion and shear flows can be observed in both. Some phenomena such as waves, shocks, kinetic temperature, and phonon heat transfer exist only in complex plasmas. The numerical results have been compared with experiments in order to make sure that simulated effects are observable in physical systems. Unfortunately due to space constraints, some effects not exclusive to complex plasmas were left out (such as lane formation) as well as some of those not extensively simulated (such as coupling of vertical and horizontal modes in monolayers). The authors hope that their choice of dynamic effects well illustrates the beauty and complexity of complex plasmas.

7. References

Aikin, A. C. & Pesnell, W. D. (1998). Uptake coefficient of charged aerosols - implications for atmospheric chemistry, *Geophys. Res. Lett.* 25(9): 1309–1312.

Allen, M. P. & Tildesley, D. J. (eds) (1987). *Computer simulations of liquids*, Oxford University Press, New York.

Arp, O., Block, D. & Piel, A. (2004). Dust Coulomb balls: Three-dimensional plasma crystals, *Phys. Rev. Lett.* 93(16): 165004.

Bonitz, M., Block, D., Arp, O., Golubnychiy, V., Baumgartner, H., Ludwig, P., Piel, A. & Filinov, A. (2006). Structural properties of screened Coulomb balls, *Phys. Rev. Lett.* 96(7): 075001.

Boufendi, L., Jouanny, M. C., Kovacevic, E., Berndt, J. & Mikikian, M. (2011). Dusty plasma for nanotechnology, *J. Phys. D-Appl. Phys.* 44:174035.

Boulon, J., Sellegri, K., Venzac, H., Picard, D., Weingartner, E., Wehrle, G., Collaud Coen, M., Bütikofer, R., Flückiger, E., Baltensperger, U. & Laj, P. (2010). New particle formation and ultrafine charged aerosol climatology at a high altitude site in the Alps (Jungfraujoch, 3580 m a.s.l., Switzerland), *Atmos. Chem. Phys.* 10(19): 9333–9349.

Bronold, F. X., Fehske, H., Kersten, H. & Deutsch, H. (2009). Towards a microscopic theory of particle charging, *Contrib. Plasma Phys.* 49(4-5): 303–315.

Cândido, L., Rino, J.-P., Studart, N. & Peeters, F. M. (1998). The structure and spectrum of the anisotropically confined two-dimensional Yukawa system, *J. Phys.: Condens. Matter.* 10: 11627–11644.

Chen, Z., Yu, M. Y. & Luo, H. (2005). Molecular dynamics simulation of dust clusters in plasmas, *Physica Scripta* 71: 638–643.

Cho, J. Y. N., Alcala, C. M., Kelley, M. C. & Swartz, W. E. (1996). Further effects of charged aerosols on summer mesospheric radar scatter, *J. Atmospher. Terrestr. Phys.* 58(6): 661–672.

Chu, J. H. & I, L. (1994). Direct observation of Coulomb crystals and liquids in strongly coupled rf dusty plasmas, *Phys. Rev. Lett.* 72(25): 4009–4012.

Couëdel, L., Zhdanov, S. K., Ivlev, A. V., Nosenko, V., Thomas, H. M. & Morfill, G. E. (2011). Wave mode coupling due to plasma wakes in two-dimensional plasma crystals: in-depth view, *Physics of plasmas* 18(8): 083707.

Crocker, J. C. & Grier, D. G. (1996). Methods of digital video microscopy for colloidal studies, *J. Colloid Interface Sci.* 179: 298–310.

Donkó, Z., Goree, J. & Hartmann, P. (2010). Viscoelastic response of Yukawa liquids, *Phys. Rev. E* 81(5): 056404.

Donkó, Z., Goree, J., Hartmann, P. & Liu, B. (2009). Time-correlation functions and transport coefficients of two-dimensional Yukawa liquids, *Phys. Rev. E* 79(2): 026401.

Donkó, Z. & Hartmann, P. (2008). Shear viscosity of strongly coupled Yukawa liquids, *Phys. Rev. E* 78(2): 026408.

Donkó, Z. & Kalman, G. J. (2001). Molecular dynamics studies of strongly coupled charged particle bilayers at finite temperatures, *Phys. Rev. E* 63(6): 061504.

Donkó, Z., Kalman, G. J. & Hartmann, P. (2008). Dynamical correlations and collective excitations of Yukawa liquids, *J. Phys.: Condens. Matter.* 20(41): 413101.

Durniak, C. & Samsonov, D. (2010). Defect dynamics in complex plasmas: Interactions between solitons and penta-hepta defects (dislocations), *Europhysics Letters* 90: 45002.

Durniak, C. & Samsonov, D. (2011). Plastic deformations in complex plasmas, *Phys. Rev. Lett.* 106(17): 175001.

Durniak, C., Samsonov, D., Oxtoby, N. P., Ralph, J. F. & Zhdanov, S. (2010). Molecular-dynamics simulations of dynamic phenomena in complex plasmas, *IEEE Transactions on Plasma Science* 38(9): 2412–2417.

Durniak, C., Samsonov, D., Zhdanov, S. & Morfill, G. (2009). Steepening of solitons (tsunami effect) in complex plasmas, *Europhysics Letters* 88: 45001.

Epstein, P. S. (1924). On the resistance experienced by spheres in their motion through gases, *Phys. Rev.* 23(6): 710–733.

Farouki, R. T. & Hamaguchi, S. (1992). Phase transitions of dense systems of charged "dust" grains in plasmas, *Appl. Phys. Lett.* 61(25): 2973–2975.

Farouki, R. T. & Hamaguchi, S. (1994). Thermodynamics of strongly-coupled Yukawa systems near the one-component-plasma limit. II. Molecular dynamics simulations, *J. Chem. Phys.* 101(11): 9885–9893.

Feng, Y., Goree, J. & Liu, B. (2007). Accurate particle position measurement from images, *Review of Scientific Instruments* 78(5): 053704.

Feng, Y., Goree, J. & Liu, B. (2008). Solid superheating observed in two-dimensional strongly coupled dusty plasma, *Phys. Rev. Lett.* 100(20): 205007.

Feng, Y., Goree, J. & Liu, B. (2011). Viscosity calculated in simulations of strongly coupled dusty plasmas with gas friction, *Physics of Plasmas* 18(5): 057301.

Fortov, V. E., Ivlev, A. V., Khrapak, S. A., Khrapak, A. G. & Morfill, G. E. (2005a). Complex (dusty) plasmas: Current status, open issues, perspectives, *Physics Reports* 421: 1–103.

Fortov, V. E., Petrov, O. F., Molotkov, V. I., Poustylnik, M. Y., Torchinsky, V. M., Naumkin, V. N. & Khrapak, A. G. (2005b). Shock wave deformation in a dc glow discharge dusty plasma, *Phys. Rev. E* 71(3): 036413.

Frenkel, D. & Smit, B. (2002). *Understanding molecular simulation: from algorithms to applications*, Academic Press.

Goedheer, W. J., Land, V. & Venema, J. (2009). Modelling of voids in complex radio frequency plasmas, *Contrib. Plasma Phys.* 49(4-5): 199–214.

Goree, J. (1994). Charging of particles in a plasma, *Plasma Sources Sci. Technol.* 3: 400–406.

Hamaguchi, S. (1999). Strongly coupled Yukawa plasmas - models for dusty plasmas and colloidal suspensions, *Plasmas & Ions* 2: 57–68.

Hamaguchi, S., Farouki, R. T. & Dubin, D. H. E. (1997). Triple point of Yukawa systems, *Phys. Rev. E* 56(4): 4671–4682.

Hartmann, P., Donkó, Z., Kalman, G. J., Kyrkos, S., Golden, K. I. & Rosenberg, M. (2009). Collective dynamics of complex plasma bilayers, *Phys. Rev. Lett.* 103(24): 245002.

Hartmann, P., Douglass, A., Reyes, J. C., Matthews, L. S., Hyde, T. W., Kovács, A. & Donkó, Z. (2010). Crystallization dynamics of a single layer complex plasma, *Phys. Rev. Lett.* 105(11): 115004.

Hartmann, P., Sándor, M. C., Kovács, A. & Donkó, Z. (2011). Static and dynamic shear viscosity of a single-layer complex plasma, *Phys. Rev. E* 84(1): 016404.

Hartquist, T. W., Havnes, O. & Morfill, G. E. (2003). The effects of charged dust on Saturn's rings, *Astron. Geophys.* 44(5): 26–30.

Harvey, P., Durniak, C., Samsonov, D. & Morfill, G. (2010). Soliton interaction in a complex plasma, *Phys. Rev. E* 81(5): 057401.

Havnes, O., Aslaksen, T., Hartquist, T. W., Li, F., Melandsø, F., Morfill, G. E. & Nitter, T. (1995). Probing the properties of planetary ring dust by the observation of Mach cones, *Journal of Geophysical Research* 100(A2): 1731–1734.

Hayashi, Y. & Tachibana, K. (1994). Observation of Coulomb-crystal formation from carbon particles grown in a methane plasma, *Jpn. J. Appl. Phys.* 33(6A): 804–806.

Hou, L.-J., Shukla, P. K., Piel, A. & Mišković, Z. L. (2009). Wave spectra of two-dimensional Yukawa solids and liquids in the presence of a magnetic field, *Physics of Plasmas* 16(7): 073704.

Hou, L.-J., Mišković, Z. L., Piel, A. & Murillo, S. (2009). Wave spectra of two-dimensional dusty plasma solids and liquids, *Phys. Rev. E* 79(4): 046412.

Hou, L.-J., Mišković, Z., Piel, A. & Shukla, P. (2009). Brownian dynamics of charged particles in a constant magnetic field, *Physics of Plasmas* 16(5): 053705.

Hou, L.-J. & Piel, A. (2009). Heat conduction in 2D strongly coupled dusty plasmas, *J. Phys. A: Math. Theor.* 42(21): 214025.

Hou, L.-J., Piel, A. & Shukla, P. K. (2009). Self-diffusion in 2D dusty-plasma liquids: numerical-simulation results, *Phys. Rev. Lett.* 102(8): 085002.

Ikezi, H. (1986). Coulomb solid of small particles in plasmas, *Physics of Fluids* 29(6): 1764–1766.

Ikkurthi, V. R., Matyash, K., Melzer, A. & Schneider, R. (2009). Computation of ion drag force on a static spherical dust grain immersed in rf discharges, *Phys. Plasmas* 16(4): 043703.

Ivanov, Y. & Melzer, A. (2007). Particle positioning techniques for dusty plasma experiments, *Review of Scientific Instruments* 78(3): 033506.

Jefferson, R. A., Cianciosa, M. & Thomas Jr., E. (2010). Simulations of one- and two-dimensional complex plasmas using a modular, oject-oriented code, *Physics of plasmas* 17(11): 113704.

Joyce, G., Lampe, M. & Ganguli, G. (2001). Particle simulation of dust structures in plasmas, *IEEE Transactions in plasma science* 29(2): 238–246.

Juan, W.-T. & I, L. (1998). Anomalous diffusion in strongly coupled quasi-2D dusty plasmas, *Phys. Rev. Lett.* 80(14): 3073–3076.

Käding, S., Block, D., Melzer, A., Piel, A., Kählert, H., Ludwig, P. & Bonitz, M. (2008). Shell transitions between metastable states of Yukawa balls, *Physics of Plasmas* 15(7): 073710.

Kählert, H., , Ludwig, P., Baumgartner, H., Bonitz, M., Block, D., Käding, S., Melzer, A. & Piel, A. (2008). Probability of metastable configurations in spherical three-dimensional Yukawa crystals, *Phys. Rev. E* 78(3): 036408.

Kennedy, R. V. & Allen, J. E. (2003). The floating potential of spherical probes and dust grains. II: Orbital motion theory, *J. Plasma Phys.* 69(6): 485–506.

Killer, C., Schella, A., Miksch, T. & Melzer, A. (2011). Vertically elongated three-dimensional Yukawa clusters in dusty plasmas, *Phys. Rev. B* 84(5): 054104.
Klumov, B. A., Rubin-Zuzic, M. & Morfill, G. E. (2006). Crystallization waves in a dusty plasma, *JETP Letters* 84(10): 542–546.
Klumov, B., Huber, P., Vladimirov, S., Thomas, H., Ivlev, A., Morfill, G., Fortov, V., Lipaev, A. & Molotkov, V. (2009). Structural properties of 3D complex plasmas: experiments versus simulations, *Plasma Phys. Control. Fusion* 51(12): 124028.
Klumov, B., Joyce, G., Räth, C., Huber, P., Thomas, H., Morfill, G. E., Molotkov, V. & Fortov, V. (2010). Structural properties of 3D complex plasmas under microgravity conditions, *Europhysics Letters* 92: 15003.
Knapek, C. A., Samsonov, D., Zhdanov, S., Konopka, U. & Morfill, G. E. (2007). Recrystallization of a 2D plasma crystal, *Phys. Rev. Lett.* 98(1): 015004.
Konopka, U., Mokler, F., Ivlev, A. V., Kretschmer, M., Morfill, G. E., Thomas, H. M., Rothermel, H., Fortov, V. E., Lipaev, A. M., Molotkov, V. I., Nefedov, A. P., Baturin, Y. M., Budarin, Y., Ivanov, A. I. & Roth, M. (2005). Charge-induced gelation of microparticles, *New J. Phys.* 7: 227.
Konopka, U., Morfill, G. E. & Ratke, L. (2000). Measurement of the interaction potential of microspheres in the sheath of a rf discharge, *Phys. Rev. Lett.* 84(5): 891–894.
Lai, Y.-J. & I, L. (2001). Defects and particle motions in the nonuniform melting of a two-dimensional Coulomb cluster, *Phys. Rev. E* 64(1): 015601(R).
Lai, Y.-J. & I, L. (1999). Packings and defects of strongly coupled two-dimensional Coulomb clusters: numerical simulation, *Phys. Rev. E* 60(4): 4743–4753.
Liu, B. & Goree, J. (2005a). Phonons in a one-dimensional Yukawa chain: dusty plasma experiment and model, *Phys. Rev. E* 71(4): 046410.
Liu, B. & Goree, J. (2005b). Shear viscosity of two-dimensional Yukawa systems in the liquid state, *Phys. Rev. Lett.* 94(18): 185002.
Liu, B. & Goree, J. (2007). Superdiffusion in two-dimensional Yukawa liquids, *Phys. Rev. E* 75(1): 016405.
Liu, B., Goree, J., Nosenko, V. & Boufendi, L. (2003). Radiation pressure and gas drag forces on a melamine-formaldehyde microsphere in a dusty plasma, *Physics of Plasmas* 10(1): 9–20.
Liu, B., Goree, J. & Vaulina, O. S. (2006). Test of the Stokes-Einstein relation in a two-dimensional Yukawa liquid, *Phys. Rev. Lett.* 96(1): 015005.
Luo, Q.-Z., D'Angelo, N. & Merlino, R. L. (1999). Experimental study of shock formation in a dusty plasma, *Physics of plasmas* 6(9): 3455–3458.
Ma, Z. W. & Bhattacharjee, A. (2002). Molecular dynamics simulations of Mach cones in two-dimensional Yukawa crystals, *Physics of Plasmas* 9(8): 3349–3354.
Melandsø, F. & Goree, J. (1995). Polarized supersonic plasma flow simulation for charged bodies such as dust particles and spacecraft, *Phys. Rev. E* 52(5): 5312–5326.
Melzer, A. (2003). Mode spectra of thermally excited two-dimensional dust Coulomb clusters, *Phys. Rev. E* 67(1): 016411.
Melzer, A., Nunomura, S., Samsonov, D., Ma, Z. W. & Goree, J. (2000). Laser-excited Mach cones in a dusty plasma crystal, *Phys. Rev. E* 62(3): 4162–4176.
Melzer, A., Trottenberg, T. & Piel, A. (1994). Experimental determination of the charge on dust particles forming Coulomb lattices, *Phys. Lett. A* 191: 301–308.

Merlino, R. L. & Goree, J. A. (2004). Dusty plasmas in the laboratory, industry, and space, *Physics Today* pp. 1–7.

Morfill, G. E. & Ivlev, A. V. (2009). Complex plasmas: an interdisciplinary research field, *Rev. Mod Phys.* 81(4): 1353–1404.

Nelissen, K., Partoens, B. & Peeters, F. M. (2007). Dynamics of topological defects and the effects of the cooling rate on finite-size two-dimensional screened Coulomb clusters, *Europhysics Letters* 79: 66001.

Nosenko, V., Goree, J., Ma, Z. W., Dubin, D. H. E. & Piel, A. (2003). Compressional and shear wakes in a two-dimensional dusty plasma crystal, *Phys. Rev. E* 68(5): 056409.

Nosenko, V., Nunomura, S. & Goree, J. (2002). Nonlinear compressional pulses in a 2D crystallized dusty plasma, *Phys. Rev. Lett.* 88(21): 215002.

Nosenko, V., Zhdanov, S., Ivlev, A. V., Morfill, G., Goree, J. & Piel, A. (2008). Heat transport in a two-dimensional complex (dusty) plasma at melting conditions, *Phys. Rev. Lett.* 100(2): 025003.

Nosenko, V. & Zhdanov, S. K. (2009). Dynamics of dislocations in a 2D plasma crystal, *Contrib. Plasma Phys.* 49(4-5): 191–198.

Nosenko, V., Zhdanov, S. & Morfill, G. (2007). Supersonic dislocations observed in a plasma crystal, *Phys. Rev. Lett.* 99(2): 025002.

Nunomura, S., Goree, J., Hu, S., Wang, X., Bhattacharjee, A. & Avinash, K. (2002). Phonon spectrum in a plasma crystal, *Phys. Rev. Lett.* 89(3): 035001.

Nunomura, S., Samsonov, D. & Goree, J. (2000). Transverse waves in a two-dimensional screened-Coulomb crystal (dusty plasma), *Phys. Rev. Lett.* 84(22): 5141–5144.

Nunomura, S., Samsonov, D., Zhdanov, S. & Morfill, G. (2006). Self-diffusion in a liquid complex plasma, *Phys. Rev. Lett.* 96(1): 015003.

Nunomura, S., Samsonov, D., Zhdanov, S. & Morfill, G. E. (2005a). Heat transfer in a two-dimensional crystalline complex (dusty) plasma, *Phys. Rev. Lett.* 95(2): 025003.

Nunomura, S., Zhdanov, S., Samsonov, D. & Morfill, G. E. (2005b). Wave spectra in solid and liquid complex (dusty) plasmas, *Phys. Rev. Lett.* 94(4): 045001.

Ohta, H. & Hamaguchi, S. (2000). Molecular dynamics evaluation of self-diffusion in Yukawa systems, *Physics of Plasmas* 7(11): 4506–4514.

Ott, T. & Bonitz, M. (2009a). Anomalous and Fickian diffusion in two-dimensional dusty plasmas, *Contrib. Plasma Phys.* 49(10): 760–764.

Ott, T. & Bonitz, M. (2009b). Is diffusion anomalous in two-dimensional Yukawa liquids?, *Phys. Rev. Lett.* 103(19): 195001.

Ott, T., Bonitz, M., Donkó, Z. & Hartmann, P. (2008). Superdiffusion in quasi-two-dimensional Yukawa liquids, *Phys. Rev. E* 78(2): 026409.

Ott, T., Stanley, M. & Bonitz, M. (2011). Non-invasive determination of the parameters of strongly coupled 2D Yukawa liquids, *Physics of Plasmas* 18(6): 063701.

Press, W., Teukolsky, S., Vetterling, W. & Flannery, B. (1992). *Numerical recipes in C: the art of scientific computing*, second edn, Cambridge University Press.

Quinn, R. A., Cui, C., Goree, J., Pieper, J. B., Thomas, H. & Morfill, G. E. (1996). Structural analysis of a Coulomb lattice in a dusty plasma, *Phys. Rev. E* 53(3): R2049.

Quinn, R. A. & Goree, J. (2001). Experimental test of two-dimensional melting through disclination unbinding, *Phys. Rev. E* 64(5): 051404.

Rapaport, D. C. (1995). *The art of molecular dynamics simulation*, Cambridge University Press.

Robbins, M. O., Kremer, K. & Grest, G. S. (1988). Phase diagram and dynamics of Yukawa systems, *J. Chem. Phys.* 88(5): 3286–3312.

Rothermel, H., Hagl, T., Morfill, G. E., Thoma, M. H. & Thomas, H. M. (2002). Gravity compensation in complex plasmas by application of a temperature gradient, *Phys. Rev. Lett.* 89(17): 175001.

Rubin-Zuzic, M., Morfill, G. E., Ivlev, A. V., Pompl, R., Klumov, B. A., Bunk, W., Thomas, H. M., Rothermel, H., Havnes, O. & Fouquét, A. (2006). Kinetic development of crystallization fronts in complex plasmas, *Nature Physics* 2: 181–185.

Samsonov, D. & Goree, J. (1999). Instabilities in a dusty plasma with ion drag and ionization, *Phys. Rev. E* 59(1): 1047–1058.

Samsonov, D., Goree, J., Ma, Z. W., Bhattacharjee, A., Thomas, H. M. & Morfill, G. E. (1999). Mach cones in a Coulomb lattice and a dusty plasma, *Phys. Rev. Lett.* 83(18): 3649–3652.

Samsonov, D., Goree, J., Thomas, H. M. & Morfill, G. E. (2000). Mach cone shocks in a two-dimensional Yukawa solid using a complex plasma, *Phys. Rev. E* 61(5): 5557–5572.

Samsonov, D., Ivlev, A. V., Quinn, R. A., Morfill, G. & Zhdanov, S. (2002). Dissipative longitudinal solitons in a two-dimensional strongly coupled complex (dusty) plasma, *Phys. Rev. Lett.* 88(9): 095004.

Samsonov, D. & Morfill, G. E. (2008). High-speed imaging of a shock in a complex plasma, *IEEE Transactions on Plasma Science* 36(4): 1020–1021.

Samsonov, D., Morfill, G., Thomas, H., Hagl, T., Rothermel, H., Fortov, V., Lipaev, A., Molotkov, V., Nefedov, A., Petrov, O., Ivanov, A., & Krikalev, S. (2003). Kinetic measurements of shock wave propagation in a three-dimensional complex (dusty) plasma, *Phys. Rev. E* 67(3): 036404.

Samsonov, D., Zhdanov, S. K., Quinn, R. A., Popel, S. I. & Morfill, G. E. (2004). Shock melting of a two-dimensional complex (dusty) plasma, *Phys. Rev. Lett.* 92(25): 255004.

Sanbonmatsu, K. Y. & Murillo, M. S. (2001). Shear viscosity of strongly coupled Yukawa systems on finite length scales, *Phys. Rev. Lett.* 86(7): 1215–1218.

Schveĭgert, I. V., Schveĭgert, V. A., Melzer, A. & Piel, A. (2000). Melting of a dust crystal with defects, *JETP Letters* 71(2): 58–61.

Seurig, R., Morfill, G., Fortov, V. & Hofmann, P. (2007). Complex plasma research on ISS past, present, and future facilities, *Acta Astronautica* 61: 940–953.

Sheridan, T. E. (2007). Criterion for bulk behavior of a Yukawa disk, *Physics of plasmas* 14(3): 032108.

Sheridan, T. E. (2009a). Dusty plasma ring model, *Phys. Scr.* 80(6): 065502.

Sheridan, T. E. (2009b). Monte Carlo study of melting in a finite two-dimensional dusty plasma, *Physics of plasmas* 16(8): 083705.

Sheridan, T. E., Wells, K. D., Garee, M. J. & Herrick, A. C. (2007). Theoretical and experimental study of elliptical Debye clusters, *J. Applied Physics* 101(11): 113309.

Sternovsky, Z., Robertson, S., Sickafoose, A., Colwell, J. & Horányi, M. (2002). Contact charging of lunar and Martian dust simulants, *J. Geophys. Res.-Planets* 107(E11): 5105.

Thomas, H. M. & Morfill, G. E. (1996). Melting dynamics of a plasma crystal, *Nature* 379: 806–809.

Thomas, H. M., Morfill, G. E., Demmel, V., Goree, J., Feuerbacher, B. & Möhlmann, D. (1994). Plasma crystal: Coulomb crystallization in a dusty plasma, *Phys. Rev. Lett.* 73(5): 652–655.

Totsuji, H., Kishimoto, T., Totsuji, C. & Tsuruta, K. (2002). Competition between two forms of ordering in finite Coulomb clusters, *Phys. Rev. Lett.* 88(12): 125002.

Tsuruta, K. & Ichimaru, S. (1993). Binding energy, microstructure, and shell model of Coulomb clusters, *Phys. Rev. A* 48(2): 1339–1344.

Vaulina, O., Khrapak, S. & Morfill, G. (2002). Universal scaling in complex (dusty) plasmas, *Phys. Rev. E* 66(1): 016404.

Vaulina, O. S., Adamovich, X. G., Petrov, O. F. & Fortov, V. E. (2008). Evolution of the mass-transfer processes in nonideal dissipative systems. I. numerical simulation, *Phys. Rev. E* 77(6): 066403.

Vaulina, O. S. & Dranzhevski, I. E. (2006). Transport of macroparticles in dissipative two-dimensional Yukawa systems, *Phys. Scr.* 73: 577–586.

Vladimirov, S. V., Maiorov, S. A. & Ishihara, O. (2003). Molecular dynamics simulation of plasma flow around two stationary dust grains, *Physics of Plasmas* 10(10): 3867–3873.

Voronoi, G. (1908). Nouvelles applications des paramètres continus à la théorie des formes quadratiques, *J. Reine Angew. Math.* 134: 198–287.

Whipple, E. C. (1981). Potentials of surfaces in space, *Rep. Prog. Phys.* 44: 1197–1250.

Yurtsever, E., Calvo, F. & Wales, D. J. (2005). Finite-size effects in the dynamics and thermodynamics of two-dimensional Coulomb clusters, *Phys. Rev. E* 72(2): 026110.

Zhdanov, S. K., Morfill, G. E., Samsonov, D., Zuzic, M. & Havnes, O. (2004). Origin of the curved nature of Mach cone wings in complex plasmas, *Phys. Rev. E* 69(2): 026407.

Zhdanov, S., Nunomura, S., Samsonov, D. & Morfill, G. (2003). Polarization of wave modes in a two-dimensional hexagonal lattice using a complex (dusty) plasma, *Phys. Rev. E* 68(3): 035401(R).

12

Application of Molecular Dynamics Simulations to Plasma Etch Damage in Advanced Metal-Oxide-Semiconductor Field-Effect Transistors

Koji Eriguchi
Graduate School of Engineering, Kyoto University
Japan

1. Introduction

According to "the international technology roadmap for semiconductors (ITRS)" (SIA, 2009), the shrinkage of silicon-based metal–oxide–semiconductor field-effect transistor (MOSFET) – an elemental device (unit) in ultra-large-scale integrated (ULSI) circuits – has been accelerating due to expanding demands for the higher performance and the lower power operation. The characteristic dimensions of current MOSFETs in mass productions are around 30 - 50 nm. Figure 1 shows the scaling trend of the key feature sizes in ULSI circuits predicted by Semiconductor Industry Association, USA. Various types of MOSFETs are designed for the specific purposes, i.e., low standby power (LSP), low operation power (LOP), and high performance (HP) operations, and built in ULSI circuits such as dynamic random access memory (DRAM) and micro-processing unit (MPU). New structured MOSFETs such as fully-depleted (FD) and metal-gate (MG) devices have been recently proposed. Since physical gate length (*Lg*) and source / drain extension depth (Ext) are the key feature sizes determining MOSFET performance (Sze & Ng, 2007), the shrinkage of *Lg* and Ext is a primal focus in the development of MOSFETs. These sizes have become a few nanometers, comparable to the scale of atomistic simulation domain.

To meet the requirements such as fabricating fine patterns with anisotropic features, plasma etching is widely used in mass production of MOSFETs. At present, the control of the pattern transfer by plasma etching needs to be within the variation of a few nanometers (SIA, 2009). In such regimes, the feature size of the region where plasma - etch reactions are occurring becomes no more negligible with respect to the scale of MOSFET. Thus, precise control of the surface reaction between plasma and device is strongly required. In plasma etch process, radicals (atoms or molecules in an excited state) react with surface material with a help of the energy of incident ions accelerated in the "sheath" between plasma and device surface. This reaction mechanism is commonly referred to as "reactive ion etching (RIE)" (Lieberman & Lichtenberg, 2005). In some plasma etch processes, an energy of the ion becomes larger than 1 keV to obtain high etch rate. In such schemes, an unexpected reaction may occur. This unexpected ("unwanted") reaction mechanism is usually called as "plasma process-induced damage" (Eriguchi & Ono, 2008; Lieberman & Lichtenberg, 2005; Oehrlein, 1989), which is bringing out many key problems in the development of MOSFETs.

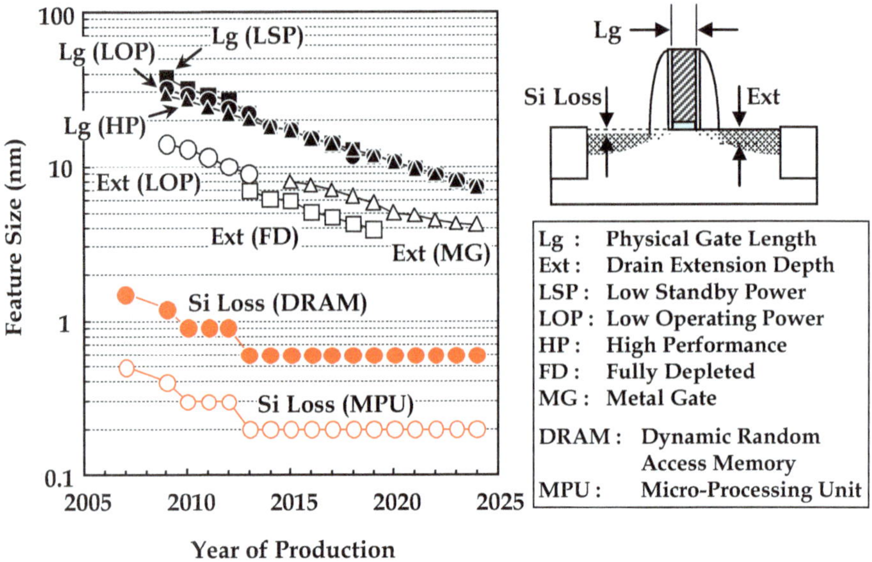

Fig. 1. Scaling trend of feature sizes in a metal–oxide–semiconductor field-effect transistor (MOSFET) in an ultra-large-scale integrated circuit.

Plasma process-induced damage (PID) is one of the serious issues causing degradation of MOSFET performance and reliability. Figure 2 illustrates an example of PID during a typical plasma etch process. Si wafer is placed on a wafer stage in a plasma chamber. Reactive plasma is generated by power supply. This figure corresponds to an inductively coupled plasma (ICP) system (Lieberman & Lichtenberg, 2005), where powers with frequencies of f_1 and f_2 are supplied to a plasma source (f_1) and a wafer stage (f_2), respectively. During plasma etching, MOSFET is exposed to a plasma and energetic ions impact on the surface. This energetic ion bombardment results in creation of defects (*ex.* displaced Si atom) in the Si surface region of MOSFET. This mechanism is one of examples of PID (Eriguchi & Ono, 2008). During more than the last two decades, PID has been studied extensively to understand the mechanisms and to solve practical problems with various approaches. In order to obtain the statistical data in mass production and to clarify the source of PID, a use of specifically designed devices called "test elementary group (TEG)" (SIA, 2009) has been a major approach. In addition to the use of TEG, physical and electrical analyses have been conducted by using a bare Si wafer to gain fundamental understanding of PID. To realize future high-performance MOSFETs, understanding and controlling (minimizing) PID is crucial because the critical dimension of reaction-layer thickness and device feature size will be in conflict with the plasma-damaged-layer thickness governed by plasma parameters. In other words, the damaged-layer thickness does not scale with device- shrinkage trends shown in Fig. 1.

Several simulation techniques have been proposed so far for plasma etch process. There are two major schemes: (1) Plasma-etch feature-profile simulations employing a small cell (Jin et al., 2002; Tsuda et al., 2011) and (2) Surface-reaction simulations based on a molecular dynamics (MD) (Graves & Humbird, 2002; Ohta & Hamaguchi, 2001a, 2001b; Sankaran & Kushner, 2004).

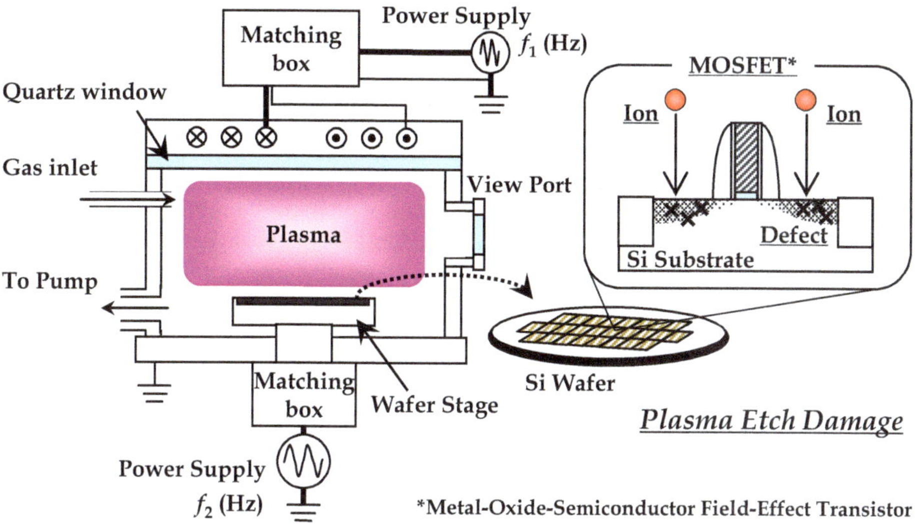

Fig. 2. Illustrations of plasma process reactor and plasma etch damage.

Since the number of particles to be simulated may be quite large (>10^{10} cm^{-3}) during plasma etching, classical MD simulations based on Newton's equation of motion (Graves & Humbird, 2002; Ohta & Hamaguchi, 2001a, 2001b) are now widely employed, compared to those using quantum mechanical calculations (Mazzarolo et al., 2001; Pelaz et al., 2009). Recently MD simulations have been used to understand formation of the surface "damaged" layer and displacement of Si atoms - PID (Graves & Humbird, 2002; Pelaz et al., 2009). However, the primary focus of these conventional MD simulations has been placed on the surface-reaction chemistry among ions, radicals, and the surface material. Since plasma etch processes are utilized for MOSFET fabrication, not only the surface reaction mechanism but also the effects of PID on MOSFET performance degradation should carefully be taken into account. Thus, MD simulations for PID should be incorporated with the prediction of electrical characteristics of MOSFETs.

There are two major challenging parts in the development of future MOSFET and plasma etch process: (1) A systematic and quantitative understanding of PID - the thickness of the damaged layer and the density of the displaced Si atoms (defects), and (2) A comprehensive design framework for future plasma by considering the effect of PID on electrical characteristics of MOSFETs. By keeping these issues in mind, this chapter discusses PID mechanisms by a classical MD simulation. We compare the simulation results with experimental data obtained by various analysis techniques. Future key issues concerning the effects of PID on MOSFET performance are provided. This chapter is organized as follows: In Sec. 2, we review the PID mechanism (Si loss mechanism - Si recess structure formation as mentioned later). In Sec. 3, the MD simulation employed in this study is briefly described. In Sec. 4, the simulation results are presented. In Sec. 5, experimental results are compared to the simulation results. Concluding remarks are in Sec. 5.

2. Ion-bombardment damage to MOSFET during plasma processing

Figure 3 illustrates a PID mechanism induced by the ion bombardment on Si surface during an offset spacer (SIA, 2009) etching (one of manufacturing steps) for MOSFET. During plasma etching, an energetic ion impinges on the Si surface with an energy of E_{ion}, leasing the energy by a series of collisions, then, it creates the defect sites under the exposed surface. This mechanism forms the damaged layer. In general, defect sites in Si substrate are referred to as displaced Si atoms, vacancies, and interstitials. As seen on the right in Fig. 3, the damaged structure consists of the surface (amorphous) and interfacial (a mixture of amorphous and crystalline structure) layers. Underneath these layers, there exist (latent) localized defect sites. (In this study, we denote these sites as "(local) defect sites".) The surface and interfacial layers can usually be monitored by an optical technique such as spectroscopic ellipsometry (SE) in production lines. The profile of defect site and the thickness of damaged-layer are determined by E_{ion} as well as the potential between Si and the incident ion.

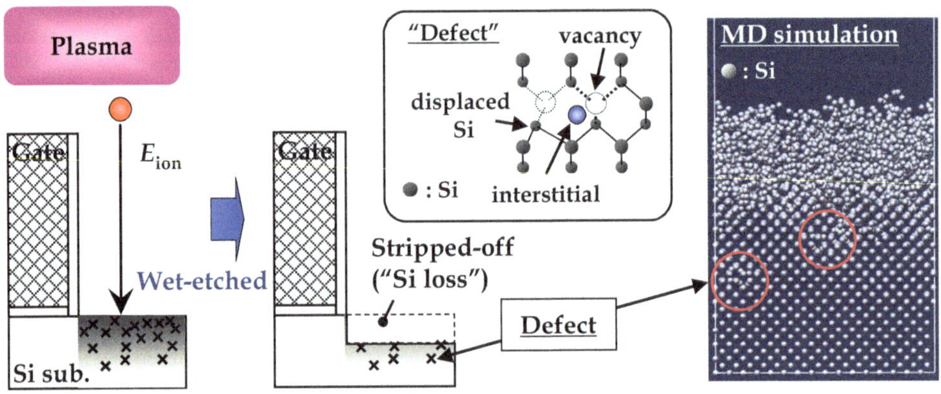

Fig. 3. Mechanism of plasma-etch damage to Si substrate and "Si loss" formation. Energetic ion bombardment creates a damaged layer underneath the Si surface. As shown on the right, localized defect sites are formed. A portion of the damaged layer with defect sites is stripped off during a subsequent wet-etch process, resulting in Si loss. (For details, see the text.)

In conventional MOSFET fabrication processes, a wet-etch step then follows the plasma etch process to remove the contaminated layer including the damaged layer. Since the damaged layer oxidizes due to an air exposure after the plasma etch, the portion is stripped off by the wet-etch. Then, the etched layer results in Si loss whose structure is observed as recessed Si surface, called "Si recess" (Ohchi et al., 2008; Petit-Etienne et al., 2010; Vitale & Smith, 2003). Si recess is formed in the source / drain extension (SDE) region in a MOSFET.

It has been reported (Eriguchi et al., 2009a; Eriguchi et al., 2008a) that the Si recess structure by PID degrades MOSFET performance, i.e., induces the shift of threshold voltage (V_{th}) for MOSFET operation. Since V_{th} (Sze, 1981; Sze & Ng, 2007) plays an important role in determining the performance, Si recess structure has become a primal problem in the present-day MOSFET development (SIA, 2009). To understand the formation of Si recess structure (PID), the damage creation mechanism should be clarified from both theoretical and experimental viewpoints. Moreover, to predict the effects of PID on the MOSFET electrical characteristics, the defect structures should be identified quantitatively with high accuracy. In this study, we performed a classical molecular dynamics simulation as well as quantitative analyses of the local defect site density. Then, we discuss the effects of PID on the electrical characteristics of MOSFET.

3. Molecular dynamics simulation for plasma etching

Regarding to classical MD simulations for plasma etch process, many papers have been focusing on the surface reactions to understand details of silicon and silicon dioxide etch characteristics by energetic halogen (Abrams & Graves, 2000; Hanson et al., 1997; Humbird & Graves, 2004; Nagaoka et al., 2009; Ohta & Hamaguchi, 2001b) and fluorocarbon (Abrams & Graves, 1999) ions. The primal focuses are placed on estimation of etch yield by incident ions and the selectivity for RIE system. Regarding the ion bombardment damage, Graves and Humbird (Graves & Humbird, 2002) have reported in detail the formation of damaged layer in crystalline Si structures by Ar ion impacts. They estimated a stopping range of ions as well as a thickness of the amorphous (amorphized) layer formed near the surface. However, the detail mechanism of local defect site formation was not discussed.

As mentioned in Sec. 1, an RIE system includes the physical and chemical reactions triggered by 10 - 10^3 eV high-energy ion impacts. Although *ab initio* MD simulations are now available, they cannot be applied in practice. This is because more than 10^3 atoms are necessary to construct a solid surface and the total number of incident ions is more than 10^{10} cm^{-2}, resulting in more than 10^3 impacts on the surface with the area of 10 nm^2 (~ commonly simulated size). Therefore, at present, the only possible candidate for atomistic RIE simulations is a classical MD, in particular, with pre-constructed interatomic potential functions.

One of the commonly used interatomic potential functions for Si-containing systems is the one proposed by Stillinger and Weber (SW) (Stillinger & Weber, 1985) wherein the total potential energy consists of two- and three-body functions. The SW potential function was originally designed for Si/F systems. Afterwards various potential sets for Si/Cl (Ohta & Hamaguchi, 2001b), Si/O (Watanabe et al., 1999), Si/O/F (Ohta & Hamaguchi, 2001a), Si/O/Cl (Ohta & Hamaguchi, 2001a), and Si/O/C/F (Ohta & Hamaguchi, 2004; Smirnov et al., 2007) systems were provided. The other widely used function was proposed by Tersoff (Tersoff, 1988a, 1988b) with bond-order parameters including multi-body interactions. This potential can be effectively applied to C-containing systems to understand the complicated behaviours by the strengths of double and triple bonds. The parameter sets were proposed for systems including Si, C, Si/C, C/H, C/F, and Si/C/F (Abrams & Graves, 1998; Tanaka et al., 2000). In addition to the SW and Tersoff potential functions, other potential functions (Biswas & Hamann, 1987; Dodson, 1987; Hanson et al., 1997) were also proposed. Although there have been many discussions on the validity of the functions (Balamane et al., 1992), all of these functions can effectively reproduce some structural and thermodynamic characteristics of the materials and the relevant structural chemistry for some selected molecules. In this study, to eliminate complicated surface reactions usually occurring in halogen-containing plasmas, we focus on Ar-Si-O system for studying PID. We used the SW potential function for the Si-Si and Si-O systems.

3.1 Interatomic potential functions used in this study

The Stillinger-Weber potential function (Stillinger & Weber, 1985) utilizes both two-body and three-body interaction terms to stabilize the diamond cubic structure of crystalline silicon. The potential function is given by

$$\Phi = \sum_{i<j} V_2(i,j) + \sum_{i<j<k} V_3(i,j,k) \tag{1}$$

where $V_2(i, j)$ is the two-body interaction term between i-th and j-th atoms expressed as

$$V_2(i,j) = A_{ij} \cdot g_{ij}(B_{ij} \cdot r_{ij}^{-p_{ij}} - r_{ij}^{-q_{ij}}) \cdot \exp\left(\frac{C_{ij}}{r_{ij} - a_{ij}}\right), \tag{2}$$

if $r_{ij} < a_{ij}$, where a_{ij} is the cut-off distance, and $V_2(i, j) = 0$ otherwise. r_{ij} is the interatomic distance between i-th and j-th atoms in the SW's length unit (0.20951 nm). The parameters A_{ij}, B_{ij}, C_{ij}, p_{ij}, and a_{ij} depend only on the species of i-th and j-th atoms. g_{ij} (< 1) is the bond-softening function introduced by Watanabe et al. (Watanabe et al., 1999), adjusting the contribution of the two-body term to reproduce the cohesive energies of Si-O bonds. The three-body interaction term has the following form.

$$V_3(i,j,k) = h(r_{ij}, r_{ik}, \theta_{jik}) + h(r_{ji}, r_{jk}, \theta_{ijk}) + h(r_{ki}, r_{kj}, \theta_{ikj}), \tag{3}$$

where θ_{ijk} is the angle between two lengths of r_{ij} and r_{jk}, etc. Given that r_{ij} and r_{ik} are less than the cut-off distance, the function h is

$$h(r_{ij}, r_{ik}, \theta_{jik}) = \lambda_{jik} \exp\left(\frac{\gamma_{jik}^{ij}}{r_{ij} - a_{jik}^{ij}} + \frac{\gamma_{jik}^{ik}}{r_{ik} - a_{jik}^{ik}}\right) \times (\cos\theta_{jik} - \cos\theta_{jik}^{0})^2, \tag{4}$$

otherwise $h = 0$. $\lambda_{jik}, \gamma_{jik}^{ij}, a_{jik}^{ij}, \gamma_{jik}^{ik}, a_{jik}^{ik}, \theta_{jik}$, and θ_{jik}^{0} are parameters for the j-i-k triplet. a_{jik}^{ij} and a_{jik}^{ik} are the cut-off distances in the three-body configuration. For the "ideal" tetrahedral angle,

$$\cos\theta_{jik}^{0} = -\frac{1}{3} \tag{5}$$

is held.

Regarding the two-body system with Ar, the Moliere-type repulsive pair potential function (Moliere, 1947; Torrens, 1972; Wilson et al., 1977) was employed. The potential function $V_2(i,j)$ includes a screening function $f_s(r_{ij})$ combined with a Coulomb potential. The function is expressed as

$$V_2(i,j) = \frac{Z_i Z_j e^2}{4\pi\varepsilon_0 r_{ij}} f_s(r_{ij}), \tag{6}$$

where e is the elementary charge, ε_0 is the permittivity of a vacuum, and Z_i and Z_j are the atomic numbers of the projectile (Ar) and target (Si, O, Ar) atoms, respectively. The expression of the screening function has been studied by many researchers (Torrens, 1972). So far, the repulsive interatomic potential between Ar and other elements has been modified (Wilson et al., 1977). In the expression of Moliere potential function, $f_s(r_{ij})$ is described as

$$f_s(r_{ij}) = \sum_{m=1}^{3} C_m \exp\left(-b_m r_{ij} / a'\right) = 0.35\exp\left(-\frac{0.3r_{ij}}{a'}\right) + 0.55\exp\left(-\frac{1.2r_{ij}}{a'}\right) + 0.10\exp\left(-\frac{6.0r_{ij}}{a'}\right) \quad (7)$$

where a' is the Firsov screening length (Firsov, 1957). The parameters proposed by Moliere (Moliere, 1947; Wilson et al., 1977) were used in this study.

3.2 Simulation procedure

We prepared a crystalline Si structure of squared Si (100) surface with a side length of 3.258 nm (six times of lattice constant ~ a squared 6-unit cell). The MD code used in this study was originally developed by Ohta and Hamaguchi (Ohta & Hamaguchi, 2001b). Each layer contained 72 atoms (= 1 monolayer (ML)). The bottom layer of the simulation cell (72 atoms) was rigidly fixed throughout the simulations. Initial depth of the simulation domain is 9-unit cells (nine times of lattice constant or 36 MLs). Periodic boundary condition was employed along the horizontal direction. Since a typical MOSFET structure has an SiO_2 layer on the Si substrate, the initial Si structure (6 × 6 × 9 cell) was "oxidized" before the Ar ion-bombardment. The oxidation was done by the 500-consecutive impacts of O atoms at 50 eV and followed by a subsequent cooling step with a set-point temperature of 400 K (See below). This step can create a surface oxidized layer corresponding to SiO_2 film formed on a source / extension region of MOSFET. Using the obtained SiO_2/Si structure, we injected Ar atoms with various incident energies. Ar atoms were injected from randomly selected horizontal locations above the surface of the cell at normal incidence. (In this study, we injected Ar ions with a constant E_{ion}, although, in practical plasma etching processes, the energy of ions obeys an ion energy distribution function (IEDF) dependent on a frequency of applied bias power (Lieberman & Lichtenberg, 2005) as illustrated in Fig. 2. However, a recent study (Eriguchi et al., 2010) showed that, in low applied bias voltages (< 500 eV), the average ion energy can be used as a primal measure for the damaged-layer thickness. Therefore, in the present MD simulation, we employed a mono-energetic ion impacts.) Note that, in plasma etching, an ion plays an important role in the reactions. However, in conventional MD simulations, charge-neutral atoms are used as incident particles. This is based on the assumption that incident ions are expected to be neutralized near the target surface due to a resonance or Auger process. Thus, in this study, we employed the above potential functions for charge-neutral Ar atom.

In the present-day RIE, plasma densities of the order of 10^9-10^{11} cm^{-3} are widely used. These densities lead to a (Bohm) flux of incident ions of 10^{13} – 10^{16} $cm^{-2}s^{-1}$, depending on the plasma density and the electron temperature (Lieberman & Lichtenberg, 2005). The interval between two successive ion impacts in the case of the present simulation domain (~ 10 nm^2) is much longer than the simulation time range. Therefore, each ion impact is thought to be an independent single event. To simulate such single events, for the first 0.7 ps after an energetic particle hits the surface, the motion of all particles in the domain are solved numerically by a classical mechanics except for those in the rigidly fixed bottom layer. Then, we applied "artificial" cooling step to all the particles for 0.3 ps using Berendsen's heat removal scheme (Berendsen et al., 1984) with a cooling constant of 2.0 x 10^{-14} s^{-1}. The set-point temperature of the simulation cell was 400 K to reproduce the practical surface-temperature range during plasma etching. After the end of cooling step, a new energetic particle was directed again to the surface, and the whole simulation cycle was repeated. Details of the present MD simulation procedure were published elsewhere (Ohta & Hamaguchi, 2001b).

To evaluate the damage creation mechanisms, the defect structure formed by the ion impacts should be identified. Since comprehensive discussions on the defect structures may be beyond the scope of this article, we focus on the displaced Si atoms from the initial lattice site and the representative defect structures obtained by the simulations. To assign the displaced atoms, the Lindemann radius (r_L) was used as a measure of the displacement threshold (Hensel & Urbassek, 1998; Nordlund et al., 1998). The Lindemann radius is defined as the vibration amplitude of Si atoms at their melting point. For the SW potential function for Si, the radius is r_L = 0.45 Å. After the MD simulations, we inspected the displacements of all the Si atoms. Then, we identified all those atoms as defects if they were outside of the cubic cell (with an edge of 2 × r_L) whose center was located at an original lattice site. The number of the displaced atoms was counted to investigate an overall trend of PID in the course of successive ion impacts.

3.3 Simulation results and discussion

Figures 4 display typical damaged-layer/Si structures after 500 impacts by Ar ions with various energies. In these figures, Ar ions are omitted from the structures to clearly show the damaged layer and Si substrate. From other simulation results, we found that Ar atoms are usually present in the tetrahedral interstitial site underneath the interfacial layer. As seen in Fig. 4, the damaged layer thickness increases with an increase in E_{ion}. The heavily damaged surface layer and mixing layer with a rough interface (interfacial layer) are observed, in particular, for higher E_{ion} cases (≥ 200 eV). The surface layer has been usually identified as "an amorphous layer" in the view of simulation as discussed in previous literatures (Graves & Humbird, 2002; Oehrlein, 1989). However, when samples (wafers) are processed to the next manufacturing step, oxidation of the surface occurs due to an air exposure. As mentioned in the next section, the partially oxidized layer is detected by an optical technique as SiO_2 layer, and the residual damaged layer is evaluated as a mixing layer. In the following discussion, based on the experimental data, we define the surface amorphous layer and the rough interfacial region (including local defect sites) as the surface and interfacial layers, respectively. One should also pay careful attention to the local defect sites underneath the rough interface as highlighted in Fig. 4.

Regarding the local defect sites observed in Fig. 4, many respective structures have been proposed so far (Baraff et al., 1980; Batra et al., 1987; Cheng & Corbett, 1974; Estreicher et al., 1997; Hastings et al., 1997; Leung et al., 1999; Schober, 1989; Tang et al., 1997). In addition to vacancy, various interstitials have been studied extensively by *ab initio* calculations using clustered Si atoms. Tetrahedral and hexagonal interstitials as well as "dumbbell" and the bond-centered interstitial are commonly proposed structures (Batra et al., 1987; Leung et al., 1999). Figure 5 shows the profile of an increase in the number of Si atoms located in each

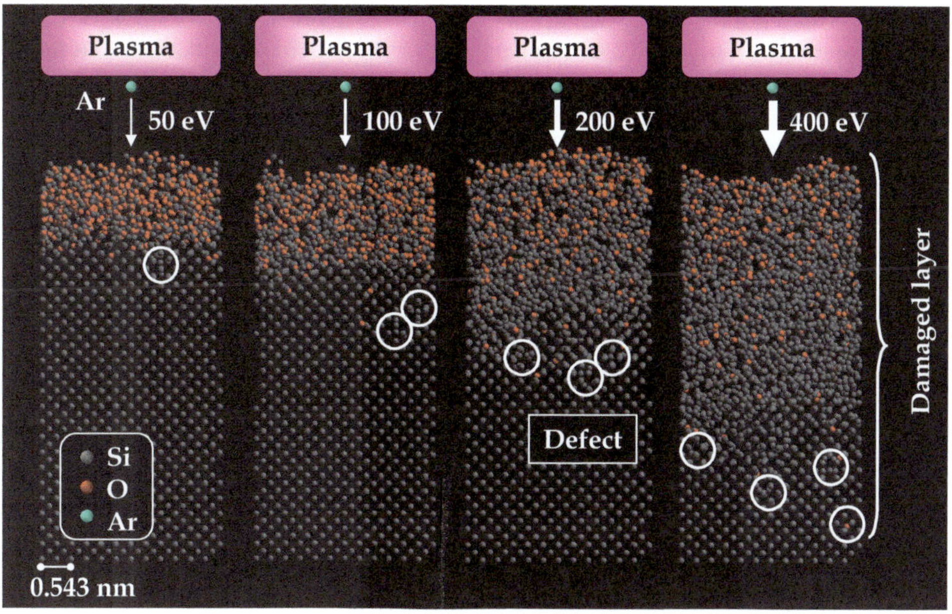

Fig. 4. SiO_2/Si structures after 500 impacts of Ar with different incident energies. The injected Ar ions are omitted from the cells for easy comparison. Local defect sites are highlighted.

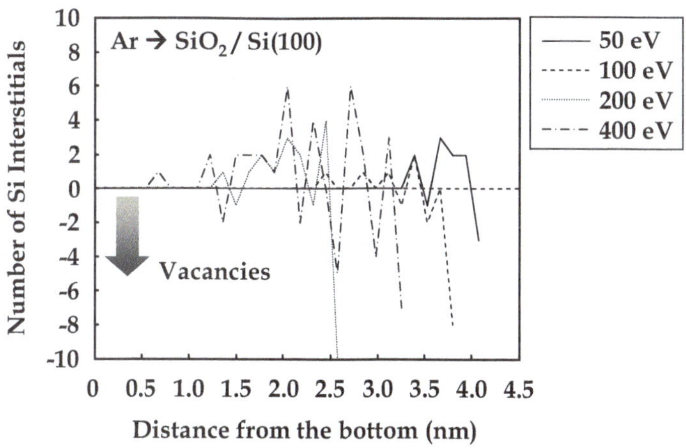

Fig. 5. Profiles of an increase in Si atoms in each atomic plane along [100].

atomic plane along the depth of [100] direction (The space between the planes is 0.13575 nm). 72-Si atoms are originally located in each plane for the present cell structure (6 × 6). In this figure, only Si atoms were inspected and counted. The difference in the interstitial structures mentioned in the above was not considered. The positive value presumably implies the presence of interstitials and the negative, that of vacancy. This means that the local defect sites are consisting of vacancies as well as Si interstitials. (Note that one can see cumulatively the net positive value when integrating the data from the bottom. This implies that the Si interstitials are more probable than vacancies in those cases.) The interstitial Si atoms were formed by knock-on process.

By keeping in mind the results in Fig. 5, we investigated in detail some of typical defects in Fig. 4 to clarify the structures of these defects formed by PID. Figures 6 present some of typical structures - we chose three representative structures. On the left are shown the views of the defects on the (100) plane (along [100]). In the middle, the bird's views to the respective defects (Si atoms) are shown. Interstitial Si atoms are highlighted in the views. From Fig. 4, we roughly categorize the defect structures (the interstitial atoms) as Type-A, Type-B, and Type-C. Type-A is like a tetrahedral interstitial, and Type-B, a hexagonal interstitial (Batra et al., 1987; Leung et al., 1999; Schober, 1989). In both cases, neighbouring

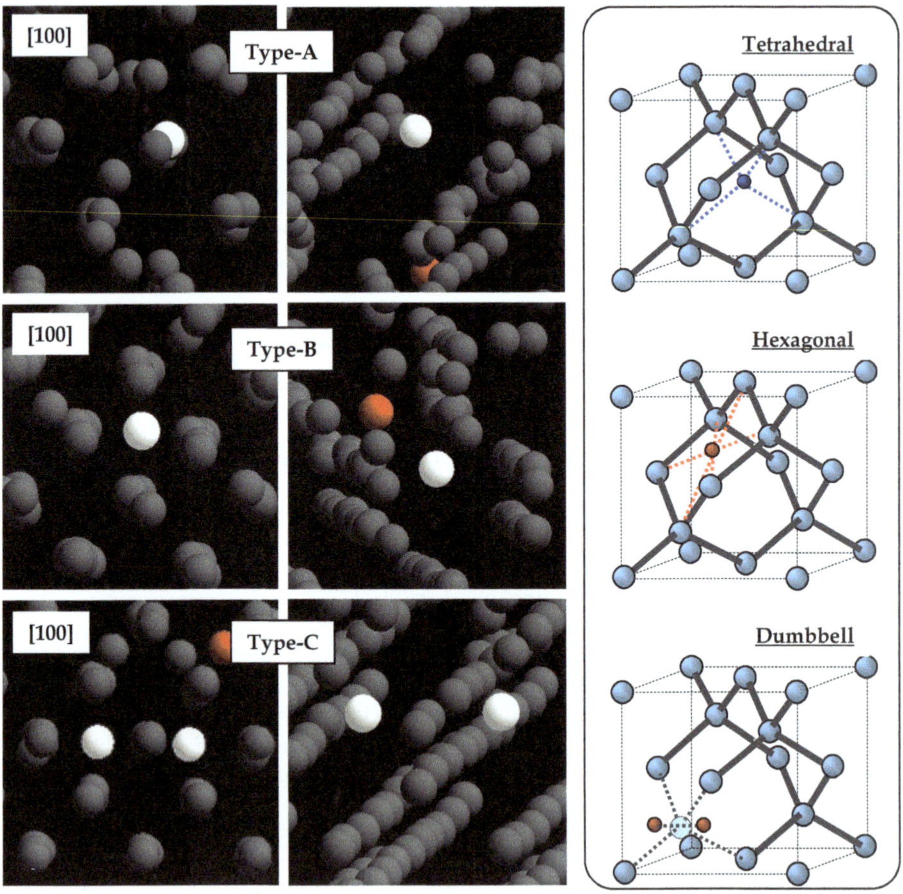

Fig. 6. Typical structures of Si-interstitial defects created by energetic ion bombardments. On the left are the views along [100] and the bird's-view are shown in the middle. On the right, typical structures assigned by previous reports are shown. (See also Fig. 4.)

Si atoms are not displaced considerably. Type-C is like a "dumbbell" structure where bonded Si atom was displaced from the lattice sites due to the presence of an interstitial Si permeating from other regions. It is widely believed that these structures are probable and found to be stable in terms of the formation energies calculated from a quantum mechanics scheme (Colombo, 2002; Leung et al., 1999; Tang et al., 1997). Typical structures reported so far are also illustrated on the right of Fig. 5. For details including other structures, please see the literatures (Batra et al., 1987; Cheng & Corbett, 1974; Leung et al., 1999). In terms of the effect of the presence of these structures on electrical characteristics of crystalline Si structure, it is also believed that these defects can create additional energy levels in the band gap ("band-gap states") (Estreicher et al., 1997; Hastings et al., 1997; Schultz, 2006). Therefore, one can speculate that the local defects by PID degrade MOSFET performance because the band-gap state plays a role as a carrier trap site inducing an increase in leakage current due to hopping conduction (Koyama et al., 1997) and/or an increase in a channel series-resistance (Eriguchi et al., 2009b) due to coulomb scattering by the trapped carrier. In general, it is difficult to identify those local defects by conventional analysis techniques such as TEM observation, in particular, to quantify the density of these defects. In the next section, by using novel analysis techniques, we quantify the density of the defects and discuss the electrical characteristics of these defects.

Since the observed defect structures are indeed derived from our MD simulations, we have to pay careful attention to the effects of the simulation procedure on our findings. In the present case, the number of defect sites (~ the density) might be dependent on both the simulation cell size and the number of Ar impacts. Conventional MD simulations for plasma etch processes employ; (1) periodic boundary condition and (2) the rigid fixed Si atoms at the bottom plane of the cell. Both technical restrictions may make the defect generation mechanism being dependent on the cell-size. It was reported (Abrams & Graves, 2000; Graves & Humbird, 2002) that the formation of amorphous layer is time-dependent, i.e., the number of ion impacts determines whether the damaged-layer formation process is in a growth or a saturation phases. Previous reports showed that the thickness of amorphous layer by Ar ion impacts became in a steady state after 1.5-monolayer (ML) impact (Graves & Humbird, 2002) and also that the reaction layer thickness by F ion impacts, after 10 MLs (Abrams & Graves, 2000).

Figure 7 shows at first the cell-size dependence of Si ion penetration depth. Si was self-implanted repeatedly with the energy of 150 eV into the initial simulation cells of various sizes. Due to a stopping process (Lindhard et al., 1963; Wilson et al., 1977), the injected Si atom loses the energy by a series of collisions and finally comes to rest in Si substrate. After 1000 impacts of Si atoms, the profiles of penetration depth ("projection range") were determined and compared for various cell sizes. As shown, one can see a small difference among the peak positions, i.e., the ion penetration depth is almost independent on the cell size. The result may suggest that one can use a smaller cell-size for plasma-etch MD simulation to reduce a "simulation cost" for estimating the ion penetration depth. However, as mentioned later, for investigating the density of local defect site, a smaller cell-size statistically contains a smaller number of local defect sites under the same number of ion impacts. Hence, although the estimation of ion penetration depth can be done by using a smaller cell size, a larger number of impacts should be conducted to investigate an overall feature of local defect site structures. In Sec. 4, we compare the number of defect sites in Fig. 4 with that obtained by the experiments.

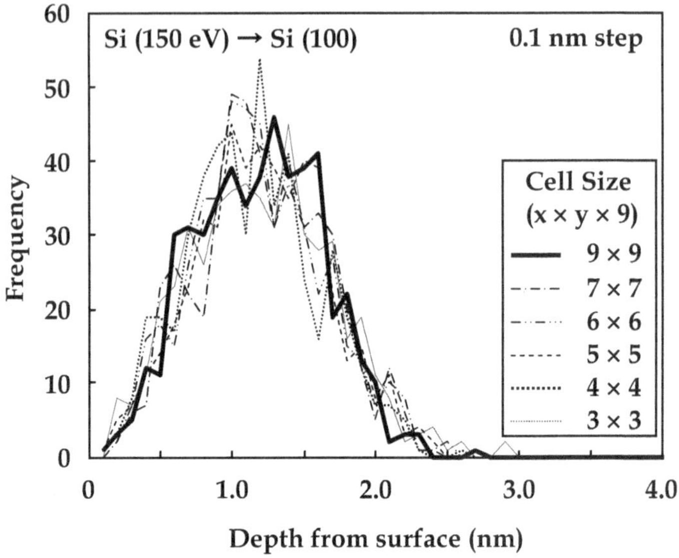

Fig. 7. Depth profiles of incident Si atoms into the cells with various sizes.

Figure 8 shows time evolutions of the number of displaced Si atoms (n_{Si}) determined from the algorithm based on the Lindemann radius, for the case of the present cell size of 6 × 6 × 9. As seen, at the initial stage, the counted n_{Si} increases with fluence until ~ 100 impacts, and then saturates after approximately 500 impacts for both E_{ion} cases. The 500-impact corresponds to ~ 7-ML injections in the case of the present (100) surface (6 × 6). Compared to previous reports (Abrams & Graves, 2000; Graves & Humbird, 2002), this saturation value is

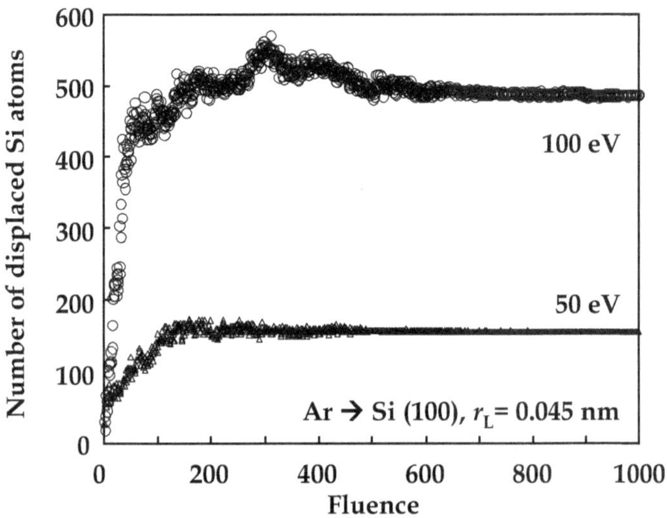

Fig. 8. Time evolutions of the number of displaced Si atoms.

in the reasonable range. Thus, it can be concluded that by considering the saturation phenomena, the number of impacts should be properly optimized in advance when performing plasma-etch MD simulations. (In Fig. 4, we have shown the results by 500 impacts for discussions on the defect sites.)

4. Comparison with experimental data

4.1 Experimental setup, test sample structure, and evaluation techniques

To verify the above MD simulation results, we carried out plasma treatments of Si wafers by using two plasma reactors. N-type (100) Si wafer with 0.02 Ωcm was exposed to an inductively coupled plasma (ICP) reactor (See Fig. 2) and a DC plasma reactor. In the ICP reactor (Eriguchi et al., 2008a), the plasma was generated by a source power supply unit. The applied power was 300 W. The plasma density can be controlled by changing this source power. On the other, a bias power applied to the wafer stage determines the energy of ion incident on the wafer surface. In the present study, the bias power was varied from 50 to 400 W. The frequencies of source and bias power were 13.56 MHz for both. Ar was used and the pressure was 20 mTorr. Unless otherwise specified, the process time was 30 s. By using an oscilloscope, we determined self dc bias (V_{dc}<0) and plasma potential (V_p), resulting in an average ion energy E_{ion} (eV) (= $e(V_p - V_{dc})$) ranging from 56 to 411 eV. From a Langmuir probe measurement, the electron density and the electron temperature were estimated to be 3.3×10^{11} cm^{-3} and 2.6 eV, respectively. Since Ar plasma is electropositive, the ion density is approximately equal to the electron density 3.3×10^{11} cm^{-3}, giving the flux of ions (Γ_{ion}) of 1.2×10^{17} cm^{-2}s^{-1}. Note that this ICP configuration results in a constant Γ_{ion} to the wafer for all bias power conditions in the present experiments.

In order to evaluate the damaged structure, we conducted several analyses – spectroscopic ellipsometry (SE) (Herman, 1996), high-resolution transmission electron microscope (HR-TEM) observation, Rutherford backscattering spectroscopy (HR-RBS), photoreflectance spectroscopy (PRS) (Herman, 1996), and capacitance – voltage (C-V) measurement. In particular, to quantify the local defect site density, we employed PRS and CV measurement.

In SE analysis, the damaged layer is assumed to consist of two regions, i.e., the surface and interfacial layers (abbreviated as SL and IL, respectively). The SL is composed of SiO_2, as a result of oxidation of heavily damaged regions by exposure to an air as well as the presence of knock-on oxygen from the surface oxide layer. (This is confirmed from the MD simulation results.) The IL is partially oxidized or disordered Si, which is identified by SE as a mixed layer consisting of crystalline Si and SiO_2 phases. Thus, an optical model assuming four layers (ambient, surface SiO_2 layer, interfacial layer, and Si substrate) was employed (Eriguchi et al., 2008a; Matsuda et al., 2010). In this analysis, the thicknesses of the surface (SL: d_{SL}) and interfacial [IL (SiO_2 + crystalline-Si): d_{IL}] layers and the composition (f_{Si}: component of crystalline-Si) were determined.

PRS is one of modulation spectroscopic techniques (Aspnes, 1973; Pollak & Shen, 1990), where the surface of the sample is modulated by a laser beam. The reflectance change by using a probe beam was monitored. In the present PRS, the amplitude of reflectance change ($\Delta R/R$) for the damaged sample was measured. A decrease in $\Delta R/R$ means the defect generation in the surface region of Si substrate (< 10 nm) (Murtagh et al., 1997;

Wada et al., 2000). Details for the basics of PRS analysis and experimental procedures were described elsewhere (Eriguchi & Ono, 2008). From the laser-power dependence of spectra, the areal defect site density (N_{dam} (cm^{-2})) can be estimated (Eriguchi & Ono, 2008; Nakakubo et al., 2010a).

For Capacitance - Voltage (C-V) analysis (Sze, 1981), we used a mercury probe system. The bias frequency was 1 MHz. To estimate the local defect site density, we investigated $1/C^2$-V curves of the damage samples (Eriguchi et al., 2008a). Details are mentioned later.

4.2 Experimental results and discussion

4.2.1 Thickness of the damaged-layer

Figure 9 shows the surface and interfacial thicknesses (d_{SL} and d_{IL}) identified by SE with an optimized four-layer model. The samples were treated by the ICP system by changing an applied bias power. The SE analysis assigns an increase in d_{IL} with the average ion energy E_{ion}, while d_{SL} does not exhibit a clear increase. This may be due to surface sputtering mechanism. Rutherford backscattering spectrometry (not shown here) identifies the presence of stoichiometric SiO_2 region in the SL, and the IL consisting of SiO_2 and *c*-Si phases (Eriguchi et al., 2008a). Figure 10 shows TEM observation results for various bias powers. Although it is difficult for the case of higher E_{ion} to assign the details of the damaged structures, one can observe the presence of interfacial layer below the surface layer. The estimated thickness was found to be consistent with the SE data (Eriguchi et al., 2008a; Matsuda et al., 2010). The roughness in the interfacial layer is confirmed to increase, in particular, for the 400-W case. The increase in roughness may agree with MD simulation results in Fig. 4. Therefore, it is concluded that the presence of both the surface and interfacial layers should be taken into account when evaluating the damaged-layer thickness by SE and/or TEM. Note that, from these TEM observation results, one can see no local defect site which was assigned by MD simulations. To identify these local defect sites, other novel techniques are required. In the subsection 4.2.3, we provide the results by these techniques.

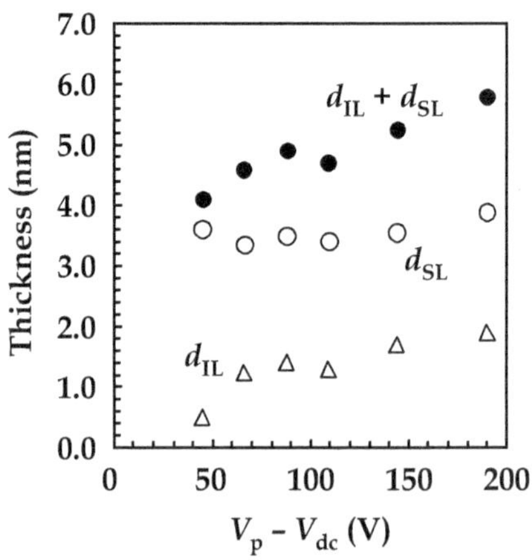

Fig. 9. Thicknesses of surface and interfacial layers obtained by spectroscopic ellipsometry with a four-layer optical model. Total optical thickness is also shown (closed circles).

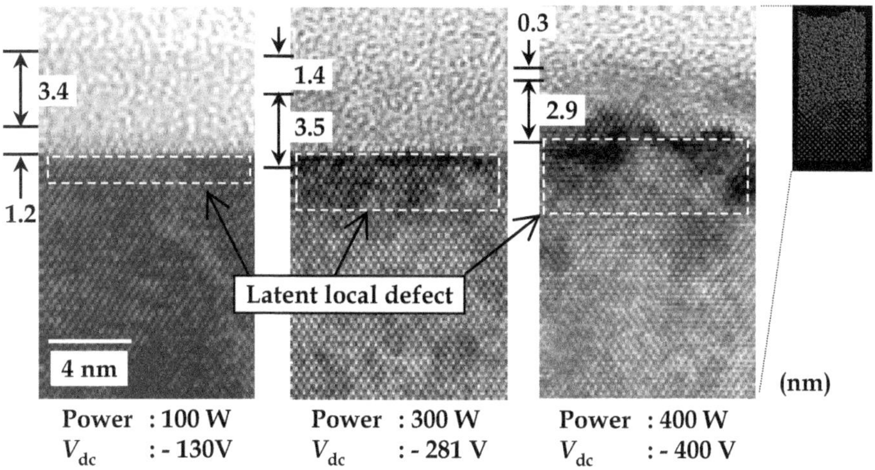

Fig. 10. TEM observation results for the Si wafer surface damaged by various bias powers.

4.2.2 Time-dependent damaged layer formation

As indicated in Fig. 8, the MD simulation predicts growth of the damaged layer in accordance with ion fluence. Some of experimental results regarding this phenomenon are shown in Fig. 11. Figure 11 indicates d_{SL} and d_{IL} as a function of process time (plasma exposure time t_{pro}). A DC plasma reactor was used to create PID for various conditions; V_{dc} = - 300 and - 350 V, respectively. Dependences of d_{SL} and d_{IL} on t_{pro} are shown. As t_{pro} increases, while the d_{SL} is almost constant (due to oxidation of surface layer by an air exposure), the d_{IL} increases for both cases. Moreover, the d_{IL} tends to saturate after a certain amount of time; 5 s for - 300 V and 10 s for - 350 V. From a Langmuir probe measurement, the electron temperature and the electron density were estimated to be 3.0 eV and ~ 10^9 cm^{-3}, respectively, giving Γ_{ion} of ~ 2 × 10^{14} cm^{-2}s^{-1} in these experiments. Thus, the total dosage (fluence) is approximately 2 × 10^{15} cm^{-2} for 10 s. Based on the results by the MD simulations in Fig. 8, 500-impact in the present cell size corresponds to a fluence of ~ 5 × 10^{15} cm^{-2}. Although the E_{ion} in the DC plasma processes was not measured precisely (only deduced from applied DC bias voltages), we can speculate that the saturating behaviour of d_{IL} in Fig. 11 corresponds to the results in Fig. 8. In other words, once Γ_{ion} is properly determined by plasma diagnostics, the MD simulation can predict the time evolution of damaged-layer formation in practical plasma etching process.

In general, the number of ion impacts combined with the cell size is a more practical and useful measure rather than MLs for MD simulations, because the density of local defect site is a key parameter determining the effects of PID on MOSFET performance degradation. Figure 12 provides the relationship among the number of ion impacts, Γ_{ion}, and t_{pro}. Figure 12 guides how many ion impacts are necessary in MD simulations to predict correctly the local defect site generation during practical plasma etching. For a given Γ_{ion}, the number of impacts in a MD simulation should be increased with an increase in t_{pro}. For example, in the case of Γ_{ion} = 10^{15} cm^{-2}s^{-1} and t_{pro} = 10^2 s, approximately a 10000–impact is needed in the MD simulation. Note that this value is dependent on the cell size. Deduced only from the time

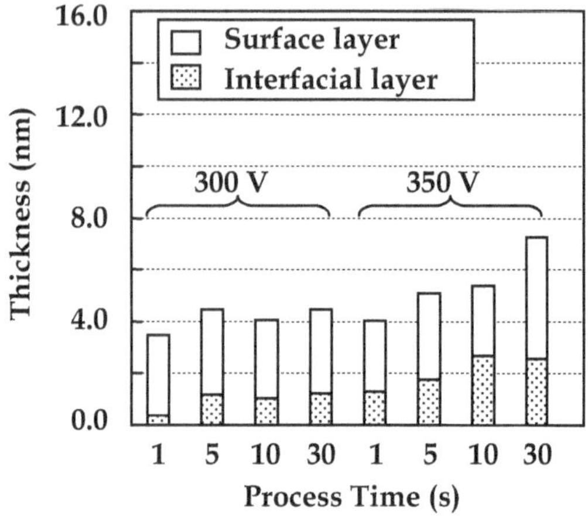

Fig. 11. Thicknesses of surface and interfacial layers as a function of process time. A DC plasma reactor was used.

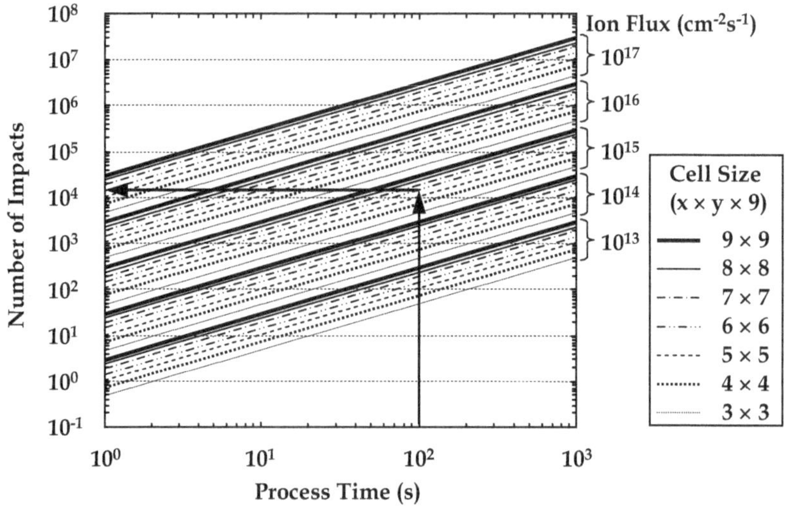

Fig. 12. The number of impacts required for MD simulation to predict practical plasma etch damage for various ion fluxes. The cell size dependence is shown by the respective lines.

evolution of damaged-layer formation as in Fig. 8, one can expect that a smaller size is better (more efficient). However, as discussed in the next, in terms of local defect site density, one has to increase a cell size as much to understand accurate pictures and overall features of the defect structures for the precise prediction of MOSFET performance degradation.

4.2.3 Density of defect sites

So far, PID has been evaluated by a wide variety of techniques (Awadelkarim et al., 1994; Egashira et al., 1998; Kokura et al., 2005; Mu et al., 1986; Oehrlein et al., 1988; Yabumoto et al., 1981). However, there have been not so many reports quantifying the density of created local defects by PID. In this study, we employed two different quantification techniques, photoreflectance spectroscopy (PRS) and capacitance-voltage (C-V) measurement. Details are presented in other literatures (Eriguchi et al., 2008a; Eriguchi & Ono, 2008).

Figure 13 shows the estimated local defect site density from photoreflectance spectra of the test structures damaged by the ICP system (closed squares, on the left axis). By using a modified PRS model for estimation of the density of defect sites (Eriguchi et al., 2008b; Eriguchi & Ono, 2008), one can determines the areal defect site density (N_{dam}) as a function of E_{ion}. From this figure, one can observe N_{dam} of the order of 10^{12} cm^{-2} for the present plasma conditions.

Figure 14 shows examples of $1/C^2$-V analysis results. Figure 14 on the right illustrates the basics of this C-V technique, where w and q are the depletion layer width and the elementary charge, respectively. By using a mercury probe system, a bias voltage applied to Si substrate was swept, and the capacitance was measured. We performed this C-V measurement for both the control (without plasma exposure) and the damaged samples. When the value $1/C^2$ is plotted along the bias voltage (V_b) for a fresh Si substrate (the control), the slope of $1/C^2$ - V_b becomes constant since the slope corresponds to the impurity (dopant) concentration n_D (Goodman, 1963; Sze & Ng, 2007). For the plasma-damaged sample, one can see the distortion of $1/C^2$ - V_b curves in the inversion region as seen on the left figure. In this experiment, we used n-type Si substrate, thus, the negative bias voltage ($V_b < 0$)) corresponds to the inversion layer formation region. The presence of the local defect site (carrier trapping site) causes doping loss (a decrease in n_D) by intermixing with subsequently-implanted ions (Kokura et al., 2005). Therefore, the distortion (a decrease in the slope of $1/C^2$ - V_b plot) indicates a defect site creation in the Si substrate. On the right in Fig. 14, the schematic view of these defect sites in the inversion scheme is also shown. From a change of the slope, we can estimate the volume density of the defect site (n_{dam} cm^{-3}) by assuming that the effect impurity concentration equals to ($n_D + n_{dam}$) (Eriguchi et al., 2008a; Nakakubo et al., 2010b). The calculated n_{dam} is plotted on the right axis in Fig. 13 (open circles). The n_{dam} on the order of 10^{18}-10^{19} cm^{-3} is assigned. Since the thickness of d_{IL} containing n_{dam} is estimated to be a few nanometers as seen in Fig. 9, the areal density calculated from the E_{ion}-dependence of n_{dam} is consistent with N_{dam} by PRS. (Exactly speaking, the depth profile of n_{dam} should be taken into account, though.) One of the other important findings in Figs. 13 and 14 is the fact that the identified local defect sites are electrically active and may induce the MOSFET performance degradation. The defect site structures observed in Fig. 6 are fatal and should be investigated with carefully attention.

The estimated range of n_{dam} or N_{dam} gives an important interpretation to the MD simulations for PID as follows. The areal density of defect site ranging from 10^{12} to 10^{13} cm^{-2} in practical samples is equivalent to the number of defect site of 0.1 to 1 in the present 6 × 6 simulation cell (~ 10 nm^2), although it depends on t_{pro} and plasma parameters. In Fig. 4, the observed local defect sites are indeed a few in the number. This number corresponds to approximately 10^{13} cm^{-2} in practical samples, which is in consistent with the experimental results in Fig. 13. If one uses a smaller simulation size such as 3 × 3 cell to reduce the simulation cost, one can not find any defect sites statistically. In other words, a simulation scheme using a smaller size (for reducing the calculation time) may bring an erroneous conclusion (no observation of local defect sites). Regarding PID by MD, an optimization of the cell size in accordance with the plasma etching parameter such as Γ_{ion} is quite important.

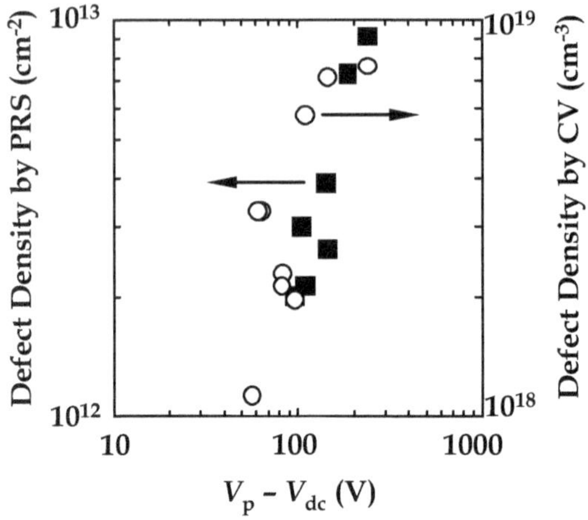

Fig. 13. Estimated defect site densities as a function of (V_p - V_{dc}) by two different analysis techniques; PRS and CV. From PRS, the areal density was determined, while, from CV, the volume density.

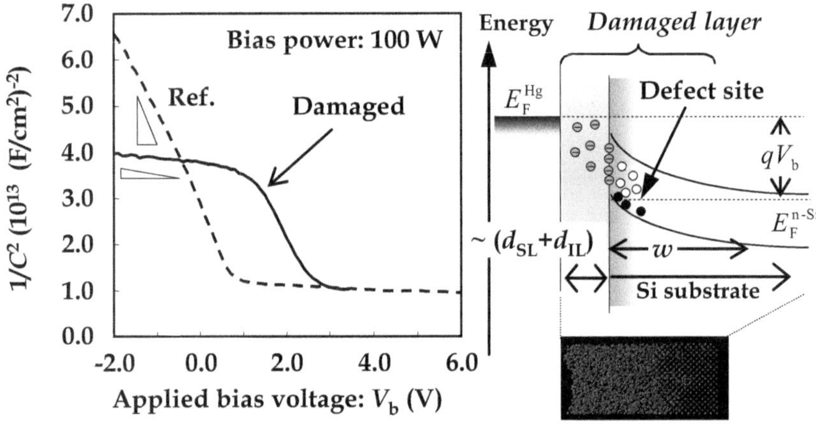

Fig. 14. Illustrations of an example for 1/C^2-V plots and the energy band diagram during the CV measurement.

4.2.4 Prediction framework for MOSFET performance degradation by MD

Finally we discuss the effects of PID observed in the MD simulations and experimental data on MOSFET performance degradation. From the findings in the above, we can summarize the following key issues:

1. The damaged layer is found to consist of two layers, the surface and the interfacial layers. By considering this oxidation process, this thickness of Si loss (Si recess depth d_R) can be predicted by MD simulations.
2. Underneath the interfacial layer, local defect sites are present. The local defect structures can be assigned from MD simulations.

3. The local defect sites (observed in the MD simulation) are confirmed to be electrically active from the experimental data.

As proposed by Eriguchi et al.(Eriguchi et al., 2009a; Eriguchi et al., 2009b), Si recess depth d_R and the local defect sites degrades MOSFET-operation parameters such as V_{th} and I_d, where I_d is drain current determining the operation speed (Sze, 1981). These mechanisms are analytically expressed as,

$$\Delta V_{th} \propto -\frac{A}{L_g} d_R , \tag{8}$$

$$I_d = I_d^0 \cdot (1 - B \cdot n_{dam}) , \tag{9}$$

where ΔV_{th} and I_d^0 are the shift of threshold voltage and the drain current of a fresh device, respectively. A and B are the process- and device-structure-dependent parameters which can be determined from Technology Computer-Aided-Design (TCAD) simulations (Eriguchi et al., 2008a). A prediction of MOSFET performance from process parameters such as E_{ion} was demonstrated by using experimental relationship between E_{ion} and d_R (or n_{dam}). Moreover, from the present studies, one presumes that d_R and n_{dam} can be determined from MD simulations. At present, the statistical distribution functions for the local defect structures by PID are not clarified yet. A future work is hoped to assign the defect generation mechanisms and the statistical distribution functions with a help of MD simulations. Since the size of MOSFET is shrinking aggressively, a prediction framework of future device design can be organized by an atomistic technique such as MD simulations.

5. Conclusion

We applied classical MD simulations to investigate plasma-etch damage mechanisms. The simulated structures were found to consist of two layers, the surface and the interfacial layers. Due to oxidation by an air exposure, these two layers were identified in practice as SiO_2 and a mixed layer consisting of crystalline Si and SiO_2 phases, respectively. Underneath the interfacial layer, we assigned local defect sites, and the structures were confirmed to include typical Si interstitials. From the experiments, these sites were found to be electrically active. Combined with plasma diagnostics and quantitative analysis techniques of the local defect site density, the number of required ion impacts and the cell size for MD simulations were optimized for PID. Finally, a prediction framework for MOSFET design was discussed. An optimized MD simulation will be a promising candidate for predicting MOSFET performance degradation by PID.

6. Acknowledgment

The author greatly thanks Dr. H. Ohta for his great support of the MD simulation in this work. Acknowledgements are also given to Mr. Y. Nakakubo, Mr. A. Matsuda, Dr. Y. Takao, and Prof. K. Ono at Kyoto University, and Drs. M. Yoshimaru, H. Hayashi, S. Hayashi, H. Kokura, T. Tatsumi, and S. Kuwabara at STARC (Semiconductor Technology Academic Research Center) for their helpful discussions. This work was financially supported in part by STARC and a Grant-in-Aid for Scientific Research (B) from the Japan Society for the Promotion of Science.

7. References

Abrams, C., & Graves, D. (1998). Energetic ion bombardment of SiO_2 surfaces: Molecular dynamics simulations, *J. Vac. Sci. Technol. A*, Vol.16, No.5, (1998), pp. 3006-3019

Abrams, C. F., & Graves, D. B. (1999). Molecular dynamics simulations of Si etching by energetic CF_3^+, *J. Appl. Phys.*, Vol.86, No.11, (1999), pp. 5938-5948

Abrams, C. F., & Graves, D. B. (2000). Molecular dynamics simulations of Si etching with energetic F^+: Sensitivity of results to the interatomic potential, *J. Appl. Phys.*, Vol.88, No.6, (2000), pp. 3734-3738

Aspnes, D. E. (1973). Third-derivative modulation spectroscopy with low-field electroreflectance, *Surf. Sci.*, Vol.37, (1973), pp. 418-442

Awadelkarim, O. O., Mikulan, P. I., Gu, T., Reinhardt, K. A., & Chan, Y. D. (1994). Electrical properties of contact etched p-Si: A comparison between magnetically enhanced and conventional reactive ion etching, *J. Appl. Phys.*, Vol.76, No.4, (1994), pp. 2270-2278

Balamane, H., Halicioglu, T., & Tiller, W. A. (1992). Comparative study of silicon empirical interatomic potentials, *Phys. Rev. B*, Vol.46, No.4, (1992), pp. 2250-2279

Baraff, G. A., Kane, E. O., & Schluter, M. (1980). Theory of the silicon vacancy: An Anderson negative-U system, *Phys. Rev. B*, Vol.21, No.12, (1980), pp. 5662-5986

Batra, I. P., Abraham, F. F., & Ciraci, S. (1987). Molecular-dynamics study of self-interstitials in silicon, *Phys. Rev. B*, Vol.35, No.18, (1987), pp. 9552-9558

Berendsen, H. J. C., Postma, J. P. M., Gunsteren, W. F. v., DiNola, A., & Haak, J. R. (1984). Molecular dynamics with coupling to an external bath, *J. Chem. Phys.*, Vol.81, No.8, (1984), pp. 3684-3690

Biswas, R., & Hamann, D. R. (1987). New classical models for silicon structural energies, *Phys. Rev. B*, Vol.36, No.12, (1987), pp. 6434-6445

Cheng, L.-J., & Corbett, J. W. (1974). Defect creation in electronic materials, *Proceedings of the IEEE*, Vol.62, No.9, (1974), pp. 1208-1214

Colombo, L. (2002). Tight-Binding Theory of Native Point Defects in Silicon, *Annual Review of Materials Research*, Vol.32, No.1, (2002), pp. 271-295

Dodson, B. W. (1987). Development of a many-body Tersoff-type potential for silicon, *Phys. Rev. B*, Vol.35, No.6, (1987), pp. 2795-2798

Egashira, K., Eriguchi, K., & Hashimoto, S. A new evaluation method of plasma process-induced Si substrate damage by the voltage shift under constant current injection at metal/Si interface, *IEDM Tech. Dig.*, pp. 563-566, San Francisco, CA, USA, Dec 06-09, 1998

Eriguchi, K., Matsuda, A., Nakakubo, Y., Kamei, M., Ohta, H., & Ono, K. (2009a). Effects of Plasma-Induced Si Recess Structure on n-MOSFET Performance Degradation, *IEEE Electron Device Lett.*, Vol.30, No.7, (2009), pp. 712-714

Eriguchi, K., Nakakubo, Y., Matsuda, A., Kamei, M., Ohta, H., Nakagawa, H., S. Hayashi, Noda, S., Ishikawa, K., Yoshimaru, M., & Ono, K. A New Framework for Performance Prediction of Advanced MOSFETs with Plasma-Induced Recess Structure and Latent Defect Site, *IEDM Tech. Dig.*, pp. 443-446, San Francisco, CA, USA, Dec 15-17, 2008

Eriguchi, K., Nakakubo, Y., Matsuda, A., Takao, Y., & Ono, K. (2009b). Plasma-Induced Defect-Site Generation in Si Substrate and Its Impact on Performance Degradation in Scaled MOSFETs, *IEEE Electron Device Lett.*, Vol.30, No.12, (2009), pp. 1275-1277

Eriguchi, K., Nakakubo, Y., Matsuda, A., Takao, Y., & Ono, K. (2010). Model for Bias Frequency Effects on Plasma-Damaged Layer Formation in Si Substrates, *Jpn. J. Appl. Phys.*, Vol.49, (2010), pp. 056203

Eriguchi, K., Ohno, A., Hamada, D., Kamei, M., & Ono, K. (2008b). Estimation of defect generation probability in thin Si surface damaged layer during plasma processing, *Thin Solid Films*, Vol.516, No.19, (2008), pp. 6604–6608

Eriguchi, K., & Ono, K. (2008). Quantitative and comparative characterizations of plasma process-induced damage in advanced metal-oxide-semiconductor devices, *J. Phys. D*, Vol.41, No.2, (2008), pp. 024002

Estreicher, S. K., Hastings, J. L., & Fedders, P. A. (1997). The ring-hexavacany in silicon: A stable and inactive defect, *Appl. Phys. Lett.*, Vol.70, No.4, (1997), pp. 432-434

Firsov, O. B. (1957). Scattering of Ions by Atoms, *Soviet Physics JETP*, Vol.34, No.7, (1957), pp. 308-311

Goodman, A. M. (1963). Metal - Semiconductor Barrier Height Measurement by the Differential Capacitance Method - One Carrier System, *J. Appl. Phys.*, Vol.34, No.2, (1963), pp. 329-338

Graves, D. B., & Humbird, D. (2002). Surface chemistry associated with plasma etching processes, *Appl. Surf. Sci.*, Vol.192, No.1-4, (2002), pp. 72-87

Hanson, D. E., Voter, A. F., & Kress, J. D. (1997). Molecular dynamics simulation of reactive ion etching of Si by energetic Cl ions, *J. Appl. Phys.*, Vol.82, No.7, (1997), pp. 3552-3559

Hastings, J. L., Estreicher, S. K., & Fedders, P. A. (1997). Vacancy aggregates in silicon, *Phys. Rev. B*, Vol.56, No.16, (1997), pp. 10215-10222

Hensel, H., & Urbassek, H. M. (1998). Implantation and damage under low-energy Si self-bombardment, *Phys. Rev. B*, Vol.57, No.8, (1998), pp. 4756-4763

Herman, I. P. (1996). *Optical Diagnostics for Thin Film Processing* (Academic Press), ISBN 0-12-342070-9, San Diego

Humbird, D., & Graves, D. B. (2004). Atomistic simulations of spontaneous etching of silicon by fluorine and chlorine, *J. Appl. Phys.*, Vol.96, No.5, (2004), pp. 791-798

Jin, W., Vitale, S. A., & Sawin, H. H. (2002). Plasma-surface kinetics and simulation of feature profile evolution in Cl_2 + HBr etching of polysilicon, *J. Vac. Sci. & Technol. A*, Vol.20, No.6, (2002), pp. 2106-2114

Kokura, H., Okabe, K., Nakaishi, M., & Miyajima, M. Plasma induced damage on ultra shallow junction in spacer etching, *Proc. Symp. Dry Process*, pp. 27-28, Jeju, Korea, Nov 28-30, 2005

Koyama, M., Cheong, C.-W., Yokoyama, K., & Ohdomari, I. (1997). Influence of Near-Surface Defects in Si Induced by Reactive Ion Etching on the Electrical Properties of the Pt/n-Si Interface, *Jpn. J. Appl. Phys.*, Vol.11, No.11, (1997), pp. 6682-6686

Leung, W. K., Needs, R. J., Rajagopal, G., Itoh, S., & Ihara, S. (1999). Calculations of Silicon Self-Interstitial Defects, *Phys. Rev. Lett.*, Vol.83, No.12, (1999), pp. 2351-2354

Lieberman, M. A., & Lichtenberg, A. J. (2005). *Principles of Plasma Discharges and Materials Processing,* (2nd ed., Wiley), ISBN 978-0-471-72001-0, New York

Lindhard, J., Scharff, M., & Schiott, H. E. (1963). Range Concepts and Heavy Ion Ranges, *Mat. Fys. Medd. K. Dan. Vidensk. Selsk.*, Vol.33, No.14, (1963), pp. 1-41

Matsuda, A., Nakakubo, Y., Takao, Y., Eriguchi, K., & Ono, K. (2010). Modeling of ion-bombardment damage on Si surfaces for in-line analysis, *Thin Solid Films*, Vol.518, No.13, (2010), pp. 3481-3486

Mazzarolo, M., Colombo, L., Lulli, G., & Albertazzi, E. (2001). Low-energy recoils in crystalline silicon: Quantum simulations, *Phys. Rev. B*, Vol.63, No.19, (2001), pp. 195207

Moliere, G. (1947). Theorie der Streuung schneller geladener Teichen I, *Z. Naturforschung*, Vol.A2, (1947), pp. 133-145

Mu, X. C., Fonash, S. J., Rohatgi, A., & Rieger, J. (1986). Comparison of the damage and contamination produced by CF_4 and CF_4/H_2 reactive ion etching: The role of hydrogen, *Appl. Phys. Lett.*, Vol.48, No.17, (1986), pp. 1147-1149

Murtagh, M., S M Lynch, Kelly, P. V., Hildebrandt, S., Herbert, P. A. F., Jeynes, C., & Crean, G. M. (1997). Photoreflectance characterization of Ar^+ ion etched and $SiCl_4$ reactive ion etched silicon (100), *Mat. Sci. Technol.*, Vol.13 (1997), pp. 961-964

Nagaoka, T., Eriguchi, K., Ono, K., & Ohta, H. (2009). Classical interatomic potential model for Si/H/Br systems and its application to atomistic Si etching simulation by HBr^+, *J. Appl. Phys.*, Vol.105, No.2, (2009), pp. 023302-023306

Nakakubo, Y., Matsuda, A., Kamei, M., Ohta, H., Eriguchi, K., & Ono, K. (2010a). Analysis of Si Substrate Damage Induced by Inductively Coupled Plasma Reactor with Various Superposed Bias Frequencies. In A. Amara & M. Belleville & T. Ea (Eds.), *Emerging Technologies and Circuits* (Vol. 66, pp. 107-120), London, (Springer)

Nakakubo, Y., Matsuda, A., Takao, Y., Eriguchi, K., & Ono, K. Study of Wet-Etch Rate of Plasma-Damaged Surface and Interface Layers and Residual Defect Sites, *Proc. Symp. Dry Process*, pp. 173-174, Tokyo, Japan, Nov 11-12, 2010

Nordlund, K., Ghaly, M., Averback, R. S., Caturla, M., Diaz de la Rubia, T., & Tarus, J. (1998). Defect production in collision cascades in elemental semiconductors and fcc metals, *Phys. Rev. B*, Vol.57, No.13, (1998), pp. 7556-7570

Oehrlein, G. S. (1989). Dry etching damage of silicon: A review, *Materials Sci. Eng. B*, Vol.4, No.1-4, (1989), pp. 441-450

Oehrlein, G. S., Bright, A. A., & Robey, S. W. (1988). X-Ray Photoemission Spectroscopy Characterization of Si Surfaces after CF_4/H_2 Magnetron Ion Etching - Comparisons to Reactive Ion Etching, *J. Vac. Sci. Technol. A*, Vol.6, No.3, (1988), pp. 1989-1993

Ohchi, T., Kobayashi, S., Fukasawa, M., Kugimiya, K., Kinoshita, T., Takizawa, T., Hamaguchi, S., Kamide, Y., & Tatsumi, T. (2008). Reducing Damage to Si Substrates during Gate Etching Processes, *Jpn. J. Appl. Phys.*, Vol.47, No.7, (2008), pp. 5324-5326

Ohta, H., & Hamaguchi, S. (2001a). Classical interatomic potentials for Si-O-F and Si-O-Cl systems, *J. Chem. Phys.*, Vol.115, No.14, (2001), pp. 6679-6690

Ohta, H., & Hamaguchi, S. (2001b). Molecular dynamics simulation of silicon and silicon dioxide etching by energetic halogen beams, *J. Vac. Sci. & Technol. A*, Vol.19, No.5, (2001), pp. 2373-2381

Ohta, H., & Hamaguchi, S. (2004). Effects of Van der Waals Interactions on SiO_2 Etching by CF_x Plasmas, *J. Plasma Fusion Res*, Vol.6, (2004), pp. 399-401

Pelaz, L., Marques, L. A., Aboy, M., Lopez, P., & Santos, I. (2009). Front-end process modeling in silicon, *The European Physical Journal B*, Vol.72, (2009), pp. 323-359

Petit-Etienne, C., Darnon, M., Vallier, L., Pargon, E., Cunge, G., Boulard, F., Joubert, O., Banna, S., & Lill, T. (2010). Reducing damage to Si substrates during gate etching processes by synchronous plasma pulsing, *J. Vac. Sci. & Technol. B*, Vol.28, No.5, (2010), pp. 926-934

Pollak, F. H., & Shen, H. (1990). Photoreflectance Characterization of Semiconductors and Semiconductor Heterostructures, *J. Electronic Materials*, Vol.19, No.5, (1990), pp. 399-406

Sankaran, A., & Kushner, M. J. (2004). Integrated feature scale modeling of plasma processing of porous and solid SiO_2. I. Fluorocarbon etching, *J. Vac. Sci. & Technol. A*, Vol.22, No.4, (2004), pp. 1242-1259

Schober, H. R. (1989). Extended interstitials in silicon and germanium, *Phys. Rev. B*, Vol.39, No.17, (1989), pp. 13013-13015

Schultz, P. A. (2006). Theory of defect levels and the "band gap problem" in silicon, *Phys. Rev. Lett.*, Vol.96, No.24, (2006), pp. 246401

SIA. (2009). *The International Technology Roadmap for Semiconductors 2009 edition, Front End Processes*

Smirnov, V. V., Stengach, A. V., Gaynullin, K. G., Pavlovsky, V. A., Rauf, S., & Ventzek, P. L. G. (2007). A molecular dynamics model for the interaction of energetic ions with SiOCH low-*k* dielectric, *J. Appl. Phys*, Vol.101, No.5, (2007), pp. 053307

Stillinger, F. H., & Weber, T. A. (1985). Computer simulation of local order in condensed phases of silicon, *Phys. Rev. B*, Vol.31, No.8, (1985), pp. 5262-5271

Sze, S. M. (1981). *Physics of Semiconductor Devices,* (2nd ed., Wiley), ISBN 0-471-05661-8, New York

Sze, S. M., & Ng, K. K. (2007). *Physics of Semiconductor Devices* (3rd ed., Wiley-Interscience), ISBN 978-0-471-14323-9, Hoboken, NJ

Tanaka, J., Abrams, C., & Graves, D. (2000). New C-F interatomic potential for molecular dynamics simulation of fluorocarbon film formation, *J. Vac. Sci. Technol. A*, Vol.18, No.3, (2000), pp. 938-945

Tang, M., Colombo, L., Zhu, J., & Diaz de la Rubia, T. (1997). Intrinsic point defects in crystalline silicon: Tight-binding molecular dynamics studies of self-diffusion, interstitial-vacancy recombination, and formation volumes, *Phys. Rev. B*, Vol.55, No.21, (1997), pp. 14279-14289

Tersoff, J. (1988a). Empirical interatomic potential for silicon with improved elastic properties, *Phys. Rev. B*, Vol.38, No.14, (1988), pp. 9902-9905

Tersoff, J. (1988b). New empirical approach for the structure and energy of covalent systems, *Phys. Rev. B*, Vol.37, No.12, (1988), pp. 6991-7000

Torrens, I. M. (1972). *Interatomic Potentials* (Academic Press, Inc.), New York

Tsuda, H., Miyata, H., Takao, Y., Eriguchi, K., & Ono, K. (2011). Three-Dimensional Atomic-Scale Cellular Model and Feature Profile Evolution during Si Etching in Chlorine-Based Plasmas: Analysis of Profile Anomalies and Surface Roughness, *Jpn. J. Appl. Phys.*, Vol.50, (2011), pp. 08JE06

Vitale, S. A., & Smith, B. A. (2003). Reduction of silicon recess caused by plasma oxidation during high-density plasma polysilicon gate etching, *J. Vac. Sci. Technol. B*, Vol.21, No.5, (2003), pp. 2205-2211

Wada, H., Agata, M., Eriguchi, K., Fujimoto, A., Kanashima, T., & Okuyama, M. (2000). Photoreflectance characterization of the plasma-induced damage in Si substrate, *J. Appl. Phys.*, Vol.88, No.5, (2000), pp. 2336-2341

Watanabe, T., Fujiwara, H., Noguchi, H., Hoshino, T., & Ohdomari, I. (1999). Novel Interatomic Potential Energy Function for Si, O Mixed Systems, *Jpn. J. Appl. Phys.*, Vol.38, No.4A, (1999), pp. L366-L369

Wilson, W. D., Haggmark, L. G., & Biersack, J. P. (1977). Calculations of nuclear stopping, ranges, and straggling in the low-energy region, *Phys. Rev. B*, Vol.15, No.5, (1977), pp. 2458-2468

Yabumoto, N., Oshima, M., Michikami, O., & Yoshii, S. (1981). Surface Damage on Si Substrates Caused by Reactive Sputter Etching, *Jpn. J. Appl. Phys.*, Vol.20, No.5, (1981), pp. 893-900

Part 4

Dynamics at the Interface

13

Simulations of Unusual Properties of Water Inside Carbon Nanotubes

Yoshimichi Nakamura and Takahisa Ohno
National Institute for Materials Science; CREST-JST
Japan

1. Introduction

Water, which is vital for all living creatures and essential to our daily lives, is one of the most researched materials on earth. It is amazing to see that even after a long history of research, this substance consisting of triatomic molecules is still rich in new discoveries on its unexpected properties. Among the hottest topics over the last decade is anomalous behavior of water molecules inside confined nanospaces, which stimulates our scientific curiosity from the viewpoint of how the tiny, polar molecules rearrange themselves when strongly confined. What the hydrogen-bond network is like? In this research field, molecular dynamics (MD) simulations have played a leading role and are expected to further increase its importance in predicting unexpected properties of water. Typical of well-defined, size-controlled nanospaces easily obtainable is the interior space of carbon nanotubes (CNTs). Figure 1 shows the relation of a single-walled CNT to the graphene honeycomb lattice. An example where the graphene lattice vector $m\mathbf{a1} + m\mathbf{a2}$ corresponds to the circumference of the CNT is shown in the figure. This type of CNT is typically referred to in this chapter and the pair of integers (m, m) is often quoted as a useful index of the CNT diameter.

When exploring a new ordered phase of the water molecules, it is quite natural and reasonable to focus on low-temperature/high-pressure conditions. MD simulation is a powerful tool for such exploration. In the CNT diameter range about 1-2 nm, the water molecules were reported to be frozen into a variety of forms of 'ice nanotube' inside the CNT (Bai et al., 2006; Koga at al., 2001; Luo et al., 2008; Mikami et al., 2009). For example, at P = 500 bar and T ≤ 240 K, 4-, 5-, and 6-gonal ice nanotubes were observed for (14, 14), (15, 15) and (16, 16) CNTs, respectively (Koga et al., 2001). All those simulations on ice-nanotubes were performed with the use of infinite or capped CNTs, which means that the confined water density can be directly controlled. In one of such MD simulation studies, a double-layered helix form was demonstrated to be possible inside the infinite (10, 10) CNT at 298K for the confined water density of '1 g/cc' (Liu et al., 2005). Here, the water density was estimated based on the number of confined molecules divided by the geometrical volume of the CNT, not by its effective inner space volume, which is at most about 60% of the geometrical volume for the (10, 10) CNT due to the hydrophobic nature of the CNT wall. The value of '1 g/cc' therefore corresponds to a very high density condition.

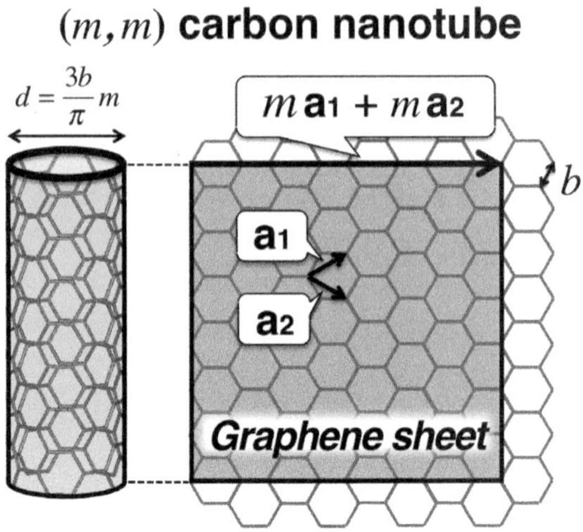

Fig. 1. Illustration of an (*m*, *m*) carbon nanotube and a graphene sheet.

Amazingly, at ambient conditions (300K, 1 bar) another ordered phase similar to the 6-gonal ice nanotube was found inside uncapped (9, 9) CNTs (Mashl. et al., 2003). The new phase reveals ice-like mobility with an amount of hydrogen bonding similar to that in the bulk liquid water. Unlike the simulations using infinite or capped CNTs, each water molecule was allowed to enter (leave) the CNTs from (for) the outside bulk water. The interior of the uncapped CNTs is therefore naturally filled with water molecules. The uncapped CNTs were embedded into wafers of neutral atoms mimicking the hydrophobic interior of a phospholipid membrane. Based on the same water model used in the above study, i.e., the SPC/E model (Berendsen et al., 1987), we have confirmed that the anomalously immobilized water at ambient conditions is also observed inside the uncapped CNTs which is NOT embedded into the membrane but directly immersed in a water reservoir (Fig. 2).

It should be noted that a more recent MD study has reported a brand new water phase called 'ferroelectric mobile water (FMW)' (Nakamura & Ohno, 2011, 2012a, 2012b). Though the FMW at first sight appears similar to the immobilized water mentioned above, they are distinct from each other both in molecular structure and in dynamics. The FMW is produced inside (8, 8) and (9, 9) CNTs immersed in a water reservoir at ambient conditions based on the TIP5P-E model (Rick, 2004). The details of the FMW will be explained later.

The H_2O/CNT simulation studies have thus revealed a variety of unusual properties of water molecules and improved our general knowledge of water and ice. Also, knowledge on influential factors such as the CNT diameter, temperature, and pressure has been increasing. For spontaneous filling of CNTs with liquid water, there are many simulations employing a similar set of these conditions. We see, however, that in some cases they do not always yield the similar results. This implies that difference in water models used is potentially another important influential factor to be considered. In this chapter, therefore, paying attention to the effect of water models, we explore the unusual behaviour of the confined water.

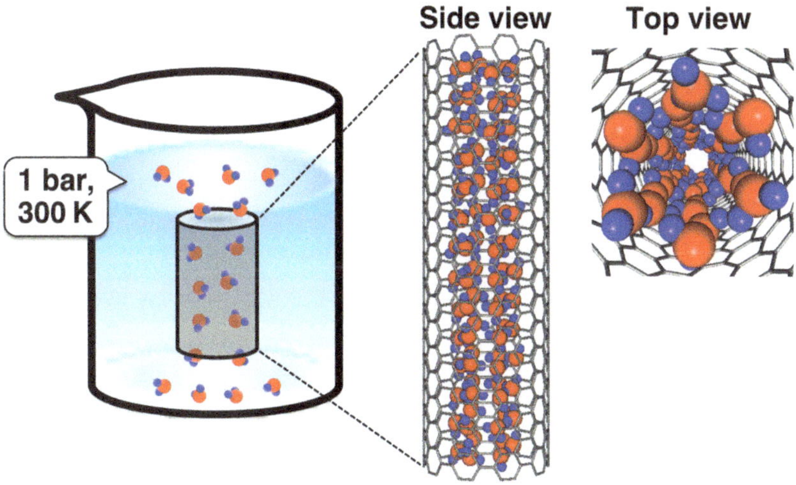

Fig. 2. Left: Illustration of an uncapped CNT in a water reservoir. Right: Snapshots of anomalously immobilized water at ambient conditions. Simulations are performed for the (9, 9) CNTs immersed in a water reservoir based on the SPC/E water model. Red for oxygen, blue for hydrogen.

2. Overview of water models

Figure 3 shows a series of the transferable interaction potentials for water molecules, TIP*n*P models (Jorgenen et al., 1983; Mohoney & Jorgensen, 2000), demonstrating how the water models are classified from the viewpoint of the number of interaction sites (denoted by n). The models for n = 3, 4, and 5 are represented in the figure. All models use the same rigid molecular structure. The van der Waals (vdW) interaction is commonly described by the 6-12 Lennard-Jones potential between the oxygen atom sites (closed black circles). Significant differences among the models are in the description of the electrostatic interaction sites (open white circles).

In the TIP3P model, the molecular charge distribution is modeled by point charges on each nuclei (Fig. 3, lower left). Together with the vdW site, the total number of interaction sites is three, as the model's name implies. In the TIP4P model, unlike in the TIP3P model, the negative charge is placed on an additional fictive site, resulting in four interaction sites overall (Fig. 3, lower middle). In the TIP5 model, the charge distribution is more realistically treated. The negative charges along the lone-pair directions are explicitly taken into account (Fig. 3, lower right).

At the expense of more computational cost than the 3- and 4-site models, the 5-site model have succeeded in reproducing more of the water properties over a range of temperatures and pressures, including the density maximum near 4 °C (Mohoney & Jorgensen, 2000) and the melting point (Fernández et al., 2006; Vega eat al., 2005). There has also been proposed a modified version of the TIP5P, called TIP5P-E (Rick, 2004), which is constructed by only modifying the Lennard-Jones parameters so as to yield more accurate results when used with Ewald sum calculations for long-ranged electrostatic interactions.

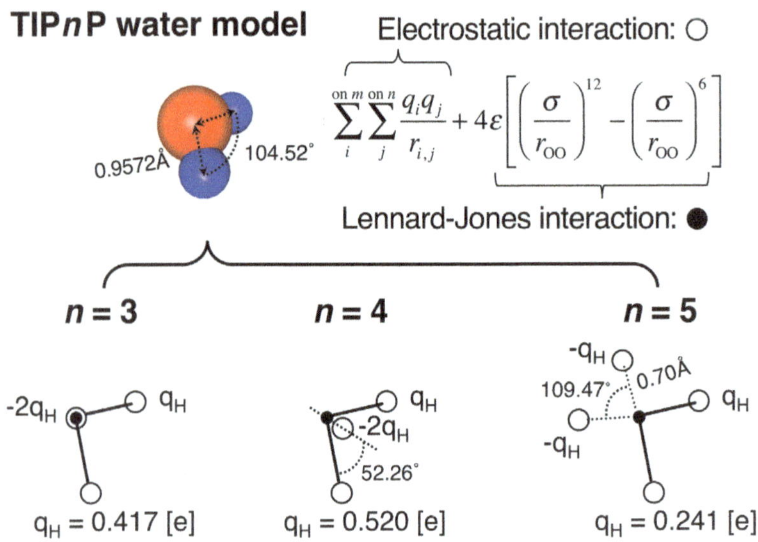

$$\sum_{i}^{\text{on } m}\sum_{j}^{\text{on } n}\frac{q_i q_j}{r_{i,j}} + 4\varepsilon\left[\left(\frac{\sigma}{r_{OO}}\right)^{12} - \left(\frac{\sigma}{r_{OO}}\right)^{6}\right]$$

Fig. 3. Illustration of TIP*n*P water models. Red for oxygen, blue for hydrogen.

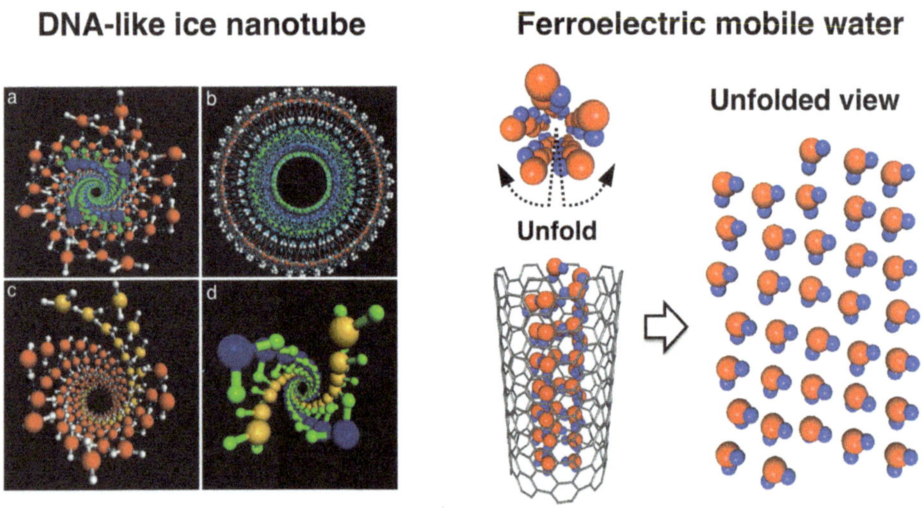

Fig. 4. Examples of H_2O/CNT simulation results based on 5-site water models (left figure is from Fig.1 of Bai et al., 2006. Copyright 2006 National Academy of Sciences.)

The TIP3P and TIP4P models have been the most commonly used water models, along with other variations of 3-site models, SPC and SPC/E (Berendsen et al., 1981, 1987). Such is also the case for H_2O/CNT simulations. Though the number of H_2O/CNT simulation studies based on the 5-site models is only a few, their findings that the water molecules are able to self-assemble into double helixes resembling DNA under high pressure (Bai at al., 2006) and that single-domain ferroelectric mobile water is possible at ambient conditions (Nakamura

& Ohno, 2011, 2012a, 2012b) are all the more outstanding, considering far more numerous studies based on the 3- and 4-site models.

3. Spontaneous filling of CNTs with liquid water

Regarding the spontaneous filling of CNTs with liquid water, various 3-site water models have been tried for a wide range of CNT diameter (Alexiadis & Kassinos, 2008a, 2008b). It was shown that the different choices of rigid/flexible model for the TIP3P, SPC, and SPC/E water molecules, along with rigid/flexible choices of CNTs, cause no significant differences. We now perform systematic simulations based on not only 3-site water models (TIP3P and SPC/E) but also 4-site (TIP4P) and 5-site (TIP5P-E) models (Nakamura & Ohno, 2012a).

A nonpolar, rigid (m, m) CNT of length L is solvated with N water molecules in a periodic box. L = 2.1 nm for m = 6, 7, 8, 9, 10, 12, 16, 20 (0.8 - 2.7 nm in diameter). L = 4.0 nm for m = 8, 9. N ranges from 2074 to 6508, depending on the CNT size. The carbon-carbon bond length is 1.4 Å. MD simulations are performed using AMBER 9.0 (Case et al., 2006). The water molecules and the CNTs are assumed to interact through the 6-12 Lennard-Jones potential between the oxygen and the sp^2 carbon atoms (AMBER force field). Based on the cross section σ_{OO} (Table 1) and σ_{CC} (3.4 Å), and the depth of the potential well ε_{OO} (Table 1) and ε_{CC} (0.086 kcal/mol), σ_{CO} and ε_{CO} are derived from the Lorentz-Berthelot combining rules,

$$\varepsilon_{CO} = \sqrt{\varepsilon_{CC} \times \varepsilon_{OO}}, \quad \sigma_{CO} = \frac{\sigma_{CC} + \sigma_{OO}}{2} . \tag{1}$$

Electrostatic interactions between the water molecules are calculated by particle-mesh Ewald method (Darden, 1993). An MD time step of 2 fsec is used. In the first runs (up to 0.5-0.6 nsec), a combination of constant volume and constant pressure simulations is performed to ensure the system is in equilibrium with the bulk water density at T K under 1 bar. In the subsequent MD time steps (10-120 nsec), the NVT-ensemble is used for statistical analysis. Snapshots are saved for analysis every 0.1 psec.

	TIP3P	SPC/E	TIP4P	TIP5P-E
σ_{OO} [Å]	3.15061	3.166	3.15365	3.097
ε_{OO} [kcal/mol]	0.1521	0.1553	0.1550	0.178

Table 1. Lennard-Jones potential parameters for each model.

Figure 5 shows the water density inside the CNTs, together with snapshots obtained by using the TIP5P-E model. As was already reported by the simulations based on the 3-site models, those based on the 4- and 5-site models also have three different filling modes, that is, 'wire', 'layered', and 'bulk' mode as the diameter size increases (Alexiadis & Kassinos, 2008a, 2008b).

In the wire mode, the water molecules form into a water wire regardless of the water model. Figure 5a shows a snapshot example of the single-file formation of the TIP5P-E molecules inside the (6, 6) CNT. It has been reported that the TIP3P water molecules in the wire mode undergo a burst-like transmission through the CNT (Hummer et al., 2001). A similar concerted and rapid axial motion is observed also for the other water models. An example for the TIP5P-E model is shown in Fig. 6. In the wire mode, significant differences among the water models are not seen in both the structure and dynamics of the confined water molecules.

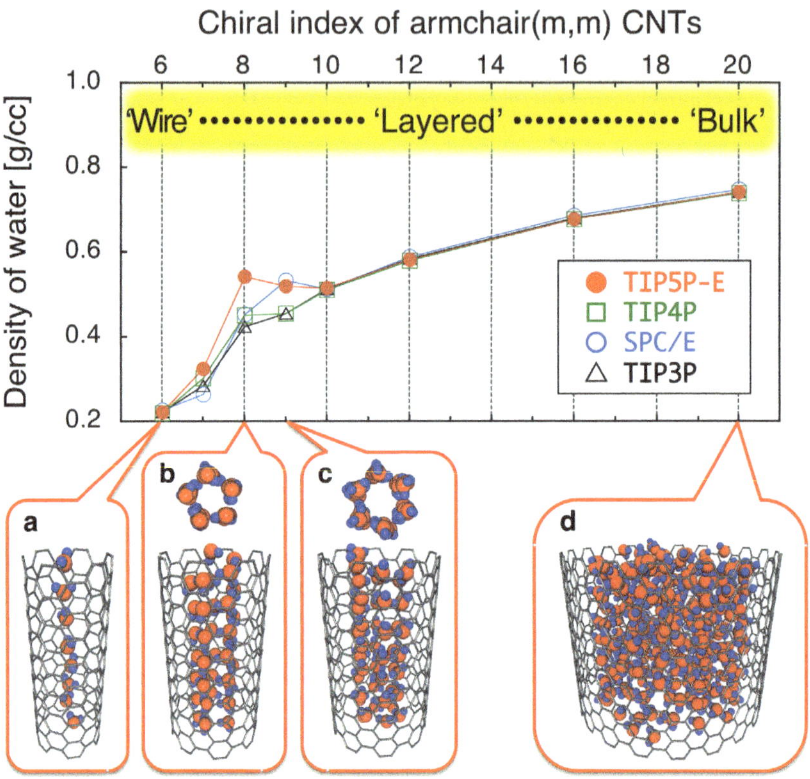

Fig. 5. Upper: Water density inside (*m, m*) CNTs of length 2.1 nm at 280 K. Data are from Fig.1b of Nakamura & Ohno, 2012a. Lower: Snapshots from MD simulations based on TIP5P-E model. (a) Single-file formation in 'wire' mode. (b) (c) Single-layered formation in 'layered' mode. (d) Example in 'Bulk' mode.

In the 'layered' mode, the larger the CNT diameter size, the more concentric layers of the confined water molecules come to appear. At the same time, those water layers become more diffuse and the bulk water structure gradually recovers. A snapshot example practically corresponding to the 'bulk' mode is shown in Fig. 5d.

The overall trend of the confined water density is that the shorter the diameter, the smaller the density. The density values for each model are practically the same except for the (8, 8) and (9, 9) CNTs (about 1.1-1.2 nm in CNT diameter). We notice that the TIP5P-E water density does not follow the overall trend for the (8, 8) and the (9, 9) CNT. The SPC/E water density does not follow the trend for the (9, 9) CNT. Inside the (8, 8) and (9, 9) CNTs, the TIP5P-E water molecules form into a single-layered water tube as shown in Fig. 5b and c.

Though they at first sight resemble the 5-gonal and 6-gonal ice nanotubes (Koga et al., 2001; Luo et al., 2008; Mikami et al., 2009), they are quite distinct from those ice nanotubes as explained in the next section.

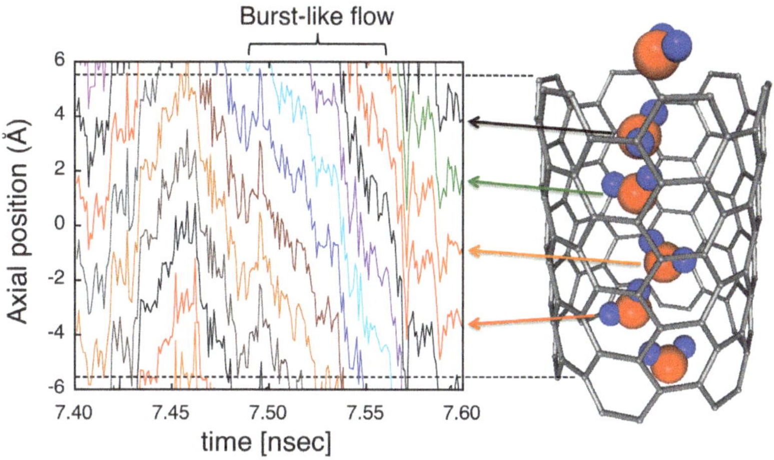

Fig. 6. Burst-like flow of the TIP5P-E water molecules along the CNT axis at 300 K.

4. Ferroelectric mobile water (FMW)

We look into the structure of the single-layered TIP5P-E water tubes inside the (8, 8) and (9, 9) CNTs. They are practically one-atom-thick water tubes as we see their top views in Fig. 5b and c. To detail the molecular structure, we unfold the water tube as shown in the upper part of Fig. 7. All the water molecules are visited by following one single helix running through the CNT. To our surprise, the dipole moment of each molecule is oriented in practically the same direction, which results in a single-domain ferroelectric arrangement of the water molecules and consequently a large spontaneous polarization along the CNT axis (Fig. 7, right-hand side). This new water phase is termed 'ferroelectric mobile water (FMW).' Its 'mobile' character will be explained later.

In Fig. 7, the direction of the axial spontaneous polarization is downward for the (8, 8) CNT and upward for the (9, 9) CNT. Each direction is observed as unchanged at least over about 100 ns. Of course, there is no reason for the net axial polarization to prefer one direction over the other. For reference, we tried other series of simulations using a different random arrangement of the water molecules, and obtained the opposite polarization direction for each. We also find that the longer the CNT length, the more stable the polarization direction. That is, employing the longer CNTs has an effect similar to reducing the temperature.

The unfolded-view snapshots in Fig. 7 reveal that a hydrogen bond network is completed, i.e., the 'ice rule' is obeyed (Bernal & Fowler, 1933). Moreover, the protons are perfectly ordered (Fig. 7, unfolded views in middle). Here, it is worth referring to proton-ordered ice (Pauling, 1935; Slater, 1941; Jackson & Whitworth, 1995; Jackson et al., 1997; Su et al., 1998; Iedema et al., 1998; Bramwell, 1999;). In ordinary ice (i.e., ice Ih), the protons are disordered though the ice rule is obeyed for its hexagonal lattice of the water molecules. The proton

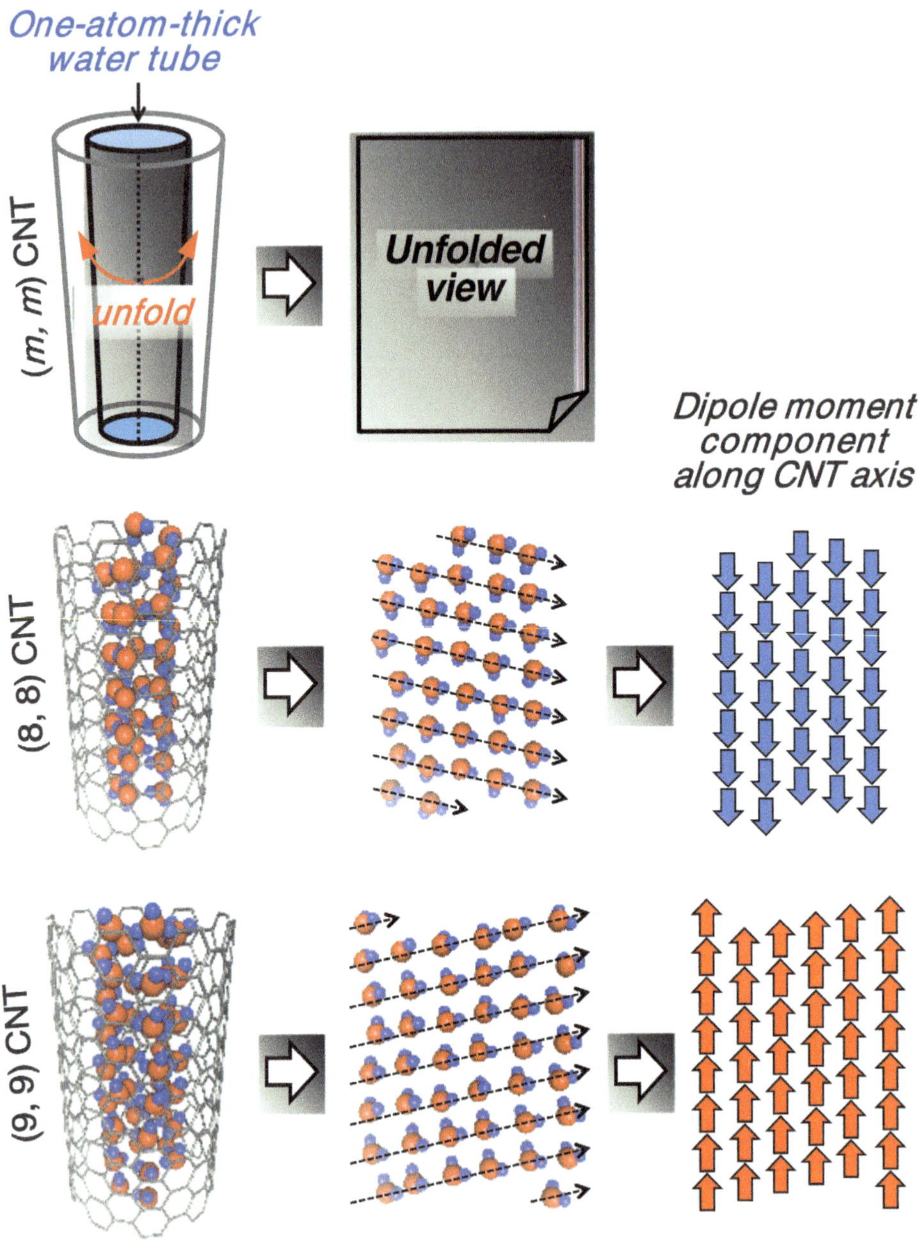

Fig. 7. Left: Side views of single-layered, one-atom-thick water tubes inside CNTs. Middle: Unfolded-views of TIP5P-E water molecules inside the (8, 8) and the (9, 9) CNTs at 280 K. Broken arrows are guide for eyes. Right: Arrangement of the dipole moment component along the CNT axis. The axial dipole moment of each water molecule is represented by red (blue) arrows when its value is zero or greater (less than zero).

ordered phase is rarely produced in pure ice because the disorder is practically frozen in before reaching the ordered phase due to the very slow kinetics of the proton reorientations at very low temperatures. Though the ordering transition can be catalyzed by doping with hydroxides, the ordered phase known as ice XI (Matsuo et al., 1986) has only been partially obtained (Tajima et al., 1982, 1984; Kawada, 1972, 1989). Researchers in this area are beginning to direct their attention to the outer solar system such as Pluto, where most water ice is considered to have existed in a crystalline phase for billions of years (Fukazawa et al., 2006). In marked contrast, we have shown that the proton-ordered water, FMW, can be produced at ambient conditions by utilizing the interior space of CNTs immersed in a reservoir water.

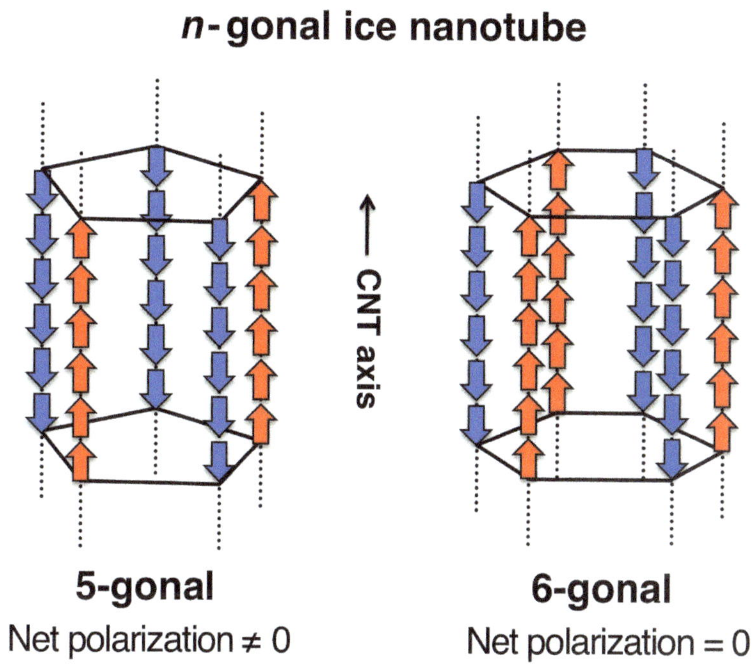

Fig. 8. Illustration of arrangement of the dipole moment component along the CNT axis for the 5- and 6-gonal ice nanotubes. Each dipole moment is represented by red (blue) arrows when its value is zero or greater (less than zero). Solid and broken lines are guide for eyes.

It is also worth referring to so far reported *n*-gonal ice-nanotubes. Those *n*-gonal ice-nanotubes have been found inside infinite or capped CNTs under high pressure/low temperature conditions with non-zero net spontaneous polarization (Luo et al., 2008; Mikami at al., 2009). Figure 8 shows schematic illustrations of the arrangement of the axial dipole moment for the 5- and 6-gonal ice nanotubes. Though each of the *n* wires of water molecules along the CNT axis is ferroelectric, the inter-wire relation is antiferroelectric. They are, therefore, not single-domain ferroelectric ice. In addition, non-zero net polarization arises only for an odd number of *n*. It should be noted that unlike the ice nanotubes, the FMW yields non-zero net polarization not only for the 5-gonal but also for the 6-gonal.

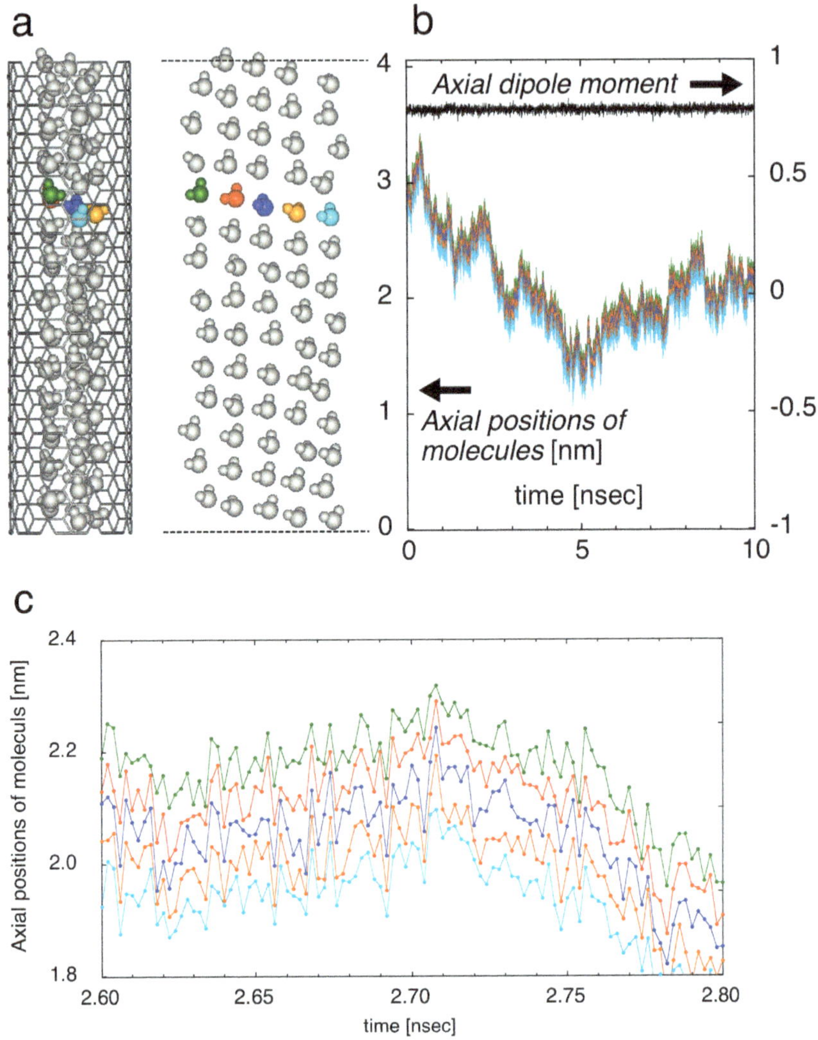

Fig. 9. Proton-ordered diffusion inside the (8, 8) CNT of length 4.0 nm. 280 K. (a) Unfolded-view snapshot. (b) Axial positions of five neighbouring molecules traced in corresponding colours. The net axial polarization of water inside the CNT traced in black (units are the magnitude of a single dipole of TIP5P-E water). a and b from Fig.2 of Nakamura & Ohno, 2011. Reproduced by permission of the PCCP Owner Societies. (c) A close-up of the motions between 2.6-2.8 nsec.

In general, a highly ordered structure such as the FMW seems suggestive of a rigid, immobile phase. Counter to our intuition, however, the FMW is dynamic. Figure 9a shows a side-view (left) and its corresponding unfolded-view snapshot obtained from the simulation employing the (8, 8) CNT. Five neighbouring molecules along the helix are coloured, and their axial positions are traced in corresponding colours (Fig. 9b, c). The net axial polarization of the water inside the CNT is also shown in Fig. 9b (black line). The molecules diffuse along the CNT axis, keeping the proton-ordered network intact. They fluctuate and gradually shift away from their initial positions like a one-dimensional Brownian motion. A nanometre-order shift over nanosecond-order time is often seen. This concerted diffusion is termed 'proton-ordered diffusion.'

Figure 10 shows a series of elementary processes of the proton-ordered diffusion. Due to the single helical chain structure of the FMW, there are 'vacant pockets' in the interior space of the CNT near its both ends. The vacant pocket near the CNT top is filled molecule by molecule with reservoir water molecules. The molecules filling the pockets enter deeper and become more stable. At the same time, another vacant pocket is created at the same place again, which makes the filling process endless. A similar filling process also takes place near the bottom of the CNT. The proton-ordered diffusion is thus deeply relevant to the structure of the FMW.

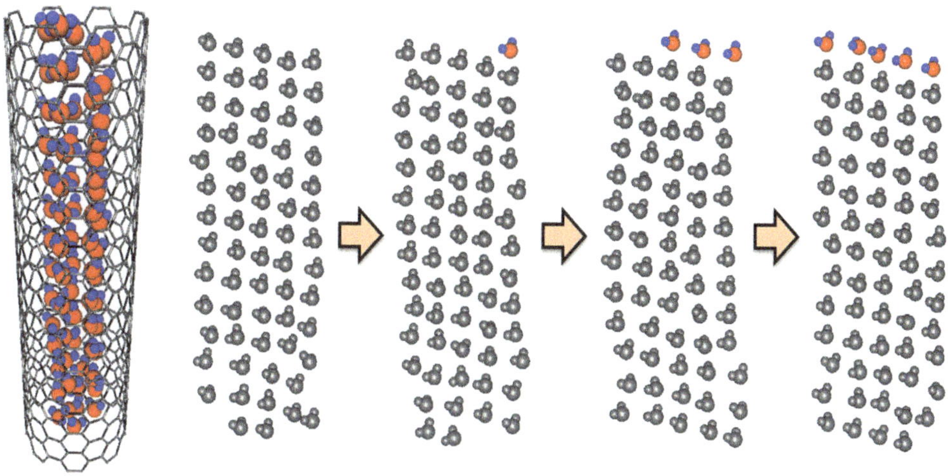

Fig. 10. Mechanism of the proton-ordered diffusion inside the (8, 8) CNT of length 4.0 nm. 280K. The oxygen (hydrogen) atoms of the water molecules entering from outside the CNT are colored red (blue).

The direction of the axial diffusion is practically dictated by the stochastic process near the CNT ends, and the probability of shifting in either directions is, on average, equal. From this viewpoint, a significant shift only in one direction seen in Fig. 9b might sound paradoxical. It is, however, understood by analogy with 'leads in a prolonged fair coin-tossing game.' When the cumulative number of heads is larger than that of tails, heads is in the lead (and vice versa). The leads are equivalent to the position shifts in the proton-ordered diffusion. When a great many series of fair coin-tossing games are conducted independently, the most likely result is for one side to be in the lead during the entire course of the game. This tendency becomes more pronounced for the games with a larger number of tosses. Contrary to popular notions, the least likely result is for the two sides to be tied for the time in the lead (Feller, 1968). The diffusion trajectory shown in Fig. 9b therefore represents a quite likely result.

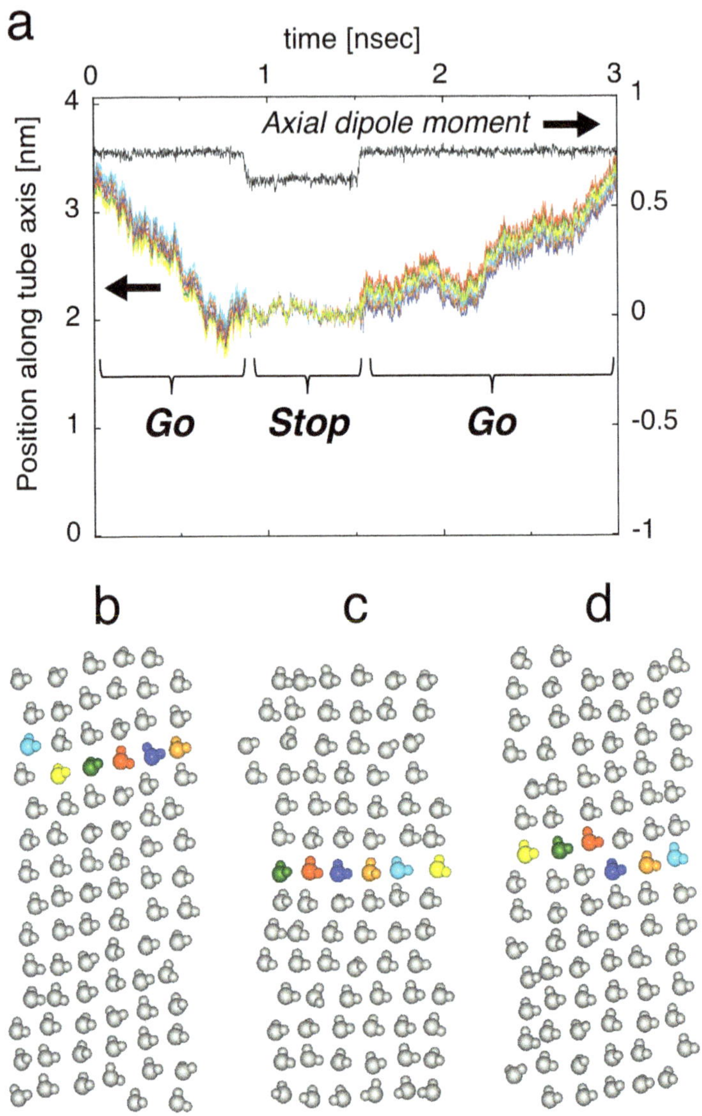

Fig. 11. 'Go–stop–go' motion of water inside the (9, 9) CNT of length 4.0 nm. 290 K. (a) Axial positions of six coloured molecules in the unfolded-view snapshots b, c, and d are traced in corresponding colours. The net axial polarization of water inside the CNT is traced in black (units are the magnitude of a single dipole of TIP5P-E water). Unfolded-view snapshot at (b) 0.072 ns, (c) 1.214 ns, and (d) 1.998 ns. From Fig.3 of Nakamura & Ohno, 2011. Reproduced by permission of the PCCP Owner Societies.

The proton-ordered diffusion is also observed for the FMW inside the (9, 9) CNTs (Fig. 11). Looking into the axial positions of six marked molecules, we notice that they are temporarily pinned at a certain axial position (at about 2.0 nm, from 0.9 to 1.5 ns in Fig. 11a). Before the pinning takes place, the confined water has a single-helix structure (b) and undergoes the axial diffusion. When pinned, however, it is no longer helical: it transforms into axially stacked layers of water molecules and consequently all the molecules become practically immobile (c). Once the pinning ends, the water recovers the FMW structure and resumes the proton-ordered diffusion (d). These motion changes literally correspond to 'Go-Stop-Go.'

In Fig. 11a, the net axial polarization of the confined water is also plotted. It should be noted that the 'Go-Stop-Go' is accompanied by step-wise changes of the net polarization. As we see the series of unfolded snapshots (b, c, d), each dipole axis comes to slightly tilt against the CNT axis when the pinning takes place, which results in a lower net polarization than that of the FMW during the pinning.

We have performed a long-time simulation with a shorter (9, 9) CNT of length 2.1 nm, in which we find a series of stepwise changes of the net polarization (Fig. 12a). Amazingly, the values fall on any one of the nine lines. Five independent phases, labeled '0' to '4', are suggested. Phase #4 corresponds to the FMW, the most stable phase (Nakamura & Ohno, 2011). Figure 12b shows unfolded-view snapshots from each phase, which is helpful to understand how the net polarization is digitized. The molecules are coloured red (blue) when their axial dipole moment P_z is zero or greater (less than zero). Each axial water wire is red or blue, i.e., ferroelectric. The ice rule is practically obeyed for all the snapshots. Under the ice rule, many other dipole arrangements are possible for each phase except #4.

Phase #0 is very similar to the reported 6-gonal ice nanotubes (Luo et al., 2008; Mikami at al., 2009) in that the inter-wire relation is antiferroelectric though the intra-wire is ferroelectric, resulting in a zero net spontaneous polarization. In phase #1 (#2), the axial water wires in red outnumber those in blue by two (four). The redundant water wires yield a non-zero net polarization. Phases #3 and #4 have no domain boundary along the CNT axis.

We detail the transition from one phase to the other, e.g., from phase #0 to #1 (at around 18 ns, indicated by a yellow arrow in Fig. 12a). A sequence of the transient snapshots is shown in Fig. 12c. At $\Delta t = 0$, the inter-wire relation is antiferroelectric. Under local perturbation from the reservoir water, the dipole orientation of the encircled molecule is changed so as to introduce a domain boundary in the axial water wire ($\Delta t = 3.5$ ps), which corresponds to a hydrogen bonding defect (Dellago et al., 2003; Köfinger et al., 2008; Köfinger & Dellago, 2009). The ice rule is consequently breached at the domain boundary, and local instability is created. Fluctuating rapidly along the CNT axis, the position of the domain boundary shifts toward the opposite CNT edge accompanied by the dipole reorientations ($\Delta t = 4.7$ ps). Eventually the whole water wire becomes red ($\Delta t = 8.5$ ps). Triggered by the local perturbation, the phase transition is thus completed in less than 10 ps. It should be noted that the perturbation from the reservoir water only works on the dipole reorientation and has little effect on their diffusion. This is the case for phases #0 to #3. In marked contrast, for phase #4, the local perturbation has little effect on the dipole reorientation: it only works on the diffusion of the molecules, that is, the proton-ordered diffusion of the FMW.

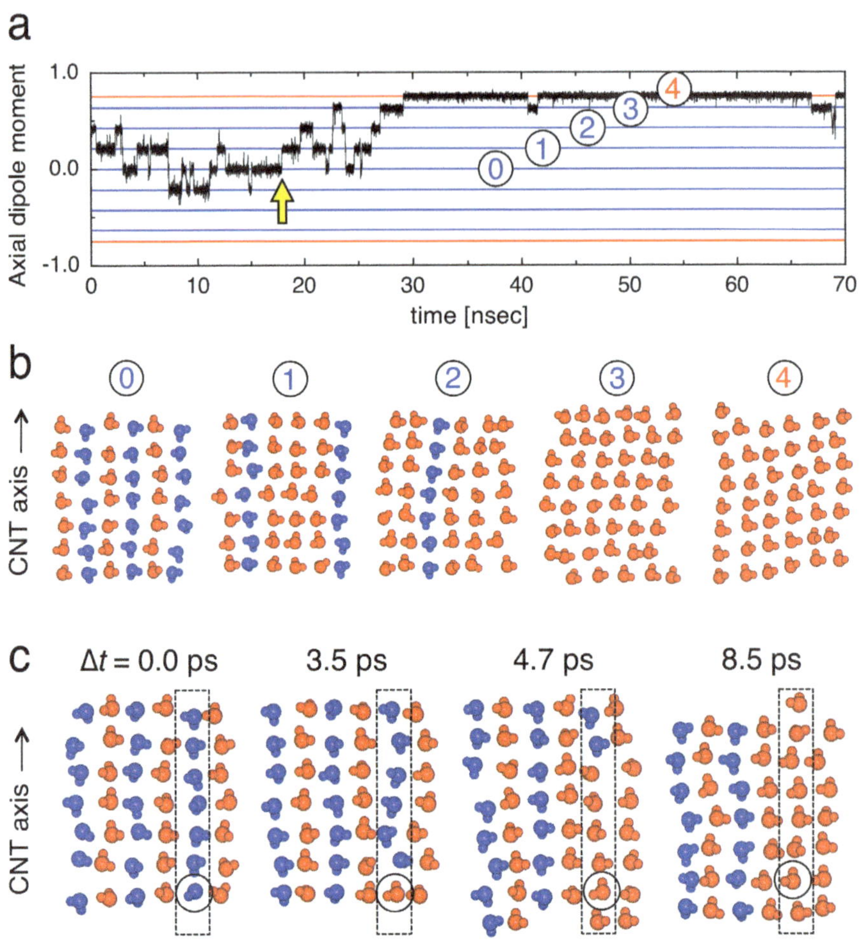

Fig. 12. Sequence of spontaneous transitions with step-wise changes of net polarization of water inside the (9, 9) CNT of length 2.1 nm. 280 K. (a) Net axial polarization of water traced in black (units are the magnitude of a single dipole of TIP5P-E water). Two red lines represent the ferroelectric mobile water. Seven blue lines the immobile water. Labels '0' to '4' on five lines represent five independent water phases. (b) Unfolded-view snapshots from each phase. Water molecules are colored red (blue) when their axial dipole moment P_z is zero or greater (less than zero). (c) Sequence of unfolded-view snapshots during the transition from phase #0 to #1, which is indicated by yellow arrow in (a). Black circles and dotted rectangles are guide for eyes. From Fig.4 of Nakamura & Ohno, 2011. Reproduced by permission of the PCCP Owner Societies.

Here, a question still remains: how the mobile phase, the FMW, is reached from the immobile phases? This process is clearly explained in Fig. 13. A schematic, representative arrangement of the confined water molecules for phase #0 to #2, all of which have axial domain boundaries, are shown in (a). In these phases, the fluctuation in directions perpendicular to the CNT axis ('lateral fluctuation') is severely restricted due to the proton-proton repulsion as shown in the yellow areas in (b). For phase #3, which has no axial domain boundary, there are two representative arrangements: domain-boundary-free (c) and lateral domain boundaries included (f).

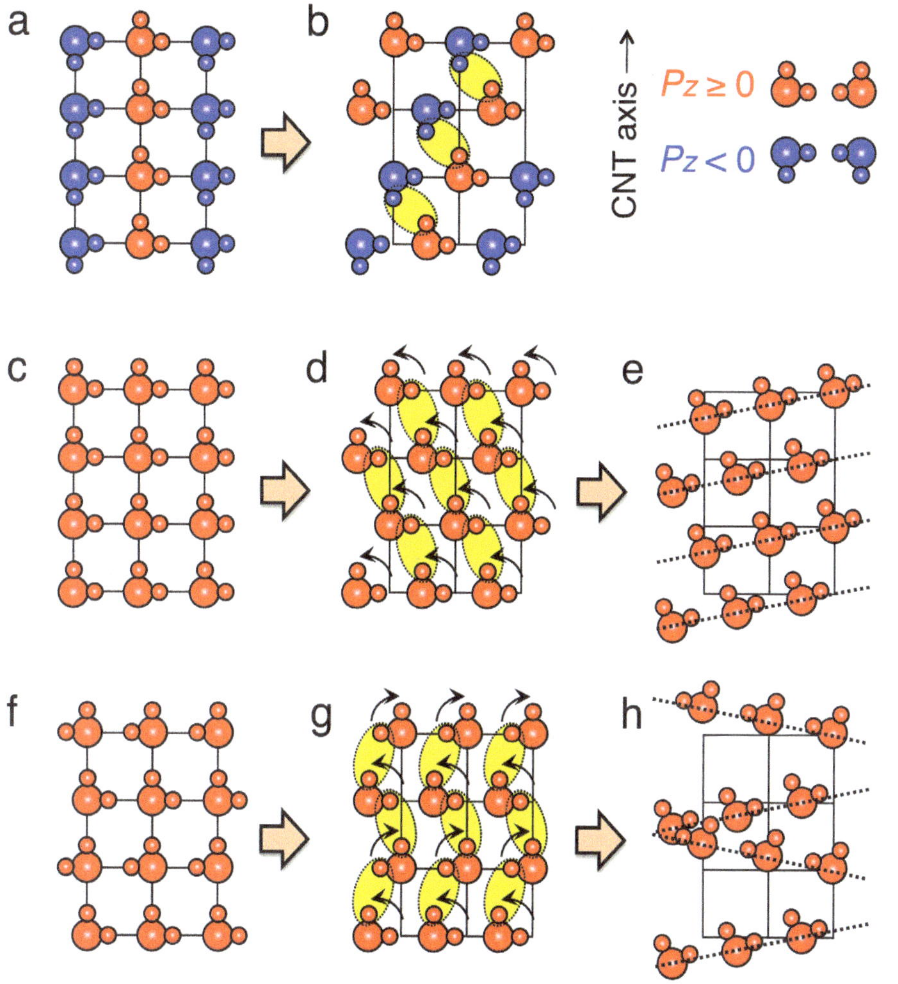

Fig. 13. Proton ordering and thermal fluctuation explained with schematic unfolded views (for simplicity, square lattice is used). Left: Without fluctuation. (a) Axial (vertical) domain boundaries included. (c) Domain-boundary-free. (f) Lateral domain boundaries included. Middle: With lateral fluctuation. (b), (d) and (g) correspond to distorted structure examples of (a), (c), and (f), respectively, under lateral fluctuation. Yellow represents repulsions induced by the lateral fluctuation. Arrows in (d) and (g) represent the rotational direction of each dipole. Right: Resultant arrangements. (e) Ferroelectric mobile water reached via (d). (h) Example of unstable arrangements reached via (g). Black dotted lines are guide for eyes.

In the domain-boundary-free case (c), the repulsion also arises under the lateral fluctuation (d). It can, however, be avoided by slightly tilting each dipole orientation toward the CNT axis (indicated by arrows) and giving the axial positions a slight downward slope along the lateral direction (e). The molecules thus align along the single helix running through the CNT without breaching the ice rule. This also gives the reason why phase #4 yeilds spontaneous polarization larger than that of phase #3. In the other case (f), however, avoiding the proton-proton repulsion induced by the lateral fluctuation only leads to instability elsewhere (g to h). More dominant the domain-boundary-free arrangement (c), the more likely the transition from phase #3 to #4 takes place.

5. Effect of different models

Based on the TIP5P-E water model, we have seen the quite unusual behaviour of water molecules inside the (8, 8) and the (9, 9) CNTs, i.e., that formation of single-domain ferroelectric water (FMW) and its concerted diffusion (proton-ordered diffusion). We have also seen in Fig. 5 that for these 'critical' CNT diameters (about 1.1-1.2 nm) there are non-trivial differences in water density among the water models, which strongly suggests that the confined water structure needs to be examined model by model under the same conditions as used for the TIP5P-E water simulations above.

Figure 14 shows the top-view snapshots and net axial polarization of water inside the (9, 9) CNT obtained for each water model, together with unfolded-view snapshot examples for the SPC/E model. The top-view snapshots for the TIP3P and the TIP4P reveal thicker water tubes compared to the SPC/E (and TIP5P-E, see Fig. 5c), clearly demonstrating model dependence. The time evolution of the net axial polarization P_z for the TIP3P and the TIP4P reveals considerable fluctuation (Fig. 14a and b, right-hand side). In contrast, a series of stepwise changes is seen for the SPC/E (Fig. 14c, right-hand side). The values fall on any of three lines, suggesting two independent phases.

We see the relation between the two phases of the SPC/E water and their corresponding molecular structure. As well as phase #0 introduced in the TIP5P-E simulations (labeled '0' in Fig. 12b), the antiferroelectric relation between the SPC/E water wires along the CNT axis yields the phase with $P_z = 0$, which is therefore named phase #0 based on the same naming scheme as used in the TIP5P-E water simulations. The transitions to the other magnitude of P_z (i.e, phase #1) take place by the flip of the water wire direction, similarly to those demonstrated in the TIP5P-E simulations (Fig. 12c). Phase #0 and #1 consist of axially stacked layers of anomalously immobile water molecules, which are essentially the same for the corresponding phases of the confined TIP5P-E water. It should be note that for the TIP5P-E, there are five independent phases (#0 to #4). For the SPC/E, however, only phase #0 and #1 are observed and the other phases including the FMW (i.e., phase #4) are not observed. The effect of the choice of model between the SPC/E and the TIP5P-E is thus clear.

Also for the (8, 8) CNT, the model dependence is significant (Fig. 15). The top-view snapshot for the TIP3P reveals a thick, less ordered tube structure (a), whereas the TIP4P and the SPC/E reveal a 4-gonal water tube (b and c). The net axial polarization P_z for the TIP3P water shows a considerable fluctuation (a, right-hand side). For the TIP4P and the SPC/E, the fluctuation is fairly suppressed and most of the values of P_z fall on any of five lines, suggesting three independent phases (b and c, right-hand side).

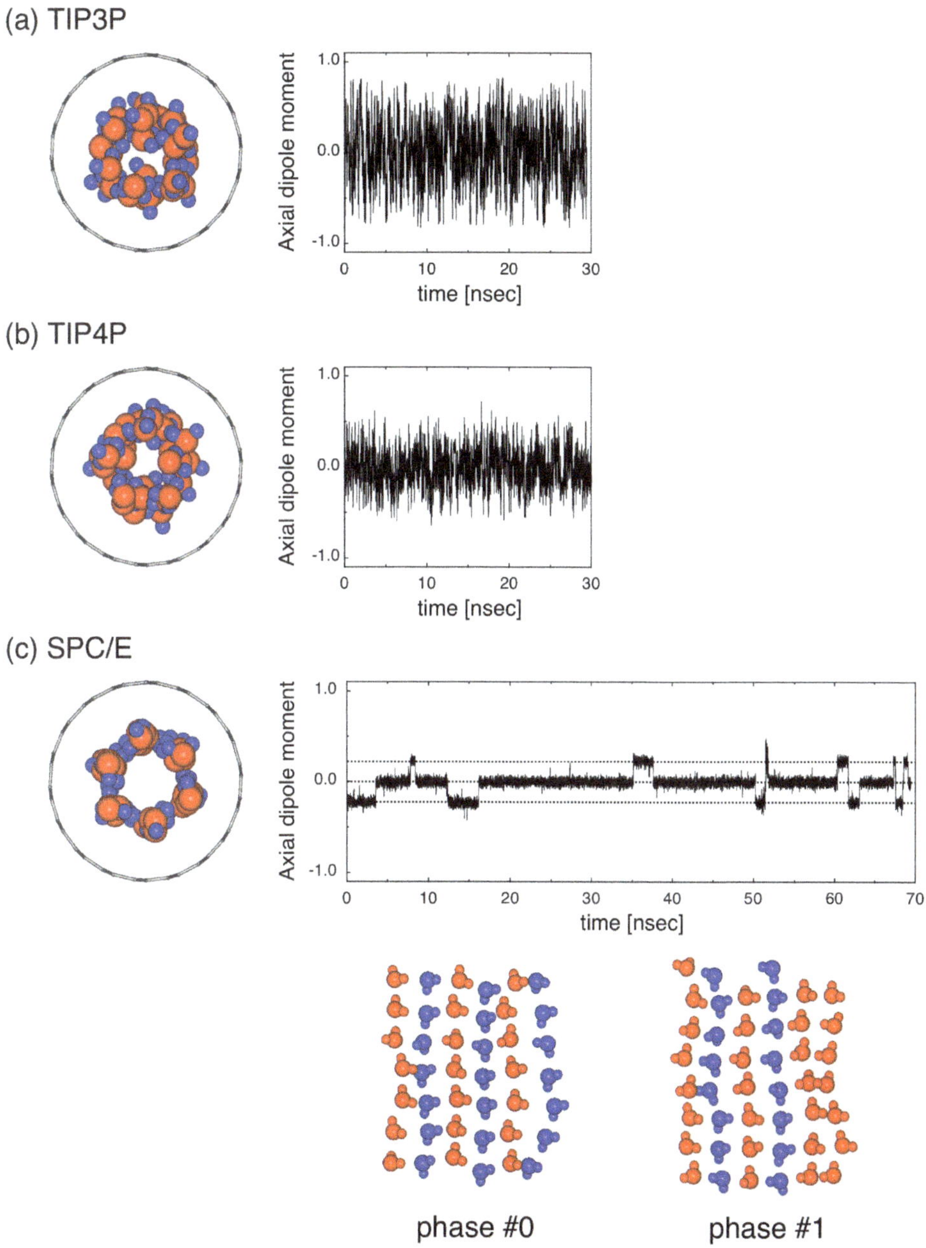

Fig. 14. Top-view snapshots and net axial polarization of water inside the (9, 9) CNT of 2.1 nm length for each water model. 280K. For SPC/E model, unfolded-view snapshot examples for phase #0 and #1 are also shown. Polarization and unfolded-view results are adapted from Fig.3 of Nakamura & Ohno, 2012a.

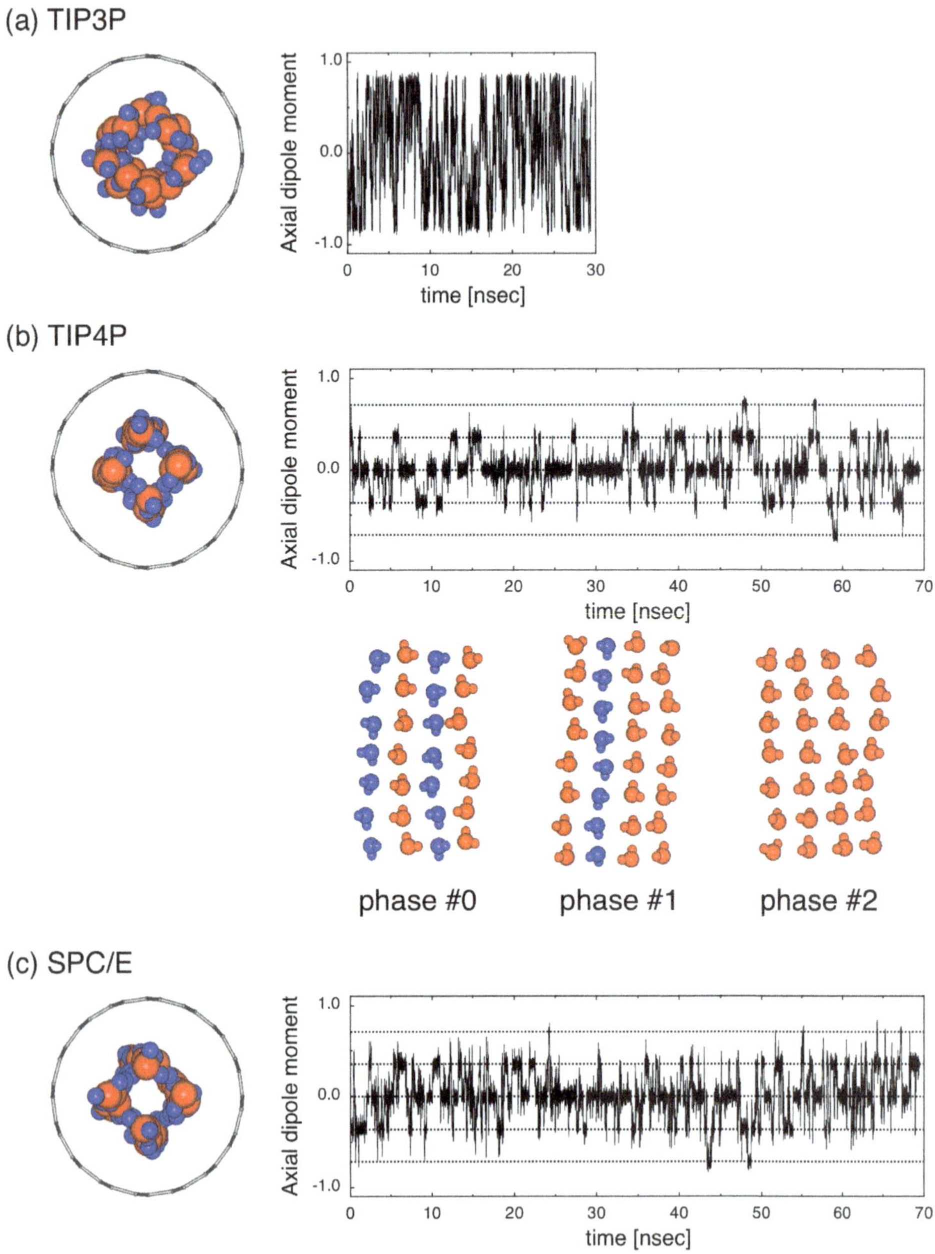

Fig. 15. Top-view snapshots and net axial polarization of water inside the (8, 8) CNT of 2.1 nm length for each water model. 280K. For TIP4P model, unfolded-view snapshot examples for phase #0 to #2 are also shown. Polarization and unfolded-view results are adapted from Fig.5 of Nakamura & Ohno, 2012a.

Examples of unfold-view snapshots corresponding to the three independent phases (phase #0 to #2) for the TIP4P are shown on the right-hand side of Fig. 15b. The transition mechanism between the phases is essentially the same as that described earlier. Unlike the confined TIP5P-E revealing the 5-gonal water tube (Fig. 5b), the TIP4P and the SPC/E reveal the 4-gonal tubes (Fig. 15b and c, left-hand side). In addition, the transition to the phase #4 (the FMW) is not observed for both the TIP4P and the SPC/E. The single helix structure of the TIP5P-E water very efficiently fills the interior space of the (8, 8) CNTs compared to the water structures of the other water models, resulting in the highest water density. We note that for the TIP5P-E water, the shorter the (8, 8) CNT length, the more often a 4-gonal single helix water structure is observed. The 4-gonal FMW is, however, rarely observed for the (8, 8) CNTs of length about 2-4 nm compared to the 5-gonal single helix water structure.

We have seen that both for the (8, 8) and the (9, 9) CNTs, the TIP3P water results in a less ordered structure compared to the other water models. The main reason for this is possibly attributed to too fast dynamics of the TIP3P water, which gives the self-diffusion constant of bulk liquid water more than double the experimental one. The reported values (25 °C, in the unit of 10^{-5} cm^2/sec) are 2.3 for experiment, 5.19 for TIP3P, 2.49 for SPC/E, 3.29 for TIP4P, and 2.80 for TIP5P-E (Mahoney & Jorgensen, 2001; Mark & Nilsson, 2001; Mills, 1973; Price et al., 1999). The structural ordering of the confined TIP3P water molecules is considered to be interrupted by the unrealistic fluctuation. Here, we briefly refer to a report on H_2O/CNT simulations employing a 'modified TIP3P' model. Though this model's diffusion constant is as high as the original TIP3P, a twisted-column shape of water inside the (7, 7), (8, 8) and (9, 9) CNTs was reported at ordinary ambient conditions (Noon et al., 2002). In the modified TIP3P model, the Lennard-Jones parameters on the hydrogen atoms were newly introduced to avoid singularities in integral equation calculations of activation free energies in complex molecular systems (Neria, 1996). The additional parameters, however, do not correct the original TIP3P's well-documented shortcomings including the very high diffusion constant (Mark & Nilsson, 2001). The effect of the additional parameters is therefore considered to be so strong that the ordered structures become possible despite the high diffusion constant.

For the rest of the water models (SPC/E, TIP4P, TIP5P-E), the model dependence has also been observed. The SPC/E and the TIP4P water molecules form into the axially stacked layers of anomalously immobile water molecules. The resultant water phases, however, correspond to the metastable phases of the confined TIP5P-E water, whose stable phase, the FMW, has not been observed for the SPC/E and the TIP4P models. What makes this difference? We pay attention to the fact that among those water models, the TIP5P-E most realistically describes the molecular charge distribution by explicitly taking into account the negative charges along the lone-pair directions of the water molecule. In the TIP4P model, the negative charge is placed on an additional fictive site other than the oxygen atom site. The number of electrostatic interaction sites, however, is three as well as the TIP3P model. Accuracy in the description of the molecular charge distribution is considered to be crucial in simulating the water inside the CNTs of the 'critical' diameters.

6. Conclusion

We have reported recent progress of H_2O/CNT simulations, focusing on the unusual behavior of the confined water molecules and the effect of the water models with different numbers of interaction sites (from three to five sites). All the models commonly show that the unusual behavior of the confined water changes in accordance with the CNT diameter size. We have found a critical CNT diameter range (about 1.1-1.2 nm), for which significant anomalous behavior of water that differs from model to model is observed. Except for this range, significant differences between the models have not been found. Based on the 5-site water models, which most realistically describes the molecular charge distribution among the water models used, single-domain ferroelectric water is produced at ambient conditions. The ferroelectric water diffuses while keeping its proton-ordered network intact. The mobile/immobile water transitions accompanied by the step-wise changes in net polarization of water have also been found. The outcome is expected to enable examining the ferroelectricity by detecting an abrupt change in mobility of water molecules, as well as calorimetry and dielectric experiments. Simulations based on the first-principles are also highly desirable to confirm the ferroelectric water. Such studies are now under way.

7. References

Alexiadis, A. & Kassinos, S. (2008a). The density ofwater in carbon nanotubes. *Chemical Engineering Science,* 63, pp. 2047 - 2056.

Alexiadis, A. & Kassinos, S. (2008b). Molecular Simulation of Water in Carbon Nanotubes. *Chemical Reviews,* 108, pp. 5014–5034.

Bai, J.; Wang, J. & Zeng, X. C. (2006). Multiwalled ice helixes and ice nanotubes. *Proceedings of the National Academy of Sciences,* 103, pp. 19664-19667.

Bernal, J. D. & Fowler, R. H. (1933). A Theory of Water and Ionic Solution, with Particular Reference to Hydrogen and Hydroxyl Ions. *Journal of Chemical Physics,* 1, pp.515-548.

Berendsen, H. J. C.; Grigera, J. R. & Straatsma, T. P. (1987). The missing term in effective pair potentials. *Journal of Physical Chemistry,* 91, pp. 6269-6271.

Berendsen, H. J. C.; Postma, J. P. M.; van Gunsteren, W. F. & Hermans, J. (1981). Interaction models for water in relation to protein hydration, In: *Intermolecular Forces,* B. Pullman, (Ed.), pp. 331-342, Reidel, Dordrecht

Bramwell, S. T. (1999). Ferroelectric ice. *Nature,* 397, pp. 212-213.

Case, D. A; Darden, T. A.; Cheatham, T. E. III; Simmerling, C. L.; Wang, J.; Duke, R. E.; Luo, R.; Merz, K. M.; Pearlman, D. A.; Crowley, M.; Walker, R. C.; Zhang, W.; Wang, B.; Hayik, S.; Roitberg, A.; Seabra, G.; Wong, K. F.; Paesani, F.; Wu, X.; Brozell, S.; Tsui, V.; Gohlke, H.; Yang, L.; Tan, C.; Mongan, J.; Hornak, V.; Cui, G.; Beroza, P.; Mathews, D. H.; Schafmeister, C.; Ross, W. S. & Kollman, P. A. (2006). *AMBER 9.* University of California, San Francisco.

Darden, T.; York, D. & Pedersen, L. (1993). Particle mesh Ewald: An *N*-log (*N*) method for Ewald sums in large systems. *Journal of Chemical Physics,*98, pp. 10089-10092.

Dellago, C.; Naor, MM. & Hummer, G. (2003). Proton Transport through Water-Filled Carbon Nanotubes. *Physical Review Letters,* 90, 105902.

Feller, W. (1968). Fluctuations in Coin Tossing and Random Walks. In: *An Introduction to Probability Theory and Its Applications,* ch. 3, Wiley, USA

Fernández, R. G.; Abascal, J. L. F. & Vega, C. (2006). The melting point of ice Ih for common water models calculated from direct coexistence of the solid-liquid interface. *Journal of Chemical Physics,* 124, 144506.

Fukazawa, H.; Hoshikawa, A.; Ishii, Y.; Chakoumakos, B. C. & Fernandez-Baca, J. A. (2006). Existence of ferroelectric ice in the universe. *The Astrophysical Journal,* 652, L57–L60.

Hummer, G.; Rasaiah, J. C. & Noworyta, J. P. (2001). Water conduction through the hydrophobic channel of a carbon nanotube. *Nature,* 414, pp. 188-109.

Iedema, M. J.; Dresser, M. J.; Doering, D. L.; Rowland, J. B.; Hess, W. P.; Tsekouras, A. A. & Cowin, J. P. (1998). Ferroelectricity in Water Ice. *Journal of Physical Chemistry B,* 102, pp. 9203-9214.

Jackson, S. M. & Whitworth, R. W. (1995). Evidence for ferroelectric ordering of ice Ih. *Journal of Chemical Physics,* 103, pp.7647-7648.

Jackson, S. M.; Nield, V. M.; Whitworth, R. W.; Oguro, M.; Wilson, C. C. (1997). Single-Crystal Neutron Diffraction Studies of the Structure of Ice XI. *Journal of Physical Chemistry B,* 101, pp. 6142-6145.

Jorgensen, W. L.; Chandrasekhar, J.; Madura, J. D.; Impey, R. W. & Klein, M. L. (1983). Comparison of simple potential functions for simulating liquid water. *Journal of Chemical Physics,* 79, pp. 926-935.

Kawada, S. (1972). Dielectric Dispersion and Phase Transition of KOH Doped Ice. *Journal of the Physical Society of Japan,* 32, pp. 1442-1442.

Kawada, S. (1989). Acceleration of dielectric relaxation by koh-doping and phase transition in ice Ih. *Journal of Physics and Chemistry of Solids,* 50, pp. 1177-1184.

Koga, K.; Gao, G. T.; Tanaka, H. & Zeng, X. C. (2001). Formation of ordered ice nanotubes inside carbon nanotubes. *Nature,* 412, pp. 802-805.

Köfinger, J.; Hummer, G. & Dellago, C. (2008). Macroscopically ordered water in nanopores. *Proceedings of the National Academy of Sciences,* 105, pp.13218–13222.

Köfinger, J. & Dellago, C. (2009). Orientational Dynamics and Dielectric Response of Nanopore Water. *Physical Review Letters,* 103, 080601.

Luo, C. F.; Fa, W.; Zhou, J.; Dong, J. M. & Zeng, X. C. (2008). Ferroelectric Ordering in Ice Nanotubes Confined in Carbon Nanotubes. *NANO LETTERS,* 8, pp. 2607-2612.

Liu, Y.; Wang, Q.; Wu, T. & Zhang, L. (2005). Fluid structure and transport properties of water inside carbon nanotubes. *Journal of Chemical Physics,* 123, 234701.

Mahoney, M. W. & Jorgensen, W. L. (2000). A five-site model for liquid water and the reproduction of the density anomaly by rigid, nonpolarizable potential functions. *Journal of Chemical Physics,* 112, pp. 8910-8922.

Mahoney, M. W. & Jorgensen, W. L. (2001). Diffusion constant of the TIP5P model of liquid water. *Journal of Chemical Physics,* 114, pp. 363-366.

Mark, P. & Nilsson, L. (2001). Structure and Dynamics of the TIP3P, SPC, and SPC/E Water Models at 298 K. *Journal of Physical Chemistry A,* 105, pp. 9954-9960.

Mashl, R. J.; Joseph, S.; Aluru, N. R. & Jakobsson, E. (2003). Anomalously Immobilized Water: A New Water Phase Induced by Confinement in Nanotubes. *NANO LETTERS,* 3, pp. 589-592.

Matsuo, T.; Tajima, Y. & Suga, H. (1986). CALORIMETRIC STUDY OF A PHASE TRANSITION IN D_2O ICE Ih DOPED WITH KOD: ICE XI. *Journal of Physics and Chemistry of Solids,* 47, pp. 165-173.

Mikami, F.; Matsuda, K.; Kataura, H. & Maniwa, Y. (2009). Dielectric Properties of Water inside Single-Walled Carbon Nanotubes. *ACS NANO,* 3, pp. 1279–1287.

Mills, R. (1973). Self-Diffusion in Normal and Heavy Water in the Range 1-45. *Journal of Physical Chemistry,* 77, pp. 685-688.

Nakamura, Y. & Ohno, T. (2011). Ferroelectric mobile water. *Physical Chemistry Chemical Physics,* 13, pp. 1064–1069.

Nakamura, Y. & Ohno, T. (2012a). Structure of water confined inside carbon nanotubes and water models. *Materials Chemistry and Physics, 132, pp. 682–687.*

Nakamura, Y. & Ohno, T. (2012b). Single-Domain Ferroelectric Water and its Concerted Diffusion in Nanotubes. Materials Science Forum, 700, pp. 108-111.

Neria, E. (1996). Simulation of activation free energies in molecular systems. *Journal of Chemical Physics,* 105, pp. 1902–1921.

Noon, W. H.; Ausman, K. D.; Smalley, R. E. & Ma, J. (2002). Helical ice-sheets inside carbon nanotubes in the physiological condition. *Chemical Physics Letters,* 355, pp. 445–448.

Pauling, L. (1935). The Structure and Entropy of Ice and of Other Crystals with Some Randomness of Atomic Arrangement. *Journal of the American Ceramic Society,* 57, pp. 2680-2684.

Price, W. L.; Ide, H. & Arata, Y. (1999). Self-Diffusion of Supercooled Water to 238 K Using PGSE NMR Diffusion Measurements. *Journal of Physical Chemistry A,* 103, pp. 448-450.

Rick, S. W. (2004). A reoptimization of the five-site water potential TIP5P for use with Ewald sums. *Journal of Chemical Physics,* 120, pp. 6085-6093.

Slater, J. C. (1941). Theory of the Transition in KH_2PO_4. *Journal of Chemical Physics,* 9, pp.16-33.

Su, X.; Lianos, L.; Shen, Y. R. & Somorjai, G. A. (1998). Surface-Induced Ferroelectric Ice on Pt(111). *Physical Review Letters,* 80, pp.1533-1536.

Tajima, Y.; Matsuo, T. & Suga, H. (1982). Phase transition in KOH-doped hexagonal ice. *Nature,* 299, pp. 810-812.

Tajima, Y.; Matsuo, T. & Suga, H. (1984). Calorimetric study of phase transition in hexagonal ice doped with alkali hydroxides. *Journal of Physics and Chemistry of Solids,* 45, pp. 1135-1145.

Vega, C.; Sanz, E. & Abascal, J. L. F. (2005). The melting temperature of the most common models of water. *Journal of Chemical Physics,* 122, 114507.

14

Studies of Cardio Toxin Protein Adsorption on Mixed Self-Assembled Monolayers Using Molecular Dynamics Simulations

Shih-Wei Hung[1], Pai-Yi Hsiao[1] and Ching-Chang Chieng[2]
[1]*Department of Engineering and System Science*
National Tsing Hua University, Hsinchu
[2]*Department of Mechanical and Biomedical Engineering*
City University of Hong Kong, Kowloon
[1]*Taiwan*
[2]*Hong Kong*

1. Introduction

To understand protein adsorption on a surface is very important in bio-related domains of technology and application such as biomaterials, implant biocompatibility, and biosensor technology. (Gray 2004) Self-assembled monolayers (SAMs) are excellent model surfaces for biology and biochemistry because they are stable, highly ordered, easy to prepare, and provide a wide range of organic functionality. (Love et al. 2005) Previous experimental studies (Ostuni et al. 2003; Prime & G.M. Whitesides 1991) have indicated that hydrophobic interactions are the major mechanism for protein adsorption on surfaces, and the dehydration of both the protein and hydrophobic SAMs provides an entropic driving force for protein adsorption. (Ostuni et al. 2003) Nonetheless, there are still many questions needed to be clarified and difficult to be investigated in experiments. Molecular dynamics (MD) simulation is a powerful tool that is able to investigate the atomic details of a molecular system. This paper summarizes the investigations of proteins adsorption on alkanethiol SAMs by means of MD simulations in order to expand the limited information that the experiments can provide. (Hung et al.2006; Hung et al.2010; Hung et al. 2011)

MD simulation has been applied to study the mechanisms of proteins adsorption on various SAMs and has provided valuable information. For example, Tobias *et al.* (Tobias et al. 1996) studied cytochrome-c (Cyt-c) covalently tethered to hydrophobic (methyl-terminated) and hydrophilic (thiol-terminated) SAMs. In their model, water molecules were not modeled explicitly. They found that Cyt-c was completely excluded from the hydrophobic SAM surface but partially dissolved on the hydrophilic SAM surface. In a follow-up study, Nordgren *et al.* (Nordgren et al. 2002) reported that the larger perturbation of Cyt-c structure occurred on hydrophilic SAM surfaces rather than on hydrophobic SAM surfaces using explicit water molecule model. They found that the protein molecule was surrounded by the water molecules, resulting in the reduction of the interaction between the protein and the surface. Zhou *et al.* (Zhou et al. 2004) investigated the orientation and conformation of Cyt-c on carboxyl-terminated SAM using a combined method of Monte Carlo (MC) and MD simulations. Their results showed that the preferable orientation of an adsorbed protein can be obtained by a strongly charged surface but the protein may lose its bioactivity due to the large conformational change. To understand surface resistance to protein adsorption, Jiang and his coworkers(Hower et al. 2006; Zheng et al. 2004; Zheng et al. 2005) studied lysozyme adsorption on various SAM surfaces with terminal methyl, hydroxyl, oligo (ethylene glycol) (OEG), mannitol, and sorbitol groups by a hybrid MC and MD simulation. They concluded that the resistance of protein adsorption to a surface was due to the tightly bound, structured water layer directly above the surface. Agashe *et al.* (Agashe et al. 2005) utilized MD simulations to investigate the adsorption of the γ-chain fragment of fibrinogen on SAM surfaces with five different terminal groups: methyl, hydroxyl, carboxyl, amino and OEG. The fibrinogen fragment did not show conformational rearrangements; rather it underwent rotational and translational motions until low-energy orientations were achieved. The above studies demonstrated that MD simulation is a useful and important tool to study protein adsorption on various SAM surfaces.

Not only the static properties, such as orientation and potential energy, but also the dynamic information about the process of CTX binding to the membrane are of interest. The dynamic biological process and the corresponding information, such as the structural change, adsorption force, interaction energy, and potential of mean force (PMF), can be investigated by steered molecular dynamics (SMD) simulation. (Isralewitz et al. 2001) The PMF is the equilibrium free energy difference along the reaction coordinate, which is an important thermodynamic quantity characterizing the dynamic process. (Kirkwood 1935) Calculation methods for PMF had been reviewed and classified as either equilibrium or non-equilibrium approaches.(Ytreberg, Swendsen, and Zuckerman 2006) The equilibrium approaches, for example, the umbrella sampling method, (Torrie & Valleau 1977) need large computer resources because they rely on fully sampled equilibrium simulations performed at each stage of the PMF calculation. The non-equilibrium approaches, for example, using Jarzynski's remarkable equality (Jarzynski 1997) from SMD simulations, has the potential to provide very rapid estimates of PMF. However, the use of Jarzynski's equality suffers from significant bias and error when the pulling velocity is too high or the number of trajectories sampled is insufficient.(Gore, Ritort, and Bustamante 2003)

CTX is a cytotoxic β-sheet basic polypeptide which is known to cause membrane leakage in many cells including human erythrocytes and phospholipid membrane vesicles.(Dufton and Hider 1988) The three-dimensional structures of various CTX homologues in both aqueous and micellar environments are available.(S C Sue et al. 2001; Dauplais et al. 1995) The interactions between CTX and lipid membranes have been widely studied by various methods, such as Fourier transform infrared spectroscopy, (Huang et al. 2003; Forouhar et al. 2003) nuclear magnetic resonance (NMR) spectroscopy, (Dubovskii et al. 2005) and computer simulation.(Levtsova et al. 2009; R G Efremov et al. 2004) CTXs do not easily adopt large conformational changes due to the existence of four disulfide bonds in their chemical structure. Therefore, they are good candidates in experimental study of protein adsorption on SAM surfaces. MD simulations can help, furthermore, in the understanding of the interaction of CTX-SAM system.

In our earlier study, (Hung et al. 2006) the binding energy of protein molecules to SAM surfaces of different mixing composition of alkanethiols chains was investigated. We found that the binding energy was enhanced due to the increasing of the hydrophobic area on the SAM surface. In that study, we focused on the enthalpic contribution of the hydrophobic interaction between a CTX molecule and SAM surfaces, and thus a solvent model of distance-dependent dielectric function was used (Ramstein & Lavery 1988) instead of explicit water molecules. However, the hydrophobic interaction is primarily driven by entropy (Chandler 2005) and it has been shown by many groups (Hower et al. 2006; Zheng et al. 2004; Zheng et al. 2005; Ostuni et al. 2003) that the water molecules on the protein-SAM interface play an important role on protein adsorption mechanism. In order to take into account both enthalpic and entropic components of hydrophobic interaction, an explicit solvent model is conducted to investigate and identify the complete physical mechanism.

Summmarizing our previous studies, MD simulations were performed to study the physical mechanism of CTX proteins adsorption on alkanethiol SAMs with different chain lengths. The dynamic information, such as structural changes and adsorption forces of CTX protein desorption from the SAM surface were investigated by means of SMD simulations. The dependence of the dynamic information on the pulling velocity was illustrated. The PMFs were calculated by the umbrella sampling method for better interpretation of the desorption process.

2. Model system and methodology

2.1 Simulation model

In order to study the adsorption of a CTX protein on the surfaces composing of mixed *alkanethiol SAM* growing on Au (111) substrate, two types of alkanethiol chains, $S(CH_2)_5CH_3$ and $S(CH_2)_9CH_3$, were composed and were denoted briefly by C5 and C9, respectively. The structure of the SAM is a $(\sqrt{3}\times\sqrt{3})R30°$ lattice on x-y plane with lattice constant equal to 0.499 nm. Each lattice point is occupied by an alkanethiol chain of either C5 or C9. Five mixing ratios were studied: $\chi_{C9} = 0, 0.25, 0.5, 0.75$ and 1, where χ_{C9} is defined as $N_{C9}/(N_{C5}+N_{C9})$ with N_{C5} and N_{C9} representing respectively the numbers of C5 and C9 chains in the SAMs. In our studies, there were 12 chains on each side of the simulation box. Periodic boundary condition was applied in x and y directions to simulate infinitely large surface of SAMs. The Au (111) substrate was modeled by a single layer of gold atoms.

The CTX protein was modeled by the nuclear magnetic resonance (NMR) structure of CTX A3 (PDB 1I02), (Sue et al. 2001) comprised of 60 amino acid residues. The interactions between CTX and lipid membrane had been well studied by Wu's group, (Huang et al. 2003; Forouhar et al. 2003) and they found that the characteristics topology of three hydrophobic fingers (Fig. 1) played a key role in binding to lipid membranes. Hence, the CTX protein was chosen as the model system to study the protein-SAM interaction, to identify the mechanism of protein adsorbing on a SAM surface, and to understand the role of the protein hydrophobicity on the adsorption in our studies.

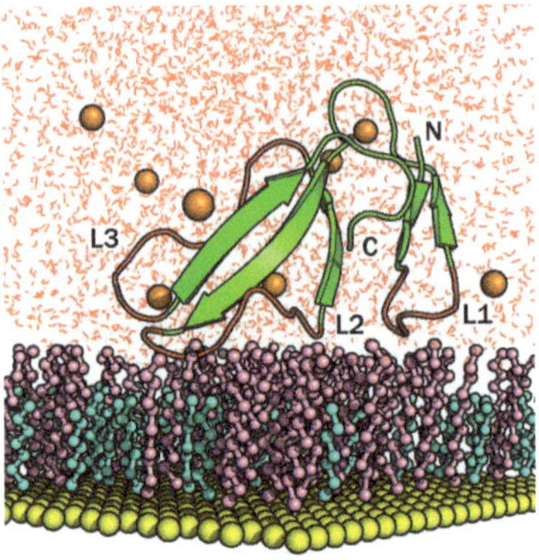

Fig. 1. Illustration of CTX adsorbed on SAM/Au (111) surface where $\chi_{C9} = 0.5$ (C5 is plotted in cyan color, C9 in pink color and gold atoms in yellow color). Water molecules are plotted in red color and the counter ions are in orange spheres. C and N indicate the C-terminus and N-terminus of the CTX protein, respectively. The three major loops, denoted by L1, L2 and L3, of CTX are colored in red. (Hung et al.2010)

2.2 Potential energy function

The potential energy function used in MD simulation included several terms describing both internal (bonded) and external (nonbonded) interactions, which ultimately determined both structure and dynamics of a molecular system. The atomic interactions inside a CTX molecule were described by the GROMOS-96(43a2) force field. (Schuler and Wilfred Van Gunsteren 2000) This force field consisted of bonded interactions, including bond, angle, dihedral and improper dihedral angle terms, and non-bonded interactions, including van der Waals (vdW) and Coulomb interactions. The complete form of this force field read

$$U = \sum_{\text{bonds}} \frac{1}{2}k_b(r-r_0)^2 + \sum_{\text{angles}} \frac{1}{2}k_\theta(\theta-\theta_0)^2 + \sum_{\text{dihedrals}} k_\varphi\left[1+\cos(n\varphi-\varphi_0)\right] + \sum_{\text{imporpers}} k_\xi(\xi-\xi_0)^2 + \sum_{\text{vdW}} 4\varepsilon_{ij}\left[\left(\frac{\sigma_{ij}}{r_{ij}}\right)^{12} - \left(\frac{\sigma_{ij}}{r_{ij}}\right)^{6}\right] + \sum_{\text{Coulombic}} \frac{q_i q_j}{4\pi\varepsilon_0 r_{ij}} \tag{1}$$

Here k_b, k_θ, k_φ, and k_ξ are the bond, angle, dihedral, and improper dihedral force constants, respectively, for the bonded interactions and r_0, θ_0, φ_0, and ξ_0 are the equilibrium values of bond length, bond angle, dihedral angle and improper dihedral angle, respectively. For the non-bonded interactions, ε is the Lennard-Jones well depth, σ is the distance at which the vdW interaction is zero, q is the charge, and r_{ij} is the distance between atoms i and j. The parameters of the vdW interaction for the cross interactions between atoms i and j were obtained by the geometry combination rule. The alkanethiol chains in SAMs were modeled by the united-atom model of Hautman and Klein model (Hautman & Klein 1989), given as

$$U = \sum_{\text{angles}} \frac{1}{2}k_\theta(\theta-\theta_0)^2 + \sum_{\text{dihedrals}} \sum_{n=0}^{5} C_n\left[\cos(\varphi-180°)\right]^n + \sum_{\text{vdW}} 4\varepsilon_{ij}\left[\left(\frac{\sigma_{ij}}{r_{ij}}\right)^{12} - \left(\frac{\sigma_{ij}}{r_{ij}}\right)^{6}\right] \tag{2}$$

In this model, the dihedral potential took the form of Ryckaert-Bellemans, (Ryckaert & Bellemans 1978) in which the potential was expanded in a series of cosine function of the dihedral angle φ with C_n = 9.28, 12.16, -13.12, -3.06, 26.24, -31.5 kJ/mole for n= 0,..., 5, respectively. The laterally interaction between SAM molecules and gold surface was described by the Lennard-Jones 12-3 potential. (Tupper & Brenner 1994)

$$U = \sum_{\text{SAM-Au}} 2.117\varepsilon_{ij}\left[\left(\frac{\sigma_{ij}}{r_{ij}}\right)^{12} - \left(\frac{\sigma_{ij}}{r_{ij}}\right)^{3}\right] \quad (3)$$

The gold atoms were restricted on the positions of the lattice points in the substrate. Water molecules were explicitly considered and were modeled by extended simple point charge (SPC/E) model. (Berendsen et al. 1987)

The hydrophobic interaction was the main interaction between CTX and SAM. The enthalpic and entropic components of the hydrophobic interaction were correctly involved and treated once the water molecules were explicitly modeled in our studies. Since the SAM molecules in the current study carry no charge, the potential energy between SAM and the other molecules was vdW interaction and the parameters were determined by the geometrical combination rule.

2.3 Initial configuration

It has been reported that phase separation can take place when SAM was composed of components of different terminal groups such as 3-mercaptopropanol and *n*-tetradecanethiol, *n*-undecanethiol and 11-mercaptoundecanoic acid, 3-mercapto-*N*-nonylpropionamide and *n*-decanethiol and so on. (Smith et al. 2004) However, for SAM composed of the components of similar terminal groups, phase separation, in general, did not happen. For example, Whitesides and coworkers had done a series of experimental studies on methyl-terminated mixed SAM systems (Laibinis et al. 1992; Folkers et al. 1992; Bain & Whitesides 1989) and shown that chain length difference in the SAM did not render the systems into the formation of macroscopic islands. Using atomic force microscope imaging, other group (Tamada et al. 1997) also reported no phase separation in the mixed SAMs composed of $S(CH_2)_3CH_3$ and $S(CH_2)_{17}CH_3$. A recent review summarized again that molecules of similar composition did not phase separate in a SAM system formed from solutions at room temperature. (Smith et al. 2004) The results indicated that the mixed SAMs were not macroscopically phase separated according to Folkers *et al.* (Folkers et al. 1992) Based upon the above information, a homogenous mixture of the simulated SAM systems was assumed and the SAM surfaces were generated by randomly placing C5 and C9 molecules on the gold substrate in the present simulations.

The initial configuration of the CTX protein was prepared in the way with the three CTX finger loops facing the SAM surface. This orientation had been confirmed as the most favorable orientation of CTX proteins binding to membrane by experiments from Wu's group. (Forouhar et al. 2003; Huang et al. 2003) Efremov *et al.* (Efremov et al. 2004) and Lomize *et al.* (Lomize et al. 2006) obtained the same result by using MC method and transfer energy minimization method respectively. Therefore, the initial configuration with the three finger loops of the CTX facing down was set to facilitate the adsorption of the system to the truly favorable orientation for the present simulations.

2.4 Simulation procedure

After generating the initial configurations, the systems were solvated in a bath of water molecules with a density of 1 g/cm^3. One Na^+ ion and ten Cl^- ions were added into the simulation box to maintain the electro-neutrality of the systems. The simulation box was a rectangular parallelepiped of 5.99 × 5.19 × 10.00 nm^3 with periodic boundary conditions applied in the x and y directions. The z direction was restricted by a wall (cf. Fig.1). The velocities were initially assigned to each atom with a Maxwell-Bolzmann distribution at 50 K. The system was then gradually heated to 300 K in a period of 400 ps to initially relax the water molecules around the protein and the SAM surface.

The annealing process (Kirkpatrick et al. 1983) was then performed to overcome local minima and search the global minimum of energy landscape of the interaction between CTX and SAM. The process was started by an initial heating stage in which the system was heated from 300 to 350 K. The temperature was then maintained at 350 K for 400 ps, followed by a slow cooling at a rate of 0.1 K/ps to 300 K. 300 K was then maintained for 4 ns until the interaction energy between CTX and SAM reached a constant value. The simulations were performed in canonical ensemble with the integrating time step equal to 1.0 fs. The temperature was controlled by Berendsen thermostat (Berendsen et al. 1984) with the time constant equal to 0.1 ps in the annealing process and then the temperature was controlled by a Nose-Hoover thermostat. (Hoover 1985; Nose 1984) Because bond vibration is very fast, all the covalent bonds were constrained by the LINCS algorithm. (Hess et al. 1997) For the nonbonded interactions, the cutoff distance was chosen to be 1.5 nm for the vdW interaction. The Coulomb interaction was cut at 2.5 nm by a force-shifting function. It has been demonstrated (Steinbach & Brooks 1994) that the cutoff method for the electrostatic interaction can correctly model the dynamics of biomolecules in solutions. The twin-range approach in neighbor searching (Wilfred F. van Gunsteren and Herman J. C. Berendsen 1990) was performed, with short range distance equal to 1.5 nm and long range distance equal to 2.5 nm. Data were saved every 1.0 ps for analysis. All the simulations were performed using the program GROMACS. (Van Der Spoel et al. 2005) The simulation snapshots were plotted using the software PyMOL. (DeLano 2002)

2.5 Steered molecular dynamic simulations

In our study (Hung et al.2011), SMD simulated the pulling process of CTX adsorbed on the SAM surface by applying an external force to the CTX molecule, and it monitored the adsorption force and structural change of the CTX molecule during the desorption process. Before the pulling process started, the model system was in equilibrium, and the CTX molecule was adsorbed onto the SAM surface. The equilibrium state was kept at 300 K for an additional 3 ns. The starting configurations of the model system for SMD simulations were generated by extracting configurations from the last nanosecond of the equilibrium stage.

In the pulling process during the SMD simulation, external forces were applied to the CTX through the center of mass to pull the CTX off of the SAM surface with a constant velocity perpendicular to the SAM surface (i.e., z direction in Fig. 1). The adsorption force f at time t can be represented by the following equation

$$f(t) = k[(Z_{CTX,0} + vt) - Z_{CTX}(t)] \tag{4}$$

where k is the force constant, v is the pulling velocity and $Z_{CTX}(t)$ and $Z_{CTX,0}$ are the z coordinates of CTX at time t and initial time 0, respectively. In our study, we defined Z as the distance from gold substrate, i.e., Z=0 is the position of the gold.

The stiff-spring approximation (Park & Schulten 2004; Park et al. 2003) with a large force constant was applied to minimize the fluctuation of reaction coordinate among different trajectories. The approximation was valid only if the force constant was sufficiently larger than the curvature of the energy landscape at its minimum, which was approximately the maximum of the second derivative of PMF profile. (Hummer & Szabo 2010) However, the fluctuations of applied force were related to k through $(k_B Tk)^{1/2}$,where k_B was the Boltzmann's constant and T was the absolute temperature, and thus the force constant cannot be arbitrarily large. (Balsera et al. 1997) In our study, the force constant was chosen to be 1209 k_BT/nm^2, which was four times as large as the maximum of the second derivative of PMF profile calculated using the umbrella sampling method, to ensure the validity of the stiff-spring approximation. The total traveling distance was up to 2 nm starting from Z_{CTX}=2.0 nm, which was the equilibrium position of CTX protein adsorbed on the SAM surface, to Z_{CTX}=4.0 nm, where the separation distance was far enough that there was no interaction between the CTX protein and SAM surface.

2.6 PMF calculation

PMF is a potential along a reaction coordinate, the gradient of which yields the negative of the average force acting on the targeted molecule over all the configurations at a given place. Physically, PMF represents the difference of free energy, ΔF, along the coordinate of reaction. In statistical physics, the difference ΔF between two coordinates z and z_0, can be calculated by the ratio of the two configurational integrals as follow:

$$\Delta F = F(z) - F(z_0) = -\frac{1}{\beta}\ln\frac{\int d^{3N-1}\mathrm{R}\exp[-\beta U(z,\mathbf{R})]}{\int d^{3N-1}\mathrm{R}\exp[-\beta U(z_0,\mathbf{R})]} \tag{5}$$

where z is the reaction coordinate, z_0 is the reference position, **R** denotes the remaining 3N-1 coordinates, U(z,**R**) is the system potential, and $\beta=1/k_BT$ with k_B being the Boltzmann's constant and T the absolute temperature.

In the circumstance of equilibrium MD simulations without applying external potentials, the regions with large PMF are not easily explored because the large difference of free energy hinders the access of the target molecule into the regions. It is hence difficult to calculate numerically the configurational integral over such regions with good accuracy. To overcome the problem, calculation methods for PMF had been devolped and classified as either equilibrium or non-equilibrium approaches.(Ytreberg et al. 2006) The equilibrium approaches, i.e. the umbrella sampling method, (Torrie & Valleau 1977) and the non-equilibrium approaches, i.e. using Jarzynski's remarkable equality (Jarzynski 1997) from SMD simulations. Further comparisons and details of these two approaches are given in our previous study. (Hung et al. 2011)

3. Results and discussions

The results and discussions were summarized in two parts. In the first part, the influence of different solvent models on the systems was investigated at first. The results demonstrated that to include water molecules explicitly was crucial in the study of the protein adsorption. Therefore, the explicit water model was implemented to study the protein adsorption on SAMs of different mixing ratios (χ_{C9}). The binding energies were obtained by means of MD simulations. The binding energy was studied by calculating the nonbonded interactions of the protein with the SAM surface. The CTX/SAM contact area and the structure of SAM molecules were examined to investigate the binding enhancement of the CTX protein adsorbing onto a SAM surface. Other physical mechanisms were discussed in our previous study. (Hung et al. 2010) In the second part, the dynamic information including the structural changes, adsorption forces a CTX protein desorbed from the pure C5 SAM surface were monitored in the SMD simulation. The thermodynamic information, PMF, was calculated using the umbrella sampling method for better interpretation of the CTX desorption process.

3.1 Static properties of CTX adsorption

3.1.1 Comparison of results using different solvent models

There have been many alternative solvent models proposed to minimize the requirement of computer resources in simulating water. (Smith & Pettitt 1994) For example, the distance-dependent dielectric function was a model widely used in many studies. To demonstrate the importance of the role of water in the adsorption of CTX on SAM, we performed the simulations using the distance-dependent dielectric model to simulate water environment. In this model, solvent molecules were not explicitly treated but implicitly considered together as a dielectric continuum. The dielectric constant was set to 1 within a cut-off distance (2.5 nm), and 78 beyond the cut-off distance. The binding energy of CTX on SAM calculated from the distance-dependent dielectric model showed the similar trend of behavior as from the explicit water model, but the error bar of data was larger for the distance-dependent dielectric model (Fig. 2). Therefore, the CTX protein in the distance-dependent dielectric model displayed a different conformation, compared to in the explicit water model. Fig. 3 showed snapshots of an equilibrium conformation of the CTX protein on the χ_{C9} = 0.5 SAM surface obtained from the two models. We observed that the CTX protein was almost lying flat on the SAM surface for the distance-dependent dielectric model. The orientation was different from the one by the explicit water model. Thus, higher binding

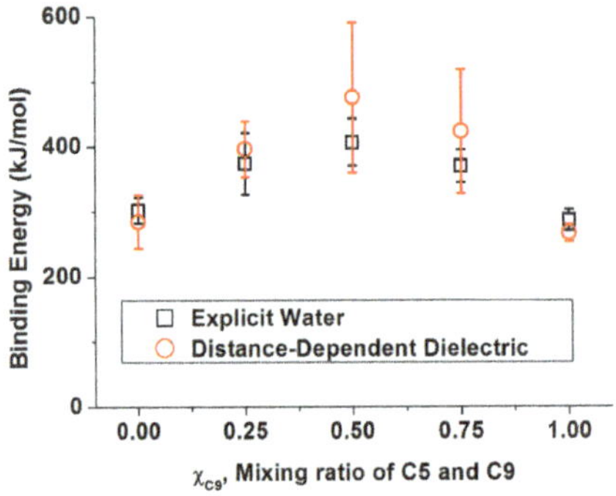

Fig. 2. Binding energy of CTX to SAM surfaces of different mixing ratios in the distance-dependent dielectric model and in the explicit solvent model. (Hung et al. 2010)

energy was obtained for the former model due to the larger contact area. In the distance-dependent dielectric model, there was no water molecule on the CTX-SAM interface to prevent direct contact of CTX and SAM and hence the interaction between them was so strong that the conformation of the CTX protein was largely deformed and the orientation was not in agreement with experimental observations. With explicit water molecules, the entropic component of the hydrophobic effect was considered and thus the hydrophobic effect induced some kinds of ordering in the surrounding water. More precisely, the hydrophilic residues of CTX were solvated by the water molecules via the hydrogen bonding. As a consequence, the CTX molecule was adsorbed on the SAM surface with specific orientation. The results demonstrated that to include water molecules explicitly in the model was crucial in the study of the protein adsorption. The binding mechanisms and protein conformation can be correctly identified only when it was considered.

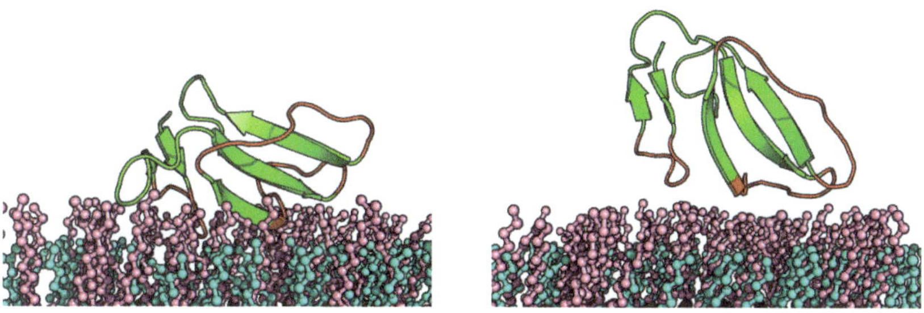

(a) (b)

Fig. 3. Snapshots of CTX on $\chi_{C9} = 0.5$ SAM surface using different solvent models: (a) distance-dependent dielectric model; (b) explicit solvent model. (Hung et al. 2010)

3.1.2 Binding energy of a CTX protein on SAM surface

The affinity of a CTX protein on SAM surface was of interest. This affinity can be quantified by calculating the binding energy between CTX and SAM, which was the sum of all the non-bonded interactions between them. Fig. 4 showed a maximum in the middle. The binding energy of the CTX protein to the pure C5 and the energy to the pure C9 SAM surfaces were similar to each other but significantly smaller than that to a mixed SAM surface. The enhancement of the binding energy on a mixed SAM surface can be as large as 34% when $\chi_{C9} = 0.5$.

3.1.3 Physical mechanisms of CTX adsorption

The enhancement of the binding energy can be explained by the equilibrium configurations of CTX landing on SAM surfaces composing of different mixing ratios as shown in Fig. 5. In the pure C5 and C9 SAM systems, the surface roughness is small. The CTX protein landed stably on the flat surface and the conformation of the CTX looked similarly to each other. On the other hand, the surface roughness was increased in the mixed SAM systems. The three-finger loops of the CTX can penetrate into the region between C5 and C9 molecules, especially when $\chi_{C9} = 0.25$ and 0.5, resulting in the increase of the CTX-SAM contact area.

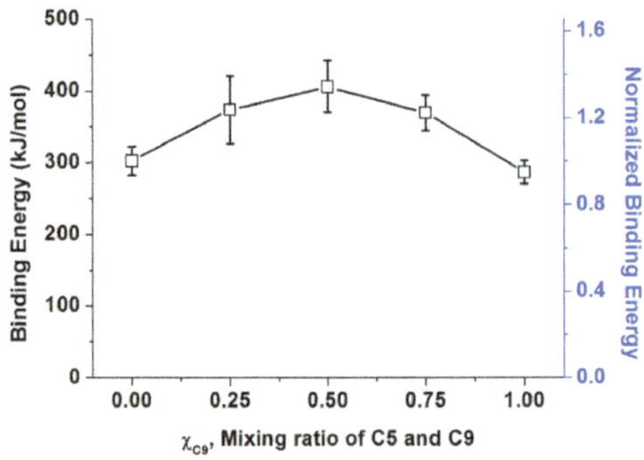

Fig. 4. Binding energy (read from left) of CTX adsorded on SAM surfaces of different mixing ratios, χ_{C9}. The normalized binding energy (read from right) is calculated by dividing the binding energy by the value on the C5 SAM surface. (Hung et al. 2010)

The CTX-SAM contact area and the structure of SAMs were described in the following sections.

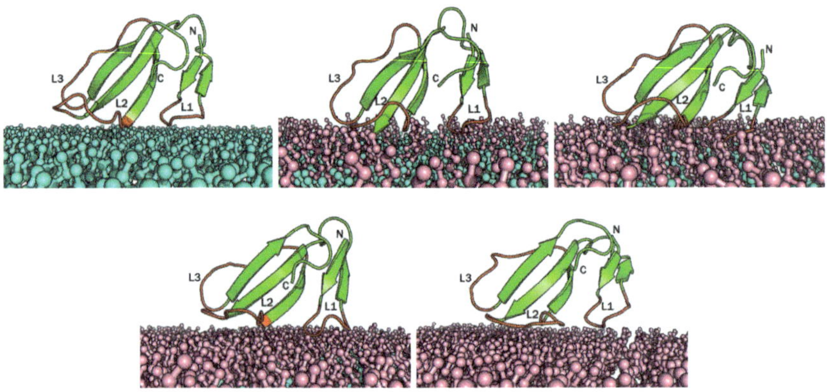

Fig. 5. Snapshots of CTX on (a) $\chi_{C9} = 0$, (b) $\chi_{C9} = 0.25$, (c) $\chi_{C9} = 0.5$, (d) $\chi_{C9} = 0.75$, and (e) $\chi_{C9} = 1$ SAM surface. Water molecules are not shown in the figures to provide clear illustrations of the protein configurations. (Hung et al. 2010)

It was commonly accepted that the free-energy change of protein from water to membrane was proportional to the change of area contacting with the surrounding water area of protein. (White & Wimley 1994; Reynolds et al. 1974; Eisenberg & McLachlan 1986) This idea was extended to our system and verified if the binding energy of CTX on SAM also followed a linear relation against the CTX-SAM contact area. The area was calculated by the method of double cubic lattice (Eisenhaber et al. 1995) with a probe radius equal to 0.14 nm. The result of the binding energy versus the contact area was plotted in Fig. 6. It showed that the binding energy satisfied a linear equation with the surface area of CTX in contact with the SAM surface. The slope of the linear equation was 36.05 kJ/mol·nm^2. Therefore, the binding energy was higher on the mixed SAMs surface than on the pure SAM. Mixed SAMs surfaces were rough surfaces which increased the contact area.

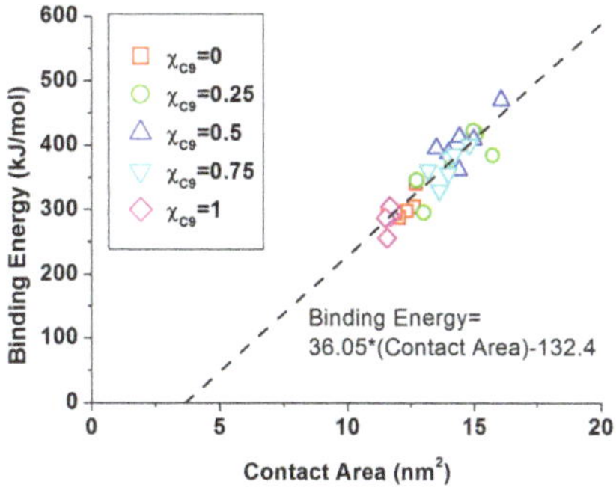

Fig. 6. Binding energy as a function of CTX-SAM contact area. (Hung et al. 2010)

Since the binding energy was strongly related to the CTX-SAM contact area, the structure of the SAM molecules can be another important factor in determining the protein adsorption. Mixed SAMs can be divided to two layer regions as observed by Bain, (Bain & Whitesides 1989) illustrated in Fig. 7(a). The first layer region was the inner region adjacent to the gold substrate and the second one was the outer layer in contact with the solution. The mobility of the SAM molecules can be calculated by the root mean square deviation (RSMD) of the alkanethiol chains in the SAM in a time interval equal to 1 ps. The larger the RMSD, the higher the mobility would be. Fig. 7(b) showed that the RMSD of the C5 surface was higher than that of the C9 surface, which showed a better ordering when the chain length was long.

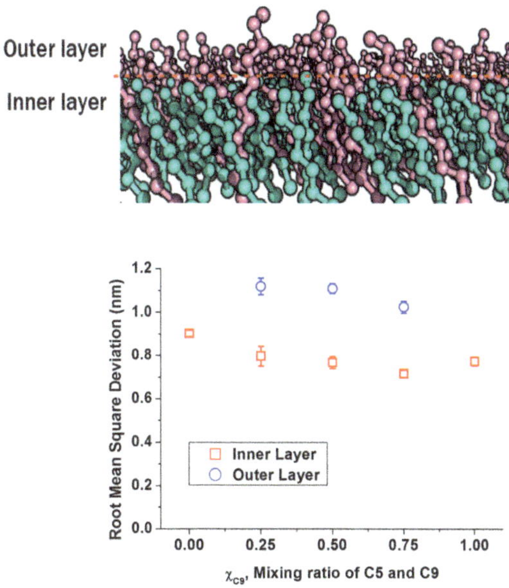

Fig. 7. (a) Illustion of inner and outer layers for mixed SAM surfaces; (b) Root mean square deviations for inner and outer layers versus mixing ratio χ_{C9} of SAM surface. (Hung et al. 2010)

The chain length dependence had been studied by experiments. (Fenter et al. 1997; Porter et al. 1987) These experiments showed that there existed distinct differences in structure between long-chained and short-chained SAMs. The long-chained SAMs formed a densely packed, crystalline-like structure while the structures of the short-chained SAMs became increasingly disordered. The results were consistent with the experiments. The large number of methylene groups in the long-chained SAM provided a strong vdW interaction to sustained an ordered structure. On the other hand, for the mixed SAM surfaces, the RMSD value was larger in the outer layer region than in the inner one. This was in agreement with the Bain's study, which showed that the inner layer packed better than the outer one.

The mixture of alkanethiols of different chain lengths in the SAM provided an additional dimension of the reaction area, especially for the hydrophobic interaction, on the limited surface. As a result, the CTX affinity was enhanced on the mixed SAM surface. In order to investigated the relation between the CTX binding and the SAM surface area, the SAS area of the mixed SAMs surface was calculated and plotted in Fig. 8. It showed that the CTX-SAM contact area was highly correlated to the SAM surface area. The result suggested that the three dimensional nanostructured morphology, due to the chain length difference of the alkanethiol chains in SAM, promoted the contact of CTX protein on the SAM surface.

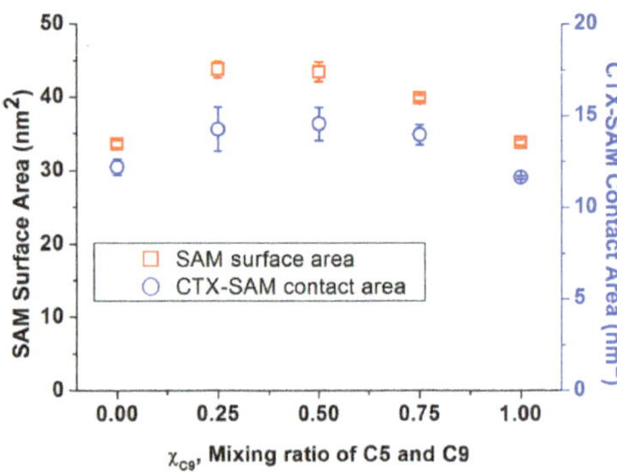

Fig. 8. SAM surface area and CTX-SAM contact area versus mixing ratio χ_{C9} of SAM surface. (Hung et al. 2010)

3.2 Dynamic information of CTX desorption process

3.2.1 Structural changes of CTX during the desorption process

CTX is a highly stable protein due to the presence of four disulfide bonds and a core of hydrophobic residues. (Sivaraman et al. 1998) As a result, the CTX protein does not undergo unfolding during desorption processes. SMD simulations provided dynamic structure histories of the CTX protein during the pulling process, and the key CTX molecular conformations were shown in Figs. 9(a)-9(d) for a pulling velocity of 0.25 nm/ns. The structural change of the CTX protein during the desorption process was due to these three loops. Starting from the equilibrium orientation of the three-finger loops facing and attached to the SAM surface (Fig. 9(a)), when CTX was pulled from the surface at a constant velocity, Loop I's tip detached from the surface first (Fig. 9(b)), and Loops II and III detached later at about the same time (Fig. 9(c)) before the entire protein was detached from the surface (Fig. 9(d)). The correlation between the trajectories of the three loops' tips ($Z_{Loop,i}$) and of the center of mass of protein (Z_{CTX}) from the gold substrate can quantitatively indicated the structural changes of the CTX protein during the desorption process, as shown in Fig. 9(e). The centers of mass of three amino acid residues, Val7, Ala28, and Leu47, were chosen to represent the tip positions of the three loops, $Z_{Loop\ I}$, $Z_{Loop\ II}$, and $Z_{Loop\ III}$, respectively. The average position of the methyl groups of the SAMs was about 1.0 nm from the gold substrate, and the vdW radius of methyl group was ~0.2 nm. (A J Li and Nussinov 1998) Thus, all three $Z_{Loop,i}$ were about 1.3~1.4 nm from the gold substrate when the loop contacted the SAM surface. It was reasonable to define that the ith loop tip detached from the SAM surface when $Z_{Loop,i}$ was larger than 1.6 nm. In general, the desorption process can be described in three stages as follows: in the first stage, CTX contacted the SAM surface with the three loops attached until Loop I detached from the SAM surface at Z_{CTX}~2.3-2.4 nm, and $Z_{Loop,i}$<1.6 nm for all three loops i=1, 2, and 3. In the second stage, Loop I rose with Z_{CTX} while Loops II and III remained in contact with the SAM surface, and $Z_{Loop,i}$<1.6 nm for i=2 and 3 only. Loops II and III detached from the SAM surface almost at the same time at Z_{CTX}~3.3 nm. The three loops were far from the SAM surface in the third stage.

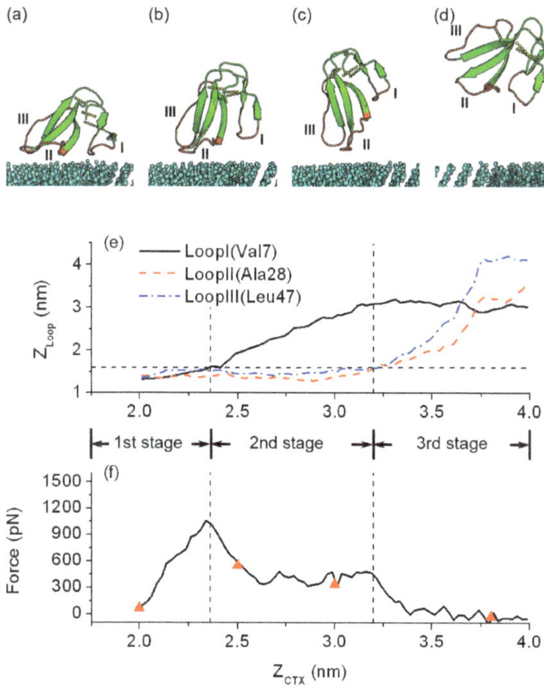

Fig. 9. (a)-(d) CTX conformations at Z_{CTX}=2.0, 2.5, 3.0, and 3.8 nm, respectively. (e) Z_{Loop}, positions of the three loop tips of CTX (black solid line: LoopI; red dashed line: LoopII; blue dashed-dotted line: LoopIII) and (f) adsorption force vs. Z_{CTX}, center of mass of CTX, when pulling with velocity v=0.25 nm/ns. Red triangles mark the positions of the snapshots in (a)-(d). (Hung et al. 2011)

3.2.2 Adsorption force during the desorption process

The adsorption forces and structural changes of the CTX protein during the desorption process from the SAM surface were the important results of the SMD simulation. Fig. 9(f) shows the adsorption force of the CTX protein as it was pulled with a constant velocity of 0.25 nm/ns. Three stages can be distinguished from the force curve. In the first stage, the force was monotonically rising until it reached a peak, i.e., the rupture force in the present study at Z_{CTX}~2.3-2.4 nm. The force then decreased to a plateau in the second stage for 2.5 nm< Z_{CTX}<3.2 nm, and the force diminishes to nearly zero in the third stage (Z_{CTX}>3.2 nm). These stages were closely correlated when the loops were either attached or detached.

Summarizing the history of the adsorption force and positions of the three loop tips, $Z_{Loop,i}$, the desorption process can be described in three stages. In the first stage, the adsorption force increased with Z_{CTX} until it was large enough to break Loop I from the SAM surface. In the second stage, the CTX-SAM was in a quasi-equilibrium state with Loops II and III contacting the SAM surface, and thus, the force remained constant in this stage. In the third stage, CTX was located far enough from the SAM surface that there was no interaction force between the CTX and SAM surface. The detachment of Loop I from the SAM surface gave the rupture force, which suggested that Loop I played an important role in the CTX desorption process from the SAM surface. This observation agreed with two-dimensional NMR spectroscopy experiments, (Sivaraman et al. 2000) which had shown that Loops II and III were more stable than Loop I. Hence, Loop I possessed the highest flexibility of the three loops, which was related to its biological activity. As it had the highest flexibility, Loop I should be the first loop to pull off from the surface, which was consistent with the SMD simulation results.

Different pulling velocities leading to different hysteresis effects (Liphardt et al. 2001) and different rupture forces (Liphardt et al. 2002) were observed in the previous studies. In the present simulations, the desorption processes at pulling velocities of v=1.0 nm/ns and v=0.25 nm/ns were qualitatively consistent with the three distinguished stages. However, three major differences were observed: the magnitude of the adsorption force at a higher pulling velocity was much higher than that at a lower pulling velocity, the quasi-equilibrium state was indeterminate and the adsorption force was not diminished when the CTX was far from the SAM surface at v=1.0 nm/ns. These distinctions were a result of the pronounced non-equilibrium phenomena, such as friction and dissipation, at high pulling velocity. Furthermore, it was noted that the orientation of CTX in the bulk solvent environment at higher pulling velocity (Fig. 10(d)) was similar to that of CTX just departing from the SAM surface (Fig. 10(c)). There was no specific orientation of CTX in the bulk solvent environment at a lower pulling velocity (Figs. 9(c) and 9(d)), which implies that CTX was not relaxed with the surrounding molecules at a high pulling velocity of v=1.0 nm/ns.

To depict the adsorption force-pulling velocity dependence, the averaged force curves at four different pulling velocities were shown in Fig. 11(a). The force curves were in qualitative agreement by shape, but the magnitude of the peak adsorption force and the width of the second plateau-like stage were different at different pulling velocities. Because the difference between the forces at v=0.125 nm/ns and at 0. 25 nm/ns was very small and the computation time for the v=0.125 nm/ns case was twice that of the v=0.25 nm/ns case, the major computations were conducted at v=0.25 nm/ns for the present study. Figure 11(b) showed that the average rupture force was related to the pulling velocity in terms of the

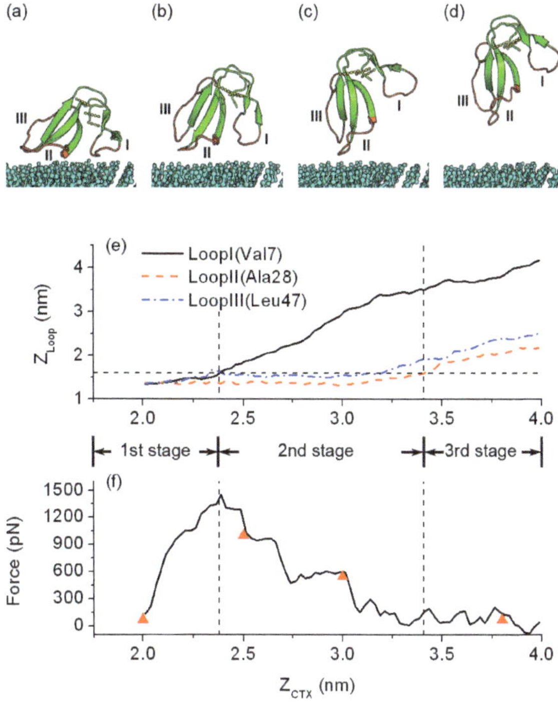

Fig. 10. (a)-(d) CTX conformations at Z_{CTX}=2.0, 2.5, 3.0, and 3.8 nm, respectively. (e) Z_{Loop}, positions of the three loop tips of CTX (black solid line: LoopI; red dashed line: LoopII; blue dashed-dotted line: LoopIII) and (f) adsorption force vs. Z_{CTX}, center of mass of CTX, when pulling with velocity v=1.0 nm/ns. Red triangles mark the positions of the snapshots in (a)-(d). (Hung et al. 2011)

linear (green dash-dotted line) and logarithmic (red dash line) fittings. Some SMD studies (Sotomayor & Schulten 2007; Gao et al. 2002) found that the rupture force values are increased logarithmically with the pulling velocities, while other studies (Heymann & Grubmüller 1999; Marrink et al. 1998) found the linear dependence on the pulling velocities for high pulling velocities. The behavior between the rupture force and pulling velocity can be described well in terms of both linear and logarithmic relationships for pulling velocity ranging from 0.125 nm/ns to 1.0 nm/ns in our SMD simulations. The near-linear correlation indicated that the desorption process was within the friction force-dominated regime for the present study.

3.2.3 PMF calculated by the umbrella sampling method

Because the pulling forces depended strongly on the pulling velocity, further information of the CTX desorption process can be obtained from the free energy landscape, i.e., PMF. The resulting PMF profile from the umbrella sampling (Fig. 12(a)) showed a sharp change of the slope at Z_{CTX}~2.5 nm, a subsequent gradual increase afterward (2.5 nm $<Z_{CTX}<$3.2 nm), and a plateau after Z_{CTX}~3.2 nm. The sharp change indicated a free energy barrier at Z_{CTX}~2.5 nm.

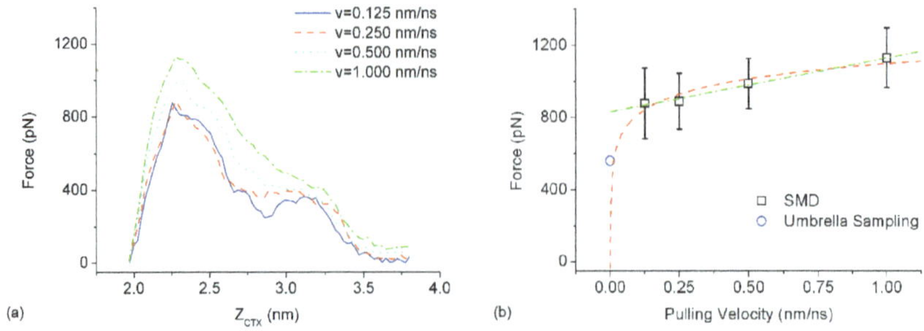

Fig. 11. (a) Average adsorption force curves with various pulling velocities, i.e., v=0.125 (blue solid line), 0.25 (red dashed line), 0.5 (cyan dotted line), and 1 nm/ns (green dashed-dotted line). (b) Average rupture force vs. pulling velocity from SMD simulations (squares). Blue circle indicates the rupture force obtained from the derivation of the PMF calculated using umbrella sampling. The green dashed-dotted and red dashed lines represent the best linear and logarithm fits to the average rupture forces, respectively. (Hung et al. 2011)

For comparison with the pulling force curves from SMD simulations, the mean force profile was obtained from the derivative of the PMF profile with respect to Z_{CTX} (Fig. 12(b)). A remarkable agreement between the mean force (Fig. 12(b)) and SMD force-distance curves (Fig. 9(f) and Fig. 10(f)) was obtained.

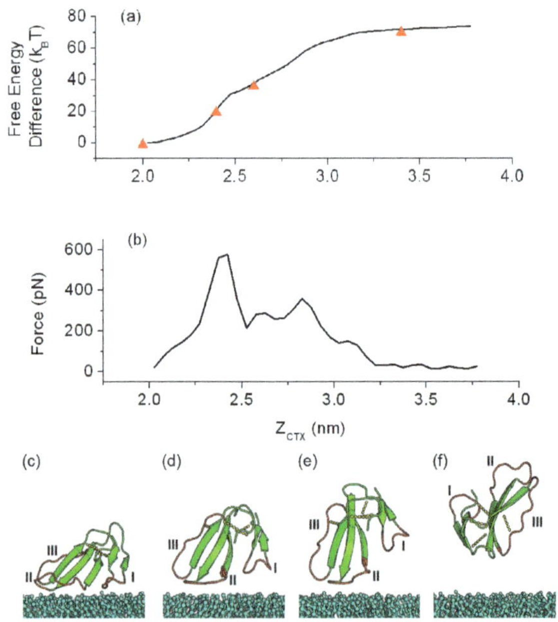

Fig. 12. (a) PMF profile calculated using umbrella sampling. Red triangles mark the positions of the snapshots in (c)-(f). (b) Mean force profile obtained from the derivative of the PMF with respect to Z_{CTX}. (c)-(f) CTX conformations at Z_{CTX} =2.0, 2.4, 2.7, and 3.4 nm, respectively. (Hung et al. 2011)

The adsorption force from the SMD simulation included the random force, friction force, and thermodynamic force, but he mean force from the umbrella sampling method represented the thermodynamic force between the CTX protein and the SAM surface. Similar force-distance curves and similar structural changes impled cross-validation and also capture key features of the process from both approaches. Furthermore, the peak forces of both force-distance curves occurred at the same departing distance, Z_{CTX}~2.5 nm, indicating that the departure of the first loop was a major rupture force in the desorption process and the use of a pulling force from SMD that was higher than the mean force from the umbrella sampling was reasonable. For the range of pulling velocities applied in SMD simulations, the friction force played a role in the adsorption force, and thus, the relationship between the adsorption force and pulling velocity was near-linear at higher pulling velocities. However, the relationship between the adsorption force and pulling velocity becomes logarithmic with lower pulling velocities for the thermodynamic force-dominated regime. (Marrink et al. 1998) As a result, the mean force (blue circle in Fig. 12(b)) cannot be achieved by extrapolating linearly to zero pulling velocity. Instead, the SMD data might approach the mean force by logarithmically extrapolating to zero pulling velocity, as shown in Fig. 12(b).

4. Conclusion and prospects

In the present study, the adsorption of CTX proteins on alkanethiol SAMs of different mixing ratios was analyzed by means of MD simulations. Different solvent models had been examined and the results demonstrated that the use of explicit water molecules was necessary to correctly take into account the enthalpic and entropic components of the hydrophobic effect when one studies protein adsorption on SAMs. The results showed that the binding energy has the highest value when χ_{C9} was 0.5 and were in good agreement with the experimental data. Moreover, the binding energy between CTX and SAM surface was proportional to the CTX-SAM contact area. The structure of SAMs molecules determined the CTX-SAM contact area, and hence the binding energy.

Dynamic information, such as structural change and adsorption force, about the desorption of a single CTX protein from a SAM surface was investigated by means of SMD simulations successfully. The simulation results indicated that CTX did not undergo unfolding during the pulling process and Loop I was the first loop to depart from the SAM surface. This observation was in good agreement with the results of the NMR spectroscopy experiment. For the pulling velocity ranging from 0.125 nm/ns to 1.0 nm/ns employed in the present study, the near-linear dependence of force on pulling velocity indicated that the friction force played a significant role in the force measured in SMD simulations. A remarkable agreement was obtained between the force-distance interaction from the umbrella sampling method and the pulling force-distance curve from SMD, which cross-validated these techniques and also captured the same key features of the process from both approaches. Furthermore, the peak forces of both force-distance curves occurred at the same departing distance, Z_{CTX}~2.5 nm, indicated that the departure of the first loop was resulted in the major rupture force in the desorption process. The results provided valuable information at atomic level toward a fundamental understanding of protein adsorption.

Future work will focus on the PMFs calculation of CTX adsorption on mixed SAMs of different mixing ratios. It will provide key features for enhacing protein adsorption or protein resistance on the designed surface by manipulating the mixing ratios.

5. Acknowledgment

The authors thank the National Center for High-Performance Computing, Taiwan for computing resources and the National Science Council, Taiwan for financial support under Grant (NSC99-2221-E007-028-MY2).

6. References

Agashe, M.; Raut, V.; Stuart, S. J. & Latour, R. A. (2005). Molecular Simulation to Characterize the Adsorption Behavior of a Fibrinogen Gamma-Chain Fragment. *Langmuir*, Vol.21, No.3, pp. 1103-1117

Bain, C.D. & Whitesides, G. M. (1989). Formation of Monolayers by the Coadsorption of Thiols on Gold: Variation in the Length of the Alkyl Chain. *Journal of the American Chemical Society*, Vol.111, No.18, pp. 7164-7175

Balsera, M.; Stepaniants, S.; Izrailev, S.; Oono, Y. & Schulten, K. (1997). Reconstructing Potential Energy Functions from Simulated Force-Induced Unbinding Processes. *Biophysical Journal*, Vol.73, No.3, pp. 1281-1287

Berendsen, H. J. C., Grigera, J. R. & Straatsma, T. P. (1987). The Missing Term in Effective Pair Potentials. *The Journal of Physical Chemistry*, Vol.91, No.24, pp. 6269-6271

Berendsen, H. J. C.; Postma, J. P. M.; Gunsteren, W. F. van; DiNola, A. & Haak, J. R. (1984). Molecular Dynamics with Coupling to an External Bath. *The Journal of Chemical Physics*, Vol.81, No.8, pp. 3684-3690

Chandler, D. (2005). Interfaces and the Driving Force of Hydrophobic Assembly. *Nature*, Vol.437, No.7059, pp. 640-647

Dauplais, M.; Neumann, J. M.; Pinkasfeld, S.; Ménez, A. & Roumestand, C. (1995). An NMR Study of the Interaction of Cardiotoxin Gamma from Naja Nigricollis with Perdeuterated Dodecylphosphocholine Micelles. *European Journal of Biochemistry / FEBS*, Vol.230, No.1, pp. 213-220

DeLano, W. L. (2002). The PyMOL Molecular Graphics System. , Available from http://www.pymol.org

Dubovskii, P. V.; Lesovoy, D. M.; Dubinnyi, M. A.; Konshina, A. G.; Utkin, Y. N.; Efremov, R. G. & Dubovskii, A. S. A. (2005). Interaction of Three-Finger toxins with Phospholipid Membranes: Comparison of S- and P-Type Cytotoxins. *The Biochemical Journal*, Vol.387, No.Pt 3, pp. 807-815

Dufton, M. J. & Hider, R. C. (1988). Structure and Pharmacology of Elapid Cytotoxins. *Pharmacology & Therapeutics*, Vol.36, No.1, pp. 1-40

Efremov, R. G.; Nolde, D. E.; Konshina, A. G.; Syrtcev, N. P. & Arseniev, A. S. (2004). Peptides and Proteins in Membranes: What can We Learn via Computer Simulations? *Current Medicinal Chemistry*, Vol.11, No.18, pp. 2421-2442

Eisenberg, D. & McLachlan, A. D. (1986). Solvation Energy in Protein Folding and Binding. *Nature*, Vol.319, No.6050, pp. 199-203

Eisenhaber, F.; Lijnzaad, P.; Argos, P.; Sander, C. & Scharf, M. (1995). The Double Cubic Lattice Method: Efficient Approaches to Numerical Integration of Surface Area and Volume and to Dot Surface Contouring of Molecular Assemblies. *Journal of Computational Chemistry*, Vol.16, No.3, pp. 273-284

Fenter, P.; Eberhardt, A.; Liang, K.-S. & Eisenberger, P. (1997). Epitaxy and Chainlength Dependent Strain in Self-Assembled Monolayers. *The Journal of Chemical Physics*, Vol.106, No.4, pp. 1600-1608

Folkers, J. P.; Laibinis, P. E. & Whitesides, G. M. (1992). Self-Assembled Monolayers of Alkanethiols on Gold: Comparisons of Monolayers Containing Mixtures of Short- and Long-Chain Constituents with Methyl and Hydroxymethyl Terminal Groups. *Langmuir*, Vol.8, No.5, pp. 1330-1341

Forouhar, F.; Huang, W.-N.; Liu, J.-H.; Chien, K.-Y.; Wu, W.-g. and Hsiao, C.-D. (2003). Structural Basis of Membrane-Induced Cardiotoxin A3 Oligomerization. *The Journal of Biological Chemistry*, Vol.278, No.24, pp. 21980-21988

Gao, M.; Lu, H. & Schulten, K. (2002). Unfolding of Titin Domains Studied by Molecular Dynamics Simulations. *Journal of Muscle Research and Cell Motility*, Vol.23, No.5-6, pp. 513-521

Gore, J.; Ritort, F. & Bustamante, C. (2003). Bias and error in estimates of equilibrium free-energy differences from nonequilibrium measurements. *Proceedings of the National Academy of Sciences of the United States of America*, Vol.100, No.22, pp. 12564-12569

Gray, J. J. (2004). The Interaction of Proteins with Solid Surfaces. *Current Opinion in Structural Biology*, Vol.14, No.1, pp. 110-115

Hautman, J. & Klein, M. L. (1989). Simulation of a Monolayer of Alkyl Thiol Chains. *The Journal of Chemical Physics*, Vol.91, No.8, pp. 4994-5001

Hess, B.; Bekker, H.; Berendsen, H. J. C. & Fraaije, J. G. E. M. (1997). LINCS: A Linear Constraint Solver for Molecular Simulations. *Journal of Computational Chemistry*, Vol.18, No.12, pp. 1463-1472

Heymann, B. & Grubmüller, H. (1999). AN02/DNP-Hapten Unbinding Forces Studied by Molecular Dynamics Atomic Force Microscopy Simulations. *Chemical Physics Letters*, Vol.303, No.1-2, pp. 1-9

Hoover, W. G. (1985). Canonical Dynamics: Equilibrium Phase-Space Distributions. *Physical Review A*, Vol.31, No.3, pp. 1695–1697

Hower, J. C.; He, Y.; Bernards, M. T. & Jiang S. (2006). Understanding the Nonfouling Mechanism of Surfaces through Molecular Simulations of Sugar-Based Self-Assembled Monolayers. *The Journal of Chemical Physics*, Vol.125, No.21, pp. 214704

Huang, W.-N.; Sue, S.-C.; Wang, D.-S.; Wu, P.-L. & Wu, W.-g. (2003). Peripheral Binding Mode and Penetration Depth of Cobra Cardiotoxin on Phospholipid Membranes as Studied by a Combined FTIR and Computer Simulation Approach. *Biochemistry*, Vol.42, No.24, pp. 7457-7466

Hummer, G. & Szabo, A. (2010). Free Energy Profiles from Single-Molecule Pulling Experiments. *Proceedings of the National Academy of Sciencess of the United States of America*, Vol.107, No.50, pp. 10-15.

Hung, S.-W.; Hsiao, P.-Y. & Chieng, C.-C. (2011). Dynamic Information for Cardiotoxin Protein Desorption from a Methyl-Terminated Self-Assembled Monolayer Using Steered Molecular Dynamics Simulation. *The Journal of Chemical Physics*, Vol.134, No.19, pp. 194705

Hung, S.-W.; Hsiao, P.-Y. & Chieng, C.-C. (2010). Mixed-SAM Surfaces Monitoring CTX-protein, part II: Analysis using Molecular Dynamics Simulations. *IEEE Transactions on Nanobioscience*, Vol.9, No.4, pp. 297-306.

Hung, S.-W.; Hwang, J.-K.; Tseng, F.; Chang, J.-M.; Chen, C.-C. & Chieng, C.-C. (2006). Molecular Dynamics Simulation of the Enhancement of Cobra Cardiotoxin and E6 Protein Binding on Mixed Self-Assembled Monolayer Molecules. *Nanotechnology*, Vol.17, No.4, pp. S8-S13.

Isralewitz, B.; Gao, M. & Schulten, K. (2001). Steered Molecular Dynamics and Mechanical Functions of Proteins. *Current Opinion in Structural Biology*, Vol.11, No.2, pp. 224-230.

Jarzynski, C. (1997). Nonequilibrium Equality for Free Energy Differences. *Physical Review Letters*, Vol.78, No.14, pp. 2690-2693

Kirkpatrick, S.; Gelatt, C. D. & Vecchi, M. P. (1983). Optimization by Simulated Annealing. *Science*, Vol.220, No.4598, pp. 671-680

Kirkwood, J. G. (1935). Statistical Mechanics of Fluid Mixtures. *The Journal of Chemical Physics*, Vol.3, No.5, pp. 300-313

Laibinis, P. E.; Nuzzo, R. G. & Whitesides, G. M. (1992). Structure of Monolayers Formed by Coadsorption of Two n-alkanethiols of Different Chain Lengths on Gold and Its Relation to Wetting. *The Journal of Physical Chemistry*, Vol.96, No.12, pp. 5097-5105

Levtsova, O. V.; Antonov, M. Y.; Mordvintsev, D. Y.; Utkin, Y. N.; Shaitan, K. V. & Kirpichnikov, M. P. (2009). Steered Molecular Dynamics Simulations of Cobra Cytotoxin Interaction with Zwitterionic Lipid Bilayer: No Penetration of Loop Tips into Membranes. *Computational Biology and Chemistry*, Vol.33, No.1, pp. 29-32

Li, A. J. & Nussinov, R. (1998). A Set of van der Waals and Coulombic Radii of Protein Atoms for Molecular and Solvent-Accessible Surface Calculation, Packing Evaluation, and Docking. *Proteins*, Vol.32, No.1, pp.111-127

Liphardt, J.; Onoa, B.; Smith, S. B.; Tinoco, I. & Bustamante, C. (2001). Reversible Unfolding of Single RNA Molecules by Mechanical Force. *Science*, Vol.292, No.5517, pp. 733-737

Liphardt, J.; Dumont, S.; Smith, S. B.; Tinoco, I. & Bustamante, C. (2002). Equilibrium Information from Nonequilibrium Measurements in an Rxperimental Test of Jarzynski's Equality. *Science*, Vol.296, No.5574, pp. 1832-1835

Lomize, M. A.; Lomize, A. L.; Pogozheva, I. D. & Mosberg, H. I. (2006). OPM: Orientations of Proteins in Membranes Database. *Bioinformatics*, Vol.22, No.5, pp. 623-625

Love, J. C.; Estroff, L. A.; Kriebel, J. K.; Nuzzo, R. G. & Whitesides, G. M. (2005). Self-Assembled Monolayers of Thiolates on Metals as a Form of Nanotechnology. *Chemical Reviews*, Vol.105, No.4, pp. 1103-1169.

Marrink, S.; Berger, O.; Tieleman, P. & Jahnig, F. (1998). Adhesion Forces of Lipids in a Phospholipid Membrane Studied by Molecular Dynamics Simulations. *Biophysical Journal*, Vol.74, No.2, pp. 931-943

Nordgren, C. E.; Tobias, D. J.; Klein, M. L. & Blasie, J. K. (2002). Molecular Dynamics Simulations of a Hydrated Protein Vectorially Oriented on Polar and Nonpolar Soft Surfaces. *Biophysical Journal*, Vol.83, No.6, pp. 2906-2917

Nose, S. (1984). A Unified Formulation of the Constant Temperature Molecular Dynamics Methods. *The Journal of Chemical Physics*, Vol.81, No.1, pp. 511-519

Ostuni, E.; Grzybowski, B. A.; Mrksich, M.; Roberts, C. S. & Whitesides G. M. (2003). Adsorption of Proteins to Hydrophobic Sites on Mixed Self-Assembled Monolayers. *Langmuir*, Vol.19, No.5, pp. 1861-1872

Park, S. & Schulten, K. (2004). Calculating Potentials of Mean Force from Steered Molecular Dynamics Simulations. *The Journal of Chemical Physics*, Vol.120, No.13, pp. 5946-5961

Park, S.; Khalili-Araghi, F.; Tajkhorshid, E. & Schulten, K. (2003). Free Energy Calculation from Steered Molecular Dynamics Simulations Using Jarzynski's Equality. *The Journal of Chemical Physics*, Vol.119, No.6, pp. 3559-3566

Porter, M. D.; Bright, T. B.; Allara, D. L. & Chidsey, C. E. D. (1987). Spontaneously organized molecular assemblies. 4. Structural characterization of n-alkyl thiol monolayers on gold by optical ellipsometry, infrared spectroscopy, and electrochemistry. *Journal of the American Chemical Society*, Vol.109, No.12, pp. 3559-3568

Prime, K.L. & Whitesides, G.M. (1991). Self-Assembled Organic Monolayers: Model Systems for Studying Adsorption of Proteins at Surfaces. *Science*, Vol.252, No.5009, pp. 1164-1167

Ramstein, J. & Lavery, R. (1988). Energetic Coupling between DNA Bending and Base Pair Opening. *Proceedings of the National Academy of Sciences of the United States of America*, Vol.85, No.19, pp. 7231-7235

Reynolds, J. A., Gilbert, D. B. & Tanford, C. (1974). Empirical Correlation between Hydrophobic Free Energy and Aqueous Cavity Surface Area. *Proceedings of the National Academy of Sciences of the United States of America*, Vol.71, No.8, pp. 2925-2927

Ryckaert, J.-P. & Bellemans, A. (1978). Molecular Dynamics of Liquid Alkanes. *Faraday Discussions of the Chemical Society*, Vol.66, pp. 95-106

Schuler, L. & Gunsteren, W. F. van (2000). On the Choice of Dihedral Angle Potential Energy Functions for n-Alkanes. *Molecular Simulation*, Vol.25, No.5, pp. 301-319

Sivaraman, T.; Kumar, T. K. S.; Chang, D.-K.; Lin, W.-Y. & Yu, C. (1998). Events in the Kinetic Folding Pathway of a Small, All Beta-Sheet Protein. *The Journal of Biological Chemistry*, Vol.273, No.17, pp. 10181-10189

Sivaraman, T.; Kumar, T. K. S.; Hung, K.-W.; & Yu, C. (2000). Comparison of the Structural Stability of Two Homologous Toxins Isolated from the Taiwan Cobra (Naja naja atra) Venom. *Biochemistry*, Vol.39, No.30, pp. 8705-8710

Smith, P. E. & Pettitt, B. M. (1994). Modeling Solvent in Biomolecular Systems. *The Journal of Physical Chemistry*, Vol.98, No.39, pp. 9700-9711

Smith, R. K., Lewis, P. A. & Weiss, P. S. (2004). Patterning Self-Assembled Monolayers. *Progress in Surface Science*, Vol.75, No.1-2, pp. 1-68

Sotomayor, M. & Schulten, K. (2007). Single-Molecule Experiments in Vitro and in Silico. *Science*, Vol.316, No.5828, pp. 1144-1148

Steinbach, P. J. & Brooks, B. R. (1994). New Spherical-Cutoff Methods for Long-Range Forces in Macromolecular Simulation. *Journal of Computational Chemistry*, Vol.15, No.7, pp. 667-683

Sue, S.-C.; Jarrell, H. C.; Brisson, J. R. & Wu. W.-g. (2001). Dynamic Characterization of the Water Binding Loop in the P-Type Cardiotoxin: Implication for the Role of the Bound Water Molecule. *Biochemistry*, Vol.40, No.43, pp. 12782–12794

Tamada, K.; Hara, M. & Sasabe, H. (1997). Surface Phase Behavior of n-alkanethiol Self-Assembled Monolayers Adsorbed on Au (111): An Atomic Force Microscope Study. *Langmuir*, Vol.13, No.6, pp. 1558-1566

Tobias, D. J.; Mar, W.; Blasie, J. K. & Klein, M. L. (1996). Molecular Dynamics Simulations of a Protein on Hydrophobic and Hydrophilic Surfaces. *Biophysical Journal*, Vol.71, No.6, pp. 2933-2941

Torrie, G. M. & Valleau, J. P. (1977). Nonphysical Sampling Distributions in Monte Carlo Free-Energy Estimation: Umbrella Sampling. *Journal of Computational Physics*, Vol.23, No.2, pp. 187-199

Tupper, K. J. & Brenner, D. W. (1994). Compression-Induced Structural Transition in a Self-Assembled Monolayer. *Langmuir*, Vol.10, No.7, pp. 2335-2338

van Gunsteren, W. F. & Berendsen, H. J. C. (1990). Computer Simulation of Molecular Dynamics: Methodology, Applications, and Perspectives in Chemistry. *Angewandte Chemie International Edition in English*, Vol.29, No.9, pp. 992-1023

van der Spoel, D.; Lindahl, E.; Hess, B.; Groenhof, G.; Mark, A. E. & Berendsen, H. J. C. (2005). GROMACS: Fast, Flexible, and Free. *Journal of Computational Chemistry*, Vol.26, No.16, pp. 1701-1718

White, S. & Wimley, W. (1994). Peptides in Lipid Bilayers: Structural and Thermodynamic Basis for Partitioning and Folding. *Current Opinion in Structural Biology*, Vol.4, No.1, pp. 79-86

Ytreberg, F. M.; Swendsen, R. H. & Zuckerman, D. M. (2006). Comparison of Free Energy Methods for Molecular Systems. *The Journal of Chemical Physics*, Vol.125, No.18, pp. 184114

Zheng, J.; Li, L.; Chen, S. & Jiang, S. (2004). Molecular Simulation Study of Water Interactions with Oligo (Ethylene Glycol)-Terminated Alkanethiol Self-Assembled Monolayers. *Langmuir*, Vol.20, No.20, pp. 8931-8938

Zheng, J.; Li, L.; Tsao, H.-K.; Sheng, Y.-J.; Chen, S. & Jiang, S. (2005). Strong Repulsive Forces between Protein and Oligo (Ethylene Glycol) Self-Assembled Monolayers: a Molecular Simulation Study. *Biophysical Journal*, Vol.89, No.1, pp. 158-166

Zhou, J.; Zheng, J. & Jiang, S. (2004). Molecular Simulation Studies of the Orientation and Conformation of Cytochrome c Adsorbed on Self-Assembled Monolayers. *The Journal of Physical Chemistry B*, Vol.108, No.45, pp. 17418-17424

15

Applications of All-Atom Molecular Dynamics to Nanofluidics

Mauro Chinappi
Department of Physics, Sapienza University of Rome, P.la Aldo Moro 5, 00185, Rome Italy

1. Introduction

In recent years, due to the progress in the fabrication of micro and nanodevices, nanofluidics has become an intense research field. The interest of the scientific community is evident from the large amount of published papers and by the issuing of dedicated journals. In this scenario, a better understanding of the key aspects of fluid motion in nanoscale systems is fundamental and hence computational techniques able to provide a deeper insight in the complex phenomena involved in nanoscale mass transport play a crucial role in nanofluidic research. The aim of this chapter is to introduce the reader to the application of classical all-atom molecular dynamics (MD) to nanofluidic problems. Nanofluidics is the study of the fluid motion in confined structures whose characteristic size is some nanometers, typically $1 - 100$ nm (Eijkel & Berg, 2005). Confined fluids in nanoscale geometries exhibit physical behaviors that, in several cases, largely differ from macroscale dynamics. The crucial difference is that in nanoscale systems the usual mathematical description for continuum fluid dynamics often fails to reproduce the correct behavior. Here we deal with simple liquids (Hansen & McDonald, 2006) and the appropriate macroscopic model is given by the incompressible Navier-Stokes equations for mass and momentum conservation

$$\nabla \cdot \mathbf{u} = 0 \tag{1}$$

$$\frac{\partial \mathbf{u}}{\partial t} + \mathbf{u} \cdot \nabla \mathbf{u} = -\frac{1}{\rho} \nabla p + \nu \nabla^2 \mathbf{u} + \mathbf{f} \tag{2}$$

where ρ is the constant fluid mass density, $\mathbf{u}$ the fluid velocity, p the pressure, ν the kinematic viscosity and $\mathbf{f}$ is the force per unit of mass due to external loads. Equations (1) and (2) are usually completed with the impermeability and the no-slip boundary conditions at solid walls that, taken together, can be written as

$$\mathbf{u} = 0 \quad \mathbf{x} \in D, \tag{3}$$

where D is the solid-liquid interface. The absence of slip at a rigid wall is largely confirmed by direct observations at the macroscale and there are only few well documented cases (see Lauga et al. (2005)) where the use of the no-slip boundary condition (3) at solid walls does not reproduce the correct fluid behavior at macroscale.

Since fluids are composed by molecules it is, in principle, always possible to investigate the flow of a fluid in a nanoconfined system by simulating the motion of each single atom via all-atom molecular dynamics (MD) simulations. This approach, that in general is inefficient and, more often, inapplicable to usual fluid dynamics problems, is crucial for nanofluidics for two main reasons i) it is not based on continuum assumption - that often fails in nanoconfined geometries - and ii) it does not require assumptions on boundary conditions at the interfaces. In recent years MD proved to be a powerful tool for the analysis of several nanofluidics problems such as, among others, the meniscus and contact line dynamics (De Coninck & Blake, 2008; Gentner et al., 2003), the role of precursor film in wetting (Chibbaro et al., 2008), the thermal exchange properties of carbon nanotube (Chiavazzo & Asinari, 2011) and the interface dynamics of a two immiscible liquid system (Orlandini et al., 2011). Stimulated by the experimental results concerning the flow rate through carbon nanotubes that, for narrow channels (a few nanometers), exceed predictions from the no-slip Poiseuille flow by up to several orders of magnitude (Majumder et al., 2005), several authors used MD simulation to investigate the liquid transport through nanopores. Both Lennard-Jones (Cannon & Hess, 2010; Chinappi et al., 2006) and more realistic models, where water and pore are modeled with the state of the art of classical force fields (Falk et al., 2010; Hummer et al., 2001; Thomas et al., 2010) have been used. Interpretations of the observed flow rate enhancement in terms of viscosity changes in the depletion region close to the wall (Myers, 2011) and change of bulk and interface properties due to confinement (Thomas et al., 2010) have been proposed and, in part, tested via MD techniques. Another nanofluidics topic that has attracted the interest of researcher is the liquid slippage on solid walls, i.e. the presence of a finite velocity at the wall and, hence, the failure of no-slip boundary condition (3). As we will see later in more details slippage is associated to both chemical and geometrical features of the solid surface. MD was largely used to explore the role of surface hydrophobicity (Chinappi & Casciola, 2010; Huang et al., 2008), of surface roughness (Zhang et al., 2011), and of the shear rate influence (Niavarani & Priezjev, 2010; Priezjev et al., 2005) on liquid slippage.

Although the large amount of applications and the increasing computational power of calculation systems, for a large number of nanofluidics phenomena the typical time and length scales accessible to MD simulations are still far from application to realistic systems. This implies that Navier Stokes equation (2) has to be used for the description of nanofluidic phenomena. Two main issues arise when using Navier-Stokes equations for nanofluidic systems, i) is the continuum assumption reasonable at the scale of the system? and ii) which boundary condition has to be applied at the wall? In this chapter we discuss these two issues for the case of liquids in nano confined geometries providing examples of how to employ MD simulations for their analysis. Concerning the former issue, we set up a model system for the estimation of the mass flux through a pore due to an external forcing and compare the results with the hydrodynamic prediction obtained via dimensional analysis of Navier-Stokes equations (2). We show that for simple-liquids the threshold for the validity of the continuum assumption is of the order of five times the molecule dimension and that, below this scale, the hydrodynamic prediction underestimates the flow rate. The second issue is addressed by estimating the slippage for a smooth solid wall using different solid surfaces. In agreement with literature results (Huang et al., 2008) we find a relationship between surface wettability and slippage. The section ends with a discussion on the rough surface case and a comparison with experimental results for a specific system of technological interest consisting in water flowing on an hydrophobic coated surface.

The chapter is structured as follows. A first brief section concerns the continuum assumption and the sub-continuum behavior for a liquid flow through a nanopore. Then we introduce the concept of liquid slippage and we present a simulation set-up for MD characterization of slippage for both smooth and rough surfaces. The final section of the chapter is dedicated to a perspective on near future applications of MD to nanofluidic problems.

2. Continuum vs single-file motion

As we discussed in the introduction, the main causes for the failure of the prediction of the standard continuum model – eq. (1), (2) and (3) – at the nanoscale are the inappropriate boundary conditions and the failure of the continuum assumption. In each MD simulation aimed at reproducing a nanoscale flow these effects are concomitant and hence it is not easy to isolate the two contributions, and, in particular, it is not possible to clearly identify a threshold for the validity of the continuum assumption and to understand how non-continuum features affect the flow. Here we introduce a MD set-up that overcomes the problem strictly controlling the boundary condition at the solid wall. The comparison of MD results with continuum model prediction is then used to estimate a threshold for the validity of the continuum assumption. As a first step in this program we need to recall a hydrodynamic prediction for the flow rate through pores.

The motion of a liquid in a macroscale system is described by the incompressible Navier-Stokes equation (2). For microscale systems the non-linear convective term $(\mathbf{u} \cdot \nabla \mathbf{u})$ is negligible respect to the viscous term $\nu \nabla^2 \mathbf{u}$ and eq. (2) reduces to the Stokes equation

$$\frac{\partial \mathbf{u}}{\partial t} = -\frac{1}{\rho} \nabla p + \nu \nabla^2 \mathbf{u} + \mathbf{f}. \tag{4}$$

With a reference length l_0, to be specified later, a reference diffusive time, $t_0 = l_0^2/\nu$, speed $u_0 = l_0/t_0 = \nu/l_0$ and pressure $p_0 = \nu^2 \rho / l_0^2$, the typical reference force for unit of mass is $f_0 = \nu^2/l_0^3$. As a result, the dimensionless parameter

$$f^* = f/f_0, \tag{5}$$

where $f = |\mathbf{f}|$, is a natural measure of the external load. With the above choices, the dimensionless form of eq. (4) reads

$$\frac{\partial \mathbf{u}^*}{\partial t^*} = -\nabla p^* + \nabla^2 \mathbf{u}^* + f^* \hat{\mathbf{f}}, \tag{6}$$

where $\hat{\mathbf{f}} = \mathbf{f}/f$ and stars indicate dimensionless units. Since we are interested in molecular flow across a pore, it is natural to identify l_0 with a length that characterizes the pore diameter and that we will indicate as effective diameter d_e. Let us use the previous formalism to obtain a hydrodynamic prediction for the mass flux through a pore. The mass flux across a surface of area S is given by

$$\Phi = \int_S \rho \mathbf{u} d\mathbf{S} = \rho l_0 \nu \Phi^* \equiv \Phi_0 \Phi^*, \tag{7}$$

with $\Phi^* = \int_{S^*} \mathbf{u}^* d\mathbf{S}^*$ the dimensionless flux. Since Equation (6) is linear, u^* (and hence Φ^*) is proportional to f^* and the scaling law

$$\Phi \propto \frac{\rho f l_0^4}{\nu} \tag{8}$$

for the mass flux is found. Equation (8) is the well-known power-four law for the mass flux of a viscous fluid in a pipe, that, in the particular case of an infinite pipe of diameter d_e with no slip boundary condition at the wall, results in the Hagen-Poiseuille expression $\phi = \pi P'(d_e/2)^4/(8\nu)$ with P' the pressure gradient. It is crucial to note that this dimensional argument is valid only if no other characteristic length scales appear in the problem and, in particular, in the boundary condition. This happens, for instance, in the two cases of no-slip (the velocity at the wall is zero) and no stress (the stress at the wall is zero) boundary conditions but not for partial slip condition where a new characteristic length (the slip length L_S, see section 3 below) is present. Equation (8) is the continuum model prediction we will compare to MD simulation results.

2.1 System set-up

The molecular dynamics set-up presented here is similar to the one presented in Chinappi et al. (2008). For the sake of the clarity we report here the main features, while the interested reader could find the details in the above cited paper. We consider a cylindrical nanopore of height h and circular section of radius r. The nanopore connects two cylindrical reservoirs of radius R (see panel a of Fig. 1). A periodic boundary condition is applied in z-direction, the box length being L_z. The liquid molecules are monoatomic and interact via a standard Lennard-Jones (LJ) potential $V_{LJ}(r) = 4\epsilon[(\sigma/r)^{12} - (\sigma/r)^6]$ truncated at distance $r_{cut} = 3.1\sigma$. In the rest of the section we will use LJ units. In all the simulations the density of the liquid is $\rho = 0.83$ and the temperature is $\theta = 0.725$. The solid wall is modeled as a continuum that occupies a volume S_w and the interaction between the wall and a fluid atom located at $\mathbf{r}$ is given by $V_w(\mathbf{r}) = \int_{S_w} n_w f(|\mathbf{r} - \mathbf{r}_w|) d\mathbf{r}_w$, where n_w is a suitable density having dimension of inverse volume, S_w is the wall domain and $f(r)$ is a LJ potential truncated at its minimum. A slab of width h endowed with the cylindrical pore separates the two reservoirs. The aspect ratio of the domain and the wall is kept constant in all simulations, namely, $L_z/r = 10$, $R/r = 4$, $h/r = 1$. The origin of reference system is set at the center of the pore with the z-coordinate running along its axis. A sketch of the simulation box is reported in panel a of Fig. 1. Due to the steepness of the wall-liquid potential $V_w(\mathbf{r})$ the isosurface $V_w = k_b\theta$ is a natural candidate for the boundary of the volume accessible by liquid atoms. Hence, we define the effective diameter d_e as the diameter of the narrow part of the pore delimited by the isosurface $V_w = k_b\theta$. At this virtual interface, no tangential forces are exerted on molecules. As a consequence of that in the hydrodynamic description this virtual interface corresponds to a free-slip (no stress) impermeable boundary. The flux across the pore is induced by a homogeneous forcing $\mathbf{f}$ acting on each liquid atom along the pore axis direction. The coupling to the heat-bath in the non-equilibrium simulations is achieved via a Berendsen thermostat (Berendsen et al., 1984) and preliminary tests provided confidence on the small sensitiveness of the presented results to changes in thermostat's characteristic time constant. All the simulations were performed with a molecular dynamics code obtained on the basis of the open-source code DL_PROTEIN-2.1 (Melchionna & Cozzini, 2001).

Simulations were performed for different values of the effective pore diameter d_e with the number of atoms ranging from $N = 435$ for $d_e = 1.83$ to 31680 for the largest one at $d_e = 9.25$. Following Zhu et al. (2004), the mass flux is expressed by means of the collective variable n, defined in differential form by its increment in the time interval dt:

$$dn(t) = \sum_i \frac{dz_i}{h} \tag{9}$$

where dz_i is the displacement of the $i-th$ molecule in a time step dt, and the sum runs over all the molecules within the channel at time t, i.e. $-h/2 < z_i(t) < h/2$. Hence each molecule crossing the channel from left/right to right/left is associated to an increase/decrease $n \to n \pm 1$. The integer part of $n(t + \Delta t) - n(t)$ measures the number of molecules which cross the channel from left-to-right, minus the ones which cross the channel from right-to-left, in a time interval Δt. In statistically stationary conditions, the flux of molecules is defined as $\Phi = \langle n(t + \Delta t) - n(t) \rangle / \Delta t$, where the average is taken by sampling the system in time. For a given pore diameter d_e simulation were performed at different forcing intensities f in order to verify that the system is in the linear response regime, i.e. $\Phi \propto f$. In the following only linear response results are considered.

2.2 Flow through a cylindrical nanopore

In order to test the validity of the continuum prediction eq. (8) we plot in panel d of Fig. 1 the quantity $\Phi d_e^{-4}/f$ as a function of the effective diameter d_e. It is apparent that for high d_e the curve tends to a constant value as expected from equation (8) that predicts a power four scaling law of the flux Φ with the diameter d_e. This hydrodynamic behavior sets in at an effective diameter in between 5 and 6 van der Waals radii. This result is coherent with existing literature (see among others Koplik & Banavar (1995) and Bocquet & Charlaix (2009)) indicating that continuum approach for simple liquids is valid on length scales one order of magnitude larger than molecule size. Snapshots of the simulated systems for $d_e = 1.83$ and $d_e = 9.25$ are provided in Fig. 1b, where typical configurations of the molecules are reported for equilibrium (no forcing) simulations. The molecules are represented as van der Waals spheres. Those inside the pore are drawn in actual size, while the radii of those outside the pore have been arbitrarily reduced for clarity.

Decreasing d_e the flux is larger than the hydrodynamic prediction and it appears to scale roughly as d_e^3. Observing the snapshot of typical configurations of the molecules for low d_e we see that in the extreme case ($d_e = 1.83$ lower panel of Fig. 1b) only a single atom can occupy the pore section. This evidence suggests that an appropriate framework for such a quasi-1D motion is the so called single-file motion, namely a sequential motion of concomitant molecules along a line, with no possibility of overtaking. Being the molecules densely packed within the nanopore, each time a molecule enters the inlet mouth, a molecule is kicked-out from the outlet (see Fig. 1c). The many-body aspects of single-file motion can be described by a single parameter, the hopping-rate (Berezhkovskii & Hummer, 2002) k, that is the inverse of the characteristic time at which molecules hop inside the inlet mouth of the nanopore, with a sufficient energy to move all the molecules inside the pore, so that the last one is expelled. As suggested by Zhu et al. (2004) in describing the flux of water through a carbon nanotube, the relevant parameter is the potential jump associated with the passage of a particle through the pore, $\Delta\mu = fL_z$. At low forcing intensity, the flux is linear in the potential jump

$$\Phi = -k\frac{\Delta\mu}{k_b\theta} \tag{10}$$

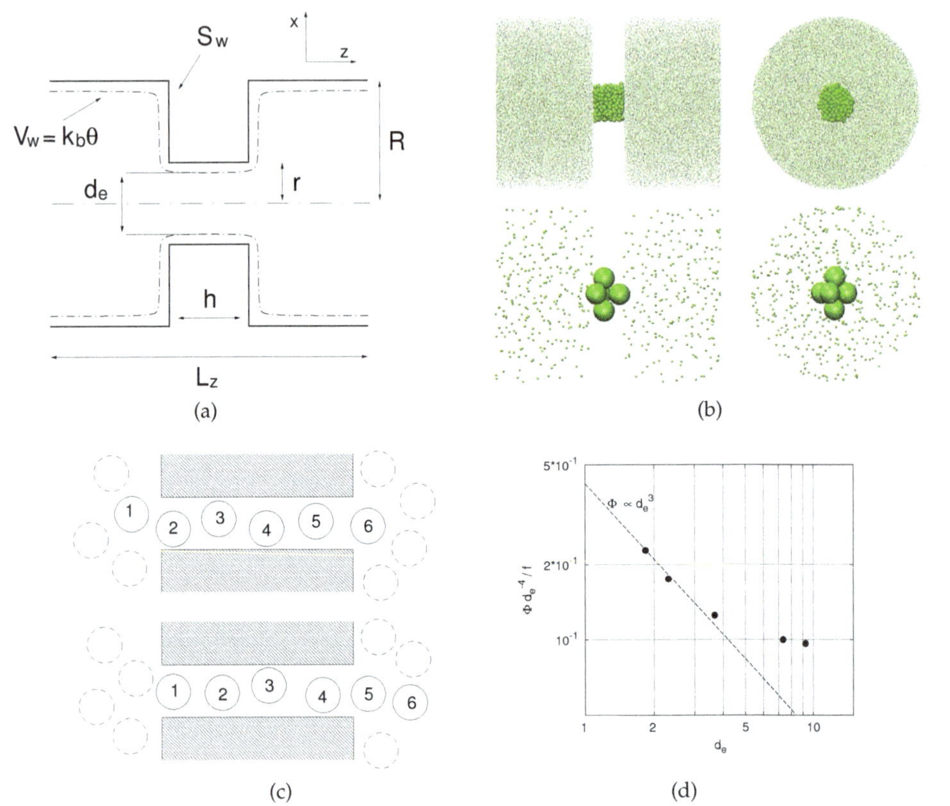

Fig. 1. a) Sketch of the system geometry. The isosurface $V_w(\mathbf{r}) = k_b\theta$ (dot-dashed line) represents a natural choice for the boundary of the volume available to the liquid atoms, hence, the effective diameter d_e is defined as the diameter of the narrow part of the pore delimited by $V_w = k_b\theta$. At this virtual interface, no tangential forces are exerted on molecules. b) Snapshots of equilibrium simulations having two different pore sizes: namely, lower panel $d_e = 1.83$, upper panel $d_e = 9.25$. For each panel the image on the left is a projection on zy plane, while the image on the right is the projection on the xy plane. For the sake of clarity particles inside the pores (i.e. $0.5h < z < 0.5h$) are drawn as spheres of diameter σ while the other particles as spheres of diameter 0.1σ. The image is realized using the VMD software (Humphrey et al., 1996). c) Schematic representation of single-file motion. When a particle enters the pore (e.g. particle 1 of the upper panel), the last particle in the pore exits from the opposite side (particle 5 in lower panel). d) Particle flux Φ divided by the forcing intensity f and the hydrodynamic scaling factor d_e^4. The dashed line represents the single-file scaling $\Phi \propto d_e^3$.

where k_b is the Boltzmann constant. In stochastic models for single-file transport (Berezhkovskii & Hummer, 2002), the hopping rate k is a phenomenological constant of the model. Roughly speaking, for slight deviations from equilibrium, the hopping rate k should depend on fluid density and temperature and, of course, on the cross section of the pore. Hence, at a given thermodynamic state, it is reasonable to argue that the hopping rate is proportional to the pore cross section, i.e. $k \propto d_e^2$. Since, given $\mathbf{f}$, $\Delta\mu$ scales with the longitudinal dimension of the pore, the above expression indicates that the molecular flux in the single-file regime scales as d_e^3, a result in agreement with the molecular dynamics finding.

The presented results could be qualitatively compared with recent MD results on mass flux through carbon nanotubes. Indeed stimulated by experimental studies of water flowing through carbon nanotubes (reporting flow rates exceeding the predictions given by the no-slip Hagen-Poiseuille flow by orders of magnitude (Holt et al., 2006; Majumder et al., 2005; Whitby et al., 2008)) several research groups studied the water flow in nanotubes via MD simulations. Falk et al. (2010) analyzed the water flux inside carbon nanotube (CNT) of different sizes and found that there is a transition in the friction coefficient when the pore diameter is smaller than a couple of nanometers. In particular they found that the water structure is affected by confinement for CNTs of radius below $1.6nm$, i.e. ~ 5 times larger that the size of water molecules, and that, below this regime, single-file motion sets in and the friction coefficient vanishes. In the system analyzed by Falk et al. (2010) both boundary (slippage) and confinement effects are present and hence their results could not be quantitatively compared to our case. However it is apparent that the threshold for the separation between the two regimes is in both cases ~ 5 times the molecule dimension. A similar threshold was found in Thomas et al. (2010) where the authors show that a continuum model is able to reproduce the MD results if slip length and water viscosity changes with the pore size. In particular, the flow enhancement for narrow pores is interpreted in the continuum model as a decrease of the viscosity and an increase of the slippage at the boundary. In this respect it is interesting to point out that recently in Myers (2011) a continuum model based on reduced viscosity in the depletion region (i.e. the viscosity is not uniform and it decreases close to the wall) and no-slip boundary was proposed. This model correctly predicts the flow enhancement in narrow nanotubes, the enhancement factor being analogous of the one obtained using a finite slip-length. The presence in literature of different theoretical models able to interpret the same phenomena remarks the role of MD as a powerful tool for nanofluidic research since proper MD set-ups allow to isolate the different causes that contribute to the observed behaviour.

3. Liquid slippage on solid walls

In the continuum framework, liquid slippage at the wall is described in terms of the Navier boundary condition, which, in the case of parallel flow over a non-moving planar wall, reads

$$v_w = L_s \frac{dv}{dz}, \tag{11}$$

where $v(z)$ is the velocity profile with z the wall-normal coordinate and v_w is the slip velocity, i.e. the value the velocity profile attains at the wall. The parameter L_s is called the slip length. It is geometrically interpreted as the distance below the wall where the extrapolated fluid velocity vanishes, see Fig. 2. Wall slippage was observed for both smooth and patterned surfaces by several groups, for a review see, among others, Lauga et al. (2005). Slippage is

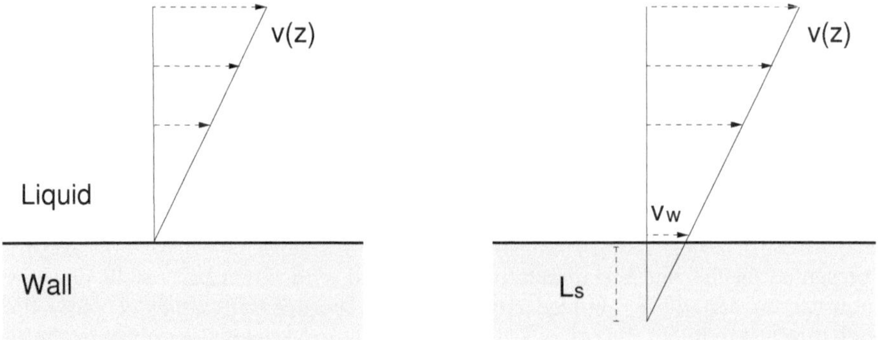

Fig. 2. Slippage on a planar wall. Left, no-slip surface: the velocity at the wall is zero. Right: partial slip surface, Navier boundary condition: the velocity at the wall v_w is proportional to the velocity gradient $dv(z)/dz$ with z directed as the wall normal toward the liquid (eq. (11)). The constant of proportionality is called the slip length (L_s).

typically classified into two broad classes: intrinsic slip, sometimes called molecular slip, and apparent slip. In the first case (intrinsic slip) a non-vanishing slip velocity at the smooth wall results from the first few layers of liquid molecules sliding on the solid surface. In the second case the slip appears at scales intermediate between the characteristic size of the system and the molecular scale. A typical case of apparent slippage is the presence of gas nano-bubbles trapped in the surface roughness elements (center panel of Fig. 5 showing the so called Cassie state), as often happens both for regularly patterned (Gogte et al., 2005) and randomly rough surfaces (Govardhan et al., 2009). The capability of a surface to trap gas bubbles is often associated to the so called superhydrophobic condition where the surfaces are characterized by a low wettability (large contact angle of a water sessile drop together with a low contact angle hysteresis) and slippage (Rothstein, 2010).

In the next section we present some results concerning slippage on smooth walls and discuss the relationship between slippage and wettability. The last section is dedicated to an example of rough surface with nanoscopic defects.

3.1 Intrinsic slippage on smooth surfaces

On smooth walls the mechanism that is responsible for liquid slippage is the so called intrinsic or molecular slippage where the first few layers of liquid molecules slide on the solid surface. Intrinsic slippage has been largely studied via MD simulations. The general picture emerging is that wettability and slippage are deeply related, in particular the larger the contact angle θ the larger the slip length L_s. In this respect the data presented by Huang et al. (2008) support the existence of a quasi-universal relationship between contact angle θ and slip length L_s. However a recent research (Ho et al., 2011) reports MD simulation results that clearly indicate that water slippage could occur also for hydrophilic surfaces suggesting that the connection between θ and L_s is purely coincidental. Here we use two simple simulation set-ups to measure the contact angle and the slip length on solid surface with different degree of hydrophobicity and discuss our results in the framework of the ongoing debate on the role of wettability in liquid slippage.

	c_{SL}	θ
A	1	30°
B	0.7	95°
C	0.5	135°
D	0.3	160°
E	0.1	non-wetting

Table 1. Summary of the contact angle measurement simulations. The second column reports the value of the parameter c_{SL} that rules the attractive part of the modified Lennard-Jones potential (12) for the solid-liquid interaction ($c_{SL} = 1$ for standard LJ potential, $c_{SL} = 0$ for repulsive potential).

3.1.1 Contact angle measurement

The system we consider is formed by a solid Lennard-Jones (LJ) wall and by a LJ liquid droplet (Fig. 4a). Each atom interacts with the others via a modified LJ potential

$$V(r_{ij}) = 4\epsilon_{ij}\left[\left(\frac{\sigma_{ij}}{r_{ij}}\right)^{12} - c_{ij}\left(\frac{\sigma_{ij}}{r_{ij}}\right)^{6}\right] \tag{12}$$

with c_{ij} the parameter used to change the affinity between atoms, indeed $c_{ij} = 0$ corresponds to a completely repulsive interaction while for $c_{ij} = 1$ the usual attractive tail of LJ potential is recovered. For liquid-liquid interaction we used standard LJ potential, i.e. $\epsilon_{LL} = 1, \sigma_{LL} = 1$ and $c_{LL} = 1$. The solid is more self attracting than the liquid, $\epsilon_{SS} = 10$, $\sigma_{SS} = 1$ and $c_{SS} = 1$ moreover solid atoms are constrained to a face centered cubic (FCC) lattice by a harmonic spring. Concerning solid-liquid interaction we have $\epsilon_{SL} = 1, \sigma_{SL} = 1$ while c_{SL} is varied from 0.1 to 1. In the initial configuration the solid atoms forms a FCC slab of dimension L_x, L_y and h in x,y, and z direction respectively, while the liquid atoms are arranged as a spherical cut, the center of the sphere being at a distance b from the last layer of solid atoms. Periodic boundary conditions are applied in the three directions being L_x, L_y and L_z the dimensions of the periodic box. Both solid and liquid atoms' initial positions are on a FCC lattice of density $\rho_w = \rho_l = 0.83$. Initial velocities are assigned to give to the system an initial kinetic energy corresponding to a temperature $T = 0.75$. During the equilibration phase a thermostat is applied to both wall and liquid atoms. The system temperature $T = 0.75$ (smaller than LJ fluid critical temperature) and the volume available to fluid atoms correspond to a two phase liquid-vapor system on the LJ phase diagram (Hansen & McDonald, 2006). During equilibration the initial FCC lattice structure used to assign the initial position to the liquid atoms disappears and some atoms leave the droplet surface and enter in the vapor phase. Moreover the droplet rearranges until reaching a steady state where the contact angle θ does not change. The system is considered at equilibrium when the time evolution of the number of fluid particles in vapor phase and the contact angle do not appreciably change. At this point the thermostat is turned off and a NVE equilibrium simulation is run. In all the cases analyzed we do not noticed further changes of the droplet contact angle during the NVE run. Fig. 3 reports snapshots of the equilibration phase of a droplet from initial configuration to equilibrium state for the cases $c_{SL} = 0.3$ and $c_{SL} = 0.7$, starting from an initial configuration where the center of the sphere (liquid phase) is located at a distance $b = 2$ from the surface. It is apparent that the equilibrium contact angle is strongly affected by the liquid-solid attraction parameter c_{SL} resulting in an almost complete dewetting condition for $c_{SL} = 0.3$ (panels a,c and e of Fig. 3) to a weak hydrophobicity for $c_{SL} = 0.7$ (panels b,d and f of Fig. 3). We observe (data not shown) that the final state is independent from the initial position of the droplet and in particular from the value of b that rules the contact angle for the initial configuration. However the time needed to reach equilibrium dramatically increase if the initial guess of contact angle is far from the equilibrium value, this effect is particularly relevant if the initial contact angle is much larger than the equilibrium one. Equilibrium contact angle is estimated as in Werder et al. (2003) and Chinappi & Casciola (2010), i.e. calculating the density profile, defining the droplet surface as the set of points for which ρ is the half of the liquid density inside the droplet and fitting the surface points to a sphere. It is known that at nanometric scale, the contact angle of a drop may significantly differ from its macroscopic value due to line-tension. For instance in the case of a liquid water droplet on a hydrophobically coated surface (Chinappi & Casciola, 2010), the equilibrium contact angle is $\theta \simeq 120°$ for a droplet of radius $r \simeq 34$Å and $\theta \simeq 112°$ for a droplet of radius $r \simeq 60$Å leading to a macroscopic contact angle $\theta_{macro} \simeq 105°$, obtained fitting the modified Young equation, $\cos\theta = \cos\theta_{macro} - \tau/(r_b \gamma_{lv})$ with τ the line tension at the three-phase line, r_b its curvature radius and γ_{lv} the surface tension at the liquid-vapor interface. For the purpose of the present section, since we are interested in the correlation between contact angle θ and slippage and not in the measurement the precise values of θ, we do not consider this systematic correction due to the line tension τ. In Table 1 the value of c_{SL} for the cases considered in this chapter and the measured contact angles θ are reported. It is apparent that θ increases when the attraction between solid and liquid decreases and, in particular, for $c_{SL} = 0.1$ (case E) the thermal agitation is able to detach the drop from the solid wall.

3.1.2 The Couette flow MD set-up and the slip length measurement

A usual way to measure intrinsic slippage by MD simulation is to induce a shear in the fluid and to estimate the slip length L_s from the extrapolation of the bulk velocity profile. To this purpose the Couette flow is a natural choice since the shear is homogeneous and, hence, the bulk velocity profile in stationary state is linear. In order to induce a Couette flow we prepared a system formed by a channel where a liquid is confined by two solid walls. Periodic boundary conditions are implemented in wall parallel directions (x and y) being L_x and L_y the box dimensions. As in the case of droplet simulations the lower wall is constrained in a FCC lattice by harmonic potential. The upper wall is constrained only in z direction and a constant force is applied to each wall atom in the x direction resulting, after a transient, in a stationary Couette flow. The atoms of the lower wall are coupled to a thermostat in order to dissipate the heat produced by viscous drag in the liquid.

3.1.3 Results, effect of liquid-solid interaction

In panel b of Fig.4 the intrinsic slip length L_s is plotted as a function of the equilibrium contact angle θ. It is apparent that the for the hydrophilic surface there is no slippage. Moreover the slippage is found to increase with the contact angle. The figure reports also the data from a previous study on water slippage on hydrophobic coatings (Chinappi & Casciola, 2010). These results qualitatively confirm the picture of Huang et al. (2008) that propose a quasi universal relationship between the contact angle θ and intrinsic slip length L_s. In particular we do not observe slippage for hydrophilic walls. As we pointed out in section 3.1 a recent research (Ho et al., 2011) reports a positive slip length also for hydrophilic surfaces. In that paper the slippage is observed for hydrophilic surface only in the case the wall lattice spacing is

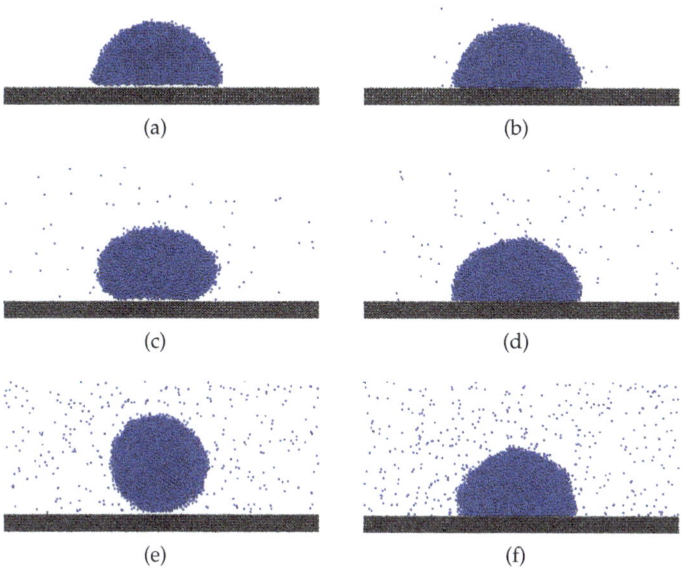

Fig. 3. Time evolution of a droplet from initial condition to steady state for a highly hydrophobic case – $c_{SL} = 0.3$ (panels a,c,e) – and a slightly hydrophobic one, $c_{SL} = 0.7$ (panels b,d,f).

significantly smaller that liquid molecule size. Here, as in Huang et al. (2008), the dimension of the liquid molecules is similar to solid lattice spacing, hence it is not surprising that our results confirm the picture that associates hydrophobicity and slippage. On the other hand this fact suggests a natural way to further investigate the issue. Indeed it is easy to systematically vary the solid lattice dimension and repeat both contact angle and slippage measurements. Such analysis, performed with a minimal model such as the LJ system, could reveal if the slippage on hydrophilic surfaces is a general phenomenon to be ascribed only to the ratio between solid and liquid molecule sizes or is due to specific choices for the force field implemented in Ho et al. (2011).

3.2 Slippage on rough surfaces: the example of OTS coatings

In order to address the role of surface nanoscale defects on slippage, we consider the case of an Octedecyltrichlorosilane (OTS) coated surface. The OTS molecule is formed by a linear alkyl of 17 carbon atoms with a methyl group on one end and a $SiCl_3$ (silane) group on the other end. OTS molecules are able to spontaneously form layers (sometimes a monolayer) where the molecules are assembled in hexagonal cells with the silane group attached to the solid surface and the methyl group exposed toward the fluid. Due to their ability to form compact layers exposing the methyl (hydrophobic) group, OTS coatings represent a promising technology for surface functionalization. Several groups quantified the slip length L_s of liquid water on smooth OTS coated surfaces (Bouzigues et al., 2008; Cottin-Bizonne et al., 2008; Li & Yoda,

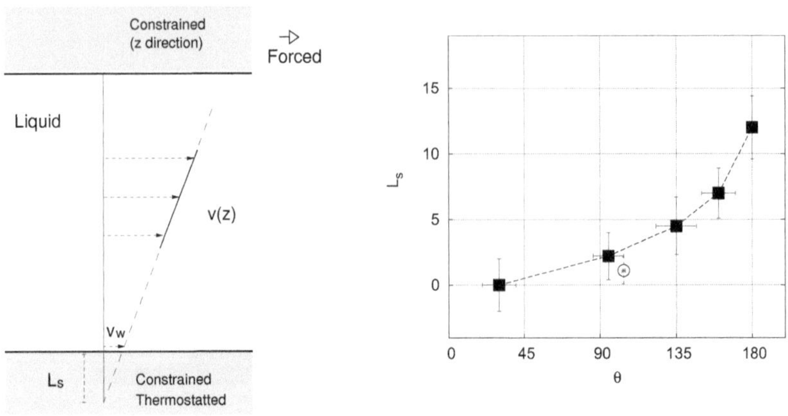

Fig. 4. a) Sketch of the MD set-up for slip length measurement. The lower wall is constrained and thermalized while a constant forcing is applied to the upper wall in wall parallel direction, moreover the z position of the upper wall is constrained. Periodic boundary conditions are applied in x and y. b) Contact angle θ vs slip length L_s for the simulations described in the section 3.1 (filled squares). The empty circle corresponds to a result of a previous study on wettability and slippage of an Octadecyltichlorosilane (OTS) coated surface Chinappi & Casciola (2010). In order to map this result in LJ units we use the value 3.3Å as van der Waals radius of the water molecule.

2010). Here our aim is to assess the effect of nanoscale roughness of the coating on the water slippage on a OTS layer.

3.2.1 Defected OTS coating: MD set-up

The typical defect considered here consists of a hole of diameter D. In the greater part of the simulations the defect is obtained removing the OTS molecules in a circle of diameter D hence exposing the LJ (hydrophilic) uncoated surface. For the largest system considered here, for reason that will be clear later, the defect is obtained in a slightly different way; using alkyl molecules of different lengths, namely 11 carbon atoms alkyl chain for the defect and 29 carbon atoms alkyl chain for other molecules. The two different length chains are represented as dark and light blue molecules in the left panel of Fig. 6. Concerning the water molecules we used the TIP3P model (Jorgensen et al., 1983). This model fails to reproduce some of the water properties, in particular the viscosity (one third of the actual water viscosity) and the surface tension (slightly smaller). More accurate models exist such as the TIP4P/2005 (Abascal & Vega, 2005; Alejandre & Chapela, 2010) but they are more computationally demanding and, for the specific case of slippage on smooth OTS surfaces, they did not lead to results qualitatively different from the TIP3P model (Chinappi et al., 2011). A summary of the simulations, performed with the NAMD software (Phillips et al., 2005), is reported in Table 2. The box dimensions L_x and L_y are reported in columns 2 and 3 while column 4 reports the diameter of the defect. Equilibration was performed at 300 K and 1 atm. Systems were equilibrated following the same procedure described in Chinappi & Casciola (2010) for water slippage on OTS ideal (not defected) coatings and here briefly reported for

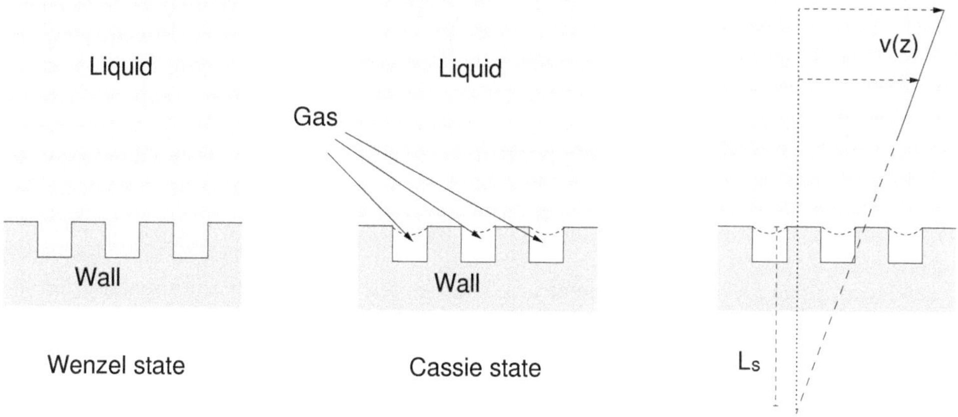

Fig. 5. Wenzel and Cassie states. In Wenzel state (left) the liquid fills all the asperities and wets the whole surface, while in the Cassie state (center) gas (or vapor) pockets are trapped in the surface grooves resulting in a patterned interface (liquid-solid zones alternated with liquid-gas zones). The right panel illustrates the concept of apparent slip on a Cassie state (superhydrophobic) surface. The negligible stresses at the liquid-vapor interface results in a bulk velocity profile that could be reproduced by a continuum model with a Navier boundary condition with a uniform slip length L_s on an effective flat surface.

the sake of clarity. The wall is modeled as a Lennard-Jones (LJ) solid with the parameters σ and ϵ selected to have a melting point at 1 atm well above the simulation temperature. The solid-liquid interface is parallel to the xy plane and corresponds to a 111 plane of the LJ FCC structure. The alkyl chain head group binds the wall and it is treated in hybrid manner, as a LJ atoms of the wall as concerning no-bonded interactions and as a carbon atom of the alkyl chains for the interactions with bonded atoms along the same chain. The equilibration phase was performed in a triply-periodic box where, by periodicity, the coated wall forms a unique bunch of solid with the upper wall. During equilibration a Langevin piston is applied in the wall-normal direction (z) in order to relax the system to the desired thermodynamic state. After equilibration the two walls were separated by inserting a void region (larger than the LJ cut-off radius of 12Å) before starting the Couette simulation. Concerning the coated wall, the atoms of its lower plane were keep fixed, while the other LJ atoms were coupled to a Langevin thermostat. Concerning the uncoated wall, the atoms of its upper plane were constrained in the wall-normal direction by a harmonic spring and, as in the case of smooth walls described in the previous section, a constant force parallel to the solid-liquid interface is applied to all the upper wall atoms.

3.2.2 Results

Water molecules did not fill the hole during the equilibration phase for all the cases we simulated with the exception of the largest hole we considered (case E in Table 2) where in the first stage of the equilibration process a great number of water molecules enter the defects and get trapped at the LJ hydrophilic surface resulting in a complete wetting (Wenzel) state (panel a of Fig. 5). In order to avoid this effect we introduced the slightly different system we discussed in the previous section where the defect is obtained by using alkyl chain of different lengths resulting in a hole where also the bottom surface is hydrophobic (see the left panel of Fig. 6). This system shows a very stable Cassie state. For all the simulations the slip length L_s is measured from the velocity profile (see right panel of Fig. 6). For the five cases considered L_s is reported in the sixth column of the Table 2. It is apparent that L_s increases with the hole diameter and with the system size. Moreover L_s is larger with respect to the smooth case ($L_s \simeq 5 - 7$Å (Chinappi & Casciola, 2010)) as expected from the vanishing friction at the liquid-vapor interface.

Following the definition introduced in section 3 the observed slippage has to be classified as apparent slippage. In this context it is interesting to compare our MD data with continuum model results that are available in literature. The simplest way to study the apparent slippage with a continuum model is to consider a patterned surfaces where the interface is composed by solid areas where the local slippage either vanishes or, alternatively, conforms to the small intrinsic slip at solid-liquid interface (L_{in}) and by gaseous areas (corresponding to the nanobubbles) where no shear stress is acting on the liquid phase (Ng & Wang, 2010; Ybert et al., 2007). Defining the solid fraction Φ_s as the ratio between the area of the liquid-solid interface and the projected area, i.e., for the present case $\Phi_s = 1 - \pi D^2/(4L_xL_y)$ reported in column 5 of Table 2), Ng & Wang (2010) found that, if no intrinsic slip is assumed ($L_{in} = 0$ at the solid-liquid interface), a continuum model based on Stokes equation leads to the relation

$$\frac{L_{s,0}}{A} = -0.134\ln(\Phi_s) - 0.023 \tag{13}$$

where A is the cell side length of a squared periodic lattice. The suffix "0" in expression (13) indicates that no-slip ($L_{in} = 0$) is assumed at the liquid-solid portion of the interface. For a partially slipping solid surface ($L_{in} > 0$) embedding the free-slip hole, the apparent slip length increases in proportion to the intrinsic slip length L_{in} and in inverse proportion to the solid fraction Φ_s (Ybert et al., 2007),

$$\Delta L_s = L_s - L_{s,0} \simeq \frac{L_{in}}{\Phi_s}, \tag{14}$$

with order one proportionality constant. Eqs. (13) and (14) allow to predict the apparent slip length for the case considered in our MD simulations, the prediction is reported in column 6 of Table 2 for three values of the intrinsic slip length, namely $L_{in} = 0, 5, 10$Å. The comparison between MD and continuum model indicates that the continuum description of the apparent slip on patterned surface is valid also at the nanoscale. Moreover the value of intrinsic slippage for L_{in} that have to be used in equation (14) to obtain a quantitative agreement is in between 5 and 10Å that is consistent with the intrinsic slip measured for smooth OTS coated surfaces (Chinappi & Casciola, 2010). A similar quantitative agreement was evidenced by the comparison of MD results for LJ fluid slippage on superhydrophobic surface by Cottin-Bizonne et al. (2003) and lattice Boltzmann simulation by Benzi et al. (2006). As in the case of simple liquid analyzed in section 2 the assessment of the capability of a continuum model to reproduce (or not) a nanoscale fluid dynamics system is a precious contribution that MD is able to provide to nanofluidic research.

Moreover the presented results concerning apparent slippage on defected OTS coatings and, in particular, the stability of the superhydrophobic (Cassie) state we observed for all the considered systems, suggest a possible explanation for an interesting issue pointed out by a careful analysis of both experimental and numerical results on water slippage on hydrophobic surfaces presented by Bocquet & Charlaix (2009). The issue is the following: while for

	L_y (Å)	L_x (Å)	D (Å)	Φ_s	L_s MD (Å)	L_s(Å) (Ng & Wang, 2010)
A	26.19	30.24	20	0.603	12.7	1.3, 9.5, 17.8
B	34.92	40.32	20	0.777	10.3	0.4, 6.8, 13.3
C	43.65	50.04	20	0.856	9.2	−−
D	34.92	40.32	30	0.497	22.4	2.6, 12.7, 22.8
E	122.22	141.12	100	0.544	24.8	7.7, 16.9, 26.1

Table 2. Summary of the molecular dynamics simulations of the Couette flow on defected OTS-SAM coatings. Columns 2 and 3 report the periodic cell dimensions, L_x and L_y respectively, column 4 the diameter D of the circular defect and column 5 the solid fraction $\Phi_s = 1 - \pi D^2/(4 L_x L_y)$. The apparent slip length L_s obtained from MD simulations is in the 6th column. The last column provides the value obtained combining expression 13 and 14 where $A = (L_x L_y)^{0.5}$ is used as characteristic length. The three values correspond to intrinsic slip on solid-liquid interface $L_{in} = 0$ (no-slip at solid-liquid interface), and $L_{in} = 5, 10$ Å. The symbol $--$ for case C with no slip condition on solid surface is due to the fact that, as pointed out by Ng & Wang (2010) expression (13) provides a good fit of their numerical data only in the range of $\Phi_s \in (0.22, 0.75)$.

smooth surfaces an increase in slip length with hydrophobicity was found for both MD and experiments, for a given contact angle θ the value of the experimental estimated L_s is larger than MD results of about one order of magnitude. In particular in the case of the OTS coatings the most credited experimental data (obtained with different techniques (Bouzigues et al., 2008; Cottin-Bizonne et al., 2008; Li & Yoda, 2010)) indicates a slip length of $\sim 20nm$ while the MD value is in the range $0.5 - 1.5nm$. Since the MD simulations were performed by different authors (Chinappi & Casciola, 2010; Chinappi et al., 2011; Huang et al., 2008; Sendner et al., 2009) using different force fields, computational codes and numerical set-ups, this discrepancy could hardly be ascribed to modeling inaccuracies affecting the MD simulations. Hence we turn our attention on the putative presence of a non-negligible amount of wall roughness in the *smooth* surface analyzed in the cited experiments. Indeed several studies on the structure of OTS coatings suggest that perfectly smooth coatings as those analyzed in ideal MD simulations do not exist in practice. Cottin-Bizonne et al. (2008) and Joseph & Tabeling (2005) indicate the peak to peak distance between asperities in their OTS coatings in a few nanometers. Additionally, spectroscopic variable angle ellipsometry, neutron reflection and atomic force microscopy by Styrkas et al. (1999) suggest that OTS films often consist of nearby multilayer domains of different thickness, ranging from 1 to 3 and even 4 times the monolayers thickness. Our conjecture is that apparent slip effects occurring on nanopatterned surfaces can, in principle, explain the difference observed between MD simulations and experiments. By being able to trap nanobubbles, such defects locally induce slippage and increase the apparent slip length. As we show, the resulting apparent L_s can be estimated with continuum model and hence, a value for the typical defect size needed to obtain the observed slippage can be estimated from equation (13) and (14). In order to validate or reject the proposed scenario, the following issues need to be addressed with care: i) Presence of suitable nanoscale surface defects able to explain the experimentally measured L_S. ii) Stability of the superhydrophobic (Cassie) state for those nanoscale defects. The stable Cassie state we observed for all the discussed MD simulations provides a first step in the assessment of the second issue, however further analysis are needed. In particular the defect size needed to reconcile MD simulation with experiment is larger that the one analyzed here. Moreover

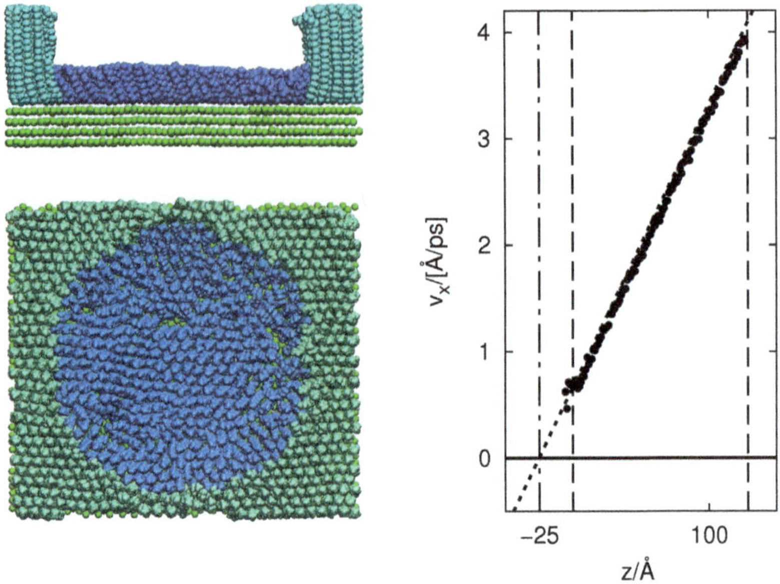

Fig. 6. Left. Snapshot of the defected OTS coated surface for simulation E of Table 2. The LJ (hydrophilic, in green) surface is coated by alkyl chains of different lengths, 11 carbon atoms (dark blue) in the circular defect and 29 carbon atoms chains otherwise. Right. Velocity profile obtained for the defect in the left panel, the observed slip length is ~ 25Å.

another crucial point to be clarified in the future is the robustness of the picture to changes in the defects geometry and the effect of meniscus curvature. All these issues could be analyzed with appropriate MD simulations.

4. Conclusion and perspectives

In this chapter we introduced the reader to some applications of classical all-atom molecular dynamics to nanofluidic problems and in particular to two main crucial issues i) the validity continuum assumption at nanoscale for simple liquids and the sub-continuum behavior and ii) the role of hydrophobicity and surface roughness on the liquid slippage at solid walls. For both problems dedicated MD set-ups have been prepared and some results were presented and compared with existing literature. Concerning the former issue we showed, in agreement with previous findings, that continuum model is appropriate for simple liquids when the characteristic size of the system is $5-6$ times larger than the liquid molecule dimension. Below this threshold the molecular flow across a pore appears to be best described by a single-file model, that predicts a more efficient power-three scaling law for the mass flux. This result, achieved with a minimal model (Lennard-Jones liquid) provides a further ingredient in the interpretation of the high flow rate measured by recent experiment for water flux through carbon nanotubes. Concerning the second issue we discussed an ongoing debate on the connection of hydrophobicity with slippage. Our results suggest that the picture proposed by Huang et al. (2008) is valid when the sizes of liquid and solid molecules are similar however further investigation are needed to assess if the slippage for hydrophilic surface recently observed by Ho et al. (2011) is a general feature that emerges when solid molecules are significantly smaller that liquid ones or it is associated to the specific choices the authors did for liquid and solid molecules. Concerning slippage we also presented results for the role of surface roughness discussed here for a specific system of technological relevance, namely, the OTS coatings, providing a further example of the usage of all-atom MD for nanofluidics research. MD not only allowed to measure the apparent slippage for the specific shape of the defect but, more interestingly, was here used to propose a different interpretation of the experimentally measured slippage for hydrophobic coatings on smooth surface. Future MD studies will provide (or not) further elements to the proposed conjecture eventually stimulating experimental validations.

Beside the results for the specific problems presented here, a secondary aim of the chapter was to use examples taken from ongoing research to give a picture of the capabilities (and the limits) of all-atom Molecular Dynamics simulation for nanofluidic applications. In general the use of MD for nanofluidics problems could divided in two broad classed. From one hand one could try to use the state of the art of the force fields to make a quantitative predictions for a specific issue. Examples are the water flow through carbon nanotubes cited in section 2 and the water slippage on a given surface (such as the OTS coating discussed in section 3.2). On the other hand MD could be used to perform in silico experiments on simple systems employing a minimal model for the interatomic interaction (e.g. Lennard-Jones liquids) aimed at isolating and analyzing new phenomena or to address specific questions such as the validity of continuum assumption 2, the role of hydrophobicity in wetting (section 3.1 presented here and many other issues discussed in the literature). Both usages will be crucial for nanofluidics research and the increase of computational power in the next years will, for sure, allow the researcher to tackle problems on length and time scales that are out of the range that can be presently explored via MD, such as macromolecules behavior in liquid flows (currently addressed via coarse grained model) and large multiphase systems.

5. Acknowledgment

We would like to thank Mr. Alberto Giacomello for useful discussion and help in LJ simulations of section 3.1 and Dr. Guido Bolognesi for useful discussions. Part of the simulations presented in section 3.2 were performed using computing resources made available by CASPUR under HPC Grant 2011.

6. References

Abascal, J. & Vega, C. (2005). A general purpose model for the condensed phases of water: TIP4P/2005, *The Journal of chemical physics* 123: 234505.

Alejandre, J. & Chapela, G. (2010). The surface tension of TIP4P/2005 water model using the ewald sums for the dispersion interactions, *The Journal of chemical physics* 132: 014701.

Benzi, R., Biferale, L., Sbragaglia, M., Succi, S. & Toschi, F. (2006). Mesoscopic modelling of heterogeneous boundary conditions for microchannel flows, *Journal of Fluid Mechanics* 548(1): 257–280.

Berendsen, H., Postma, J., Van Gunsteren, W., DiNola, A. & Haak, J. (1984). Molecular dynamics with coupling to an external bath, *The Journal of Chemical Physics* 81: 3684.

Berezhkovskii, A. & Hummer, G. (2002). Single-file transport of water molecules through a carbon nanotube, *Physical review letters* 89(6): 64503.

Bocquet, L. & Charlaix, E. (2009). Nanofluidics, from bulk to interfaces, *Chem. Soc. Rev.* 39(3): 1073–1095.

Bouzigues, C., Bocquet, L., Charlaix, E., Cottin-Bizonne, C., Cross, B., Joly, L., Steinberger, A., Ybert, C. & Tabeling, P. (2008). Using surface force apparatus, diffusion and velocimetry to measure slip lengths, *Philosophical Transactions of the Royal Society A: Mathematical, Physical and Engineering Sciences* 366(1869): 1455.

Cannon, J. & Hess, O. (2010). Fundamental dynamics of flow through carbon nanotube membranes, *Microfluidics and Nanofluidics* 8(1): 21–31.

Chiavazzo, E. & Asinari, P. (2011). Enhancing surface heat transfer by carbon nanofins: towards an alternative to nanofluids?, *Nanoscale Research Letters* 6(1): 1–13.

Chibbaro, S., Biferale, L., Diotallevi, F., Succi, S., Binder, K., Dimitrov, D., Milchev, A., Girardo, S. & Pisignano, D. (2008). Evidence of thin-film precursors formation in hydrokinetic and atomistic simulations of nano-channel capillary filling, *EPL (Europhysics Letters)* 84: 44003.

Chinappi, M. & Casciola, C. (2010). Intrinsic slip on hydrophobic self-assembled monolayer coatings, *Physics of Fluids* 22: 042003.

Chinappi, M., De Angelis, E., Melchionna, S., Casciola, C., Succi, S. & Piva, R. (2006). Molecular dynamics simulation of ratchet motion in an asymmetric nanochannel, *Physical review letters* 97(14): 144509.

Chinappi, M., Gala, F., Zollo, G. & Casciola, C. (2011). Tilting angle and water slippage over hydrophobic coatings, *Philosophical Transactions of the Royal Society A: Mathematical, Physical and Engineering Sciences* 369(1945): 2537.

Chinappi, M., Melchionna, S., Casciola, C. & Succi, S. (2008). Mass flux through asymmetric nanopores: microscopic versus hydrodynamic motion, *The Journal of Chemical Physics* 129: 124717.

Cottin-Bizonne, C., Barrat, J., Bocquet, L. & Charlaix, E. (2003). Low-friction flows of liquid at nanopatterned interfaces, *Nature Materials* 2: 237–240.

Cottin-Bizonne, C., Steinberger, A., Cross, B., Raccurt, O. & Charlaix, E. (2008). Nanohydrodynamics: The Intrinsic Flow Boundary Condition on Smooth Surfaces, *Langmuir* 24(4): 1165–1172.

De Coninck, J. & Blake, T. (2008). Wetting and molecular dynamics simulations of simple liquids, *Annu. Rev. Mater. Res.* 38: 1–22.

Eijkel, J. & Berg, A. (2005). Nanofluidics: what is it and what can we expect from it?, *Microfluidics and Nanofluidics* 1(3): 249–267.

Falk, K., Sedlmeier, F., Joly, L., Netz, R. & Bocquet, L. (2010). Molecular origin of fast water transport in carbon nanotube membranes: Superlubricity versus curvature dependent friction, *Nano letters* .

Gentner, F., Ogonowski, G. & De Coninck, J. (2003). Forced wetting dynamics: a molecular dynamics study, *Langmuir* 19(9): 3996–4003.

Gogte, S., Vorobieff, P., Truesdell, R., Mammoli, A., van Swol, F., Shah, P. & Brinker, C. (2005). Effective slip on textured superhydrophobic surfaces, *Physics of Fluids* 17: 051701.

Govardhan, R., Srinivas, G., Asthana, A. & Bobji, M. (2009). Time dependence of effective slip on textured hydrophobic surfaces, *Physics of Fluids* 21: 052001.

Hansen, J. & McDonald, I. (2006). *Theory of simple liquids*, Academic press.

Ho, T., Papavassiliou, D., Lee, L. & Striolo, A. (2011). Liquid water can slip on a hydrophilic surface, *Proceedings of the National Academy of Sciences* 108(39): 16170–16175.

Holt, J., Park, H., Wang, Y., Stadermann, M., Artyukhin, A., Grigoropoulos, C., Noy, A. & Bakajin, O. (2006). Fast mass transport through sub-2-nanometer carbon nanotubes, *Science* 312(5776): 1034.

Huang, D., Sendner, C., Horinek, D., Netz, R. & Bocquet, L. (2008). Water slippage versus contact angle: A quasiuniversal relationship, *Physical review letters* 101(22): 226101.

Hummer, G., Rasaiah, J. & Noworyta, J. (2001). Water conduction through the hydrophobic channel of a carbon nanotube, *Nature* 414(6860): 188–190.

Humphrey, W., Dalke, A. & Schulten, K. (1996). VMD: visual molecular dynamics, *Journal of molecular graphics* 14(1): 33–38.

Jorgensen, W., Chandrasekhar, J., Madura, J., Impey, R. & Klein, M. (1983). Comparison of simple potential functions for simulating liquid water, *The Journal of chemical physics* 79: 926.

Joseph, P. & Tabeling, P. (2005). Direct measurement of the apparent slip length, *Physical Review E* 71(3): 35303.

Koplik, J. & Banavar, J. (1995). Continuum deductions from molecular hydrodynamics, *Annual Review of Fluid Mechanics* 27(1): 257–292.

Lauga, E., Brenner, M. & Stone, H. (2005). Microfluidics: the no-slip boundary condition, *Handbook of Experimental Fluid Mechanics, chap. 15* .

Li, H. & Yoda, M. (2010). An experimental study of slip considering the effects of non-uniform colloidal tracer distributions, *Journal of Fluid Mechanics* 662(1): 269–287.

Majumder, M., Chopra, N., Andrews, R. & Hinds, B. (2005). Nanoscale hydrodynamics: Enhanced flow in carbon nanotubes, *Nature* 438(7064): 44–44.

Melchionna, S. & Cozzini, S. (2001). DLprotein user manual, *University of Rome* .

Myers, T. (2011). Why are slip lengths so large in carbon nanotubes?, *Microfluidics and nanofluidics* pp. 1–5.

Ng, C. & Wang, C. (2010). Apparent slip arising from Stokes shear flow over a bidimensional patterned surface, *Microfluidics and Nanofluidics* 8(3): 361–371.

Niavarani, A. & Priezjev, N. (2010). Modeling the combined effect of surface roughness and shear rate on slip flow of simple fluids, *Physical Review E* 81(1): 011606.

Orlandini, S., Meloni, S. & Ciccotti, G. (2011). Hydrodynamics from statistical mechanics: combined dynamical-nemd and conditional sampling to relax an interface between two immiscible liquids, *Phys. Chem. Chem. Phys.* .

Phillips, J., Braun, R., Wang, W., Gumbart, J., Tajkhorshid, E., Villa, E., Chipot, C., Skeel, R., Kale, L. & Schulten, K. (2005). Scalable molecular dynamics with namd, *Journal of computational chemistry* 26(16): 1781.

Priezjev, N., Darhuber, A. & Troian, S. (2005). Slip behavior in liquid films on surfaces of patterned wettability: Comparison between continuum and molecular dynamics simulations, *Physical Review E* 71(4): 041608.

Rothstein, J. (2010). Slip on superhydrophobic surfaces, *Annual Review of Fluid Mechanics* 42: 89–109.

Sendner, C., Horinek, D., Bocquet, L. & Netz, R. (2009). Interfacial water at hydrophobic and hydrophilic surfaces: Slip, viscosity, and diffusion, *Langmuir* 25(18): 10768–10781.

Styrkas, D., Keddie, J., Lu, J., Su, T. & Zhdan, P. (1999). Structure of self-assembled layers on silicon: Combined use of spectroscopic variable angle ellipsometry, neutron reflection, and atomic force microscopy, *Journal of Applied Physics* 85: 868.

Thomas, J., McGaughey, A. & Kuter-Arnebeck, O. (2010). Pressure-driven water flow through carbon nanotubes: Insights from molecular dynamics simulation, *International Journal of Thermal Sciences* 49(2): 281–289.

Werder, T., Walther, J., Jaffe, R., Halicioglu, T. & Koumoutsakos, P. (2003). On the water-carbon interaction for use in molecular dynamics simulations of graphite and carbon nanotubes, *The Journal of Physical Chemistry B* 107(6): 1345–1352.

Whitby, M., Cagnon, L., Thanou, M. & Quirke, N. (2008). Enhanced fluid flow through nanoscale carbon pipes, *Nano letters* 8(9): 2632–2637.

Ybert, C., Barentin, C., Cottin-Bizonne, C., Joseph, P. & Bocquet, L. (2007). Achieving large slip with superhydrophobic surfaces: Scaling laws for generic geometries, *Physics of fluids* 19: 123601.

Zhang, H., Zhang, Z. & Ye, H. (2011). Molecular dynamics-based prediction of boundary slip of fluids in nanochannels, *Microfluidics and Nanofluidics* pp. 1–9.

Zhu, F., Tajkhorshid, E. & Schulten, K. (2004). Collective diffusion model for water permeation through microscopic channels, *Physical review letters* 93(22): 224501.

Part 5

Dynamics of Nanomachines

16

Analysis of the Atomization Process by Molecular Dynamics Simulation

Yeh Chun-Lang
Department of Aeronautical Engineering, National Formosa University
Taiwan, ROC

1. Introduction

Investigation of the atomizer flows has received a lot of attentions due to their broad applications in many fields, e.g. industrial processes, agriculture, meteorology, medicine, etc. Historically, less attention has been devoted to the internal flow development in atomizers because of the small size of practical atomizers, which makes measurements difficult. However, internal flows in atomizers are of interest due to their potential effect on the subsequent atomization process that takes place external to the atomizers.

Owing to the difficulties inherent in the simulation of the atomizer internal and external flows, no systematic numerical researches have been carried out on this subject. One way to study internal flows in atomizers is to conduct experimental investigations by the use of large-scale models. Another approach is to do detailed numerical simulations of the flow. An advantage of the numerical simulation is that practical-scale atomizers can be studied as easily as large scale ones. The methods of analyzing atomizer flows can be divided into two categories, namely, the macroscopic analysis and the microscopic analysis, each with its own advantages and disadvantages. The former approach is usually adopted as an aided tool for atomizer design due to its relatively lower computational cost. However, when atomization occurs, the liquid breaks into ultra-fine nano-scale droplets. Under such situation, the Navier-Stokes equation based on continuum concept is not suitable and the microscopic analysis should be used for subsequent simulation of the atomization process. In the past decade, the author has carried out a series of researches on the macroscopic analysis of the atomizer internal and external flows (Yeh, 2002, 2003, 2004, 2005, 2007). The macroscopic behavior and characteristics of the atomizer flows, such as discharge coefficient, spray angle, film thickness, atomizer geometry, etc. were analyzed. However, as mentioned above, the macroscopic analysis is not suitable when atomization occurs. The author's previous studies were therefore initially confined to the investigation of the early stage of the atomization process, during which the liquid had not yet ruptured to ultra-fine nano-scale droplets. In order to improve this inherent deficiency of the macroscopic analysis, the author has employed molecular dynamics (MD) simulation to investigate the microscopic evolution of the atomizer internal and external flows. In the past few years, the author has studied the atomization processes of a nano-scale liquid thread and a nanojet. Liquid thread and nanojet are two of the most fundamental and important phenomena during atomization process. Although the size of a nano-scale atomizer is much smaller than that of a practical-scale atomizer, analysis of the nano-scale atomizer flow can provide a preliminary understanding of the detailed process of a practical-scale atomizer flow, especially when the available computing resources can not directly facilitate the microscopic analysis of a practical-scale atomizer flow.

2. Analysis by the molecular dynamics simulation

From one of the author's previous studies (Yeh, 2005) for atomizer flow by macroscopic analysis, it can be seen that the liquid evolves into threads after leaving the atomizer. In the author's recent papers (Yeh, 2009a, 2009b, 2010), the atomization processes of the nano-scale liquid thread and nanojet have been discussed in detail by molecular dynamics simulation.

2.1 Molecular dynamics analysis of the instability for a nano-scale liquid thread

In the scope of fluid flow researches, the instability has received much attention due to its potential influence on the flow development. Previous instability analyses were focused mainly on large scale flow fields. The applicabilities of these theories to micro- or nano-scale flow fields are still uncertain. Lord Rayleigh (1879) studied the instability of cylindrical thin films. He analyzed an inviscid liquid cylinder and a viscous one. Later on, Weber (1931) and Tomotika (1935) considered more realistic cases of liquid threads in unbounded domains. Goren (1962) studied the annular liquid films in contact with a solid, i.e., supported on a wire or lining the interior walls of a capillary. He used linear stability analysis to determine the fastest growing mode when either inertia or viscous forces are negligible. Koplik and Banavar (1993) studied the Rayleigh's instability of a cylindrical liquid thread in vacuum by three-dimensional MD simulation. The maximum number of molecules they used consists of 8,192 liquid argon Lennard-Jones molecules for a cylindrical liquid thread of non-dimensionalized radius of 7.5 in a box of non-dimensionalized length of 54.7. For this simulation condition, only one liquid particle was formed. If a smaller computational domain was used instead, no liquid particle was found from their study. Kawano (1998) applied 10,278 Lennard-Jones molecules of liquid and vapor coexisting argon in three dimensions to analyze the interfacial motion of a cylindrical liquid thread of non-dimensionalized radii of 2.0 to 4.0 in a box of non-dimensionalized length up to 120. For this condition of larger computational domain, a maximum number of 8 to 9 liquid particles were observed. Min and Wong (2006) studied the Rayleigh's instability of nanometer scale Lennard-Jones liquid threads by MD simulation and concluded that Rayleigh's continuum prediction holds down to the molecular scale. Kim, Lee, Han and Park (2006) applied MD simulation to investigate the thermodynamic properties and stability characteristics of the nano-scale liquid thread. They found that the overall trends of the simulation results agree with the classical stability theory. However, the classical theory overpredicts the region of stable domain compared to the MD results as the radius decreases.

In the following discussion, the instability of a liquid thread is investigated by MD simulation. The influences of liquid thread radius, fundamental cell length, and temperature will be discussed. Snapshots of molecules, number of liquid particles formed, and density field are analyzed. Two linear stability criteria, namely Rayleigh's stability criterion and Kim's stability criterion, are accessed for their validity in molecular scale. This can provide insights into the mechanism and prediction of the atomization process.

2.1.1 Molecilar dynamics simulation method

In this study, the vaporization process of a liquid thread is investigated by MD simulation. The inter-atomic potential is one of the most important parts of MD simulation. Many possible potential models exist, such as hard sphere, soft sphere, square well, etc (Haile, 1992). In this research, the Lennard-Jones 12-6 potential model, which is widely used, is adopted for calculation. It is

$$\phi(r) = 4\varepsilon\left[\left(\frac{\sigma}{r}\right)^{12} - \left(\frac{\sigma}{r}\right)^{6}\right] \tag{1}$$

where r denotes the distance between two molecules, ε and σ are the representative scales of energy and length, respectively. The Lennard-Jones fluid in this research is taken to be argon for its ease of physical understanding. The parameters for argon are as follows (Kawano, 1998) : the length parameter σ=0.354 nm, the energy parameter ε/k_B=93.3K, and the molecular weight m=6.64×10^{-26} kg, where k_B=1.38×10^{-23} J/K denotes the Boltzmann constant. The cut-off radius r_c beyond which the intermolecular interaction is neglected is 5.0σ.

The simulation domain is schematically shown in Fig.1, with periodic boundary conditions applied in all three directions. Simulation domain dimensions, temperatures and number of molecules, are listed in Table 1, together with some simulation results. The time integration of motion is performed by Gear's fifth predictor-corrector method (Haile, 1992) with a time step of Δt^*=0.001 (i.e. 2.5 fs). Note that all quantities with an asterisk in this paper, such as L*, R^*, ρ^*, Δt^*, etc., are non-dimensionalized in terms of σ, ε, and m, i.e. L*=L/σ, $R^*=R/\sigma$, $\rho^*=N\sigma^3/V$, $\Delta t^*=\Delta t\,(\varepsilon/m)^{1/2}/\sigma$, $T^*= k_BT/\varepsilon$.

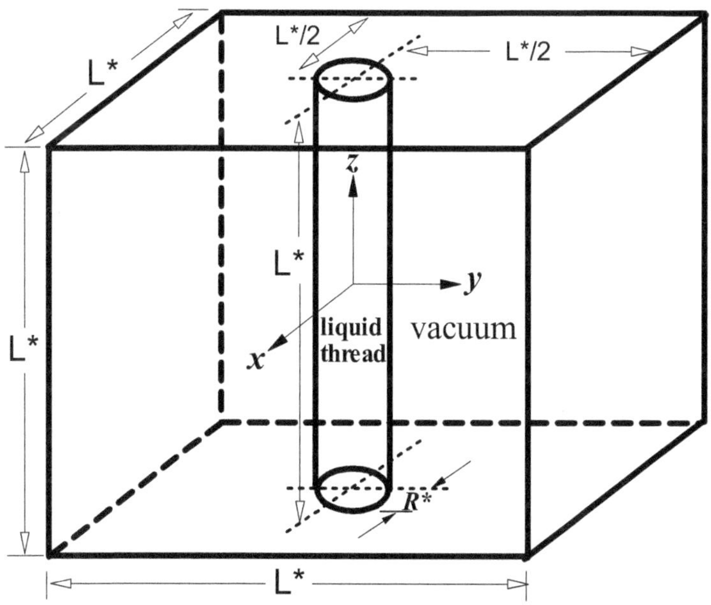

Fig. 1. Illustration of the computational domain for the simulation of a liquid thread

Case No.	L*	R^*	T*	N_{mol}	$\overline{f_\rho}$	Np
1	10	2	0.75	109	0.286	1
2	20	2	0.75	210	0.289	1
3	30	2	0.75	303	0.356	1
4	60	2	0.75	638	0.542	1
5	120	2	0.75	1221	0.635	2
6	240	2	0.75	2476	0.688	5
7	480	2	0.75	4929	0.690	10
8	10	3	0.75	235	0.341	intact
9	16	3	0.75	370	0.492	1
10	20	3	0.75	468	0.496	1
11	30	3	0.75	690	0.503	1
12	60	3	0.75	1395	0.616	1
13	120	3	0.75	2787	0.752	1
14	240	3	0.75	5590	0.805	2
15	480	3	0.75	11090	0.820	5
16	10	4	0.75	410	0.206	intact
17	20	4	0.75	824	0.555	intact
18	24	4	0.75	990	0.598	1
19	30	4	0.75	1225	0.611	1
20	60	4	0.75	2467	0.652	1
21	120	4	0.75	4943	0.741	1
22	240	4	0.75	9867	0.839	1
23	480	4	0.75	19709	0.857	3
24	10	4	1.0	407	0.001	intact
25	20	4	1.0	825	0.165	1
26	30	4	1.0	1235	0.195	1
27	60	4	1.0	2470	0.411	1
28	120	4	1.0	4927	0.487	1
29	240	4	1.0	9888	0.698	2
30	480	4	1.0	19797	0.736	4

Table 1. Simulation domain dimensions, temperatures, number of molecules, and simulation results

In this research, a cylindrical liquid thread of length L* and radius R^* is placed at the center of the computational domain and the remaining space is vacuum. The initial density of the liquid argon is $\rho_L^*=0.819$. The system temperature is kept at $T^*=0.75$ or 1.0. These dimensionless values correspond to $\rho_L=1223$ kg/m^3 for argon and T=70K or 93.3K. Note that these two temperatures are below the critical temperature（150K）of argon.

The procedure for MD simulation includes three stages : initialization, equilibration and production. Initially, equilibration is performed for liquid argon molecules in a rectangular parallelepiped with length and width equal to the liquid thread diameter ($D^*=2R^*=4, 6, 8$) and with height equal to the side length of the computational domain (L*=10, 16, 20, 24, 30, 60, 120, 240, 480). The initial velocities of molecules are decided by normal random numbers.

Velocity rescaling is performed at each time step by Eq.(2) to make sure that the molecules are at the desired temperature T^* :

$$v_i^{new} = v_i^{old} \sqrt{\frac{T_D}{T_A}} \tag{2}$$

where v_i^{new} and v_i^{old} are the velocities of molecule i after and before correction, respectively, and T_D and T_A are the desired and the actual molecular temperatures, respectively. The liquid molecules are equilibrated for 10^6 time steps at the desired temperature T^*. The achievement of equilibrium state is confirmed by obtaining the radial distribution function. After the liquid molecules are equilibrated, the rectangular parallelepiped for the liquid molecules is truncated to the desired cylindrical liquid thread by removing unwanted regions. The cylindrical liquid thread then is put into the computational domain and the production stage proceeds. A minimum image method and the Verlet neighbor list scheme (Haile, 1992) to keep track of which molecules are actually interacting at a given time interval of 0.005 are used in the equilibration and the production stages.

2.1.2 Liquid thread vaporization process

In the following discussion, a cylindrical liquid thread of length L* and radius R^* is placed at the center of the simulation domain and the remaining space is in vacuum, as illustrated in Fig.1. Simulation domain dimensions, temperatures and number of molecules, are listed in Table 1.

Figures 2~4 show the vaporization processes of liquid threads at T*=0.75 (cases 1~23 in Table 1) . Note that L*=120 and R^*=3 correspond to L=42.5nm and R=1.06nm, respectively. It is found that when R^*=2, the liquid thread breaks up into drops for all fundamental cell lengths. For L*=10, 20, 30 and 60, the liquid thread ruptures only from its two ends, i.e. the top and bottom surfaces of the fundamental cell, and gets shorter due to the contraction motion in its axial direction. Only one liquid particle is formed. On the other hand, for L*=120, 240 and 480, the thread ruptures not only from the top and bottom surfaces of the fundamental cell but also from its interior section. The number of liquid particles produced for L*=120, 240 and 480 is 2, 5 and 10, respectively. During the vaporization process, collision and coalescence of the liquid particles occur. The two liquid particles produced for L*=120 coalesce into one liquid particle. The number of liquid particles at t*=1000 for L*=240 and 480 is 3 and 5, respectively.

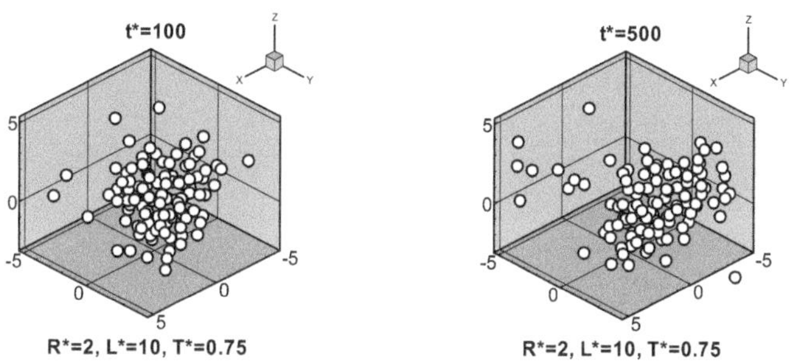

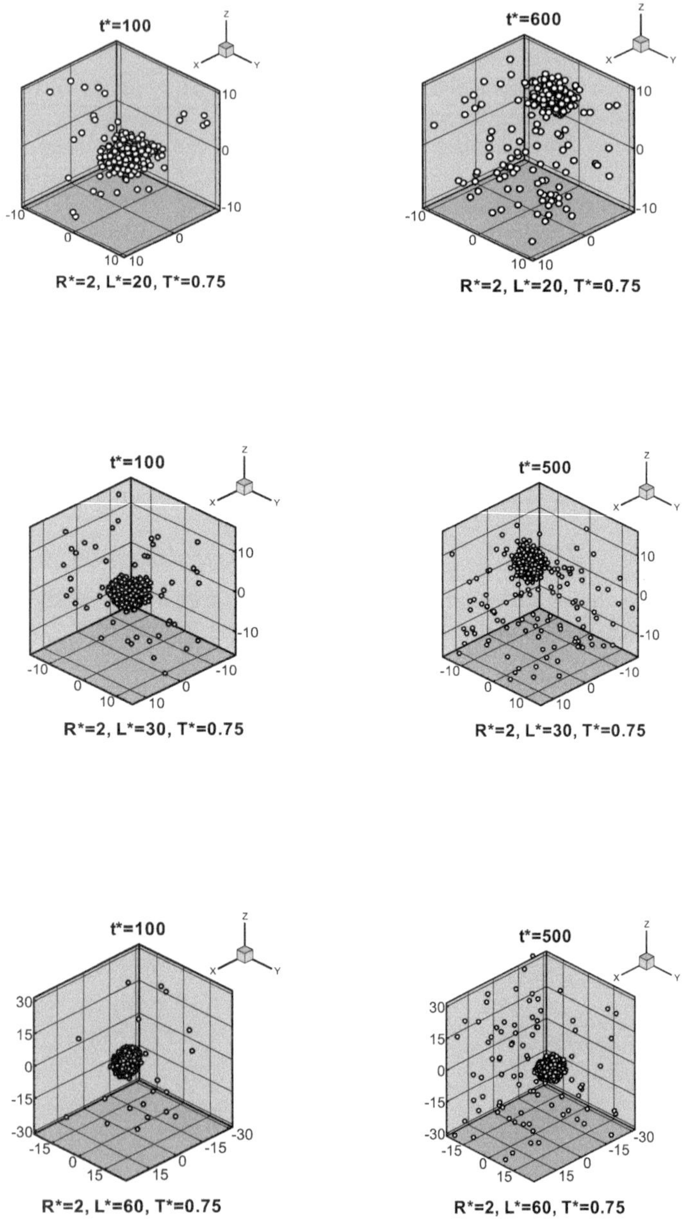
t*=100
R*=2, L*=20, T*=0.75
t*=600
R*=2, L*=20, T*=0.75
t*=100
R*=2, L*=30, T*=0.75
t*=500
R*=2, L*=30, T*=0.75
t*=100
R*=2, L*=60, T*=0.75
t*=500
R*=2, L*=60, T*=0.75

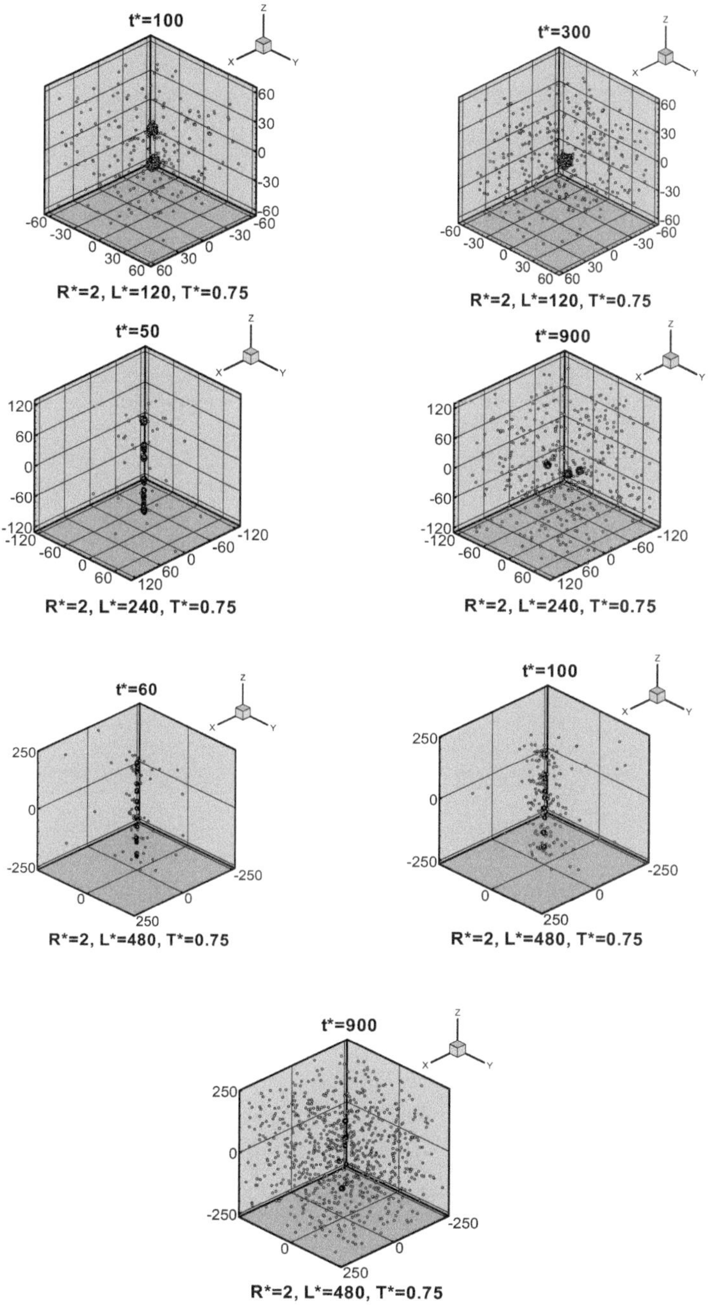

Fig. 2. Vaporization processes of liquid threads of R^*=2 at T*=0.75

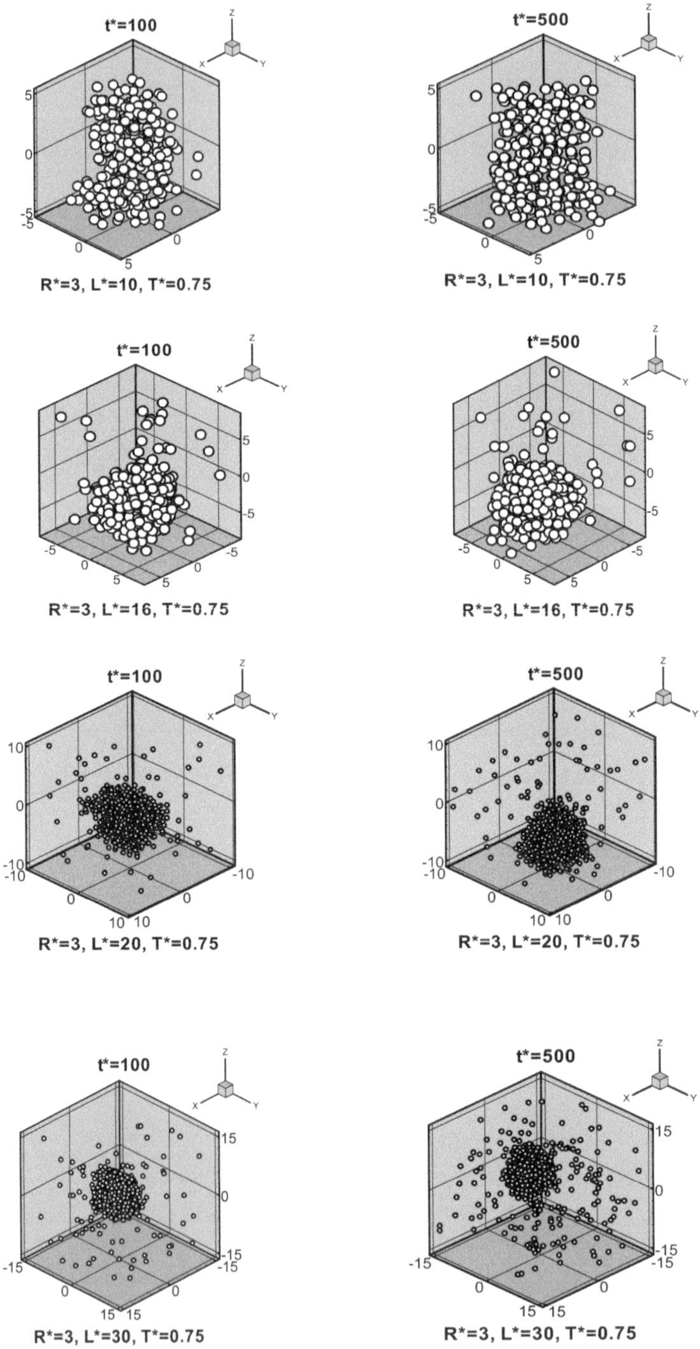
t*=100
t*=500
R*=3, L*=10, T*=0.75
R*=3, L*=10, T*=0.75
t*=100
t*=500
R*=3, L*=16, T*=0.75
R*=3, L*=16, T*=0.75
t*=100
t*=500
R*=3, L*=20, T*=0.75
R*=3, L*=20, T*=0.75
t*=100
t*=500
R*=3, L*=30, T*=0.75
R*=3, L*=30, T*=0.75

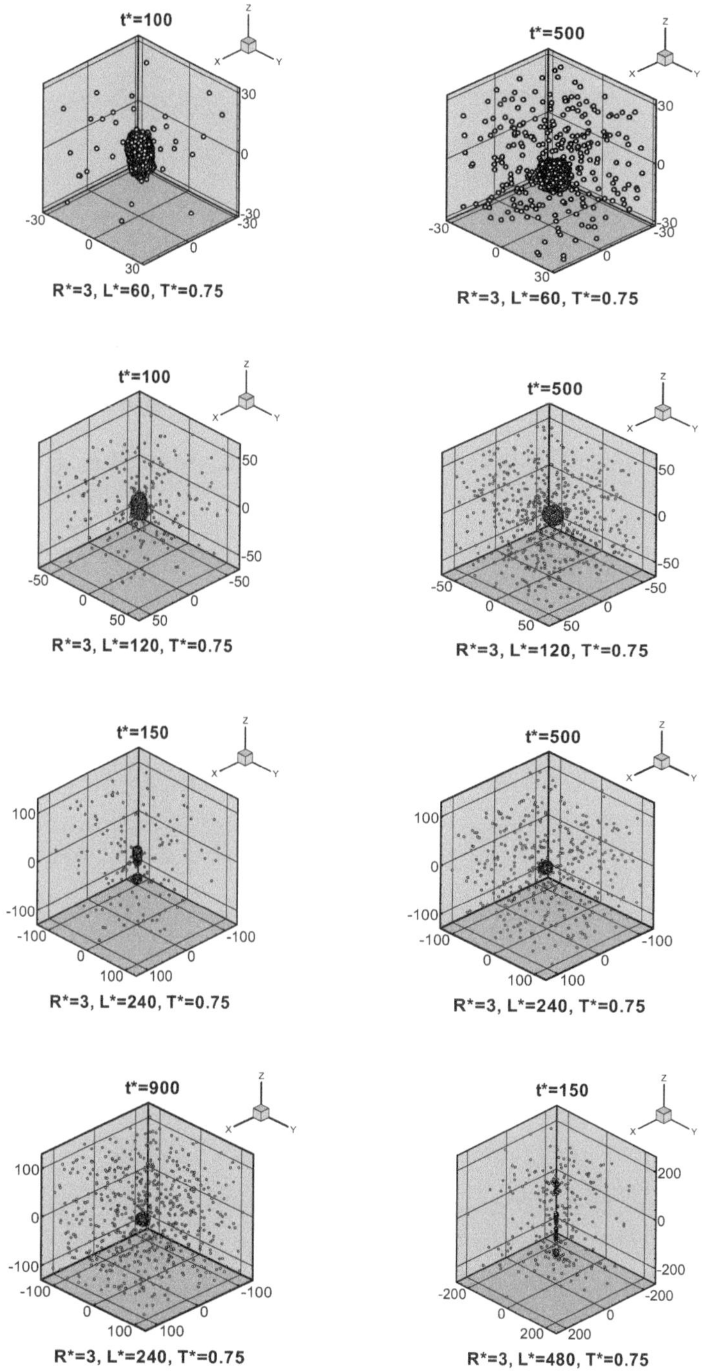
t*=100
R*=3, L*=60, T*=0.75
t*=500
R*=3, L*=60, T*=0.75
t*=100
R*=3, L*=120, T*=0.75
t*=500
R*=3, L*=120, T*=0.75
t*=150
R*=3, L*=240, T*=0.75
t*=500
R*=3, L*=240, T*=0.75
t*=900
R*=3, L*=240, T*=0.75
t*=150
R*=3, L*=480, T*=0.75

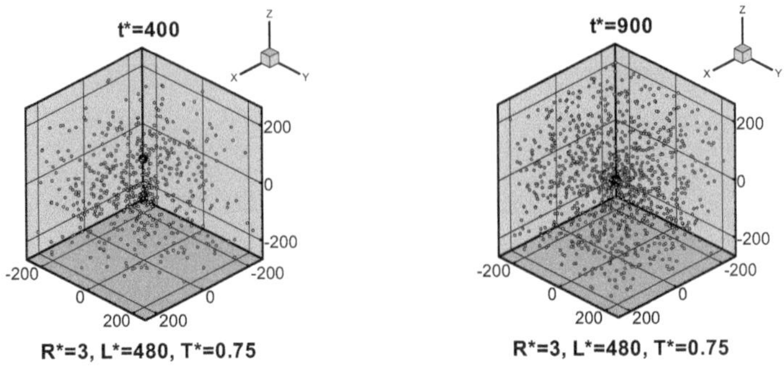

Fig. 3. Vaporization processes of liquid threads of $R^*=3$ at $T^*=0.75$

When $R^*=4$, the liquid thread remains intact for $L^*=10$ and 20. For $L^*=24$, 30, 60, 120 and 240, the liquid thread ruptures merely from its two ends and only one liquid particle is formed. For $L^*=480$, the thread ruptures not only from the top and bottom surfaces of the fundamental cell but also from its interior section. Three liquid particles are produced. The liquid particles eventually coalesce into one liquid particle.

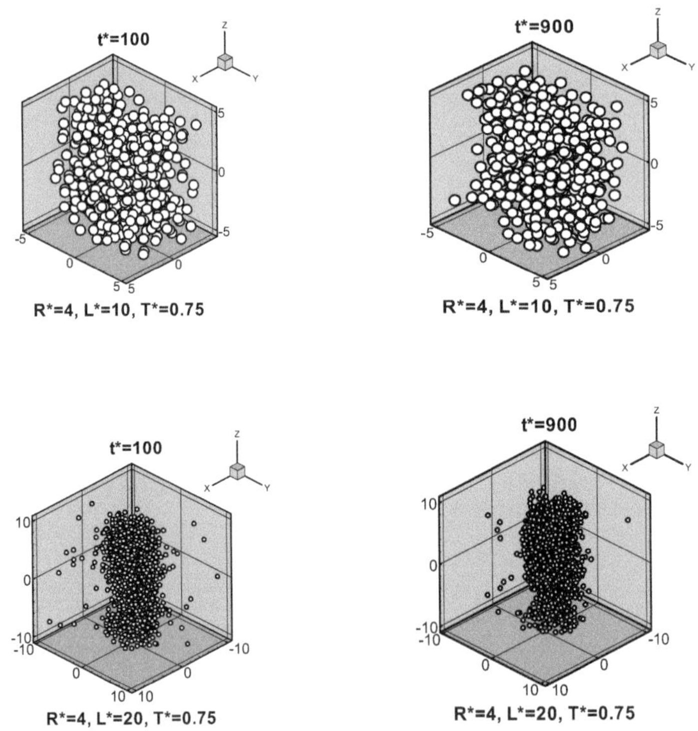

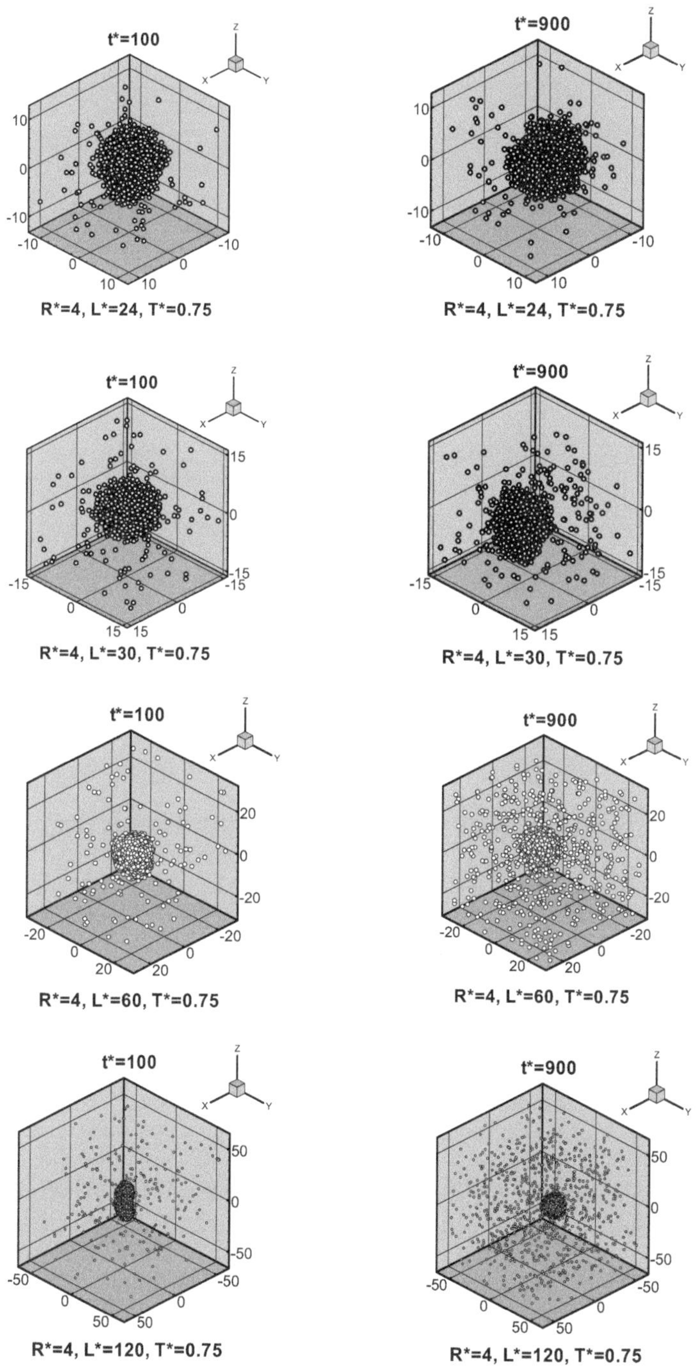
t*=100
t*=900
R*=4, L*=24, T*=0.75
R*=4, L*=24, T*=0.75
t*=100
t*=900
R*=4, L*=30, T*=0.75
R*=4, L*=30, T*=0.75
t*=100
t*=900
R*=4, L*=60, T*=0.75
R*=4, L*=60, T*=0.75
t*=100
t*=900
R*=4, L*=120, T*=0.75
R*=4, L*=120, T*=0.75

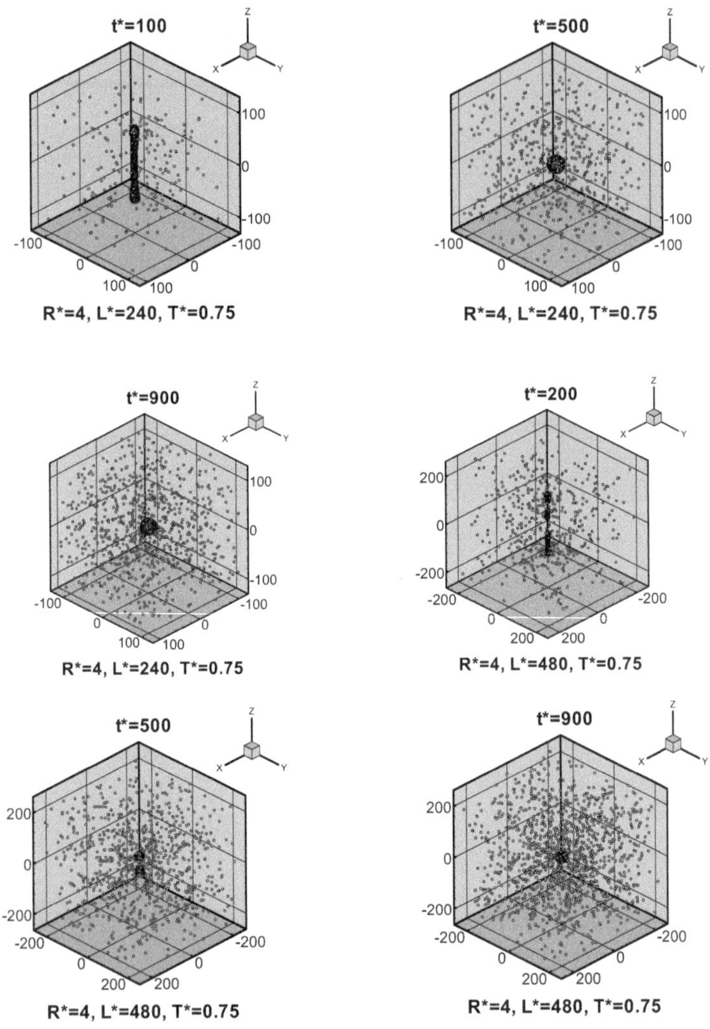

Fig. 4. Vaporization processes of liquid threads of $R^*=4$ at T*=0.75

Figure 5 shows the vaporization processes for liquid threads of $R^*=4$ at T*=1.0 (cases 24~30 in Table 1). Unlike the liquid threads of the same radius $R^*=4$ at a lower temperature T*=0.75 (cases 16~23 in Table 1), the liquid thread remains intact only for L*=10. For L*=20, 30, 60 and 120, the liquid thread ruptures from its two ends, i.e. the top and bottom surfaces of the fundamental cell, and only one liquid particle is formed. For L*=240 and 480, the thread ruptures not only from the top and bottom surfaces of the fundamental cell but also from its interior section. The number of liquid particles produced for L*=240 and 480 is 2 and 4, respectively. During the vaporization process, collision and coalescence of the liquid particles occur. The liquid particles eventually coalesce into one liquid particle.

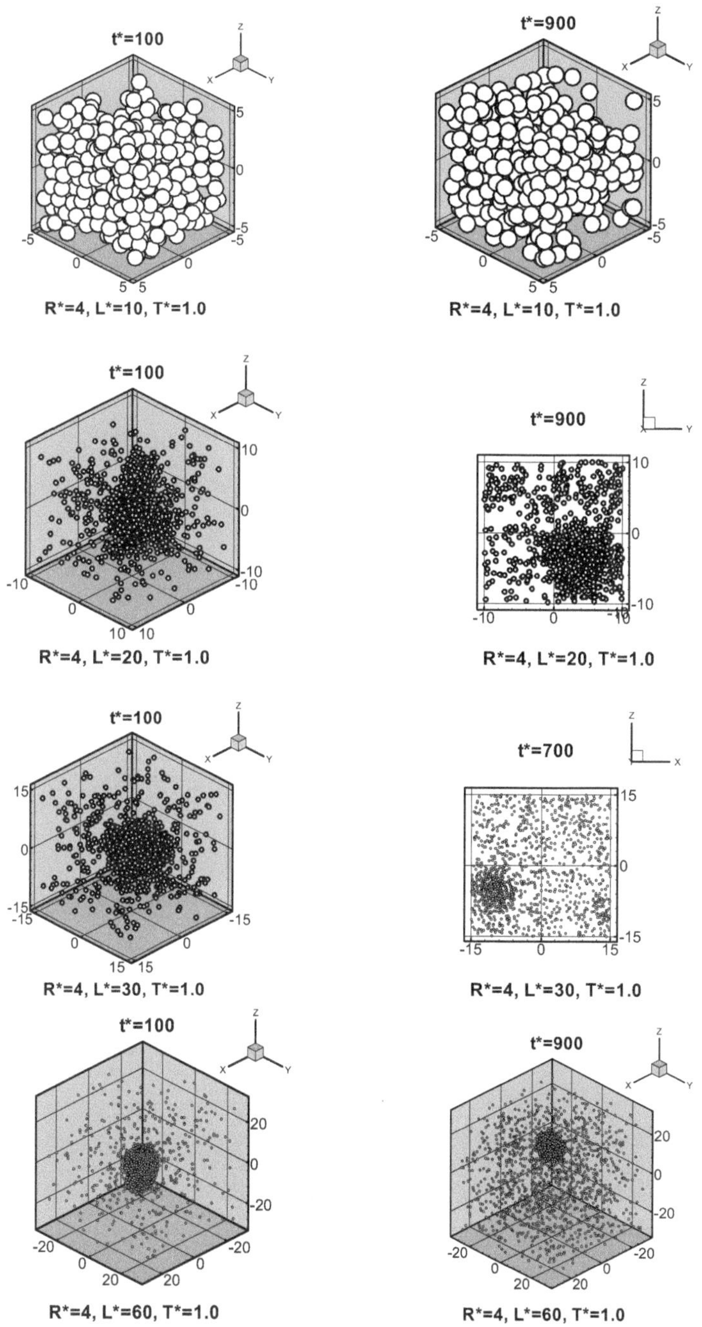
t*=100
R*=4, L*=10, T*=1.0
t*=900
R*=4, L*=10, T*=1.0
t*=100
R*=4, L*=20, T*=1.0
t*=900
R*=4, L*=20, T*=1.0
t*=100
R*=4, L*=30, T*=1.0
t*=700
R*=4, L*=30, T*=1.0
t*=100
R*=4, L*=60, T*=1.0
t*=900
R*=4, L*=60, T*=1.0

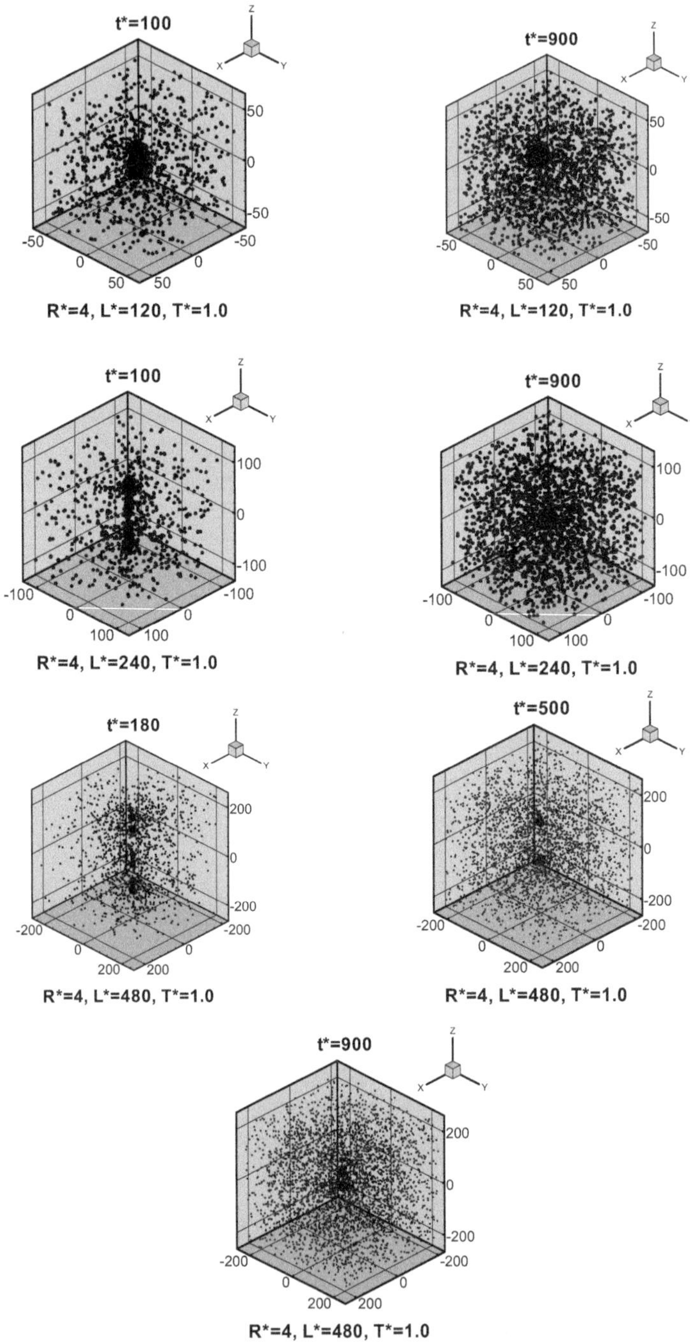

Fig. 5. Vaporization processes of liquid threads of $R^*=4$ at $T^*=1.0$

From the above discussion for Figs.2~5, the following observations can be obtained.

First, if the fundamental cell length is long enough, more than one liquid particle may be produced in the cell. Under such situation, the liquid thread ruptures not only from its two ends but also from the interior of the fundamental cell. On the other hand, if the fundamental cell length is small, the liquid thread may remain intact or produce only one liquid particle in the cell. If the thread breaks up, it ruptures from its two ends only, i.e. the top and bottom surfaces of the fundamental cell, but not from its interior.

Second, a liquid thread with a longer fundamental cell length may produce more liquid particles in the cell.

Third, a thinner liquid thread may produce more liquid particles in the cell.

Fourth, a liquid thread at a higher temperature may produce more liquid particles.

2.1.3 Stability analysis

According to the classical theory by Rayleigh (1879), the radius of the liquid thread, R, is related to the critical wavelength of perturbation, λ_c, as

$$\lambda_c = 2\pi R \tag{3}$$

A liquid thread will break up into drops if the axial wavelength of the surface perturbation $L > \lambda_c$. If $L < \lambda_c$, the thread is stable and will remain intact. Owing to the periodic boundary conditions in this study, the fundamental cell size L^* can be regarded as the longest wavelength of the perturbation.

For $R^*=2$, λ_c^* is 12.6. According to Rayleigh's stability criterion, a liquid thread of radius $R^*=2$ and length $L^*=10$ is expected to be stable and remain intact. However, from Fig.21 or Table 1 (Case 1) , it is seen that the liquid thread of $R^*=2$ and $L^*=10$ is unstable, which implies that Rayleigh's stability criterion overpredicts the stable domain as compared to the MD simulation results. Rayleigh's stability criterion holds for the remaining liquid threads with $L^* > \lambda_c^*$ ($L^*=20\sim480$) , which are unstable and break up into drop(s).

For $R^*=3$, λ_c^* is 18.9. From Fig.3 or Table 1 (Cases 8 and 9) , the liquid thread of $R^*=3$ and $L^*=10$ remains intact while the liquid thread of $R^*=3$ and $L^*=16$ is unstable. Rayleigh's stability criterion holds for the former case but violates the MD simulation results for the latter case. The remaining liquid threads with $L^* > \lambda_c^*$ ($L^*=20\sim480$) are unstable and break up into drop(s). Rayleigh's stability criterion holds for these cases.

For $R^*=4$, λ_c^* is 25.2. From Fig.4 or Table 1 (Cases 16~23) , the liquid threads of $L^*=10$ and 20 remain intact while the liquid thread of $L^*=24$ is unstable. Rayleigh's stability criterion holds for the first two cases but violates the MD simulation results for the third case. The remaining liquid threads with $L^* > \lambda_c^*$ ($L^*=30\sim480$) are unstable and break up into drop(s). Rayleigh's stability criterion holds for these cases.

If the temperature of the liquid threads of $R^*=4$ is increased to $T^*=1.0$, from Fig.5 or Table 1 (Cases 24~30) , only the liquid thread of $L^*=10$ remains intact while, in contrast to the situation of $T^*=0.75$, the liquid thread of $L^*=20$ is unstable. Rayleigh's stability criterion holds for the former case but violates the MD simulation results for the latter case. The remaining liquid threads with $L^*>\lambda_c^*$ (L^*=30~480) are unstable and break up into drop(s). Rayleigh's stability criterion holds for these cases.

Recently, Kim, Lee, Han and Park (2006) proposed a new linear relation from their MD simulation results. From their study, Rayleigh's linear relation can be modified as

$$\lambda_c=2.44\pi(R\text{-}1.23) \tag{4}$$

From Eq.(4), λ_c^* for R^* of 2, 3 and 4 are 5.9, 13.6 and 21.2, respectively. Then, from Table 1 (Cases 1~23) , it is seen that Kim's stability criterion holds for the liquid threads at T*=0.75. On the other hand, at T*=1.0 (Cases 24~30) , the liquid thread of R^*=4 and L^*=10 remains intact while the liquid thread of R^*=4 and L^*=20 is unstable. Kim's stability criterion holds for the former case but violates the MD simulation results for the latter case. The remaining liquid threads with $L^*>\lambda_c^*$ (L^*=30~480) are unstable and break up into drop(s). Kim's stability criterion holds for these cases.

From the above discussion, the following observations can be made.

First, a liquid thread with a longer fundamental cell length is more unstable.

Second, a thinner liquid thread is more unstable.

Third, a liquid thread at a higher temperature is more unstable.

Fourth, the trends of linear stability theories agree with MD simulation results. However, Rayleigh's stability criterion overpredicts stable domain as compared to the MD simulation results. Kim's stability criterion gives more accurate predictions. However, it overpredicts the stable domain at a higher temperature as compared to the MD simulation results.

2.1.4 Density distribution

In practical applications, e.g. combustor or printer, faster vaporization is usually desirable. Criteria have to be made to quantify the discussion regarding the vaporization process of a liquid thread. In this research, a liquid thread is considered to vaporize faster if the distribution of molecules reaches uniform state quicker during the vaporization process. This criterion essentially concerns with the evolution of the density distribution. The density at a specified point in the fundamental cell can be defined as

$$\rho = \lim_{\delta V \to 0} \frac{\delta N}{\delta V} \tag{5}$$

where δV is a small volume surrounding the point considered and δN is the number of molecules inside the volume δV. The density defined by Eq.(5) is actually an averaged density of a small volume surrounding the point considered. The value will approach the density of a specified point if the volume δV shrinks to that point. However, for a meaningful density field, the volume δV can not be too small because when δV becomes too small, it is difficult to obtain a definite value for $\delta N/\delta V$. In this study, the volume δV is taken to be a sphere with non-dimensionalized radius R^*=2 and with its center located at the point considered. This is an optimal choice after numerical test.

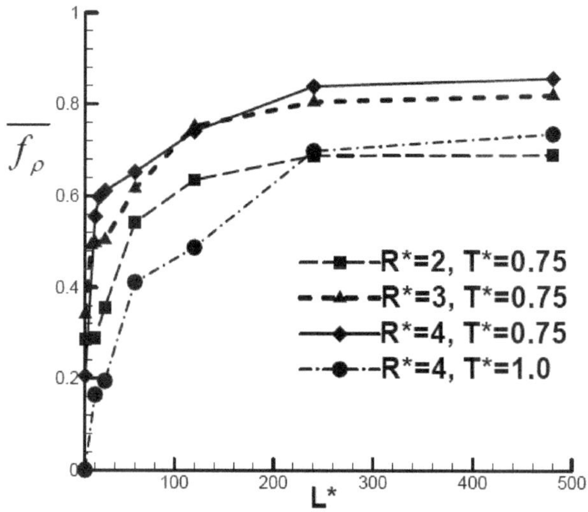

Fig. 6. Comparison of time averaged density uniformity factor, $\overline{f_\rho}$, for liquid threads of different radius, length and temperature

The time averaged density uniformity factor, $\overline{f_\rho}$, in a time interval of t^*=0 to 1000, as shown in Fig.6 or listed in Table 1, can be used to indicate the vaporization speed of a liquid thread. The density uniformity factor, f_ρ, is defined as

$$f_\rho = \frac{\sum_N \left(\rho^* - \rho^*_{eq}\right)_{t^*} \Delta V}{\sum_N \left(\rho^* - \rho^*_{eq}\right)_{t^*=0} \Delta V} \tag{6}$$

where N is the total number of molecules in the fundamental cell, i.e. N_{mol} in Table 1, ρ^* and ΔV are the density and volume of molecule i, respectively, as defined by Eq.(5), and ρ^*_{eq} is the density value when the molecules are uniformly distributed, i.e. $\rho^*_{eq} \equiv N_{mol}$ / Vol, where Vol is the volume of the fundamental cell. The density uniformity factor, f_ρ, as defined by Eq.(6) represents the deviation from uniform state. From Fig.6 or Table 1, the following observations can be drawn.

First, a liquid thread with a shorter fundamental cell length evaporates quicker. Take the liquid threads of R^*=2 (Cases 1~7) as an example. The liquid thread of L*=10 (Case 1) produces one liquid particle while the liquid thread of L*=480 (Case 7) produces ten liquid particles. The length of the latter liquid thread is forty-eight times of the former one. This implies that if forty-eight liquid threads of L*=10 are connected in series, there will be forty-eight liquid particles. It is known that molecular interaction plays an important role in the vaporization process. More liquid particles can provide more molecular interactions and this is conducive to vaporization.

Second, a liquid thread evaporates quicker at a higher temperature.

Third, a liquid thread with a higher $\overline{f_\rho}$ is more unstable and produces more liquid particles in the fundamental cell.

2.2 Molecular dynamics analysis of the vaporization process for two nano-scale liquid threads coexisting in a periodic fundamental cell

As mentioned in section 2.1, previous studies of nano-scale liquid threads have mostly been devoted to the investigation of a single liquid thread in a periodic fundamental cell. Because of the interaction between the two nano-scale liquid threads coexisting in a periodic fundamental cell, the vaporization process is different from that of a single liquid thread in a periodic fundamental cell. This section discusses the influences of the liquid thread radius, fundamental cell length, and relative position of the two threads. Snapshots of molecules, the number of liquid particles formed, and density field are analyzed. Two linear stability criteria, namely, Rayleigh's stability criterion and Kim's stability criterion, are accessed for their validity in molecular scale. This approach will be helpful for the understanding and prediction of the atomization process.

In this study, two cylindrical liquid threads of length L* and radius R^* are placed in the computational domain and the remaining space is a vacuum. The relative position of the two threads is controlled by L_1^*, L_2^* and L_4^*, as shown in Fig.7. L_3^* is set to be zero, i.e., the two liquid threads are kept at x=0 initially and are shifted only in the y direction. The initial density of the liquid argon is $\rho_L^*=0.819$ and the system temperature is kept at $T^*=0.75$. These dimensionless values correspond to $\rho_L=1223$ kg/m^3 for argon and T=70K, which is below the critical temperature (150K) of argon. Simulation domain dimensions, the relative position of the two threads and number of molecules are listed in Table 2, together with some simulation results.

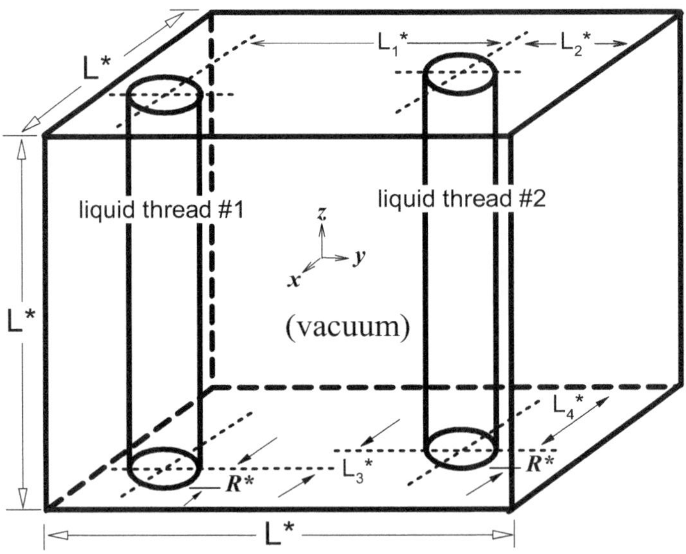

(The origin of the coordinates is located at the center of the fundamental cell.)

Fig. 7. Illustration of the computational domain for the simulation of two nano-scale liquid threads coexisting in a periodic fundamental cell

2.2.1 Liquid thread vaporization process

Figure 8 shows the vaporization process of two liquid threads of $R^*=4$, $L^*=30$, $L_1^*=10$, $L_2^*=10$ and $L_4^*=15$ (case 1 in Table 2) . Note that $L^*=30$ and $R^*=4$ correspond to L=10.6nm and R=1.41nm, respectively. It can be seen that the two liquid threads quickly coalesce into a single thread and remain intact. No liquid particles are formed during the vaporization process. Figure 9 shows the vaporization process of two liquid threads of smaller radius $R^*=2$ but with the same length and relative position : $L^*=30$, $L_1^*=10$, $L_2^*=10$ and $L_4^*=15$ (case 2 in Table 2) . For this case in which the liquid threads are thinner, it is seen that the two threads rupture from their two ends, i.e., the top and bottom surfaces of the fundamental cell, and get shorter due to the contraction motion in their axial directions. Two liquid particles are formed and they quickly coalesce into a single particle by collision and coalescence. The coalesced liquid particle prevails during the subsequent vaporization process.

Case No.	Corresponding Figure No.	R^*	L^*	L_1^*	L_2^*	L_4^*	N_{mol}	$\overline{f_\rho}$	Np
1	2	4	30	10	10	15	2450	0.649	0
2	3	2	30	10	10	15	606	0.465	2(1)※
3	4	4	60	30	15	30	4934	0.633	2
4	5	4	60	10	25	30	4934	0.683	1
5		4	60				2467	0.652	1
6	6	2	60	30	15	30	1276	0.519	2
7	7	2	60	10	25	30	1276	0.614	2(1)
8		2	60				638	0.542	1
9	8	4	120	60	30	60	9886	0.706	2
10	9	4	120	40	40	60	9886	0.708	2
11	10	4	120	10	55	60	9886	0.745	1
12		4	120				4943	0.741	1
13	11	2	120	60	30	60	2442	0.531	10(2)
14	12	2	120	40	40	60	2442	0.539	9(2)
15	13	2	120	10	55	60	2442	0.568	9(1)
16		2	120				1221	0.635	2(1)
17	14	4	240	120	60	120	19734	0.807	2
18	15	4	240	40	100	120	19734	0.811	2
19	16	4	240	10	115	120	19734	0.824	1
20		4	240				9867	0.839	1
21	17	4	480	160	160	240	39418	0.829	8(4)
22	18	4	480	40	220	240	39418	0.847	6(2)
23	19	4	480	10	235	240	39418	0.875	2(1)
24		4	480				19709	0.857	3(1)

* The number in the parenthesis in the Np column denotes the number of liquid particles formed at time $t^*=1000$, while the number outside the parenthesis in the Np column denotes the maximum number of liquid particles formed during the vaporization process. If no parenthesis is denoted, it means that the maximum number of liquid particles formed prevails over the vaporization process.

Table 2. Simulation domain dimensions, relative position of the two threads, number of molecules, and simulation results ($L_3^*=0$)

From Table 2, it is seen that the value of $\overline{f_\rho}$ is smaller for the thinner liquid threads. As stated in the previous section, this implies that the thinner liquid threads evaporate more quickly. In addition, it can be seen from Table 2 that the thinner liquid threads produce more liquid particles. In our previous study (Yeh, 2009b), we found that a single liquid thread is more unstable and produces more liquid particles when it is thinner. The present results for two liquid threads coexisting in a fundamental cell corroborate the findings of the previous study of a single liquid thread.

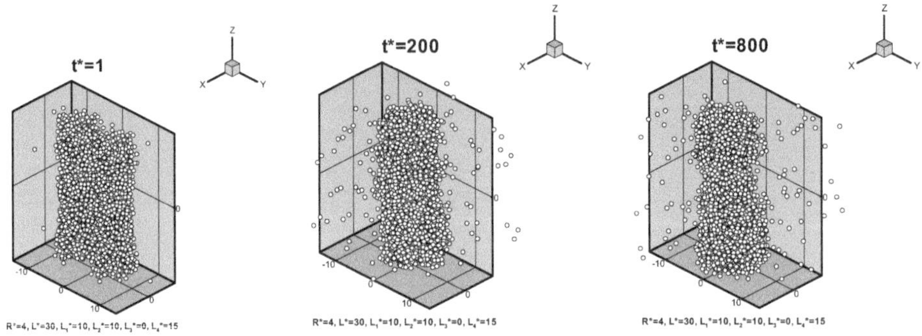

Fig. 8. Vaporization process of two liquid threads of $R^*=4$, $L^*=30$, $L_1^*=10$, $L_2^*=10$, $L_3^*=0$, $L_4^*=15$

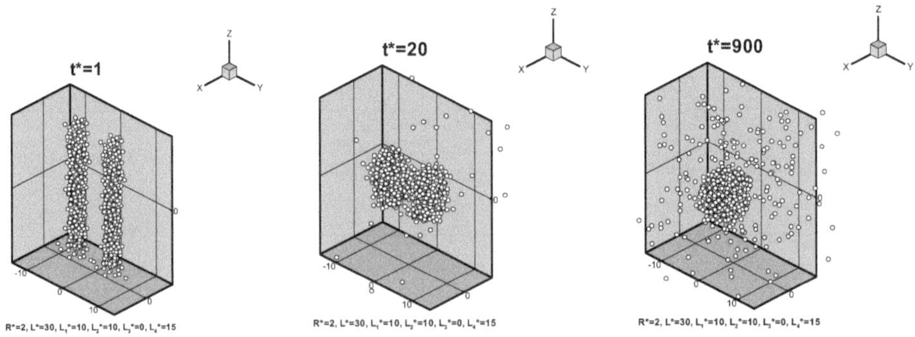

Fig. 9. Vaporization process of two liquid threads of $R^*=2$, $L^*=30$, $L_1^*=10$, $L_2^*=10$, $L_3^*=0$, $L_4^*=15$

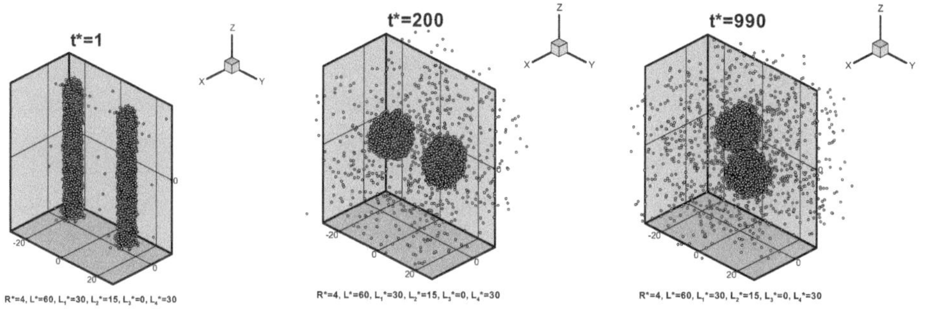

Fig. 10. Vaporization process of two liquid threads of R^*=4, L*=60, L_1*=30, L_2*=15, L_3*=0, L_4*=30

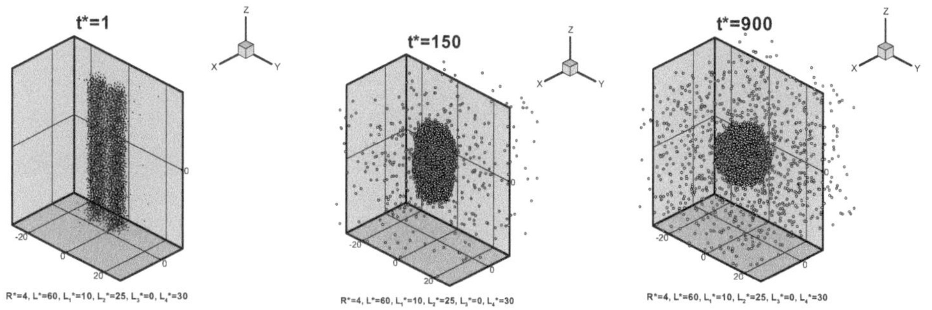

Fig. 11. Vaporization process of two liquid threads of R^*=4, L*=60, L_1*=10, L_2*=25, L_3*=0, L_4*=30

Figure 10 shows the vaporization process of two liquid threads of R^*=4, L*=60, L_1*=30, L_2*=15 and L_4*=30 (case 3 in Table 2) . It is observed that the two threads rupture from their two ends, i.e., the top and bottom surfaces of the fundamental cell, and get shorter due to the contraction motion in their axial directions. Two liquid particles are formed and prevail during the subsequent vaporization process. In contrast to cases 1 and 2 for shorter liquid threads, the two liquid particles do not coalesce. If the two liquid threads get closer to each other, e.g., R^*=4, L*=60, L_1*=10, L_2*=25 and L_4*=30 (case 4 in Table 2) , the vaporization process, which is shown in Fig.11, would be very different from case 3. From Fig.11, it is seen that because the two threads are very close to each other, they quickly coalesce into a single thread. The coalesced liquid thread gets shorter due to the contraction motion in its axial direction and finally evolves into a single liquid particle.

From Table 2, it can be seen that the value of $\overline{f_\rho}$ is larger for case 4. This implies that vaporization is slower when the two liquid threads are close to each other. In addition, Table 2 shows that more liquid particles are formed when the separation of the two threads is larger (case 3 in Table 2) . In our previous study (Yeh, 2009b), we found that molecular interaction plays an important role in the vaporization process. More liquid particles can provide more molecular interactions, which is conducive to vaporization. Comparison of cases 3, 4 and 5 also reveals this tendency. Case 5, which was investigated in our previous study (Yeh, 2009b), denotes the situation for a single liquid thread of the same radius and length as in cases 3 and 4. From Table 2, it is observed that only one liquid particle is formed for case 5 and the vaporization speed for case 5 is slower than case 3, for which two liquid particles are formed. The present results for two liquid threads coexisting in a fundamental cell corroborate the findings of our previous study for a single liquid thread.

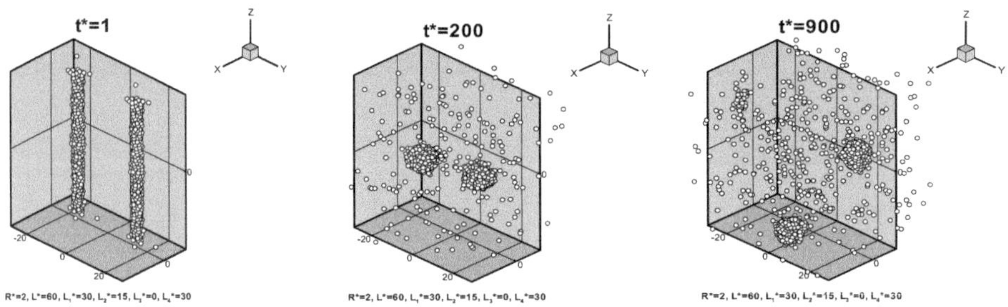

Fig. 12. Vaporization process of two liquid threads of R^*=2, L^*=60, L_1^*=30, L_2^*=15, L_3^*=0, L_4^*=30

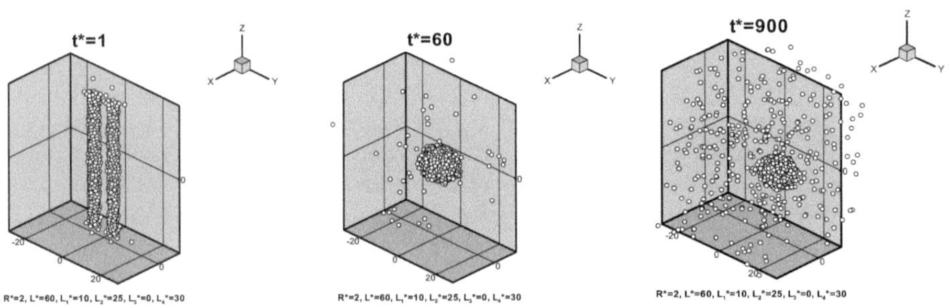

Fig. 13. Vaporization process of two liquid threads of R^*=2, L^*=60, L_1^*=10, L_2^*=25, L_3^*=0, L_4^*=30

Figure 12 shows the vaporization process of two liquid threads of R^*=2, L^*=60, L_1^*=30, L_2^*=15 and L_4^*=30 (case 6 in Table 2) . Similar to case 3, the two threads rupture from their two ends and get shorter due to the contraction motion in their axial directions. Two liquid particles are formed and prevail during the subsequent vaporization process. For two closer liquid threads, e.g., $R^*=2$, $L^*=60$, $L_1^*=10$, $L_2^*=25$ and $L_4^*=30$ (case 7 in Table 2) , the vaporization process, which is shown in Fig.13, is different from case 6. The two threads also rupture from their two ends and get shorter due to the contraction motion in their axial directions. Two liquid particles are formed but they quickly coalesce into a single particle and prevail during the subsequent vaporization process. This vaporization process is similar to but slightly different from that for thicker liquid threads (case 4) . The thinner liquid threads (case 7) evolve into two liquid particles and then quickly coalesce into a single particle while the thicker liquid threads (case 4) quickly coalesce into a single liquid thread and then evolve into a single liquid particle.

From Table 2, it is seen that the value of $\overline{f_\rho}$ is larger for case 7, i.e., vaporization is slower when the two liquid threads are close to each other. In addition, Table 2 shows that more liquid particles are produced when the separation of the two threads is larger (case 6) . Case 8, which was investigated in our previous study (2009b), denotes the situation for a single liquid thread of the same radius and length as those in cases 6 and 7. From Table 2, it is seen that only one liquid particle is formed for case 8 and the vaporization speed for case 8 is slower than case 6, for which two liquid particles are formed. This corroborates the previous results for thicker liquid threads (cases 3, 4 and 5) . Furthermore, comparing cases 6~8 for thinner liquid threads ($R^*=2$) with cases 3~5 for thicker liquid threads ($R^*=4$) , it can be seen that the value of $\overline{f_\rho}$ is smaller for thinner liquid threads, which implies that thinner liquid threads evaporate more quickly. This also corroborates the previous results for shorter liquid threads (cases 1 and 2).

Figure 14 shows the vaporization process of two liquid threads of $R^*=4$, $L^*=120$, $L_1^*=60$, $L_2^*=30$ and $L_4^*=60$ (case 9 in Table 2) . Similar to case 3 for shorter liquid threads, the two threads rupture from their two ends and get shorter due to the contraction motion in their axial directions. Two liquid particles are formed and prevail during the subsequent vaporization process. For two closer liquid threads, e.g., $R^*=4$, $L^*=120$, $L_1^*=40$, $L_2^*=40$ and $L_4^*=60$ (case 10 in Table 2) , the vaporization process, which is shown in Fig.15, is similar to case 9. Two liquid particles are formed and prevail during the subsequent vaporization process. If the two liquid threads get even closer, e.g., $R^*=4$, $L^*=120$, $L_1^*=10$, $L_2^*=55$ and $L_4^*=60$ (case 11 in Table 2) , the vaporization process, which is shown in Fig.16, is quite different from cases 9 and 10. Similar to case 4 for shorter liquid threads, it is observed that because the two threads in case 11 are very close to each other, they quickly coalesce into a single thread before they separately evolve into liquid particles. The coalesced liquid thread then gets shorter due to the contraction motion in its axial direction and finally evolves into a single liquid particle.

Similar to cases 4 and 7, Table 2 shows that the value of $\overline{f_\rho}$ is larger for case 11, i.e., vaporization is slower when the two liquid threads are very close to each other. Similarly, it is seen that more liquid particles are produced when the separation of the two threads is larger (cases 9 and 10) . Case 12, which was investigated in our previous study (Yeh, 2009b), denotes the situation for a single liquid thread of the same radius and length as those in cases 9~11. From Table 2, it is seen that only one liquid particle is formed for case 12 and the vaporization speed for case 12 is slower than in cases 9 and 10, for which two liquid particles are formed. This corroborates the previous results for shorter liquid threads (cases 3~8) .

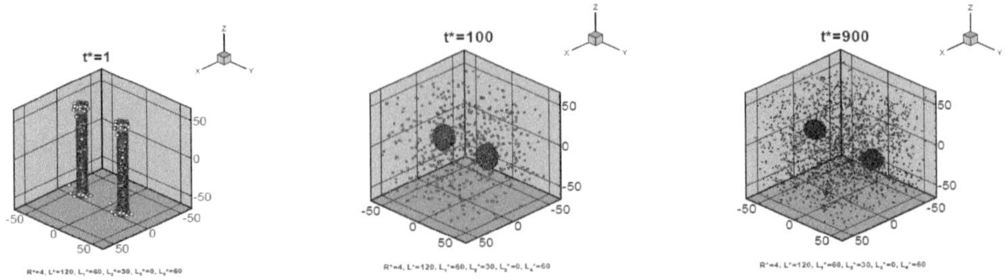

Fig. 14. Vaporization process of two liquid threads of $R^*=4$, $L^*=120$, $L_1^*=60$, $L_2^*=30$, $L_3^*=0$, $L_4^*=60$

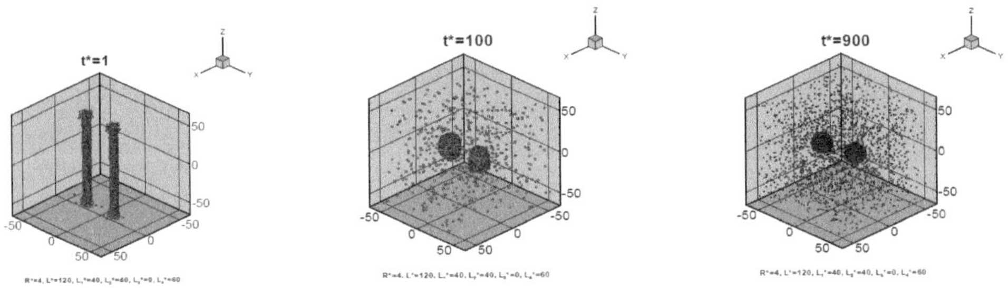

Fig. 15. Vaporization process of two liquid threads of $R^*=4$, $L^*=120$, $L_1^*=40$, $L_2^*=40$, $L_3^*=0$, $L_4^*=60$

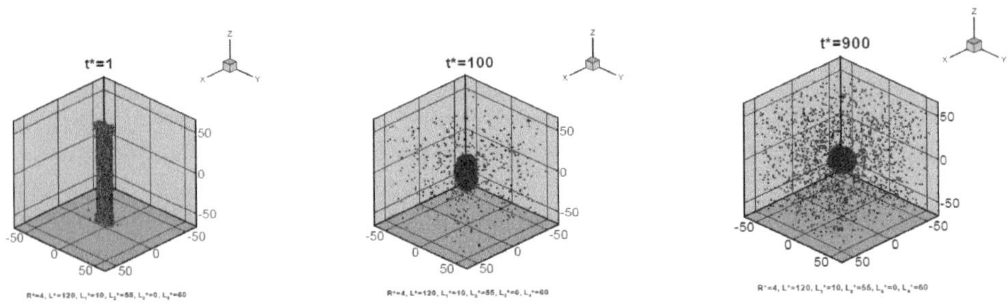

Fig. 16. Vaporization process of two liquid threads of $R^*=4$, $L^*=120$, $L_1^*=10$, $L_2^*=55$, $L_3^*=0$, $L_4^*=60$

Figure 17 shows the vaporization process of two liquid threads of $R^*=2$, $L^*=120$, $L_1^*=60$, $L_2^*=30$ and $L_4^*=60$ (case 13 in Table 2) . For this case of thinner liquid threads, it is seen that the two threads rupture from their two ends and quickly evolve into ten liquid particles. During the vaporization process, collision and coalescence of the liquid particles occur and finally two liquid particles are formed. For two closer liquid threads, e.g., $R^*=2$, $L^*=120$, $L_1^*=40$, $L_2^*=40$ and $L_4^*=60$ (case 14 in Table 2) , the vaporization process, which is shown in Fig.18, is similar to but slightly different from that of case 13. Nine liquid particles are formed initially and they finally evolve into two liquid particles by collision and coalescence. If the two liquid threads get even closer, e.g., $R^*=2$, $L^*=120$, $L_1^*=10$, $L_2^*=55$ and $L_4^*=60$ (case 15 in Table 2) , the vaporization process, which is shown in Fig.19, is different from those in cases 13 and 14. Nine liquid particles are formed, but they finally evolve into a single particle by collision and coalescence.

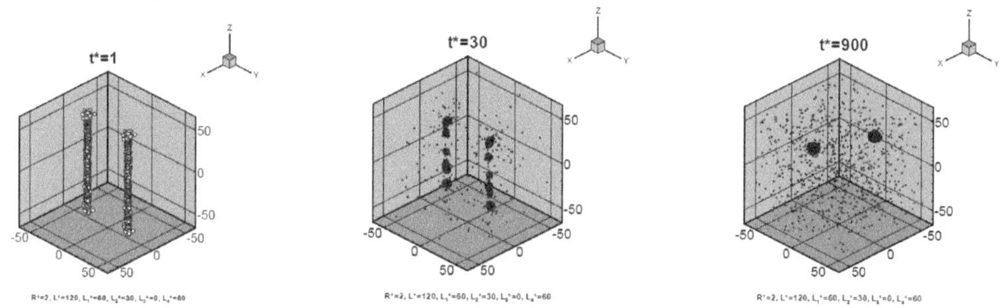

Fig. 17. Vaporization process of two liquid threads of R^*=2, L^*=120, L_1^*=60, L_2^*=30, L_3^*=0, L_4^*=60

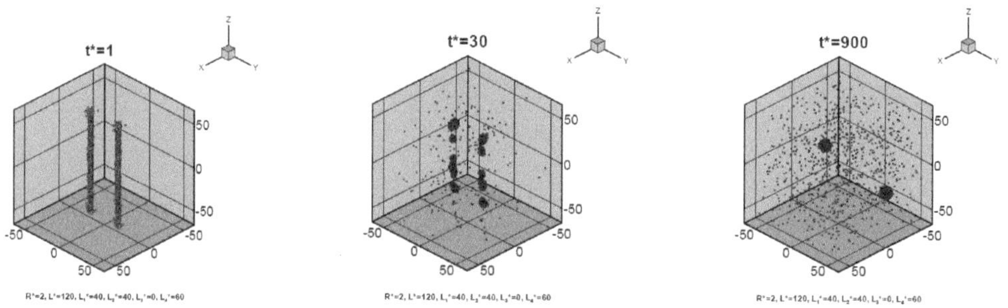

Fig. 18. Vaporization process of two liquid threads of R^*=2, L^*=120, L_1^*=40, L_2^*=40, L_3^*=0, L_4^*=60

Similar to cases 4, 7 and 11, Table 2 shows that the value of $\overline{f_\rho}$ is larger for case 15, i.e., vaporization is slower when the two liquid threads are very close to each other. Similarly, it can also be seen that more liquid particles are formed when the separation of the two threads is larger (cases 13 and 14) . Case 16, which was investigated in our previous study (Yeh, 2009b), denotes the situation for a single liquid thread of the same radius and length as those in cases 13~15. From Table 2, it is seen that less liquid particles are formed for case 16 and the vaporization speed for case 16 is slower than in cases 13 and 14, in which more liquid particles are formed. This corroborates the previous results (cases 3~12) . Furthermore, comparing cases 13~16 for thinner liquid threads (R^*=2) with cases 9~12 for thicker liquid threads (R^*=4) , it is seen that the value of $\overline{f_\rho}$ is smaller for thinner liquid threads, which implies that thinner liquid threads evaporate more quickly. This also corroborates the previous results for shorter liquid threads (cases 1~8).

Figure 20 shows the vaporization process of two liquid threads of R^*=4, L^*=240, L_1^*=120, L_2^*=60 and L_4^*=120 (case 17 in Table 2) . Similar to cases 3 and 9 for shorter liquid threads, the two threads rupture from their two ends and get shorter due to the contraction motion in their axial directions. Two liquid particles are formed and prevail during the subsequent vaporization process. For two closer liquid threads, e.g., R^*=4, L^*=240, L_1^*=40, L_2^*=100 and L_4^*=120 (case 18 in Table 2) , the vaporization process, which is shown in Fig.21, is similar to that in case 17. If the two liquid threads get even closer, e.g., $R^*=4$, $L^*=240$, $L_1^*=10$, $L_2^*=115$ and $L_4^*=120$ (case 19 in Table 2) , the vaporization process, which is shown in Fig.22, is quite different from those in cases 17 and 18. Similar to cases 4 and 11 for shorter liquid threads, it is seen that because the two threads in case 19 are very close to each other, they quickly coalesce into a single thread before they separately evolve into liquid particles. The coalesced liquid thread then becomes shorter due to the contraction motion in its axial direction and finally evolves into a single liquid particle.

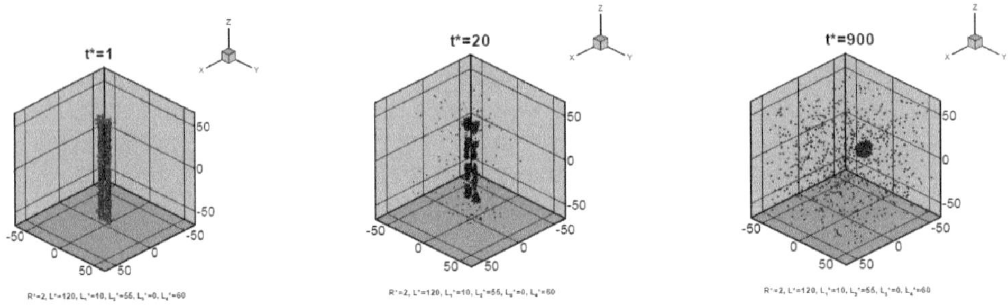

Fig. 19. Vaporization process of two liquid threads of $R^*=2$, $L^*=120$, $L_1^*=10$, $L_2^*=55$, $L_3^*=0$, $L_4^*=60$

From Table 2, it is observed that the value of $\overline{f_\rho}$ is larger for case 19, i.e., vaporization is slower when the two liquid threads are very close to each other. Similarly, it can be seen that more liquid particles are produced when the separation of the two threads is larger (cases 17 and 18) . Case 20, which was investigated in our previous study (Yeh, 2009b), denotes the situation for a single liquid thread of the same radius and length as those in cases 17~19. From Table 2, it is seen that only one liquid particle is formed for case 20 and the vaporization speed for case 20 is slower than those in cases 17 and 18, for which two liquid particles are formed. This corroborates the previous results for shorter liquid threads (cases 3~16) .

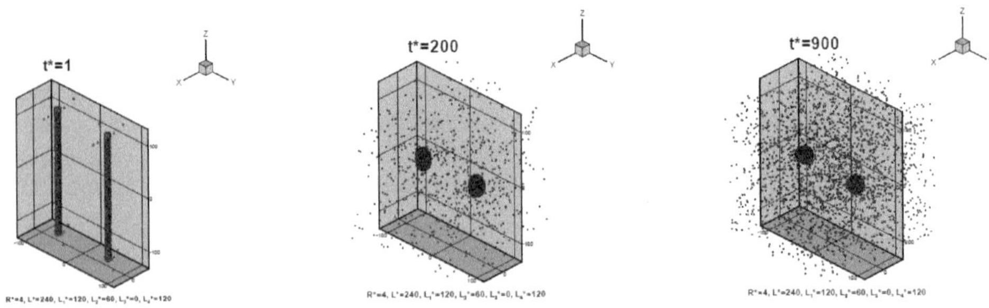

Fig. 20. Vaporization process of two liquid threads of $R^*=4$, $L^*=240$, $L_1^*=120$, $L_2^*=60$, $L_3^*=0$, $L_4^*=120$

If the length of the liquid thread is further increased, e.g. $R^*=4$ and $L^*=480$, more liquid particles are formed. Figure 23 shows the vaporization process of two liquid threads of $R^*=4$, $L^*=480$, $L_1^*=160$, $L_2^*=160$ and $L_4^*=240$ (case 21 in Table 2) . Eight liquid particles are formed and they evolve into four liquid particles by collision and coalescence. For two closer liquid threads, e.g., $R^*=4$, $L^*=480$, $L_1^*=40$, $L_2^*=220$ and $L_4^*=240$ (case 22 in Table 2) , the vaporization process, which is shown in Fig.24, is similar to but slightly different from that in case 21. Six liquid particles are formed initially and they finally evolve into two liquid particles by collision and coalescence. If the two liquid threads get even closer, e.g., $R^*=4$, $L^*=480$, $L_1^*=10$, $L_2^*=235$ and $L_4^*=240$ (case 23 in Table 2) , the vaporization process, which is shown in Fig.25, is very different from those in cases 21 and 22. Because the two threads in case 23 are very close to each other, they quickly coalesce into a single thread before they separately evolve into liquid particles. The coalesced liquid thread ruptures from its ends and interior and gets shorter due to the contraction motion in its axial direction and finally evolves into a single liquid particle by collision and coalescence. The vaporization process for case 23 (Fig.25) is similar to but slightly different from that in case 19 (Fig.22) . For both cases, the two threads quickly coalesce into a single thread because they are very close to each other. However, the coalesced liquid thread in case 19 evolves into a single liquid particle without rupturing in its interior while the coalesced liquid thread for case 23 ruptures in its interior before it finally evolves into a single liquid particle.

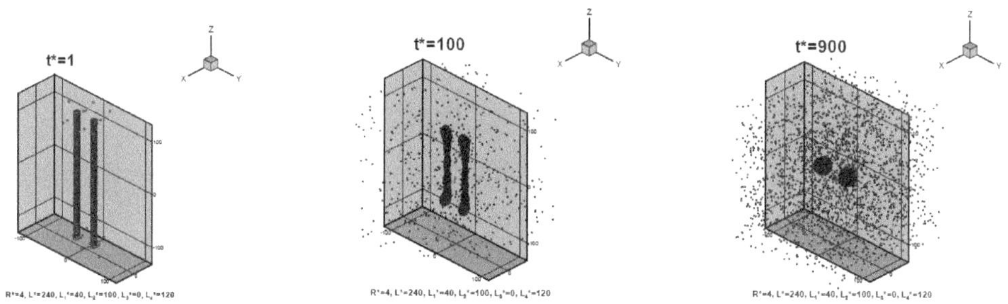

Fig. 21. Vaporization process of two liquid threads of $R^*=4$, $L^*=240$, $L_1^*=40$, $L_2^*=100$, $L_3^*=0$, $L_4^*=120$

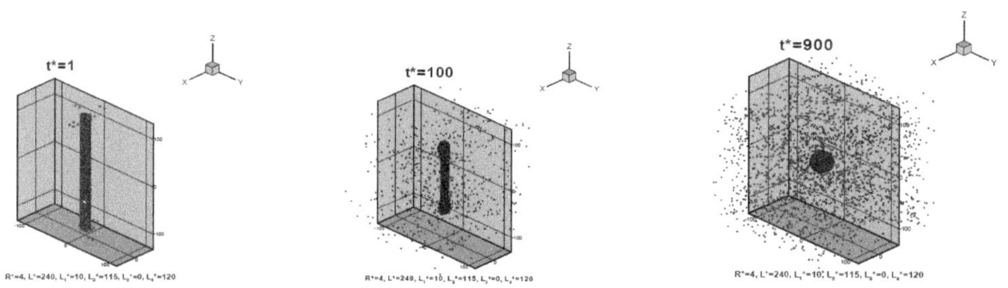

Fig. 22. Vaporization process of two liquid threads of $R^*=4$, $L^*=240$, $L_1^*=10$, $L_2^*=115$, $L_3^*=0$, $L_4^*=120$

From Table 2, it is observed that the value of $\overline{f_\rho}$ is larger for case 23, i.e., vaporization is slower when the two liquid threads are very close to each other. Similarly, it can be seen that more liquid particles are produced when the separation of the two threads is larger (cases 21 and 22) . Case 24, which was investigated in our previous study (Yeh, 2009b), denotes the situation for a single liquid thread of the same radius and length as those in cases 21~23. From Table 2, it can be seen that less liquid particles are formed for case 24 and the vaporization speed for case 24 is slower than those in cases 21 and 22, in which more liquid particles are formed. This corroborates the previous results for shorter liquid threads (cases 3~20).

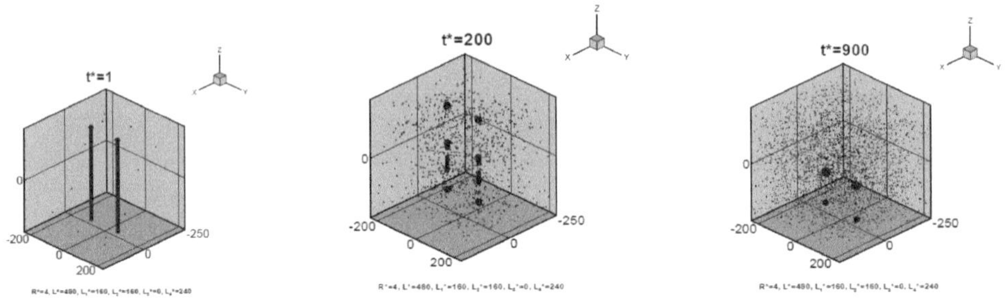

Fig. 23. Vaporization process of two liquid threads of R^*=4, L^*=480, L_1^*=160, L_2^*=160, L_3^*=0, L_4^*=240

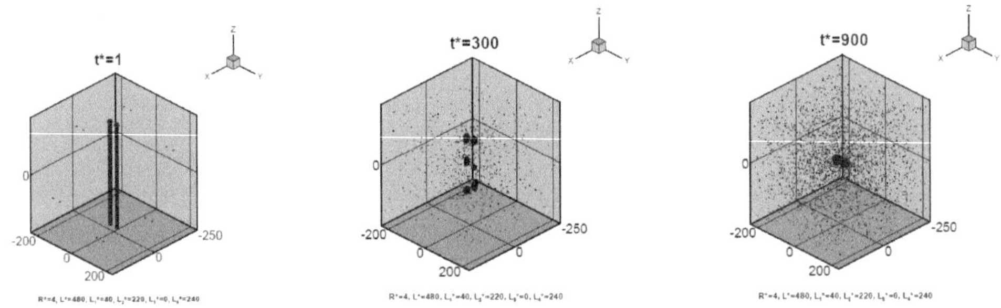

Fig. 24. Vaporization process of two liquid threads of R^*=4, L^*=480, L_1^*=40, L_2^*=220, L_3^*=0, L_4^*=240

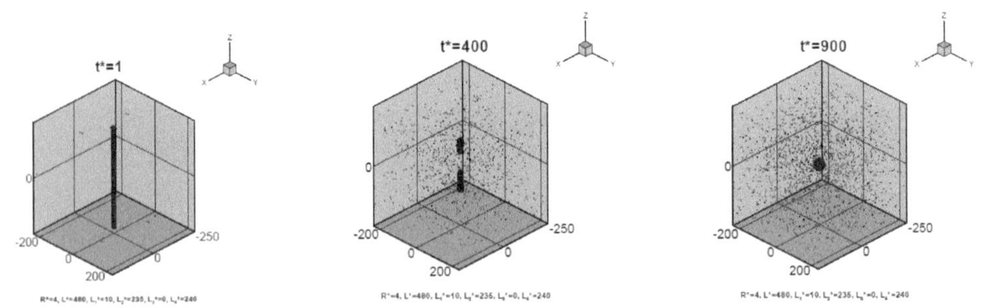

Fig. 25. Vaporization process of two liquid threads of R^*=4, L^*=480, L_1^*=10, L_2^*=235, L_3^*=0, L_4^*=240

From the above discussion, the following observations can be made.

First, if the fundamental cell length is small, the liquid threads may remain intact or evolve into only one liquid particle in the cell. If the threads break up in this case, they rupture from their ends only, i.e. the top and bottom surfaces of the fundamental cell, but not from their interiors. On the other hand, if the fundamental cell length is larger, more than one liquid particle may be produced in the cell and the liquid threads may rupture not only from their ends but also from their interiors.

Second, thinner liquid threads may produce more liquid particles in the cell and evaporate more quickly.

Third, more liquid particles are formed when the separation of the two threads is larger. Moreover, vaporization is slower when the two liquid threads are close to each other.

Fourth, on the basis of identical liquid thread radius and length, liquid threads that produce more liquid particles evaporate more quickly.

2.2.2 Stability analysis

In the following discussion, two linear stability criteria, namely Rayleigh's stability criterion (1879) and Kim's stability criterion (2006), are accessed for their validity in molecular scale. This is the first study to investigate the instability of two nano-scale liquid threads coexisting in a periodic fundamental cell by MD simulation.

According to classical theory by Rayleigh (1879), the radius of a liquid thread, R, is related to the critical wavelength of perturbation, λ_c , by Eq.(3) in section 2.1.3. A liquid thread will break up into drops if the axial wavelength of the surface perturbation $L>\lambda_c$. If $L<\lambda_c$, the thread is stable and will remain intact. Owing to the periodic boundary conditions in this study, the fundamental cell size L^* can be regarded as the longest wavelength of the perturbation.

For $R^*=2$, λ_c^* is 12.6. According to Rayleigh's stability criterion, liquid threads of radius $R^*=2$ and length $L^*=30$, 60 and 120 are expected to be unstable and will break up into drop(s). From Figs.9, 12, 13, 17, 19 or Table 2 (cases 2, 6~8, 13~16) , Rayleigh's stability criterion holds for these cases.

For $R^*=4$, λ_c^* is 25.2. According to Rayleigh's stability criterion, liquid threads of radius $R^*=4$ and length $L^*=30$, 60, 120, 240 and 480 are expected to be unstable and will break up into drop(s). However, from Figs.8, 10, 11, 14~16, 20~25 or Table 2 (cases 1, 3~5, 9~12, 17~24) , the liquid threads of $R^*=4$ and $L^*=30$ (case 1) remain intact while the remaining liquid threads are unstable and break up into drop(s). Rayleigh's stability criterion violates the MD simulation result for case 1 but holds for the remaining cases. The invalidity of Rayleigh's stability criterion for case 1 can be interpreted according to the vaporization process shown in Fig.8. It is seen that the two liquid threads quickly coalesce into a single thread and remain intact. This is somewhat like combining the two threads to form a thread of a larger radius, which can increase the critical wavelength of perturbation, λ_c, according to Eq.(3) and, therefore, broaden the stable domain.

Recently, Kim, Lee, Han and Park (2006) proposed a new linear relation from their MD simulation results. From their study, Rayleigh's linear relation can be modified by Eq.(4) in section 2.1.3. From Eq.(4), λ_c^* for R^* of 2 and 4 are 5.9 and 21.2, respectively. Then, from Table 2, it is seen that Kim's stability criterion violates the MD simulation result for case 1 while it holds for the remaining cases. The validity of Kim's stability criterion is similar to that of Rayleigh's stability criterion.

From the above discussion, it is found that the trends of Rayleigh's stability criterion and Kim's stability criterion agree with MD simulation results. However, when the two liquid threads coalesce into a single thread and remain intact, the critical wavelength of perturbation may be increased and the stable domain is broadened. In such a situation, Rayleigh's stability criterion and Kim's stability criterion underpredict the stable domain.

2.3 Molecular dynamics simulation for the atomization process of a nanojet

In this section, liquid argon nanojets made of 44000 Lennard-Jones molecules are investigated under various simulation parameters to examine their influence on the nanojet atomization process. Snapshots of the molecules, evolution of the density field, and evolution of the intermolecular force are analyzed. This can provide insight into the fundamental mechanism of the atomization process and will be helpful for the design of nanojet devices such as nano-printer or nano-sprayer.

The nano-atomizer is schematically shown in Fig.26. The simulation domain comprises a cubical box of side length 360σ, with periodic boundary conditions applied in all three directions. The nano-atomizer is placed at the center of the box. Simulation parameters are listed in Table 3, which include nano-atomizer dimensions, temperatures, number of molecules and simulation results.

The argon molecules inside the nano-atomizer are liquid and the nano-atomizer is made of rigid argon molecules. A push panel composed of 600 argon molecules is constructed with a downward velocity of 120m/s. In this study, the interactions among liquid argon molecules, nano-atomizer and push panel are taken into account.

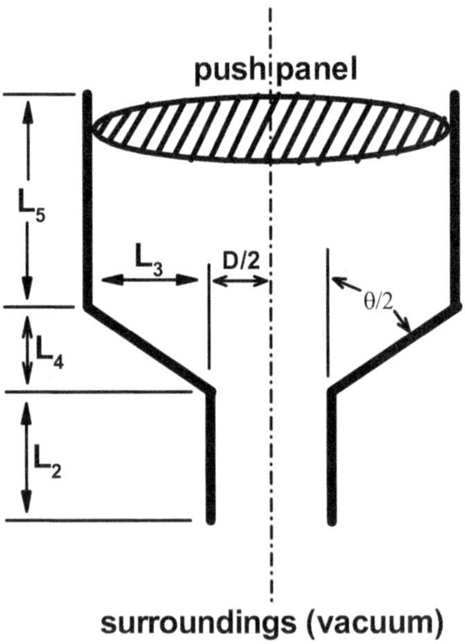

Fig. 26. Illustration of the nano-atomizer configuration and dimensions

Case	L_2^*	L_3^*	L_4^*	L_5^*	$D^*/2$	T_D^*	N	$\overline{f_\rho}$
1	5.73	5.73	5.73	76.7	8.81	0.75	43970	1.34
2	5.73	5.73	5.73	76.7	8.81	1.5	43970	1.08
3	5.73	5.73	5.73	76.7	8.81	2.0	43970	0.94
4	5.73	5.73	5.73	76.7	8.81	3.0	43970	0.74
5	5.73	5.73	5.73	76.7	8.81	4.5	43970	0.59
6	8.66	5.73	2.79	76.7	8.81	2.0	43989	0.95
7	2.79	5.73	8.66	76.7	8.81	2.0	43949	0.92
8	5.73	8.67	5.73	76.7	5.87	2.0	43934	1.27
9	8.66	5.73	2.79	76.7	8.81	3.0	43989	0.75
10	2.79	5.73	8.66	76.7	8.81	3.0	43949	0.72
11	5.73	8.67	5.73	76.7	5.87	3.0	43934	1.10

Table 3. Nano-atomizer dimensions, temperatures, number of molecules and simulation results

In the following discussion, a liquid argon nanojet of length L_5^* and diameter $2L_3^*+D^*$ is pushed by a panel into vacuum through a nano-nozzle of orifice diameter D^*, as illustrated in Fig.26. Simulation conditions are listed in Table 3.

2.3.1 Nanojet atomization process

Figure 27 shows the atomization process for a nanojet of $L_2^*=L_3^*=L_4^*=5.73$, $L_5^*=76.7$, $D^*/2=8.81$ and $T^*=0.75$, which corresponds to a nanojet of length 26.2 nm and diameter 10 nm, and a nano-nozzle of orifice length 2nm, diameter 6 nm, as well as an actual temperature of 70 K. The dot in Fig.27 indicates the center of the molecule. From the figure it is found that the nanojet does not break up. Owing to the low temperature, the molecular kinetic energies are so low that the molecules congregate near the orifice exit. Very few liquid molecules evaporate at this low temperature. At a higher temperature $T^*=1.5$ (140 K) , as depicted in Fig.28, the molecules leave the orifice exit earlier than at $T^*=0.75$, due to their higher molecular kinetic energies. More liquid molecules evaporate at this higher temperature. However, like at $T^*=0.75$, the nanojet has not broken up before $t^*=80$. If the temperature is further increased to $T^*=2.0$ (187 K) , as shown in Fig.29, evident evaporation is observed. Many evaporated molecules are produced and the non-evaporated liquid molecules concentrate within the central region. Figure 30 depicts the snapshots for temperature $T^*=3.0$ (278 K) . It is observed that breakup of the nanojet occurs and its spray angle is larger than at $T^*=2.0$. The spurted molecules from the nano-atomizer are more evenly distributed at this temperature. If the temperature is further increased to $T^*=4.5$ (420 K) , as shown in Fig.31, the spray angle is even larger than at $T^*=3.0$ and the spurted molecules from the nano-atomizer are much more uniformly distributed as compared to the lower temperature cases. Comparison of Figs.27, 28, 29, 30 and 31 reveals that the liquid nanojet evaporates quicker at higher temperatures. This will be further illustrated in later sections discussing the density distribution and the intermolecular force.

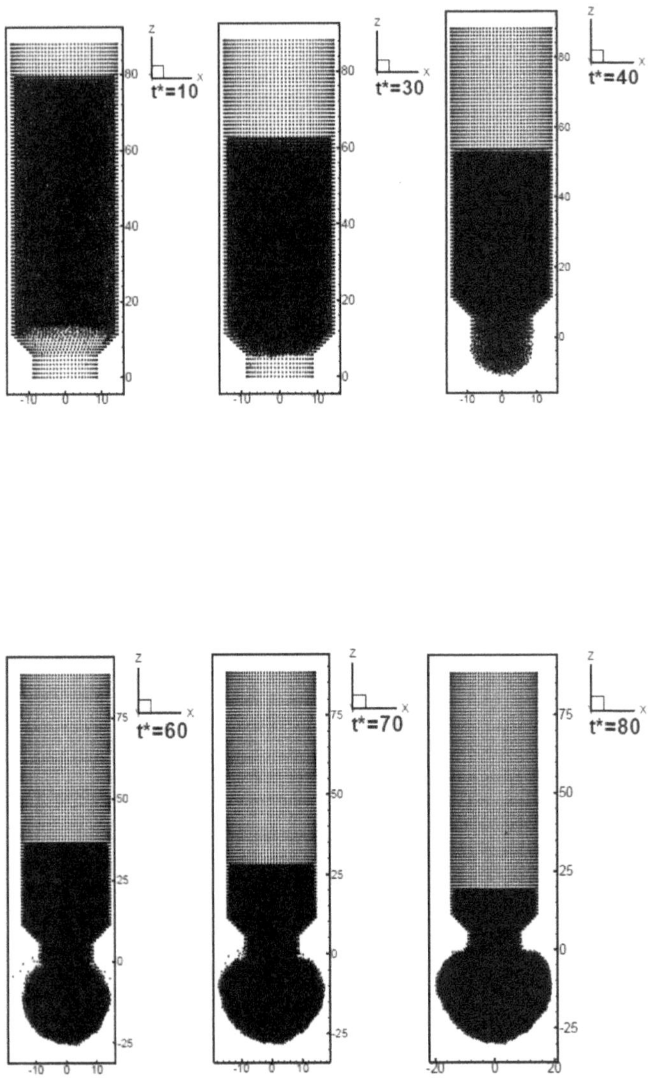

Fig. 27. Atomization process for case 1 in Table 3 (L_2^* =5.73, L_3^* =5.73, L_4^* =5.73, L_5^* =76.7, $D^*/2$=8.81, T_D^*=0.75)

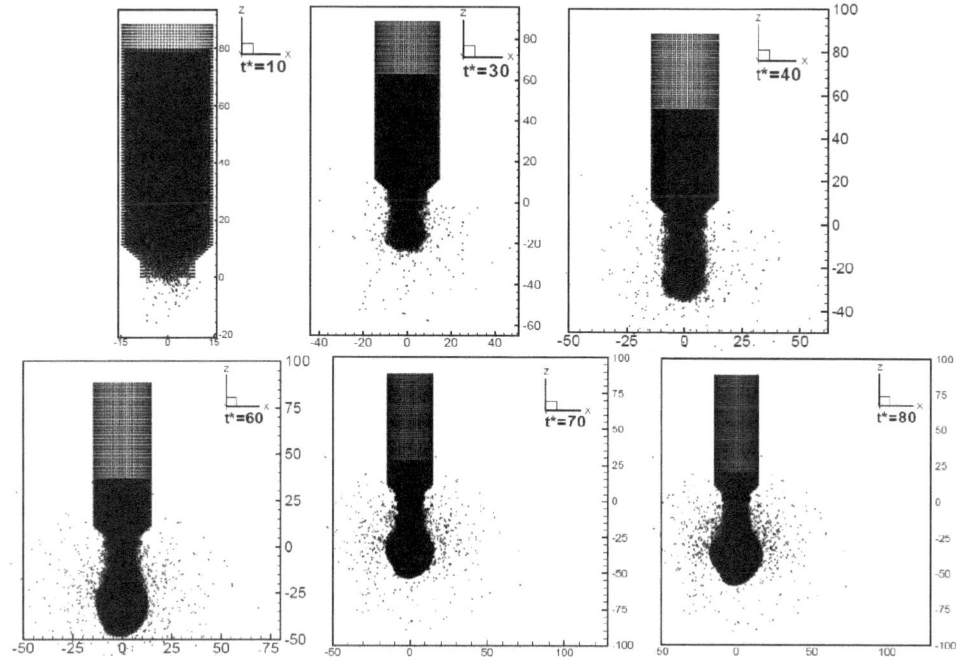

Fig. 28. Atomization process for case 2 in Table 3 (L_2^* =5.73, L_3^* =5.73, L_4^* =5.73, L_5^* =76.7, $D^*/2$=8.81, T_D^*=1.5)

To investigate the influence of nozzle geometry on the nanojet atomization, comparison of the snapshots at t*=80 for four different nozzle geometries and T*=2.0 or 3.0 is shown in Figs.32 and 33. Note that Figs.32(a), (b), (c) and (d) correspond to cases 6, 3, 7 and 8, respectively, in Table 3; while Figs.33(a), (b), (c) and (d) correspond to cases 9, 4, 10 and 11, respectively, in Table 3. In Figs.32(a), (b), (c) and 9(a), (b), (c), the nozzle orifice diameters are equal (6 nm) but the nozzle orifice lengths are varied; while in Figs.32(d) and 33(d), the nozzle orifice length is the same as for Figs.32(b) and 33(b) (2 nm) but the nozzle orifice diameter is smaller (4 nm) . By a careful comparison of Figs.32(a), (b) and (c), it can be observed that, on the basis of identical nozzle orifice diameter, a nanojet from a nozzle with a shorter orifice length (L_2) moves farther. On the other hand, from Figs.32(b) and (d), on the basis of identical nozzle orifice length, a nanojet from a nozzle with a larger orifice diameter moves farther. Figure 52 reveals similar tendency. Note that the nano-atomizer in this research is basically a plain-orifice atomizer. As pointed out by Lefebvre (1989) , in a practical plain-orifice atomizer, resistance increases with nozzle orifice length/diameter ratio. Therefore, a nanojet from a nozzle with a smaller orifice length/diameter ratio moves farther due to its smaller resistance. This will be further illustrated in later sections discussing the density distribution and the intermolecular force.

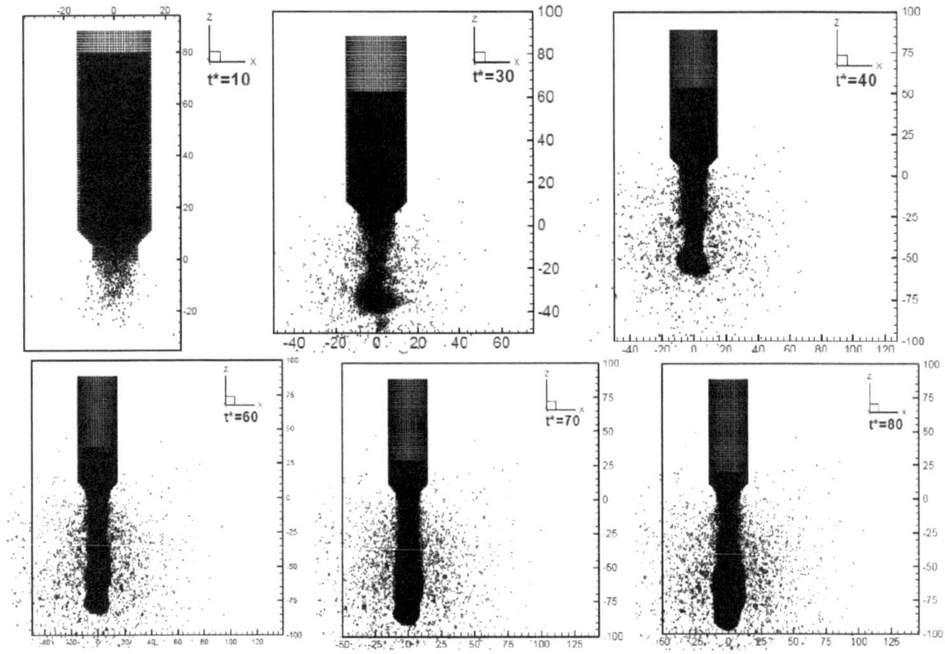

Fig. 29. Atomization process for case 3 in Table 3 (L_2^* =5.73, L_3^* =5.73, L_4^* =5.73, L_5^* =76.7, $D^*/2$=8.81, T_D^*=2.0)

2.3.2 Density distribution

It is important that the system be in equilibrium state before statistical values of the local properties can be taken. However, owing to the computational capacity limitations, the MD simulation can not proceed to a macroscopically long period. Nevertheless, the purpose of this paper is not to discuss statistical values of the local properties but to investigate the atomization process of a nanojet, which is important and conducive to the understanding of the fundamental mechanism of the atomization process. Criteria have to be made to quantify the discussion regarding the nanojet atomization process. Unfortunately, such criteria are still arbitrary in the literature. Because the system temperature in this study is kept at the desired temperature, a constant temperature criterion is not suitable for the discussion of the atomization process. In this research, a nanojet is considered to vaporize faster if the distribution of molecules reaches a uniform state quicker during the atomization process. This criterion essentially concerns with the evolution of the density distribution. The density at a specified point in the fundamental cell can be defined as Eq.(5). In this study, the volume δV is taken to be a sphere with non-dimensionalized radius $R^*=2$ and with its center located at the point considered. This is an optimal choice after numerical test.

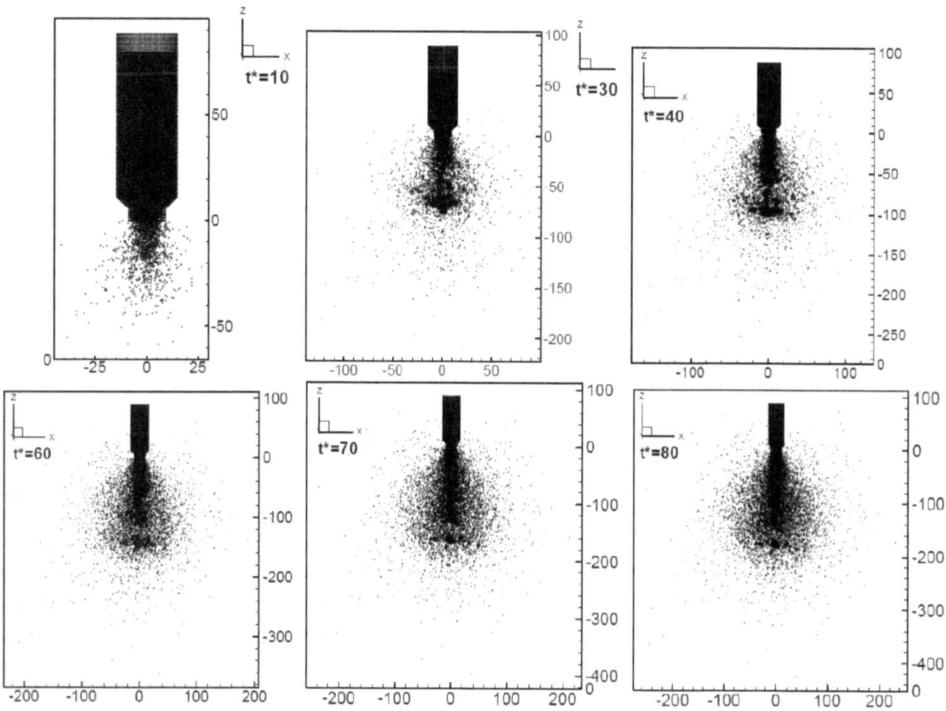

Fig. 30. Atomization process for case 4 in Table 3 (L_2^* =5.73, L_3^* =5.73, L_4^* =5.73, L_5^* =76.7, $D^*/2$=8.81, T_D^*=3.0)

Figure 34 shows the evolution of density uniformity factor for nanojets at different temperatures and the conditions of $L_2^*=L_3^*=L_4^*=5.73$, $L_5^*=76.7$, $D^*/2=8.81$ (cases 1~5 in Table 3). The density uniformity factor is defined as Eq.(6). From Fig.34 it is observed that a higher temperature nanojet evaporates faster than a lower temperature one and this corroborates the results of Figs.27~31 as discussed in section 2.3.1. The time averaged value of the density uniformity factor, $\overline{f_\rho}$, in a time interval of t*=0 to 80, as listed in Table 3, also reveals this observation. In Fig.34, it is noted that at lower temperatures

(T*=0.75 and 1.5) , the density uniformity factor increases first and then decreases. For a lower temperature nanojet, the momenta of the liquid molecules away from the push panel in the nano-atomizer are low while the molecules near the push panel have relatively higher momenta due to the action of the push panel. This results in a compression effect that leads to the increase of the density uniformity factor at the earlier stage of the atomization process; while at a later stage, the density uniformity factor drops because of the ejection of the molecules.

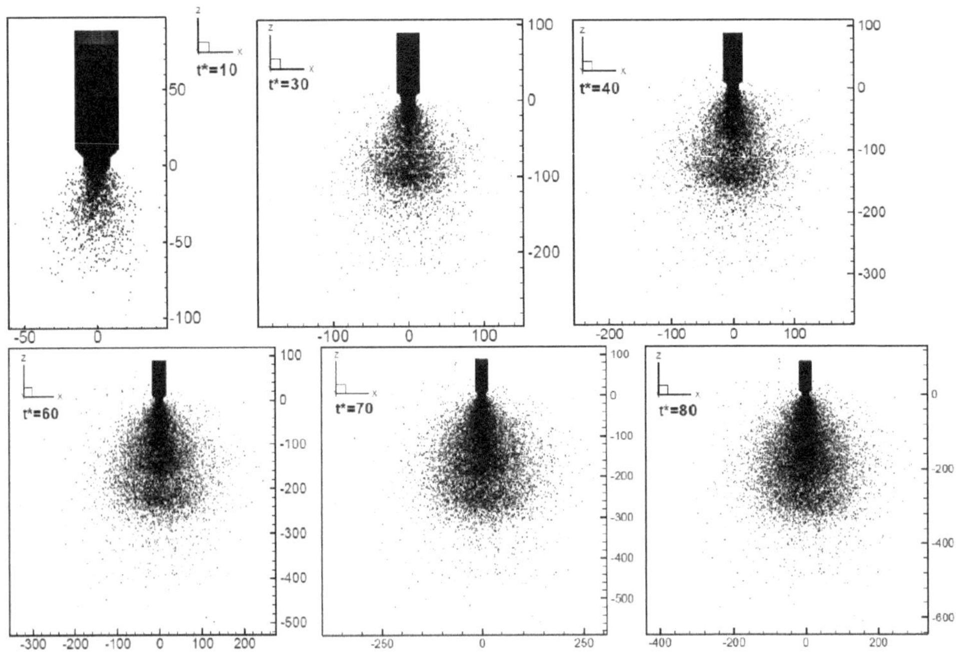

Fig. 31. Atomization process for case 5 in Table 3 (L_2^* =5.73, L_3^* =5.73, L_4^* =5.73, L_5^* =76.7, $D^*/2$=8.81, T_D^*=4.5)

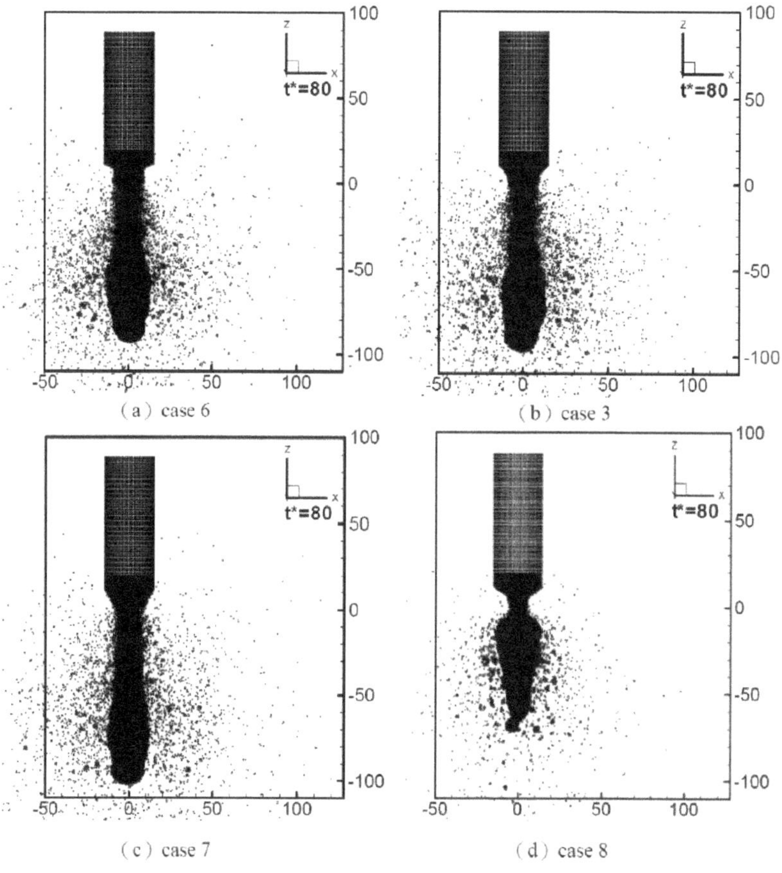

Fig. 32. Comparison of the snapshots at t*=80 and T*=2.0 for four different nozzle geometries

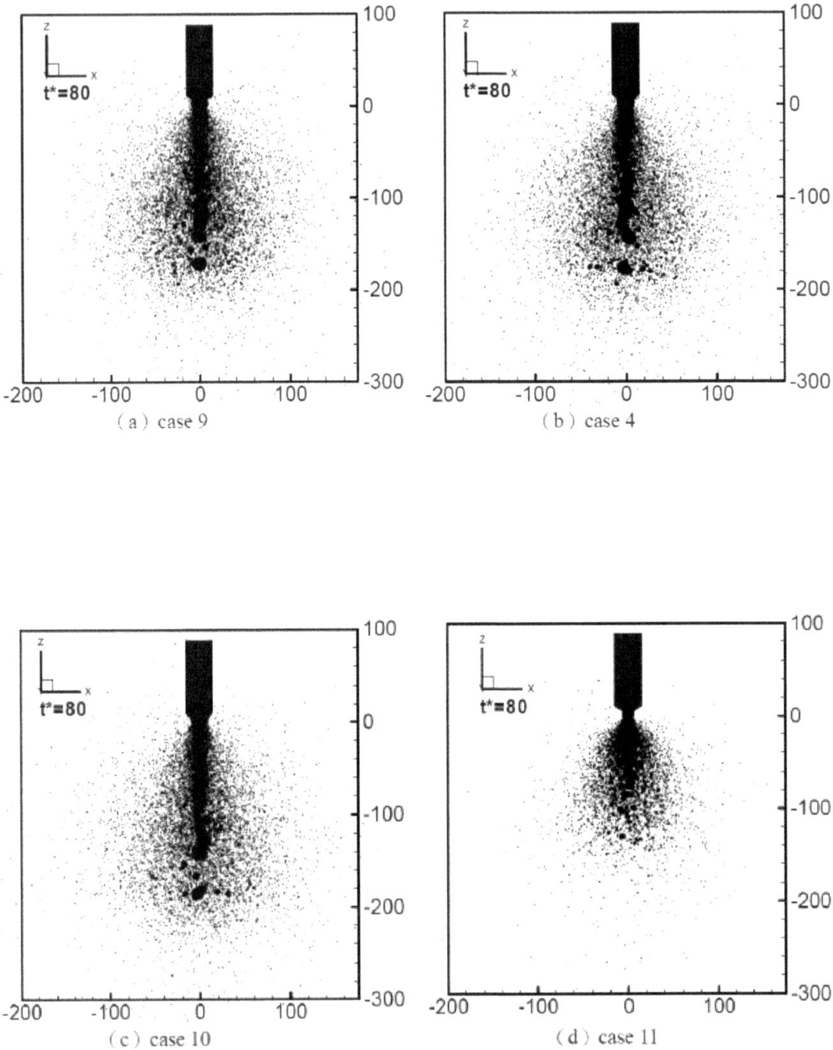

Fig. 33. Comparison of the snapshots at t*=80 and T*=3.0 for four different nozzle geometries

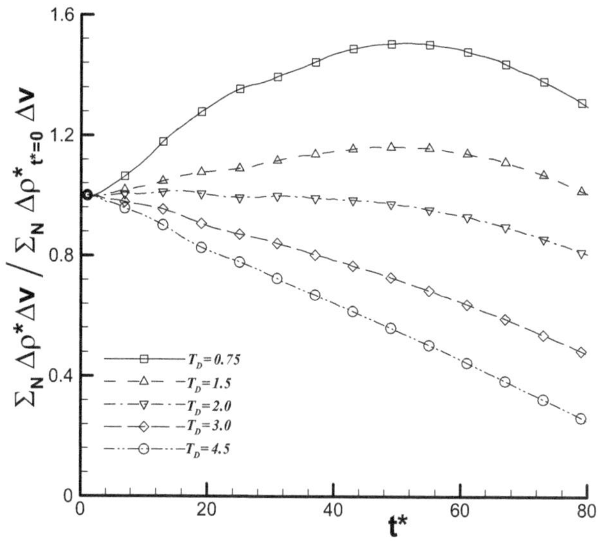

Fig. 34. Evolution of the density uniformity factor for different temperatures (cases 1~5 in Table 3)

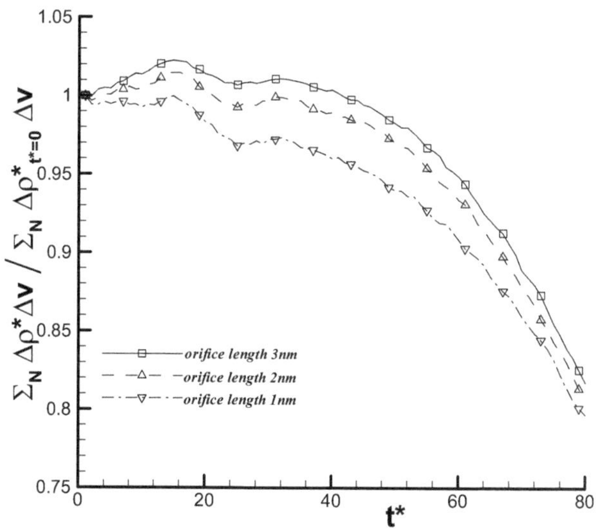

Fig. 35. Evolution of the density uniformity for different orifice lengths at (cases 3, 6 and 7 in Table 3)

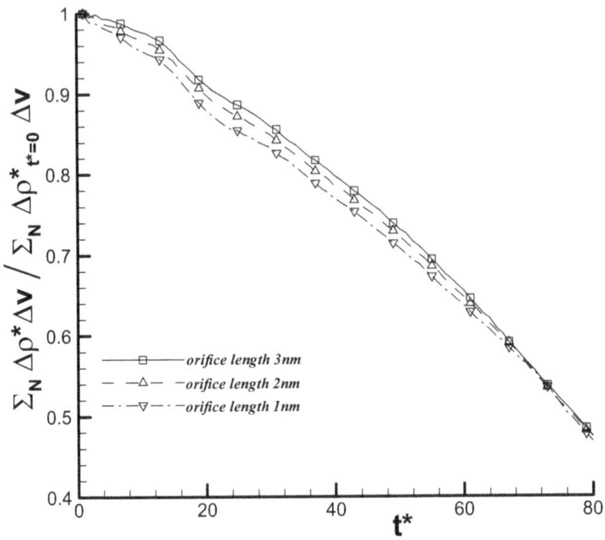

Fig. 36. Evolution of the density uniformity factor for different orifice lengths at T*=3.0 (cases 4, 9 and 10 in Table 3)

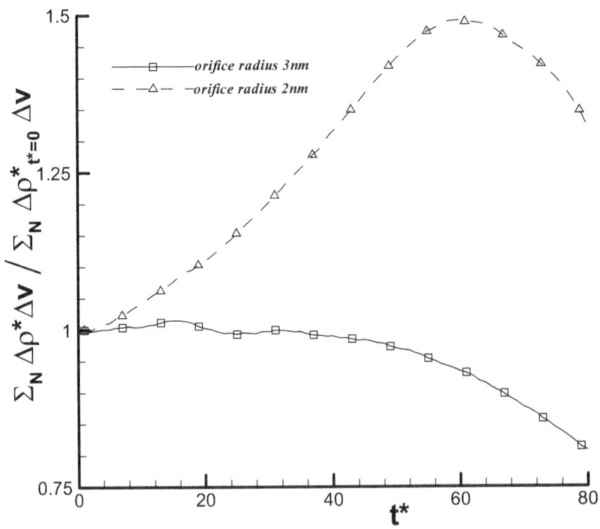

Fig. 37. Evolution of the density uniformity factor for different orifice radii at T*=2.0 (cases 3 and 8 in Table 3)

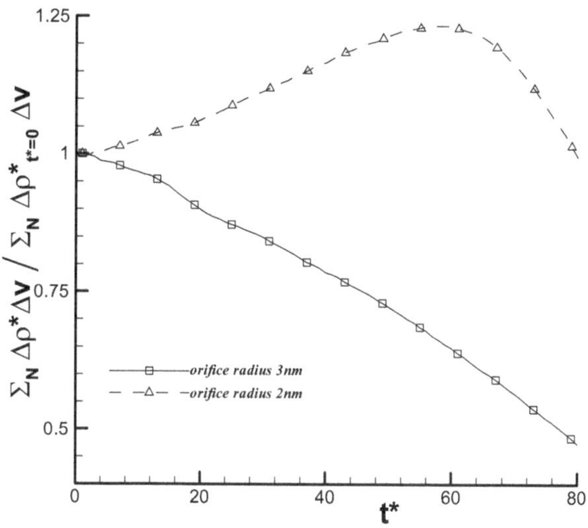

Fig. 38. Evolution of the density uniformity factor for different orifice radii at T*=3.0 (cases 4 and 11 in Table 3)

Figure 35 shows the evolution of density uniformity factor for different orifice lengths on the basis of identical nozzle orifice diameter (6 nm) and the conditions of L_3^*=5.73, L_5^*=76.7, T*=2.0 (cases 3, 6 and 7 in Table 3). It is observed that a nanojet with a shorter orifice length evaporates quicker. The time averaged value of the density uniformity factor, $\overline{f_\rho}$, in a time interval of t^*=0 to 80, as listed in Table 3, also reveals this observation. This corroborates the results of Figs.32(a), (b) and (c) as discussed in section 2.3.1. In addition, it is also observed that, as time elapsed, the influence of the orifice length mitigates. This is because as time elapsed, more and more molecules spurt from the atomizer and hence the interaction between the liquid molecules and the rigid atomizer molecules mitigates due to the decrease of number of molecules inside the atomizer. Figure 36 also reveals this tendency. However, as can be observed from Fig.36, the influence of the orifice length becomes less pronounced at a higher temperature because of the higher molecular kinetic energy to overcome the resistance caused by the orifice. Figure 37 shows the evolution of density uniformity factor for different orifice diameters on the basis of identical nozzle orifice length (2 nm) and the conditions of L_4^*=5.73, L_5^*=76.7, T*=2.0 (cases 3 and 8 in Table 3). It is observed that a nanojet with a larger orifice diameter evaporates quicker. The time averaged value of the density uniformity factor, $\overline{f_\rho}$, in a time interval of t^*=0 to 80, as listed in Table 3, also reveals this observation. This corroborates the results of Figs.32(a) and (d) and also reveals previous observation that a nanojet from a nozzle with a smaller orifice length/diameter ratio evaporates quicker. Similar tendency is obtained from Fig.38 for a higher temperature nanojet.

2.3.3 Intermolecular force

The intermolecular force is an indication of the surface tension experienced by the liquid particles and has a great effect upon the atomization process. Lefebvre (1989), Chigier (1999) and Hiroyasu (2000) pointed out that surface tension and interfacial force are the major controlling mechanisms for atomization. Owing to the vacuum environment, the aerodynamic effect on the atomization process is negligible in this study. Thus, the surface tension becomes the major controlling mechanism for the atomization process. Figure 39 shows the evolution of averaged non-dimensionalized intermolecular force for nanojets with different temperatures and the conditions of $L_2^*=L_3^*=L_4^*=5.73$, $L_5^*=76.7$, $D^*/2=8.81$ (cases 1~5 in Table 3). The averaged non-dimensionalized intermolecular force at time t* is defined as

$$\overline{F_{t^*}^*} = \frac{\sum_{i=1}^{N} F_{i,t^*}^*}{N} \qquad (7)$$

where N is the total number of molecules in the fundamental cell and F^*_{i,t^*} is the resultant force of the non-dimensionalized intermolecular force vector acting on molecule i at time t* , i.e. $F^*_{i,t^*} = (F^*_{x,i,t^*}{}^2 + F^*_{y,i,t^*}{}^2 + F^*_{z,i,t^*}{}^2)^{1/2}$, where F^*_{x,i,t^*}, F^*_{y,i,t^*} and F^*_{z,i,t^*} are the components of the intermolecular force vector at the x, y and z directions, respectively, acting on molecule i at time t*. Note that in the above definition of $\overline{F_{t^*}^*}$, N is the total number of molecules in the fundamental cell, which includes liquid, vapor and solid molecules (atomizer and push panel) ; while in the definition of density uniformity factor, Eq.(6), N is only the initial number of liquid molecules in the fundamental cell, i.e. the solid molecules are excluded. The intermolecular force diminishes with time because of the increase of distances between molecules as the nanojet vaporizes. From Fig.39, it is observed that a higher temperature nanojet evaporates faster than a lower temperature one. This corroborates the results of Figs.27~31 discussed in section 2.3.1 and Fig.34 in section 2.3.2. In Fig.39, it is also noted that although a higher temperature nanojet has a larger intermolecular force at the earlier stage of the atomization process due to its higher momentum, it evaporates faster and therefore the intermolecular force decays quicker. Figure 40 shows the time averaged value of the averaged non-dimensionalized intermolecular force for different orifice lengths on the basis of identical nozzle orifice diameter (6 nm) and the conditions of $L_3^*=5.73$, $L_5^*=76.7$, $T^*=2.0$ (cases 3, 6 and 7 in Table 3). It is observed that a nanojet with a shorter orifice length evaporates quicker. This corroborates the results of Figs.32(a), (b) and (c) as discussed in section 2.3.1 and Fig.35 as discussed in section 2.3.2. In addition, it is also observed that, as time elapsed, the influence of the orifice length mitigates. As explained in section 2.3.2, more and more molecules spurt from the atomizer as time elapsed. This causes the interaction between the liquid molecules and the rigid atomizer molecules to mitigate due to the decrease of number of molecules inside the atomizer. Figure 41 also reveals this tendency. However, as can be observed from Fig.41, the influence of the orifice length becomes less pronounced at a higher temperature because of the higher molecular kinetic energy to overcome the resistance caused by the orifice. Figure 42 shows the evolution of averaged non-dimensionalized intermolecular force for different orifice diameters on the basis of identical nozzle orifice length (2 nm) and the conditions of L_4^*=5.73, L_5^*=76.7, T*=2.0 (cases 3 and 8 in Table 3). It is observed that a nanojet with a larger orifice diameter evaporates quicker. This corroborates the results of Figs.51(a) and (d) as discussed in section 2.3.1 and Fig.37 as discussed in section 2.3.2 and also reveals that a nanojet from a nozzle with a smaller orifice length/diameter ratio evaporates quicker. Similar tendency can be observed from Fig.43 for a higher temperature nanojet.

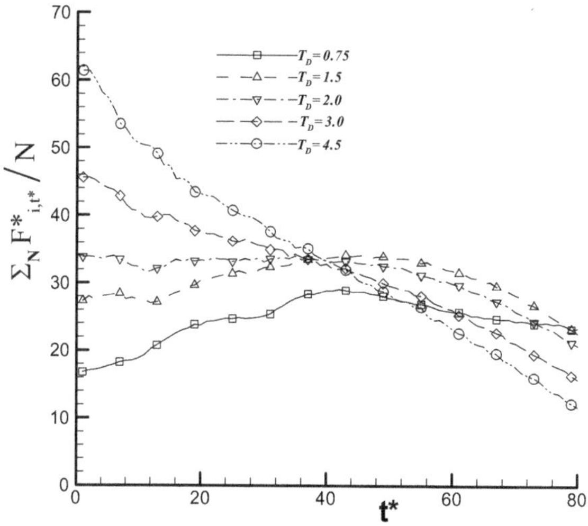

Fig. 39. Evolution of the averaged non-dimensionalized intermolecular force for different temperatures (cases 1~5 in Table 3)

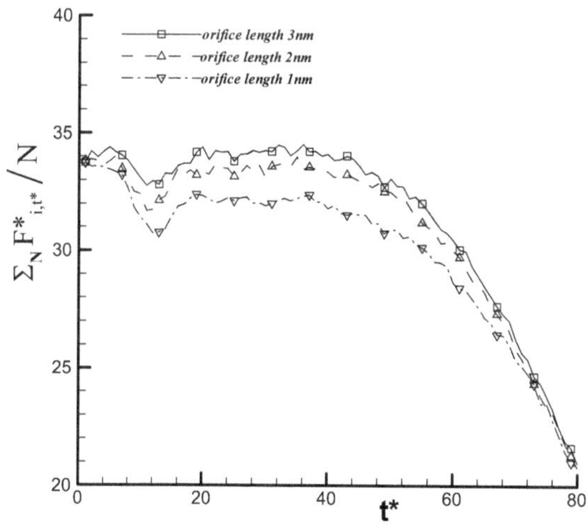

Fig. 40. Evolution of the averaged non-dimensionalized intermolecular force for different orifice lengths at T*=2.0 (cases 3, 6 and 7 in Table 3)

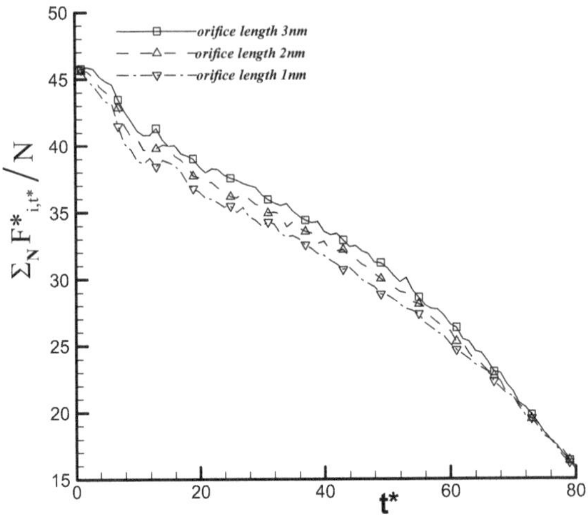

Fig. 41. Evolution of the averaged non-intermolecular force for different orifice lengthsdimensionalized at T*=3.0 (cases 4, 9 and 10 in Table 3)

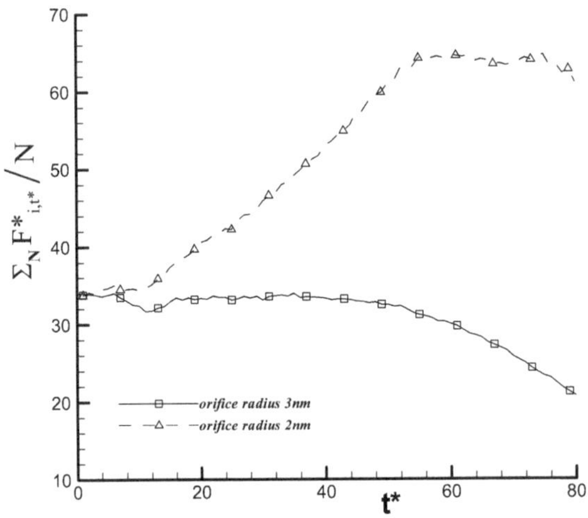

Fig. 42. Evolution of the averaged non-dimens ionalized intermolecular force for different orifice radii at T*=2.0 (cases 3 and 8 in Table 3)

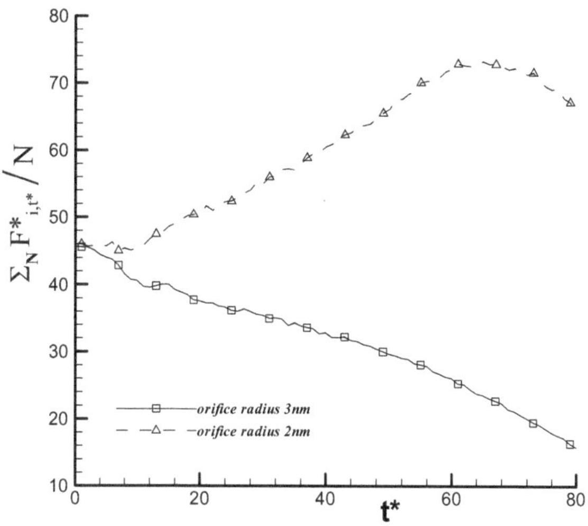

Fig. 43. Evolution of the averaged non-dimensionalized intermolecular force for different orifice radii at T*=3.0 (cases 4 and 11 in Table 3)

3. Conclusions

The purpose of this study is to provide a thorough investigation and understanding of the microscopic evolution and detailed mechanism of the flow inside and outside the atomizers. The main findings of this study are listed below.

1. For a single nano-scale liquid thread in a periodic fundamental cell, the liquid thread is more unstable and produces more liquid particles in the fundamental cell when it is thinner or at a higher temperature. In addition, a liquid thread with a longer fundamental cell length is also more unstable and produces more liquid particles in the fundamental cell, but it evaporates slower. The trends of linear stability theories agree with MD simulation results. However, Rayleigh's stability criterion overpredicts stable domain as compared to the MD simulation results. Kim's stability criterion gives more accurate predictions but overpredicts the stable domain at a higher temperature. Finally, a liquid thread with a higher time averaged density uniformity factor, $\overline{f_\rho}$, is more unstable and produces more liquid particles in the fundamental cell.
2. For two nano-scale liquid threads coexisting in a periodic fundamental cell, the two liquid threads may remain intact or evolve into only one liquid particle if the fundamental cell length is small. If the threads break up in this case, they rupture from their ends only, i.e., the top and bottom surfaces of the fundamental cell, but not from their interiors. On the other hand, if the fundamental cell length is larger, more than one liquid particle may be produced in the cell and the liquid threads rupture not only from their ends but also from their interiors. It is also found that thinner liquid threads may produce more liquid particles in the cell and evaporate more quickly. In addition, more

liquid particles are formed when the separation of the two threads is larger. Moreover, vaporization is slower when the two liquid threads are close to each other. On the basis of identical liquid thread radius and length, liquid threads that produce more liquid particles evaporate more quickly. Finally, the trends of Rayleigh's stability criterion and Kim's stability criterion agree with MD simulation results. However, when the two threads coalesce into a single one and remain intact, the critical wavelength of perturbation may be increased and the stable domain is broadened. In such a situation, Rayleigh's stability criterion and Kim's stability criterion underpredict the stable domain.

3. For the atomization process of a nanojet, it is found that a liquid nanojet evaporates faster at a higher temperature. On the basis of identical nozzle orifice diameter, a nanojet from a nozzle with a shorter orifice length evaporates quicker. However, the influence of the orifice length mitigates as time elapsed. In addition, the influence of the orifice length becomes less pronounced at a higher temperature. On the other hand, on the basis of identical nozzle orifice length, a nanojet from a nozzle with a larger orifice diameter evaporates quicker. The present simulation results reveal that a nozzle with a smaller orifice length/diameter ratio produces better atomization. This corroborates the results from conventional macroscopic analysis.

4. Acknowledgment

The author gratefully acknowledges the grant support from the National Science Council, Taiwan, R.O.C., under the contract NSC100-2221-E-150-047.

5. References

Ashgriz, N. & Poo, J. Y. (1991). FLAIR : Flux Line-Segment Model for Advection and Interface Reconstruction. *Journal of Computational Physics*, Vol.93, pp.449-468.

Brackbill, J. U., Kothe, D. B. & Zemach, C. (1992). A Continuum Method for Modeling Surface Tension. *Journal of Computational Physics*, Vol.100, pp.335-354.

Chen, S., Johnson, D. B. & Raad, P. E. (1995). Velocity Boundary Conditions for the Simulation of Free Surface Fluid Flow. *Journal of Computational Physics*, Vol.116, pp.262-276.

Chen, S. K. & Lefebvre, A. H. (1994). Discharge Coefficients for Plain-Orifice Effervescent Atomizers. *Atomization and Sprays*, Vol.4, pp.275-290.

Chigier, N. (1999). Breakup of Liquid Sheets and Jet. *AIAA paper* 99-3640.

Crank, J. (1984). *Free and Moving Boundary Problems*, Oxford University Press, New York.

Floryan, J. M. & Rasmussen, H. (1989). Numerical Methods for Viscous Flows with Moving Boundaries. *Appl. Mech. Rev.*, Vol.42, pp.323-340.

Goren, S. (1962). The Instability of an Annular Thread of Fluid. *Journal of Fluid Mechanics*, Vol.12, pp.309-319.

Haile, J. M. (1992). *Molecular Dynamics Simulation*, John Wiley & Sons, New York, Chap.5.

Harlow, F. H. & Welch, J. E. (1965). Numerical Calculation of Time-Dependent Viscous Incompressible Flow of Fluid with Free Surface. *The Physics of Fluids*, Vol.8, pp.2182-2189.

Hayase, T., Humphrey, J. A. C. & Grief, R. (1992). A Consistently Formulated QUICK Scheme for Fast and Stable Convergence Using Finite-Volume Iterative Calculation Procedures. *Journal of Computational Physics*, Vol.98, pp.108-118.

Hiroyasu, H. (2000). Spray Breakup Mechanism from the Hole-type Nozzle and its Applications. *Atomization and Sprays*, Vol.10, pp.511-527.

Hirt, C. W. & Nichols, B. D. (1981). Volume of Fluid (VOF) Method for the Dynamics of Free Boundaries. *Journal of Computational Physics*, Vol.39, pp.201-225.

Jeng, S. M., Jog, M. A. & Benjamin, M. A. (1998). Computational and Experimental Study of Liquid Sheet Emanating from Simplex Fuel Nozzle. *AIAA Journal*, Vol.36, pp.201-207.

Kawano, S. (1998), Molecular Dynamics of Rupture Phenomena in a Liquid Thread. *PHYSICAL REVIEW E*, Vol.58, No.4, pp.4468-4472.

Kawano, S., Hashimoto, H., Togari, H., Ihara, A., Suzuki, T. & Harada, T. (1997). Deformation and Breakup of an Annular Liquid Sheet in a Gas Stream. *Atomization and Sprays*, Vol.7, pp.359-374.

Kim, B. G., Lee, J. S., Han, M. & Park, S. (2006). A Molecular Dynamics Study on Stability and Thermophysical Properties of Nano-scale Liquid Threads. *Nano-scale and Micro-scale Thermophysical Engineering*, Vol.10, pp.283-304.

Koo, J. Y. & Martin, J. K. (1995). Near-Nozzle Characteristics of a Transient Fuel Spray. *Atomization and Sprays*, Vol.5, pp.107-121.

Koplik, J. & Banavar, J. R. (1993). Molecular Dynamics of Interface Rupture. *Physics of Fluids*, A, Vol.5, No.3, pp.521-536.

Launder, B. E. & Spalding, D. B. (1974). The Numerical Computations of Turbulent Flows. *Computer Methods in Applied Mechanics and Engineering*, Vol.3, pp.269-281.

Lefebvre, A. H. (1989). *Atomization and Sprays*, Hemisphere, New York, Chap.2.

Lord Rayleigh (1879). On the Stability of Jets, *Proceedings of the London Mathematical Society*, Vol.10, pp.4-13.

Min, D. & Wong, H. (2006). Rayleigh's Instability of Lennard-Jones Liquid Nanothreads Simulated by Molecular Dynamics. *Physics of Fluids*, Vol.18, 024103.

Moseler, M. & Landman, U. (2000). Formation, Stability, and Breakup of Nano-jets. *SCIENCE*, Vol.289, pp.1165-1169.

Ohrn, T. R., Senser, D. W. & Lefebvre, A. H. (1991). Geometrical Effects on Discharge Coefficients for Plain-Orifice Atomizers. *Atomization and Sprays*, Vol.1, pp.137-153.

Sakman, A. T., Jog, M. A., Jeng, S. M. & Benjamin, M. A. (2000). Parametric Study of Simplex Fuel Nozzle Internal Flow and Performance. *AIAA Journal*, Vol.38, pp.1214-1218.

Shyy, W., Udaykumar, H. S., Rao, M. M. & Smith, R. W. (1996). *Computational Fluid Dynamics with Moving Boundaries*, Taylor & Francis, Washington, D.C.

Spikes, R. H. & Pennington, G. A. (1959). Discharge Coefficient of Small Submerged Orifices, *Proceedings of the Institution of Mechanical Engineers*, Vol.173, pp.661-665.

Steinthorsson, E. & Lee, D. M. (2000). Numerical Simulations of Internal Flow in a Simplex Atomizer, *Eighth International Conference on Liquid Atomization and Spray Systems*, Pasadena, CA, USA, pp.324-331.

Thompson, J. F., Warsi, Z. U. A. & Mastin, C. W. (1985). VI. Elliptic Generation Systems, *Numerical Grid Generation*, North-Holland, pp.188-271.

Tomotika, S. (1935). On the Instability of a Cylindrical Thread of a Viscous Liquid Surrounded by Another Viscous Liquid. *Proceedings of the Royal Society of London, Series A, Mathematical and Physical Sciences*, Vol.150, No.870, pp.322-337.

Van Doormaal, J. P. & Raithby, G. D. (1984). Enhancements of the SIMPLE Method for Predicting Incompressible Fluid Flows. *Numerical Heat Transfer*, Vol.7, pp.147-163.

Vicelli, J. A. (1969). A Method for Including Arbitrary External Boundaries in the MAC Incompressible Fluid Computing Technique. *Journal of Computational Physics*, Vol.4, pp.543-551.

Weber, C. (1931). Zum zerfall eines ussigkeitsstrahles. *Zeitschrift fur Angewandte Mathematik und Mechanik*, Vol.11, No.2, pp.136-154.

Yeh, Chun-Lang (2010). Molecular Dynamics Analysis of the Vaporization Process for Two Nano-Scale Liquid Threads Coexisting in a Periodic Fundamental Cell. *CMES: Computer Modeling in Engineering & Sciences*, Vol.67, No.3, pp.175-209.

Yeh, Chun-Lang (2009a). Molecular Dynamics Simulation for the Atomization Process of a Nanojet. *CMES: Computer Modeling in Engineering & Sciences*, Vol.39, No.2, pp.179-200.

Yeh, Chun-Lang (2009b). Molecular Dynamics Analysis of the Instability for a Nano-Scale Liquid Thread. *CMES: Computer Modeling in Engineering & Sciences*, Vol.50, No.3, pp.253-283.

Yeh, Chun-Lang (2007). Numerical Simulation of Turbulent Liquid Jet Emanating from Plain-Orifice Atomizer and Pressure-Swirl Atomizer. *Numerical Heat Transfer, Part A*, Vol.51, pp.1187-1212.

Yeh, Chun-Lang (2005). Turbulent Flow Investigation inside and outside Plain-Orifice Atomizers with Rounded Orifice Inlets. *Heat and Mass Transfer*, Vol.41, No.9, pp. 810-823.

Yeh, Chun-Lang (2004). Numerical Investigation of Liquid Jet Emanating from Plain-Orifice Atomizers with Chamfered or Rounded Orifice Inlets. *JSME International Journal, Series B*, Vol.47, No.1, pp.37-47.

Yeh, Chun-Lang (2003). Effect of Inlet Turbulence Intensity on Discharge Coefficients for Liquid Jet Emanating from a Plain-Orifice Atomizer：a Numerical Study. *Journal of Aeronautics Astronautics and Aviation*, Vol.35, No.3, pp.299-306.

Yeh, Chun-Lang (2002). Numerical Study of Inlet and Geometry Effects on Discharge Coefficients for Liquid Jet Emanating from a Plain-Orifice Atomizer. *Journal of Mechanics, Series A*, Vol.18, No.3, pp.153-161.

17

High-Throughput Simulations of Protein Dynamics in Molecular Machines: The 'Link' Domain of RNA Polymerase

Robert O. J. Weinzierl
Imperial College London
United Kingdom

1. Introduction

Complex molecular machines are involved in all information-processing steps in cells. The handling of information-bearing macromolecules (predominantly nucleic acids and proteins) requires a range of catalytic capabilities, such as the sequence-specific synthesis of macromolecular entities from smaller precursors and further processing by breaking/joining of pre-existing molecular components. These catalytic activities have to be spatially and temporally precisely controlled to ensure accuracy and efficiency levels that are commensurate with the associated biological functions (Lyubimov et al., 2011). Unlike metabolic enzymes, which are often medium-sized enzymes with freely accessible active sites, the enzymes associated with replication, transcription, translation, recombination are typically large multi-subunit protein complexes capable of a complex conformational spectrum. This conformational spectrum manifests itself through the dynamic features of individual domains within these molecular machines: many of the domains act explicitly as nanomechanical elements by serving as allosteric sensors, molecular hinges and motor units (Bustamante et al., 2011; Heindl et al., 2011). A deeper understanding of the functions of molecular machines is currently a high-priority topic and ultimate success will strongly depend on a combination of 'wet-lab' experimental techniques (X-ray crystallography, NMR, spectroscopic methods, site-directed mutagenesis and chemical cross-linking) with sophisticated computational simulations. Fully atomistic molecular dynamics (MD) simulations offer the most comprehensive and detailed insights into the behavior of molecular entities and are therefore the method of choice for attempts to unravel the structural basis of molecular mechanisms (reviewed in Karplus & McCammon, 2002; Freddolino et al., 2010; McGeagh et al. 2011; Schlick et al., 2011).

The ultimate goal of MD studies is the complete and comprehensive simulation of molecular machines operating within a relevant time span. Since the functions of molecular machines critically depend on a series of tightly orchestrated conformational changes that typically occur in the micro- and milliseconds range, such a goal is not yet routinely achievable. Currently available computational resources are still insufficient for simulating the molecular dynamics of macromolecular entities for the length required for complex conformational changes to take place. We are therefore focusing on the more practicable goal of studying the dynamic properties of individual components of molecular machines in order to obtain a better understanding of their nanomechanical properties. In any natural or man-made machine, the mechanical properties of individual components determine both their functional capabilities, as well as their interactions with other parts. In many cases a description of these properties allows either a clear-cut deduction of the role of such components within the overall mechanism, or offers at least valuable suggestions for further experimental approaches. Most importantly, once the nanomechanical properties of individual components are understood in sufficient detail, their mode of interactions within defined assemblies can be modelled with increased accuracy and confidence. Such approaches result in insights into complex molecular mechanisms generated from the coordinated interplay between multiple protein domains.

Efforts in my laboratory are predominantly focused on understanding the intricate mechanisms of an enzyme playing a key role in gene expression: RNA polymerase (RNAP). Cellular RNAPs are highly conserved in evolution (ranging from bacteria to archaea and eukaryotes) and consist of a number of different subunits (reviewed in Werner, 2008). The large size and intrinsic complexity of these enzymes presents a particular challenge to understand the various mechanisms involved, such as separation and re-annealing of the DNA strands and the template-directed synthesis of an RNA transcript. Although structures of RNAPs in various states/conformations have been determined (e.g. Zhang et al., 1999; Cramer et al., 2001; Gnatt et al., 2001; Vassylyev, 2002), we are currently still far from having a complete catalogue of the most important conformations. Maybe even more importantly, the intermediate structural transitions between stable structural states are likely to be kinetically short-lived and are therefore very unlikely to be captured in the crystallized state required for high-resolution structural studies (Kaplan & Kornberg, 2008). Having a detailed understanding of such short-lived and metastable intermediate structures will, however, be critical for piecing together the conformational changes required for the various biological activities of RNAPs. Within the context of a larger strategy aimed at investigating the nanomechanical mechanisms underlying RNAP function, we are pursuing a two-pronged strategy: a laboratory-based robotic high-throughput approach capable of studying the functional consequences of a large number of site-directed mutants (the "RNAP Factory"; Nottebaum et al., 2008), and a computer-based in-depth investigation of nanomechanical properties of domains (or domain assemblies) using MD methods. The combination of these two complementary methods has already been proven to be highly effective for obtaining further insights into the nanomechanical 'motor' of RNAPs, as represented by the catalytic site and its surrounding protein domains (Fig. 1).

Our most recent studies focused on a prominent component of the catalytic site, the Bridge Helix. This structure is a 35 amino acid long α-helix which spans the catalytic site and multiple lines of evidence suggest that the Bridge Helix is, in conjunction with the Trigger Loop, predominantly involved in the translocation of nucleic acid substrates through the catalytic site. High-throughput mutagenesis experiments, combined with systematic molecular dynamics simulations, have demonstrated the presence of two hinges ('BH-H_N' and 'BH-H_C') which bend during the nucleotide addition cycle to propagate the DNA template strand and the nascent transcript through the active site (Klug, 2001; Bar-Nahum et al., 2005; Tan et al., 2008; Brueckner et al., 2009; Seibold et al., 2010; Weinzierl, 2010a and 2010b; Heindl et al., 2011).

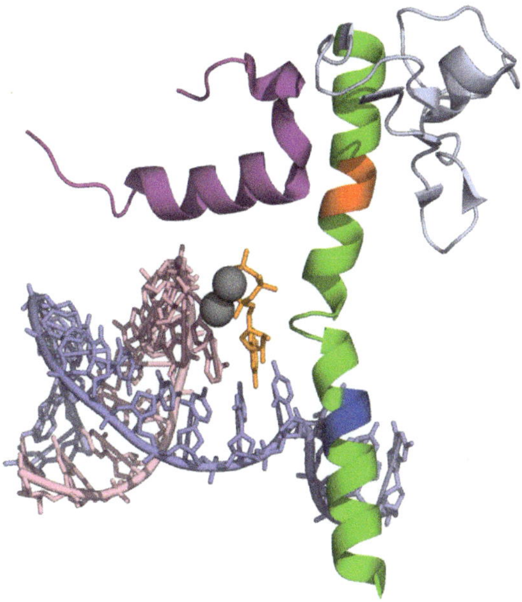

Fig. 1. Part of the nanomechanical 'motor unit' in the catalytic center of RNA polymerase. The Bridge Helix is shown in green (the two hinges BH-H_N and BH-H_C are red and blue, respectively), the F-Loop in grey and the Link domain in purple. The nucleic acid substrates are shown as stick models (DNA light blue, RNA pink, rNTP orange).

While the C-terminal hinge ('BH-H_C') may be actively involved in pushing the nascent RNA out of the active site to create space for the next incoming ribonucleotide triphosphate (rNTP; Klug, 2001; Bar-Nahum et al., 2005; Tan et al., 2008; Brueckner et al., 2009), the role of the more recently discovered N-terminal hinge (BH-H_N) is still poorly understood. In contrast to BH-H_C, the BH-H_N region is not in direct contact with any of the substrates, suggesting that BH-H_N interacts with the catalytic site more indirectly via allosteric contacts (Weinzierl, 2010b). Of the domains surrounding BH-H_N, the 'Link' domain emerges as the most likely candidate for playing such a communication role. As can be seen in Fig. 1, the Link domain consists of a short α-helix which contacts directly the N-terminal portion of the Bridge. One particular residue present in the α-helical portion of the Link domain (arginine 766 in the RPB2 subunit of *S. cerevisiae* RNAPII) is evolutionarily invariant and structural studies suggest that it plays a key role in contacting the γ-phosphate of the rNTP in the active site (Fig. 2).

A simple working hypothesis therefore suggests that the Bridge Helix N-terminal hinge can 'sense' - via the Link domain - whether the catalytic site is loaded with a correctly positioned rNTP before it is joined to the nascent transcript. When the β-γ pyrophosphate group is cleaved off during nucleotide addition and diffuses out of the catalytic site, this particular contact is lost and may result in kinking of BH-H_N. This model thus proposes that the Link domain acts as a conformational sensor that communicates the state of the catalytic site (empty, rNTP loaded, nucleotide addition and pyrophosphate release) in order to influence the conformational state of Bridge Helix. In mechanistic terms, completion of the nucleotide addition reaction signals to the Bridge Helix to initiate translocation of the DNA-RNA hybrid to create space for the next rNTP (Fig. 3)

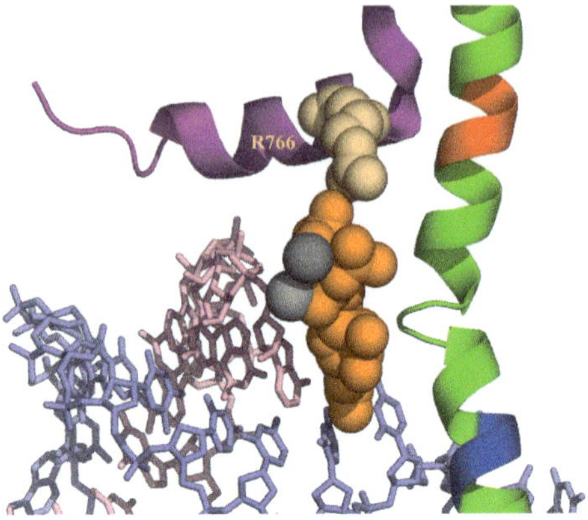

Fig. 2. Arginine 766 (R766) in the RPB2 subunit of *S. cerevisiae* RNAPII contacts the rNTP loaded in the catalytic site. The R766 residue (light brown), rNTP (orange) and the two catalytic metal ions (grey) are shown in space-filling mode. The remainder of the color scheme is as in Fig. 1.

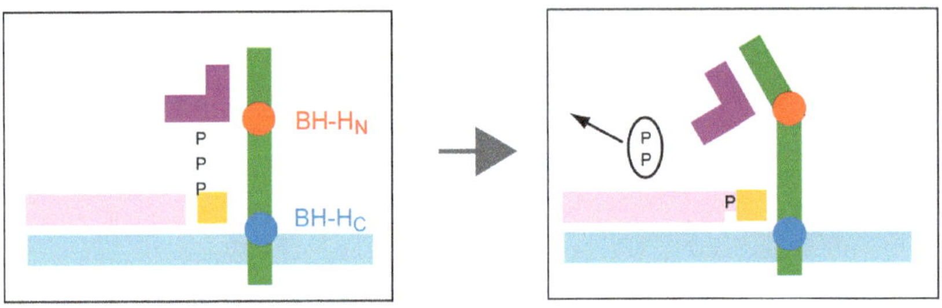

Fig. 3. Role of the Link domain as a conformational sensor for the detection of an rNTP in the catalytic site of RNAP.

Taking into account the possible role of the Link domain in communicating the rNTP loading state of the catalytic site to the Bridge Helix, we decided to investigate the mechanistic properties of this system in more detail. In particular, we are interested in studying the conformational spectrum of the Link domain to gain as much information as possible before initiating an experimental high-throughput mutagenesis approach. As pointed out earlier, it seems highly probable that molecular dynamics simulations are nowadays sufficiently reliable to deduce key nanomechanical properties inherent in a protein domain. We specifically wanted to simulate the Link domain extensively under highly standardized conditions to capture a broad spectrum of conformational changes the domain is capable of. We expected that the data obtained in this way would provide quantitative evidence for the relative stability of different parts of the domain and highlight regions that could potentially be involved in allosteric switching processes.

2. Implementation

The implementation of the strategy outlined above requires a considerable computational and organisational infrastructure. Here I will discuss the various issues that need to be considered and will provide an overview of the methods we developed to map out an universally applicable approach that can also be used by other researchers to study different systems.

2.1 What do we want to achieve?

Before proceeding further, we need to define the desirable outcome of our MD simulation strategy. In order to 'get to know' the dynamic properties of a protein domain, two key requirements must be met: First and most obvious, we need to simulate a domain for a sufficient amount of time so that we can gather quantitative data regarding the overall features of the domains, such as conformational spectrum and overall stability, local variations in secondary structure RMSD values and other information of interest. The simulation time that is considered to be 'sufficient' has increased over the last few years with the steady increase in computational resources available, but is nowadays generally considered to be in the 10-100 nanosecond range. For many stably folded domains, simulation over such a time-frame will allow data collection sufficient to quantitate, for example, the half-life of secondary structure elements. The second key requirement is the need to run multiple independent simulations. In previous work on the Bridge Helix domain (Weinzierl, 2010b; Heindl et al., 2011) we encountered structural features that behaved stochastically, i.e. not every simulation run displayed the full conformational spectrum. We found that kinking of the Bridge Helix hinges observed in some molecular dynamics production runs resulted in very stable alternative conformations that were essentially irreversible within the tens of nanosecond simulation time. Such 'trapped' conformations play almost certainly an important role in molecular switching processes occurring in the millisecond range, but would result in a biased picture of the conformational spectrum of a domain if only a small number of simulations was carried out. For practical purposes (and based on practical experience) we therefore typically perform approximately 50 independent simulations of 20 nanoseconds each to maximize the chance of detecting and quantitating relevant conformational changes occurring within small protein domains of fewer than 30 amino acids (this results in an overall simulation timeof one microsecond).

2.2 Constructing a personal high-performance MD simulation environment

Although we have used national grid computer resources in the past to carry out molecular dynamics simulations of protein domains (Weinzierl, 2010b; Heindl et al., 2011), we switched for our most recent work to high-performance personal computer systems based on workstations. The availability of multi-core CPUs, powerful graphics cards and affordable high-capacity mass storage media has over the last few years enabled the construction of workstation computing systems that are capable of providing the processing power for executing molecular dynamics simulations on a useful time scale. The custom-built workstation is based on liquid cooled dual Xeon 5690 Westmere CPUs clocked at 3.47 GHz with 24 GByte of DDR-3 RAM and running 64-bit Windows 7 Ultimate as operating system (more technical details are available on request). This 12 core (24 hyperthread) platform is capable of delivering 'raw' computing power consistently in excess of 120 GFlops in Linpack output. The MD simulation software of choice, 'YASARA Structure' (www.yasara.org), enables real-time simulations with standard (e.g. AMBER-03; Duan et al., 2003), or various custom force fields (NOVA, YAMBER, YASARA; Krieger et al., 2002; 2004; 2009). All structures were energy-minimized in pre-equilibrated simulation boxes filled with TP3 water, and sodium and chloride ions added to a final concentration of approximately 150 mM. The simulation runs were executed using a 2 femtosecond time step in a fully solvated atomistic production mode without restraints employing an AMBER03 force field (Duan et al., 2003) at 300 K (27°C) at pH 7.0. The solvent density was kept constant at 0.997 g/l. Long-range Coulomb interactions were calculated using the Particle Mesh Ewald algorithm with periodic boundary conditions (cut-off 7.86Å; Essman et al., 1995).

In practical terms, the MD workstation allows the atomistic simulation of complex macromolecular assemblies (e.g. complete RNAPs in fully hydrated simulation cells [~300, 000 atoms] to be carried out at a rate of 1 nanosecond per 28 hours computation (average CPU usage ~75%), which compares favourably with resources typically allocated to average users from supercomputing facilities.

The resulting MD trajectories were analyzed with a series of analysis packages to study the spatial and temporal stability of secondary structures, distance relationships, and correlated/concerted conformational movements using YANACONDA macro scripts (http://www.yasara.org/yanaconda.htm). The structures shown in this text were generated using PyMOL (The PyMOL Molecular Graphics System, Version 1.2r3pre, Schrödinger, LLC).

2.3 High-throughput molecular dynamics simulations of the RNAP link domain

On high-performance computational platforms the simulation of small domains is relatively inefficient because the multi-core CPU processing power is not used to the fullest extent under such circumstances. On the other hand, running multiple simulations at the same time in parallel applications allows more effective use of processing power, but the reading in and out of data for the different simulations wastes time that otherwise would be available for simulations. We have come up with a simple solution to this problem which allows simulations of multiple domains to run in parallel, but within a single application. Construction of a simulation cell that contains multiple copies of the target structure as a regular three-dimensional array allows an increased number of small domains to be simulated as a large structure (Fig. 4). The linear dimensions of simulation cells in any particular direction are often limited by the simulation software used, so we typically use three-dimensional clusters that contain the simulated domains in a crystal-like array or cube structure.

The domains within such an array can either be identical copies, thus allowing the systematic sampling of the conformational properties of a particular domain within a single run, or could represent different (e.g. containing various mutations) versions. The presence of multiple copies within the same simulation cell automatically guarantees comparability because the same simulation parameters (e.g. ionic concentration, temperature, pressure and force field) will be applied to all domains in a consistent manner. A particular concern applying to such an arrangement is that the individual domains must be spaced sufficiently far apart to avoid artefactual contacts between the different molecules. Since the domains move randomly during MD simulation (diffusion), we automatically reposition the domains to their original position at fixed intervals (typically after every nanosecond simulation time), thus avoiding any 'clashes' within the simulation cell.

Fig. 4. A three-dimensional array of copies of the Link domain arranged for high-throughput MD simulation.

We will illustrate how this procedure works in practice using the Link domain described above. For the experiments described below we ran three independent repeats of arrays containing 18 distinct copies of the Link domain (as illustrated in Fig. 4) from the yeast RNAPII structure (PDB 2E2H; Wang et al., 2006) for 20 nanoseconds simulation time each, thus resulting in data representing just over one microsecond of total simulation data. Snapshots of the simulation cell were stored at ten picosecond intervals and stored for further processing and analysis.

2.3.1 The link domain contains a surprisingly stable α-helical core

When subjecting individual domains to molecular dynamics simulations, a common concern is that a domain may not be intrinsically stable, resulting in rapid unfolding and loss of defined structure. We expected this to happen in the case of the isolated Link domain, which consists only of a relatively short α-helical region and irregularly structured N- and C-termini (Fig. 5).

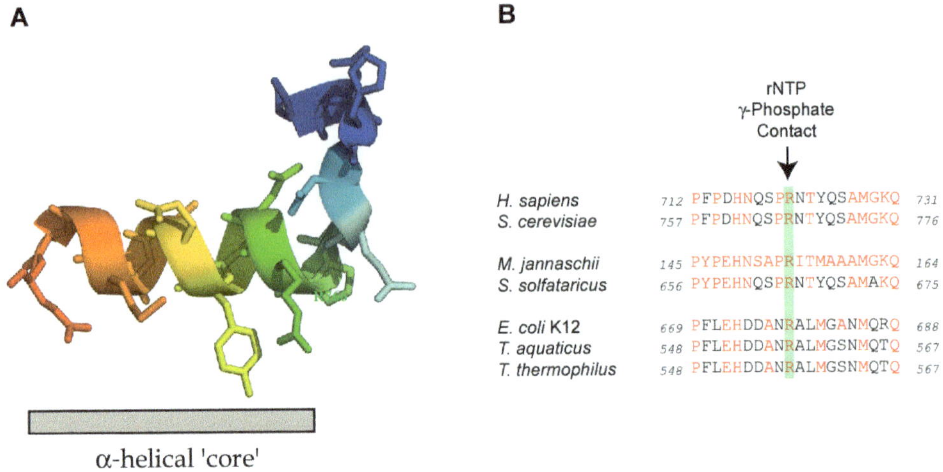

Fig. 5. Detailed structure (A) and evolutionary conservation (B) of the Link domain. *H. sapiens* (human) and *S. cerevisae* (yeast) represent eukaryotes, *M. jannaschii* and *S. solfataricus* are archaea and *E.coli*, *T. aquaticus* and *T. thermophilus* signify bacteria. Residues identical to the *M. jannaschii* sequence are shown in red and the position of the invariant arginine is highlighted with a pale green bar.

Surprisingly, this small domain turned out to be exceptionally stable, with an essentially invariant α-helical 'core' (amino acid residues spanning serine 764 to alanine 772) remaining in constant α-helical conformation throughout most, or even all (asparagine 767 to glutamine 770) of the 20 nanosecond simulation periods (Fig. 6). In contrast, the irregularly structured regions, encompassing proline 757 to glutamine 763 undergoes a variety of conformations, including (random) coil, short-lived α and 3_{10} helix and turns (no β-sheets were observed at any stage).

From this initial analysis we can therefore conclude that the Link domain consist of a nanomechanically robust portion and a highly flexible part (Fig. 7). A strategically placed proline reside (proline 765) delimits the N-terminal boundary of the α-helix (Cochran et al., 2001; Rey et al., 2010; Tendulkar & Wangikar, 2011) and thus determines the position of the sharp kink present in the L-shaped Link domain (Figs. 5 & 7).

It is remarkable that, despite the apparently random movement of the N-terminal portion of the Link domain (proline 757 to glutamine 763), the overall 'L-shape' of the domain is nevertheless maintained in more than 90% of all the simulations carried out so far (Fig. 7). This observation suggests the presence of interacting side-chains that stabilize the intramolecular positioning of the mobile N-terminus relative to the rigid C-terminal α-helix. A detailed investigation of the trajectories highlights such interactions involving histidine 761 ('H761'), glutamine 763 ('Q763'), arginine 766 ('R766') and - to a lesser extent - glutamine 770 ('Q770'; Fig. 8).

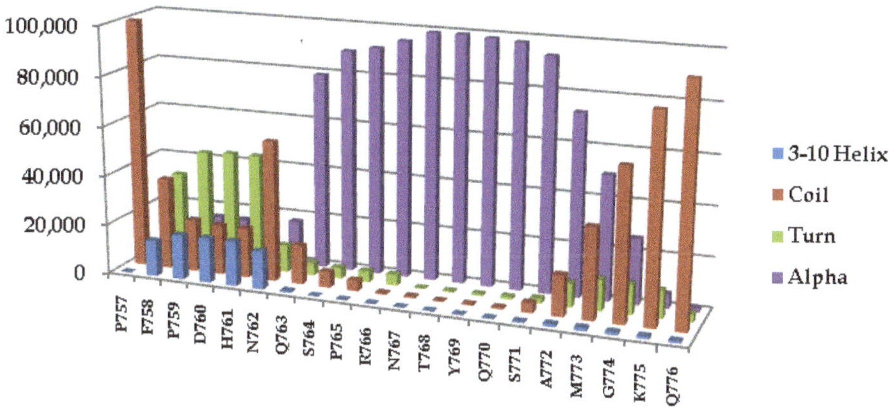

Fig. 6. Secondary structure analysis of the Link domain conformation based on 50 independent 20 nanosecond simulation runs. The proportion of four separate secondary structures adopted is shown for each residue (P757 to Q776).

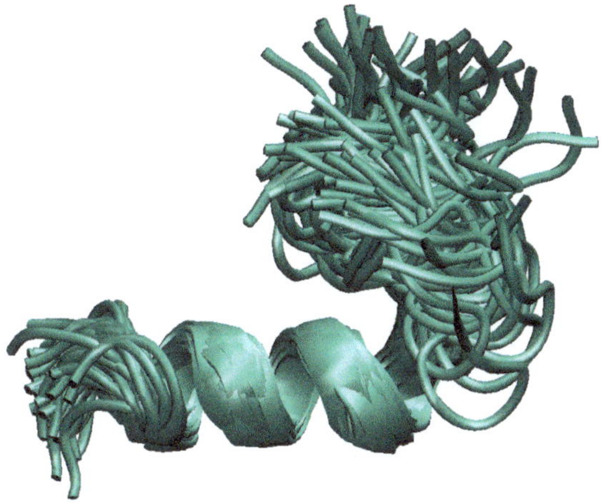

Fig. 7. Superimposition of trajectory traces of a representative 20 nanosecond MD simulation of the Link domain. The α-helical part (serine 764 to alanine 772) is structurally exceptionally stable, whereas the N-terminal portion undergoes significant conformational change.

At the beginning of the simulation (Fig. 8, t=0), the R766 side-chain engages in extensive van der Waals and/or hydrophobic interactions with Q763, which - due to the α-helical repeat, occupies a similar orientation. The side-chains appear tightly lined up relative to each other and even move synchronously (Fig. 8, compare t=0 and t=30 ps). R766 can, however, make alternative contacts with Q770, whose side-chain (similar to Q763) occupies the same side of the α-helix (Fig. 8, t=120 ps). While H761 is not initially involved in these interactions of R766 with Q763 or Q770, there is a clear tendency at a later stage of the simulation for it to

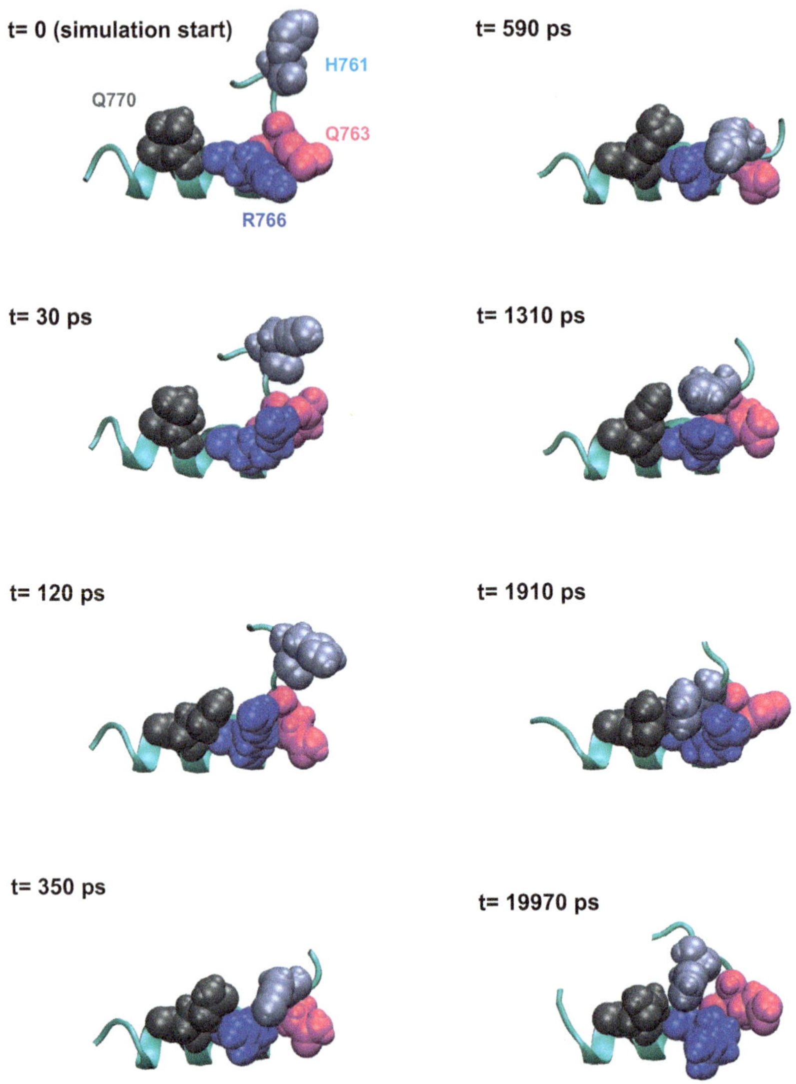

Fig. 8. Snapshots of the Link domain during 20 nanoseconds of MD simulation. See text for further information.

approach R766 (Fig. 8, t=350 ps). A conformational change moving Q763 out of the way appears necessary for the kink between the N-terminal flexible part and the α-helix to become more distinct. This is followed by a tightening of the contacts between H761 and R766 (Fig. 8, t=590 ps). Strong interactions involving side-chains of histidine and arginine have been documented previously (Heyda et al., 2010) and it therefore appears that this observation is based on similar principles. In the final stages we observe additional stabilization of the H761-R766 interaction through the adjacent glutamine side-chains that were already previously identified as R 766 interaction partners (Fig. 8, t=1310 ps). This association, however, appears to be reversible within the nanosecond time frame of the MD simulation since we observe a distinct break-up of the side-chain cluster towards the end of the run (Fig. 8, t=1910 ps and t= 19970 ps).

3. Conclusion

The gap between reality (conformational changes occurring in complex molecular machines in the millisecond to second range) and currently existing methodology (atomistic simulation of macromolecular structures in the nano- to microsecond range) is large and prevents us from obtaining a comprehensive understanding of the detailed functions of molecular machines that play fundamental role in key cellular processes. While it is very likely that this gap will narrow with the further development of computing power, viable intermediate solutions have to be found that can bridge the gap in the meantime. The approach outline here is aimed at breaking up complex macromolecular structures into components that provide distinctly identifiable nanomechanical functions which can be extensively studied using MD simulations. By gaining an understanding of the conformational spectrum of the individual components we can obtain two categories of complementary insights: the structural changes that a component is naturally capable of and the conformations that are either only achievable through energetic input or are unlikely to occur at all (this concept is already familiar to many molecular biologists in simpler systems, such as the conformational space allowed to a amino acids as represented by Ramachandran plots).

The work presented here has provided many new insights into the nanomechanical property of a newly identified structure of RNAP, the Link domain. As briefly outlined in the working model presented in Fig. 3, the Link domain appears to be strategically placed to act as an rNTP sensor, capable of transmitting the occupancy state of the catalytic site to the Bridge Helix. The Bridge Helix itself is a conformationally dynamic domain and almost certainly plays a critical role in translocating the nucleic acid substrate through the active site after successful rNTP incorporation (Weinzierl 2010a; 2010b). It thus appears essential for the translocation mechanism (Bridge Helix) to be allosterically linked to a sensor capable of reporting the successful completion of the incorporation of a nucleotide building block.

Although the initial identification of the Link domain was made entirely for structural reasons (i.e. its proximity to the N-terminal hinge region of the Bridge Helix), many of the observations emerging from the extensive molecular dynamic characterizations described here strengthen various aspects of the working model. The MD results show very clearly that the C-terminal α-helical portion of the Link domain is very stable and conformationally essentially invariant. A proline residue, P765, separates this rigid portion of the domain from a much more flexible part, which also contacts the N-terminal portion of the Bridge Helix (probably with a key involvement of a portion of the F-Loop (Fig. 1; Miropolskaya et al., 2009; 2010). An arginine residue, R766, appears to play multiple critical roles. Apart from contacting the γ-phosphate of the rNTP in the catalytic site (Fig. 2), our MD results suggest that R766 may have a decisive influence on the conformation of the Link domain. Strong contacts (Heyda et al., 2010) between H761 and R766 determine the kinking of the L-shaped Link domain (Fig. 8), suggesting that the R766 side-chain makes exclusive contacts with the rNTP or H761, thus providing a plausible model for the transmission of the sensor information to the Bridge Helix (Fig. 3).

The exploration and definition of conformational space accessible to an individual protein domain provides new insights and better-informed working hypotheses for future experimental work. We will shortly initiate a high-throughput mutagenesis screen aimed at replacing each of the residues of the Link domain in the structurally highly orthologous archaeal RNAP (Fig. 5) in order to test the preliminary insights obtained from the computer modelling.

In our MD simulation work we have up to now focused on characterizing the nanomechanical properties of the wildtype domain. The approach illustrated here can also be expanded to include *in silico* studies of mutant variants to aid the interpretation of experimental results. MD simulations of domain assemblies of increasing size and complexity (e.g. Link domain, F-Loop, Bridge Helix and nucleic acid substrates) will also make additional contributions towards achieving a better understanding of inter-domain communication through allosteric signalling mechanisms in the foreseeable future.

4. Acknowledgment

The research presented here was funded by MRC project grant G1100057 to ROJW.

5. References

Bar-Nahum, G., Epshtein, V., Ruckenstein, A.E., Rafikov, R., Mustaev, A. & Nudler, E. (2005). A ratchet mechanism of transcription elongation and its control. *Cell*, Vol. 120, No. 2, pp. 183-193, ISSN 0092-8674

Brueckner, F., Ortiz, J. & Cramer, P. (2009). A movie of the RNA polymerase nucleotide addition cycle. *Current Opinions in Structural Biology*, Vol. 19, No. 3, pp. 294-299, ISSN 0959-440X

Bustamante, C., Cheng, W. & Mejia, Y.X. (2011). Revisiting the central dogma one molecule at a time. *Cell*, Vol. 144, No. 4, pp. 480-497, ISSN 0092-8674

Cochran, D.A., Penel, S. & Doig A.J. (2001). Effect of the N1 residue on the stability of the alpha-helix for all 20 amino acids. *Protein Science*, Vol. 10, No. 3, pp. 463-70, ISSN 1469-896X

Cramer, P., Bushnell, D.A. & Kornberg, R.D. (2001). Structural basis of transcription: RNA polymerase II at 2.8 angstrom resolution. *Science*, Vol. 292, No. 5523, pp. 1863-1876, ISSN 0036-8075

Duan, Y., Wu, C., Chowdhury, S., Lee, M.C., Xiong, G., Zhang, W., Yang, R., Cieplak, P., Luo, R., Lee, T.; Caldwell, J., Wang, J. & Kollman, P. (2003). A point-charge force field for molecular mechanics simulations of proteins based on condensed-phase quantum mechanical calculations. *Journal of Computational Chemistry*, Vol. 24, No. 16, pp. 1999–2012, ISSN 1096-987X

Essmann, U., Perera, L., Berkowitz, M.L., Darden, T., Lee, H. & Pedersen, L.G. (1995). A smooth particle mesh Ewald method. *The Journal of Chemical Physics*, Vol. 103, No. 19, pp. 8577-8593, ISSN 0003-6951

Freddolino, P.L., Harrison, C.B.; Liu, Y.& Schulten, K. (2010). Challenges in protein folding simulations: Timescale, representation, and analysis. *Nature Physics* Vol. 6, No. 10, pp. 751-758, ISSN 1745-2473

Gnatt, A.L., Cramer, P., Fu, J.; Bushnell, D.A. & Kornberg, R.D. (2001). Structural basis of transcription: an RNA polymerase II elongation complex at 3.3 A resolution. *Science*, Vol. 292, No. 5523, pp.1876-1882, ISSN 0036-8075

Heindl, H., Greenwell, P., Weingarten, N., Kiss, T., Terstyanszky, G. & Weinzierl, R.O. (2011). Cation-π interactions induce kinking of a molecular hinge in the RNA polymerase bridge-helix domain. *Biochemical Society Transactions,* Vol. 39, No. 1, pp. 31-35, ISSN 0300-5127

Heyda, J., Mason, P.E. & Jungwirth, P. (2010). Attractive interactions between side chains of histidine-histidine and histidine-arginine-based cationic dipeptides in water. Journal of Physical Chemistry B, Vol. 114, No. 26, pp. 8744-8749, ISSN 1520-6106

Kaplan, C.D. & Kornberg, R.D. (2008). A bridge to transcription by RNA polymerase. *Journal of Biology*, Vol. 7, No. 10:39, ISSN 1741-7007

Karplus, M. & McCammon, J.A. (2002). Molecular dynamics simulations of biomolecules., *Nature Structural Biology*, Vol. 9, pp. 646-652, ISSN: 1545-9993

Klug, A. (2001). Structural biology. A marvellous machine for making messages. *Science,* Vol. 292, No. 5523, pp. 1844-1846, ISSN 0036-8075

Krieger, E., Koraimann, G. & Vriend, G. (2002). Increasing the precision of comparative models with YASARA NOVA – a self-parameterizing force field. *Proteins*, Vol. 47, No. 3, pp. 393-402, ISSN 1097-0134

Krieger, E., Darden, T., Nabuurs, S.B., Finkelstein, A. & Vriend, G. (2004). Making optimal use of empirical energy functions: Force-field parameterization in crystal space. *Proteins,* Vol. 57, No. 4, pp. 678-683, ISSN 1097-0134

Krieger, E.; Joo, K., Lee, J., Lee, J., Raman, S., Thompson, J., Tyka, M., Baker, D. & Karplus, K. (2009). Improving physical realism, stereochemistry, and side-chain accuracy in homology modeling: Four approaches that performed well in CASP8. *Proteins*, Vol. 77, Supplement 9, pp. 114-122, ISSN 1097-0134

Lyubimov, A.Y., Strycharska, M. & Berger, J.M. (2011). The nuts and bolts of ring-translocase structure and mechanism. *Current Opinion in Structural Biology,* Vol. 21, No. 2, pp. 240-248, ISSN 0959-440X

McGeagh, J.D., Ranaghan, K.E. & Mulholland, A.J. (2011). Protein dynamics and enzyme catalysis: insights from simulations. *Biochimica et Biophysica Acta*, Vol. 1814, No. 8, pp. 1077-1092, ISSN 0006-3002

Miropolskaya, N., Artsimovitch, I., Klimasauskas, S., Nikiforov, V. & Kulbachinskiy, A. (2009). Allosteric control of catalysis by the F loop of RNA polymerase. *Proceedings of the National Academy of Science USA*, Vol. 106, No. 45, pp. 18942-18947, ISSN-0027-8424, ISSN: 2154-1272

Miropolskaya, N., Nikiforov, V., Klimašauskas, S., Artsimovitch, I. & Kulbachinskiy, A. (2010). Modulation of RNA polymerase activity through trigger loop folding. *Transcription*, Vol. 1, No. 2, pp. 89-94

Nottebaum, S., Tan, L., Trzaska, D., Carney, H.C. & Weinzierl, R.O. (2008). The RNA polymerase factory: a robotic in vitro assembly platform for high-throughput production of recombinant protein complexes. *Nucleic Acids Research*, Vol. 36, No. 1, pp. 245-252, ISSN 0305-1048

Rey, J., Deville, J. & Chabbert, M. (2010). Structural determinants stabilizing helical distortions related to proline. *Journal of Structural Biology*, Vol. 171, No. 3, pp. 266-276, ISSN: 1047-8477

Seibold, S.A., Singh, B.N., Zhang, C., Kireeva, M., Domecq, C., Bouchard, A., Nazione, A.M., Feig, M., Cukier, R.I., Coulombe, B., Kashlev, M., Hampsey, M. & Burton, Z.F. (2010). Conformational coupling, bridge helix dynamics and active site dehydration in catalysis by RNA polymerase. *Biochimica et Biophysica Acta*, Vol. 1799, No. 8, pp. 575-587, ISSN 0006-3002

Schlick, T., Collepardo-Guevara, R., Halvorsen, L.A., Jung, S. & Xiao X. (2011). Biomolecular modeling and simulation: a field coming of age. *Quarterly Reviews in Biophysics*, Vol. 44, No. 2, pp. 191-228, ISSN 0033-5835

Tan, L., Wiesler, S., Trzaska, D., Carney, H.C. & Weinzierl, R.O. (2008). Bridge helix and trigger loop perturbations generate superactive RNA polymerases. *Journal of Biology*, Vol. 7, No. 10:40, 1741-7007, ISSN 1741-7007

Tendulkar, A.V. & Wangikar, P.P. (2011). Characterization and sequence prediction of structural variations in α-helix. *BMC Bioinformatics*, Vol. 12, Suppl 1:S20, ISSN 1471-2105

Vassylyev, D.G., Sekine, S.; Laptenko, O.; Lee, J.; Vassylyeva, M.N., Borukhov, S. & Yokoyama, S. (2002). Crystal structure of a bacterial RNA polymerase holoenzyme at 2.6 Å resolution. *Nature*, Vol. 417, No. 6890, pp. 712-719, ISSN 0028-0836

Wang, D., Bushnell, D.A., Westover, K.D., Kaplan, C.D. & Kornberg, R.D. (2006). Structural basis of transcription: role of the trigger loop in substrate specificity and catalysis. *Cell*, Vol. 127, No. 5, pp. 941-54, ISSN 0092-8674

Weinzierl, R.O. (2010a). Nanomechanical constraints acting on the catalytic site of cellular RNA polymerases. *Biochemical Society Transactions*, Vol. 38, No. 2, pp. 428-432, ISSN (printed) 0300-5127

Weinzierl, R.O. (2010b). The nucleotide addition cycle of RNA polymerase is controlled by two molecular hinges in the Bridge Helix domain. *BMC Biology*, Vol. 8:134, ISSN 1741-7007

Werner, F. (2008). Structural evolution of multisubunit RNA polymerases. *Trends in Microbiology*, Vol. 16, No. 6, pp. 247-250, ISSN 0966-842X

Zhang, G., Campbell, E.A., Minakhin, L., Richter, C., Severinov, K. & Darst, S.A. (1999). Crystal structure of Thermus aquaticus core RNA polymerase at 3.3 Å resolution. *Cell*, Vol. 98, No. 6, pp. 811-824, ISSN 0092-8674

18

Molecular Dynamics Simulation of Nanoscale Machining

Akinjide Oluwajobi
Obafemi Awolowo University, Ile-Ife
Nigeria

1. Introduction

Product miniaturization is a major motivation for the development of ultra-precision technologies and processes which can achieve high form and excellent surface finish. Of all the available manufacturing approaches, mechanical nanometric machining is still a good option for machining complex 3D devices in a controllable way (Jackson, 2008). As the dimension goes down to the nanoscale, the machining phenomena take place in a limited region of tool-workpiece interface. At this length scale and interface, the material removal mechanisms are not fully understood, so more insight is needed, which on the long run will help to achieve high precision manufacturing with predictability, repeatability and productivity (Luo, 2004). At present, it is very difficult to observe the diverse microscopic physical phenomena occurring through experiments at the nanoscale (Rentsch, 2008). Subsequently, the other alternative is to explore available simulation techniques. Continuum mechanics approach is not adequate, as the point of interest/interface cannot be assumed to be homogeneous, but rather discrete, so, atomistic simulation methods are the suitable techniques for modelling at the nanoscale.

The aim of this chapter is to review the use of classical molecular dynamics (MD) method for nanoscale machining. The steps on how the MD method can be applied to model nanometric machining and the importance of choosing appropriate interatomic potentials for the simulation are considered. Also, some examples of MD simulation of nanoscale machining are presented.

2. The MD method

Molecular dynamics (MD) is a computer simulation technique used in the study of the motions of a set of particles - atoms and molecules (Allen & Tildesley, 1988; Haile, 1997; Field, 1999; Frenkel & Smit, 2001; Leach, 2001; Schlick, 2002; Rapaport, 2004). The technique works by following the time evolution of a set of interacting atoms while integrating the equations of their motion. It is employed to study the classical motion of a system and then to extract experimental observables from the dynamics (Turkerman & Martyna, 2000). The MD is deterministic; once the positions, velocities and accelerations of the particles are known, the state of the system can be predicted. The method is also based on statistical mechanics - a way to obtain a set of configurations distributed according to some statistical distribution functions (Ercolessi, 1997, Hernandez, 2008).

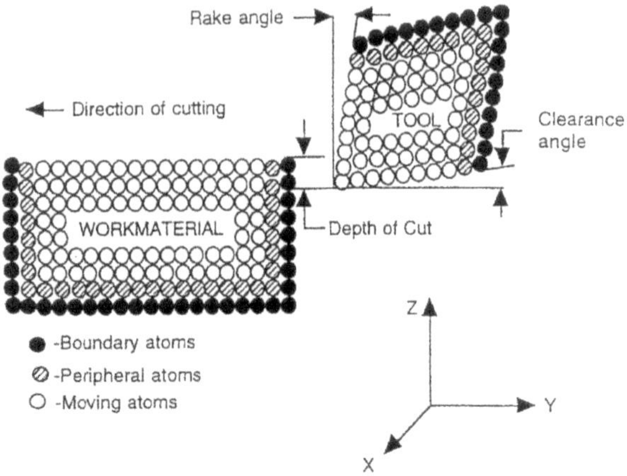

Fig. 1. Schematic of the MD Simulation of Nanometric Cutting (2D) (Komanduri & Raff, 2001)

The MD method was initiated in the late 1950s at Lawrence Radiation Laboratory in the US by Alder and Wainwright in the study of interactions of hard spheres (Alder & Wainwright, 1957, 1959). Gibson et al., 1961, carried out probably the first MD calculations with a continuous potential based on a finite difference time integration method. In 1964, Rahman carried out the simulation of liquid argon by using a realistic potential (Rahman, 1964). Later on, the first MD simulation of a realistic system was conducted by (Stillinger & Rahman, 1974) in 1974. Since then, the use of the simulation method has spread from Physics to Materials Science and now to Mechanical Engineering and other disciplines.

In the field of nanometric cutting, Belak pioneered work on the study of cutting copper with a diamond tool (Belak & Stowers, 1990). Initially, the method was used extensively to model indentation and cutting. Figure 1 shows the schematic of the MD simulation of nanoscale cutting. In 1991, Belak and Stowers first applied the MD to abrasive processes (Belak & Stowers, 1991) and Rentsch and Inasaki's study later presented the first results of simulations targeted on the pile-up phenomenon in abrasive machining (Rentsch & Inasaki, 1994). The MD method has been used in a variety of other application areas, namely; indentation (Kenny & Mulliah, 2005, Smith et al., 2003), wear and friction (Cheng et al., 2003, Shimizu et al., 1998), void growth (Potirniche et al., 2006, Zhao et al., 2009), etcetera.

The MD simulation is based on Newton's second law of motion. It consists of the numerical step-by-step integration of the classical equations of motion. For a set of N atoms,

$$F_i = m_i a_i \tag{1}$$

Where m_i is the mass of atom i, $a_i = \dfrac{d^2 r_i}{dt^2}$ is the acceleration of the atom i and F_i is the resultant force acting on atom i. These forces should be balanced by the potential energy between atoms, which are usually presented as the gradient of a potential energy function.

2.1 Thermodynamic ensembles

In MD simulations, various thermodynamic ensembles are employed. An ensemble is a large group of atoms or systems which are in different microscopic states, but have the same macroscopic or thermodynamic states. If a system of N atoms in a given macrostate is defined in terms of thermodynamic quantities, number of atoms, N; pressure, P; temperature, T; entropy, S; volume, V etc, there are many configurations at the atomic scale, which will lead to the same macrostate. The microstate of a system, defined by the atomistic positions and momenta cannot be known in a deterministic manner, because of the uncertainty principle of quantum mechanics. To avoid this problem, statistical mechanics approach is used for the atomic description.

The commonly used ensembles are the following, namely;

Microcanonical Ensemble (NVE)

This is an isolated system, with N atoms, which occupies a constant volume and the overall energy, E of the system is constant.

Canonical Ensemble (NVT)

This is a system in a temperature bath, with N atoms; and the volume, V and the temperature, T of the system are kept constant.

Isobaric Isothermal Ensemble (NPT)

This is a system in a temperature and pressure bath, with N atoms; and the pressure, P and the temperature, T of the system are kept constant.

Grand Canonical Ensemble (μ VT)

This is a system with a constant chemical potential and temperature T.

2.2 Steps in MD simulation

The outline of the MD simulation is as follows:

- Select the model
- Choose an appropriate interatomic potential
- Choose the algorithm for the integration of the equations of motion
- Initialize the model
- Relax the model from its initial state to its dynamically equilibrium condition
- Run the simulation and analyze the results

2.2.1 Select the model

The model should be selected correctly to reproduce what is expected (whether 2D/3D). The level of details and accuracy needed should guide this decision.

2.2.2 Choose an appropriate interatomic potential

Then, the next thing is a major task in MD simulation, which is the selection of the potential function for the atomic interactions. This is a function, $V(r_i, ...r_N)$ of the position of the nuclei which represents the potential energy of the system. Where $(r_i,.........r_N)$ represents the complete co-ordinate position of the atoms. Then forces are derived from it as,

$$F_i = -\frac{\partial}{\partial r_i} V(r_i,..........r_N) \tag{2}$$

Strictly speaking, the problem of modelling a material is that of finding a potential function for that material. If the potential function doesn't model the atomic behaviour correctly, the results produced from the simulation would be wrong. Consider the energy of N interacting particles, which can be written as (Tersoff, 1988);

$$E = \sum_i V_1(r_i) + \sum_{i<j} V_2(r_i,r_j) + \sum_{i<j<k} V_3(r_i,r_j,r_k) + ... + \sum_{i<j<k.......<N} V_N(r_i,r_j,r_k......r_N) \tag{3}$$

Where r_i, r_j, r_k are the positions of the particles and the functions $V_1, V_2, V_3,....V_N$ are the m-body potentials.

From equation (3), the second term, $\sum_{i<j} V_2(r_i,r_j)$ is the two-body or pair potential and the third term is the three-body potential and so on. The pair potentials are the simplest interatomic potentials used for the interaction of a set of particles. The most popularly used is the Lennard-Jones potential; others are Morse potential, Born-Mayer potential and et cetera. Apart from the pair potentials, there are multi-body potentials, like Tersoff and Embedded-Atom Method (EAM) potentials. The following is a consideration of some widely used interatomic potentials.

Lennard-Jones Potential (Lennard-Jones, 1924)

$$V_{ij} = 4\varepsilon\left[\left(\frac{\sigma}{r}\right)^{12} - \left(\frac{\sigma}{r}\right)^{6}\right] \tag{4}$$

Where ε and σ are constants which are dependent on the interacting particles. The LJ potential is ideal for rare gases, where the interactions between the non- bonded and uncharged atoms are due to weak van der Waal forces.

Morse Potential (Morse, 1929)

$$V_{ij} = D\{\exp[-2\alpha(r_{ij} - r_e)] - 2\exp[-\alpha(r_{ij} - r_e]\} \tag{5}$$

Where, r_{ij} and r_e are instantaneous and equilibrium distances between atoms i and j respectively; α and D are constants determined on the basis of the physical properties of the material. The Morse potential is suitable for cubic metals and they can be used to model the interactions between an atom and a surface.

Born-Mayer Potential

Born and Mayer suggested that the repulsion between the atoms would have a roughly exponential dependence on distance (Born & Mayer, 1932).

$$V_{ij} = A\{\exp[-\frac{r}{a_{BM}})]\} \tag{6}$$

Where A and a_{BM} are constants dependent on the material. This potential is used for metals, Group III-V semiconductors and ceramics.

Tersoff Potential

Tersoff modelled the total energy of the system as a sum of pair like interactions and as a function of the atomic coordinates, given as equation (7). The potential is based on the concept that the strength of a bond between two atoms is not a constant, but depends on the local environment (Tersoff, 1988a ,1988b).

$$E = \sum_i E_i = \frac{1}{2}\sum_i \sum_{i \neq j} V_{ij} \tag{7}$$

and

$$V_{ij} = f_C(r_{ij})[a_{ij} f_R(r_{ij}) + b_{ij} f_A(r_{ij})] \tag{8}$$

$$
\begin{aligned}
&\textit{where} \\
&f_R(r) = A\exp(-\lambda_1 r), \\
&f_A(r) = -B\exp(-\lambda_2 r), \\
&f_C(r) = \begin{cases} 1, \; r < R - D \\ \frac{1}{2} - \frac{1}{2}\sin\left[\frac{\pi}{2}(r-R)/D\right], \; R - D < r < R + D, \\ 0, \; r > R + D \end{cases} \\
&b_{ij} = (1 + \beta^n \zeta_{ij}^n)^{-1/2n}, \\
&\zeta_{ij} = \sum_{k(\neq i,j)} f_C(r_{ik}) g(\theta_{ijk}) \exp\left[\lambda_3^3 (r_{ij} - r_{ik})^3\right], \\
&g(\theta) = 1 + \frac{p^2}{q^2} - \frac{p^2}{\left[q^2 + (h - \cos\theta)^2\right]}, \\
&a_{ij} = (1 + \alpha^n \eta_{ij}^n)^{-1/2n}, \\
&\eta_{ij} = \sum_{k(\neq i,j)} f_C(r_{ik}) \exp\left[\lambda_3^3 (r_{ij} - r_{ik})^3\right]
\end{aligned}
$$

Where R and D are cutoff parameters; $A, B, \lambda_1, \lambda_2, \lambda_3, \alpha, \beta, n, p, q, h$ are fitting parameters of the Tersoff potential.

The Tersoff potential is used for covalently bonded materials like silicon and carbon atoms.

Embedded-Atom Method (EAM) Potential

For the EAM potential, the total energy of the system can be written as (Foiles ,1985, Foiles et al., 1986).

$$E_{tot} = \sum_i G_i(\rho_{h,i}) + \frac{1}{2}\sum_{i,j} V_{ij}(r_{ij}) \quad (9)$$

$\rho_{h,i}$ is the total electron density at atom i due to the rest of the atoms in the system.
G_i is the embedding energy for placing an atom into the electron density
$V_{i,j}$ is the short range pair interaction representing the core-core repulsion
r_{ij} is the separation of atoms i and j

The EAM potential was developed for a wide range of metals. It incorporates an approximation for the many-atom interactions neglected by the pair-potential scheme.

Modified Embedded-Atom Method (MEAM) Potential

The MEAM is theoretically an extension of the EAM potential (Baskes, 1992) with modifications to include the directionality of the bonding. The bond-angle was explicitly handled so as to accommodate covalent systems. The total energy is given by (Equation 9). The MEAM is suitable for modelling metals and alloys with fcc, bcc, hcp and cubic structures, and also for covalent materials such as silicon and carbon (Oluwajobi & Chen, 2010a).

The list of the potentials is not exhaustive; and there are other potentials which are modified forms of the ones already discussed. To obtain physically meaningful results from atomistic simulations, it is essential that reliable interatomic potentials are used. A reliable potential would reproduce various physical properties of the relevant elements or alloys, including the elastic, structural, defect, surface and thermal properties etc (Lee et al., 2005). For example, Tersoff potential was designed for the description of covalent materials like silicon, germanium, carbon, silicon carbide etc. and it cannot adequately model metals. Also, the EAM potential was designed for metals, as it describes the bonding in metals more satisfactorily, but the MEAM potential can be used for the modelling of both metallic systems and covalently bonded materials. Table 1 shows some interatomic potentials and their suitability (Oluwajobi & Chen, 2010).

It should be stated that EAM and MEAM potentials should be used for metals, where appropriate; Tersoff and MEAM potentials should be used for covalent materials, and for the interface of materials where suitable potentials are not available, appropriate existing potentials like LJ and Morse can be used with caution.

2.2.3 Algorithms for the Integration of the Equations of Motion

The next step, after choosing the interatomic potential/s is to select an appropriate algorithm for the integration of the Newton's equations of motions. This is to numerically solve for positions and velocities at a time, t and at a later time, t+Δt. (Generally, the algorithm is already chosen in the MD software that one uses for the simulation). MD simulations use time steps from a few femto seconds (10^{-15} s) (Shimada et al., 1993). If a large time step is used, the computational time is reduced, but when the time step is too large, this can cause instability and inaccuracy in the solutions.

Name	Model	Application
Lennard-Jones Potential	$V_{ij} = 4\varepsilon\left[\left(\frac{\sigma}{r}\right)^{12} - \left(\frac{\sigma}{r}\right)^{6}\right]$, σ and ε are constants which are dependent on the physical property of the materials. (Lennard-Jones, 1924)	Mostly suitable for rare gases
Morse Potential	$V_{ij} = D\{\exp[-2\alpha(r_{ij} - r_e)] - 2\exp[-\alpha(r_{ij} - r_e]\}$ r_{ij} and r_e are instantaneous and equilibrium distances between atoms *i* and *j* respectively α and D are constants determined on the basis of the physical properties of the material (Morse, 1929)	Mostly suitable for cubic metals
Born-Mayer Potential	$V_{ij} = A\{\exp[-2\alpha(r_{ij} - r_0)]\}$ A and r_0 are constants dependent on the material (Born & Mayer, 1932)	Mostly suitable for ceramics
Tersoff Potential	$V_{ij} = V_r(r_{ij}) - B_{ij}V_a(r_{ij})$ V_r and V_a are the potentials due to repulsive and attractive forces between atoms *i* and *j* and B_{ij} is a parameter that provides the information for the direction and the length of the bond. (Tersoff 1988a, 1988b)	Mostly suitable for covalently bonded materials
Embedded-Atom Potential (EAM)	$E_{tot} = \sum_i G_i(\rho_{h,i}) + \frac{1}{2}\sum_{i,j} V_{ij}(r_{ij})$ $\rho_{h,i}$ is the total electron density at atom i due to the rest of the atoms in the system. G_i is the embedding energy for placing an atom into the electron density $V_{i,j}$ is the short range pair interaction representing the core-core repulsion r_{ij} is the separation of atoms i and j (Foiles 1985, Foiles et al., 1986)	Mostly suitable for a wide range of metals

Table 1. Comparison of the Interatomic Potentials (Oluwajobi & Chen, 2010a)

Many numerical schemes have been developed for the integration of the equations of motions. Some of these are the Verlet algorithm (Verlet, 1967), the predictor-corrector algorithm (Rahman, 1964, Gear & Tu, 1974, Gear & Watanabe, 1974) and the Beeman's algorithm (Beeman, 1976). All the above integration schemes make the assumption, that the positions, velocities and accelerations can be approximated using a Taylor series expansion:

$$\begin{aligned} r(t+\Delta t) &= r(t) + \dot{r}(t)\Delta t + \frac{1}{2}\ddot{r}(t)\Delta t^2 + \ldots \\ v(t+\Delta t) &= \dot{r}(t) + \ddot{r}(t)\Delta t + \frac{1}{2}\dddot{r}(t)\Delta t^2 + \ldots \\ a(t+\Delta t) &= \ddot{r}(t) + \dddot{r}(t)\Delta t + \frac{1}{2}\ddddot{r}(t)\Delta t^2 + \ldots \end{aligned} \tag{10}$$

Where Δt is a finite time step, r is the position, v and $\dot{r}$ are the velocity, a and $\ddot{r}$ are the acceleration; $\dddot{r}$ and $\ddddot{r}$ are the third and the fourth derivative of position with time.

The Basic Verlet Algorithm (Verlet, 1967)

Using this method, the next step of position can be predicted as follows;

$$r(t+\Delta t) = r(t) + \dot{r}(t)\Delta t + \frac{1}{2}\ddot{r}(t)\Delta t^2 + \ldots \tag{11}$$

In the same way, the previous step;

$$r(t-\Delta t) = r(t) - \dot{r}(t)\Delta t + \frac{1}{2}\ddot{r}(t)\Delta t^2 - \ldots \tag{12}$$

Summing equations (11) and (12), we have

$$r(t+\Delta t) = 2r(t) - r(t-\Delta t) + \ddot{r}(t)\Delta t^2 \tag{13}$$

The method uses no explicit velocities, it is quite straightforward, easy to implement and its storage requirements are modest, but it is not too accurate.

The Verlet Leapfrog Algorithm

In this algorithm, the velocities are calculated by taking the average value halfway between position steps. The equations are as follows;

$$r(t+\Delta t) = r(t) + \dot{r}(t+\frac{1}{2}\Delta t)\Delta t \tag{14}$$

$$v(t+\frac{1}{2}\Delta t) = \dot{r}(t-\frac{1}{2}\Delta t) + \ddot{r}(t)\Delta t \tag{15}$$

In this method, the velocities leap over the positions and then, the positions leap over the velocities. Consequently, the positions and the velocities are not known simultaneously, but the velocities are calculated explicitly.

The Velocity Verlet Algorithm

Using equations deduced from equations (10) and (13), and ignoring infinitesimals, this algorithm calculates new positions, velocities and accelerations using their values at time, t.

$$r(t+\Delta t) = r(t) + \dot{r}(t)\Delta t + \frac{1}{2}\ddot{r}(t)\Delta t^2 \tag{16}$$

$$v(t+\Delta t) = \dot{r}(t) + \frac{1}{2}[\ddot{r}(t) + \ddot{r}(t+\Delta t)]\Delta t \tag{17}$$

The velocity Verlet algorithm requires low memory.

The Predictor-Corrector Algorithm (Rahman, 1964; Gear & Tu, 1975; Gear & Watanabe, 1974)

Velocities at time t+Δt are predicted and the forces calculated, and then the corrected forms of the velocities are later calculated. Combining the *r(t +Δt)* and *a(t +Δt)* in equation 10, the position can be expressed as:

$$r(t+\Delta t) = r(t) + \dot{r}(t)\Delta t + \frac{2}{3}\ddot{r}(t)\Delta t^2 - \frac{1}{6}\ddot{r}(t-\Delta t)\Delta t^2 + O(\Delta t^4) \tag{18}$$

The velocities at time t = t+Δt are then calculated from the positions. Combining the *r(t + Δt)* and *a(t +Δt)* in equation (18), the position can be expressed as:

$$\text{(Predicted)}\ \ v(t+\Delta t) = \dot{r}(t) + \frac{3}{2}\ddot{r}(t)\Delta t - \frac{1}{2}\ddot{r}(t-\Delta t)\Delta t \tag{19}$$

The acceleration at t = t+Δt are calculated from the positions and the predicted velocities.

$$\text{(Corrected)}\ \ v(t+\Delta t) = \dot{r}(t) + \frac{1}{3}\ddot{r}(t+\Delta t)\Delta t + \frac{5}{6}\ddot{r}(t)\Delta t - \frac{1}{6}\ddot{r}(t-\Delta t)\Delta t \tag{20}$$

This algorithm has the advantage that, by comparing the predicted and the corrected values of parameters, it is possible to perform a self-check on the algorithm for accuracy.

The Beeman's Algorithm

The Beeman's algorithm (Beeman, 1976) is based on equations (21) and (22), which can be deduced from the equations (18) and (20). The algorithm is more complicated and requires more memory than the velocity Verlet, but it provides more accurate expression for the velocities and better energy conservation.

$$r(t+\Delta t) = r(t) + \dot{r}(t)\Delta t + \frac{2}{3}\ddot{r}(t)\Delta t^2 - \frac{1}{6}\ddot{r}(t-\Delta t)\Delta t^2 \tag{21}$$

$$v(t+\Delta t) = \dot{r}(t) + \dot{r}(t)\Delta t + \frac{7}{6}\ddot{r}(t)\Delta t - \frac{1}{6}\ddot{r}(t-\Delta t)\Delta t \tag{22}$$

When choosing an integration algorithm following factors need to be considered:

- Accuracy – it should produce an approximate result close to the exact solution
- Memory requirement – it should require little memory
- Efficiency – it should be fast and computationally efficient
- Time Step – it should permit long time step Δt for the integration

2.2.4 Initialize the model

To initialize the simulation, the MD 'box' (the control volume) must be defined; then initial positions and velocities of the atoms must be assigned - this is a kind of initial randomization. Positions of the atoms can be defined, by assuming certain crystal structure and the initial velocities can be randomly assigned.

2.2.5 Relax the model from its initial state to its dynamically equilibrium condition

The model has to be relaxed from its artificially assigned initial conditions to its natural, dynamically equilibrium condition, to mimic real materials. This can be achieved by running the MD program under constant temperature for a pre- determined time steps, so that the velocities of atoms that are initially assigned randomly or based on a normal distribution, will gradually reach equilibrium at the specified temperature of the simulation (Cheong et al., 2001). This relaxation time steps will vary, depending on the time needed for the model to reach its natural, dynamically equilibrium state, that is consistent with the environmental temperature.

2.2.6 Run the simulation and analyze the results

The simulation is then run and the results analyzed. Using the MD, the effect of such variables as edge effect, depth of cut, velocity and other machining parameters can be defined and the simulations conducted accordingly. It is also easy to vary the properties of the work materials and the cutting tools in MD simulations (Komanduri & Raff, 2001).

From the results of the simulation, there are various quantities can that be derived for nano machining predictions. Some of which are cutting forces, cutting temperature, potential and kinetic energies, pressure etc.

2.3 Cutting forces

In nanometric machining, the cutting forces are the interatomic forces (Luo et al., 2003). These forces are the superposition of the interaction forces between the cutting tool and the workpiece atoms. A low cutting force is as a result of fine cutting conditions, which will in turn decrease the vibration of the cutting system and then result in better surface roughness (Jackson, 2008).

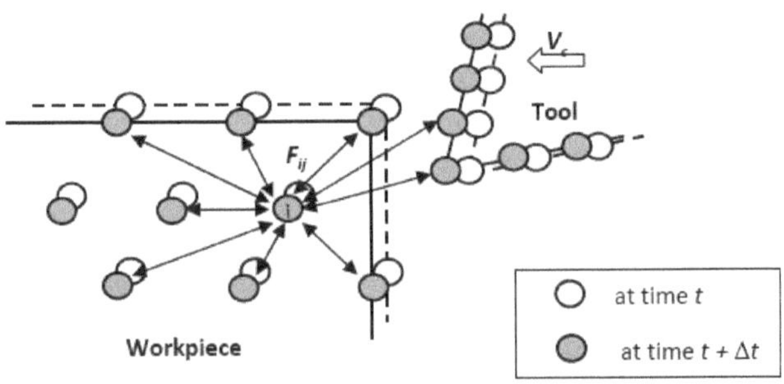

(a) (b)

Fig. 2. Atomistic Interaction in Nanometric Machining (Ikawa et al., 1991, Promyoo et al., 2008)

Figure 2 shows the time increment (Δt), in which every atom changes its position and interacts with its surrounding neighbour atoms in a manner that is determined from the interatomic potential function.

E.g. for atoms described by Lennard Jones potential, V_{ij}, the interatomic force between atoms i and j is

$$F(r_{ij}) = -\frac{dV_{ij}}{dr_{ij}} \tag{23}$$

The force acting on the i-th workpiece atom is thus a summation of the interaction with the surrounding atoms (Luo et al., 2003);

$$F_{wi} = \sum_{j\neq 1}^{N_t} F_{wtij} + \sum_{j\neq 1}^{N_w} F_{wwij} = \sum_{j\neq 1}^{N_t} -\frac{dV(r_{wtij})}{dr_{wtij}} + \sum_{j\neq 1}^{N_w} -\frac{dV(r_{wwij})}{dr_{wwij}} \tag{24}$$

Similarly, the force on each tool atom is

$$F_{ti} = \sum_{j\neq 1}^{N_t} F_{ttij} + \sum_{j\neq 1}^{N_w} F_{twij} = \sum_{j\neq 1}^{N_t} -\frac{dV(r_{ttij})}{dr_{ttij}} + \sum_{j\neq 1}^{N_w} -\frac{dV(r_{twij})}{dr_{twij}} \tag{25}$$

2.4 Cutting temperature

In MD simulations, it is assumed that the cutting energy is totally transferred into cutting heat and this results in an increase of the cutting temperature and kinetic energy of the system.

2.5 Minimum depth cut

The Minimum Depth of Cut (MDC) is defined as the minimum undeformed chip thickness that can be removed stably from a work surface at a cutting edge under perfect performance of a machine tool (Ikawa et al., 1992). The concept of MDC is that the depth of cut must be over a certain critical thickness before any chip is formed. This phenomenon of MDC leads to a rising of slipping forces, burr formation and surface roughness (Ducobu et al., 2009). Conventionally, the tool - workpiece material interface has been considered to be homogeneous and continuum mechanics are used in the analysis of the MDC. In nanomachining, analysis is based on discrete atoms whose interactions are governed by appropriate interatomic potentials. The understanding and the accurate prediction of the MDC is very crucial in improving the ultra-precision metal removal technologies, as this would assist in the selection of appropriate machining parameters and optimal geometry design (Oluwajobi & Chen, 2010b).

2.6 Nanometric cutting versus conventional cutting mechanics

Nanometric cutting is based on a very small region of the tool-workpiece interface, which contains few atoms and so discrete mechanics apply. On the other hand, conventional cutting is based on continuum mechanics. The cutting forces are based on the interatomic potentials in nanometric cutting, but they are dependent on the plastic deformation in conventional cutting. Table 2 shows the comparison of nanometric cutting and conventional cutting mechanics, and it should be noted that the basic physics of the cutting mechanism on the nanoscale is quite different from that on the macro/conventional scale.

	Nanometric Cutting	Conventional Cutting
Fundamental Cutting Principles	Discrete Molecular Mechanics	Continuum Mechanics
Workpiece Material	Heterogeneous	Homogeneous
Cutting Physics	Atomic Cluster Model	Shear Plane Model
Energy Consideration	Interatomic Potential Functional	Shear/Friction Power
Cutting Force	Interatomic Forces	Plastic Deformation
Chip Formation	Inner Crystal Deformation (Point Defects or Dislocation)	Inter Crystal Deformation (Grain Boundary Void)
Deformation and Stress	Discontinuous	Continuous
Cutting Tool Edge Radius	Significant	Ignored
Cutting Tool Wear	Cutting Face and Cutting Edge	Rake Face

Table 2. Comparison of Nanometric Cutting and Conventional Cutting Mechanics (Adapted from Luo et al., 2003)

3. The simulation set-up and procedure

There are many issues to address for the MD simulation set-up, namely; the software and the hardware. For the hardware, suitable computer system configuration should be utilized depending on whether serial or parallel processing is required. For the software, there are many open source MD simulation software that are available. Two of the most popular software are DLPOLY and LAMMPS. (Another option is to develop one's own MD software from the scratch.) There are also open source software for the visualization of the simulation like the VMD. A typical structure can be seen in Figure 3, which comprises the pre-processing, the main MD processing and the post-processing parts.

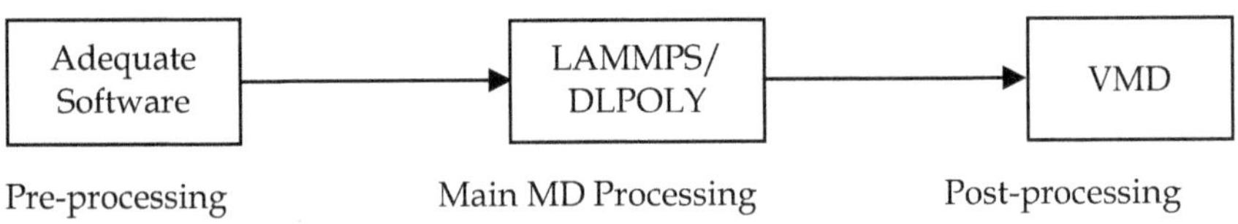

Fig. 3. Software Methodology Flowchart

3.1 Pre-processing

A suitable software has to be used to set-up the simulation model. A way to do this is to place the atoms on a perfect crystal lattice structure that represents the atomic structure of the material to be studied. The positions of atoms from an earlier simulation can also be used to initialize a new simulation (Cheong et al., 2001).

3.2 MD processing

The main MD software accepts the input data of the atomic configuration and then integrates the equation of motion for the interacting atoms.

LAMMPS– Large-Scale Atomic/Molecular Massively Parallel Simulator

This is a classical MD software that models an ensemble of atoms using a variety of empirical potentials and boundary conditions. It runs on single and parallel processor computers. It is an open-source code and it is distributed under the terms of the GNU public licence. The LAMMPS computes the Newton's equation of motion for a system of interacting atoms. It requires as its input; the types of atoms and the list of their initial co-ordinates, molecular topology information and the empirical potentials assigned to all the atoms.

DL_POLY

This is a general purpose MD software for the simulation of a wide range of atomic and molecular systems. It affords the use of standard and user defined interatomic potentials for the selected system. It also accommodates many boundary conditions, namely; cubic periodic, slab, orthorhombic periodic, parallelepiped periodic etc. (Smith & Todorov, 2005).

3.3 Post-processing

VMD – Visual molecular dynamics

The VMD is a molecular visualization software for displaying and analyzing atomic systems by using 3-D graphics technology. The VMD can be employed for the visual display of the LAMMPS/DL_POLY MD simulation results.

3.4 The simulation configuration

Various configurations are used for the simulation of nanoscale machining, to optimize computational resources. All the configuration share the same features in that they consist of the workpiece and the tool. Generally, the workpiece can be divided into boundary, Newtonian and thermostat atoms (See Figure 4).

Parameters	Values
Bulk Temperature	*293 K*
Cutting Direction	*[100]- Along the x-axis*
Cutting Speed	*150m/s*
Cutting Depth	*1.0nm*
Time Step	*0.3fs*
Run	*100000steps*

Table 3. A Typical MD Simulation Parameters

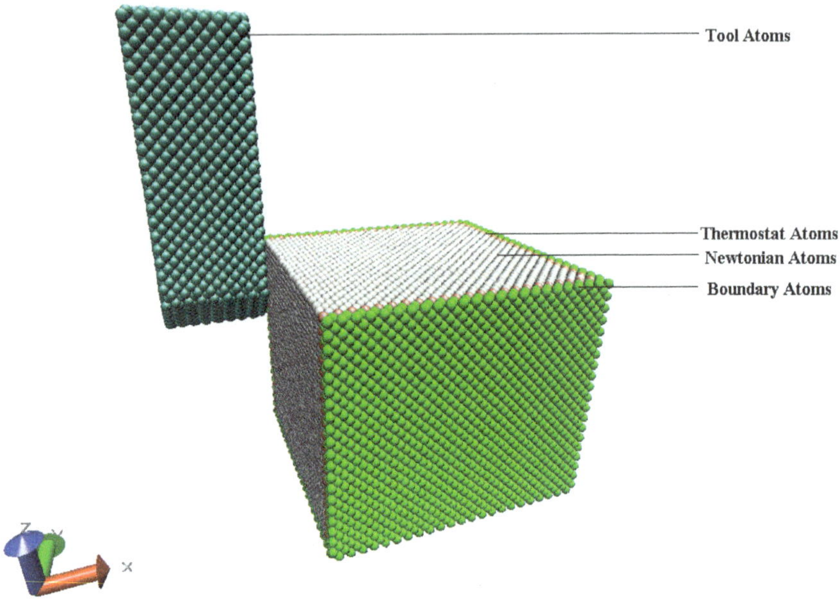

Fig. 4. A Simulation Model

Table 3 shows the simulation conditions that can be applied in a study. The workpiece is crystalline copper and the tool is also crystalline diamond. The configuration has a total of 54232 atoms. The workpiece consists of 43240 copper atoms with FCC lattice It includes 3 kinds of atoms namely; boundary atoms, thermostat atoms and Newtonian atoms. The boundary atoms are kept fixed to reduce edge effects. The thermostat atoms conduct the heat generated during the cutting process out of the cutting region. This is achieved by the velocity scaling of the thermostat atoms, (with the conversion between the kinetic energy (KE) and temperature) via equation (26),

$$\sum_i \frac{1}{2} m_i v_i^2 = \frac{3}{2} N k_B T_i \tag{26}$$

Where m_i is the mass of the ith atom, v_i is the resultant velocity of the ith atom, N is the number of the thermostat atoms, T_i is the temperature of the ith atom and k_B is the Boltzmann constant (1.3806504 x10^{-23} JK^{-1})

Whenever the temperature of the thermostat atoms exceeds the preset bulk temperature of 293K, their velocities are scaled by using equation (27),

$$v_{i,new} = v_i \sqrt{\frac{T_{desired}}{T_{current}}} \tag{27}$$

Where $T_{current}$ is the current temperature that is calculated from the KE and the $T_{desired}$ is the desired temperature.

The Newtonian atoms obey the Newton's equation of motion. The cutting tool consists of 10992 carbon atoms with diamond lattice structure. In this case, the cutting tool is pointed shaped and it can be modelled as a rigid or as a non-rigid body.

The atomic interactions in the simulation are the following, namely;
Cu-Cu : interactions between copper atoms
Cu-C : interactions between copper atoms and diamond atoms
C-C : interactions between the diamond atoms

Suitable interatomic potentials are required for all the above atomic interactions.

4. Some examples of MD simulation of nanomachining

Belak and Stowers pioneered work on the MD simulation of nanometric machining (See Figure 5). Copper was used as the workpiece with 10^3 - 10^6 atoms at room temperature. The simulations were in 2D and the focus was on the chip formation process, mechanisms of plastic deformation, flow of material and the flow of energy into the chip and workpiece. They modelled the workpiece by using the EAM potential and they used a shifted and truncated LJ potential for the tool-workpiece interface. Shimada et al 1992 also carried out MD machining simulations using copper as the workpiece (Figure 6). They used the Morse potential for the atomic interactions and showed the chip formation for the simulation model of 6076 atoms at the cutting speed of 200m/s.

(Rentsch & Inasaki, 1994) modelled a copper work material and a diamond tool for their study. They used the Lennard-Jones potential function for the copper atom interactions, but kept the boundaries and the tool stiff. A total of 11476 atoms in 13 horizontal (1,1,1) - layers of fcc-lattice were used for the copper, and the tool was shaped from a diamond lattice block by clearing on the four (1,1,1) -planes. Using a cutting speed of 100m/s, they observed a pronounced build-up phenomenon after 25000 time steps (see Figure 7).

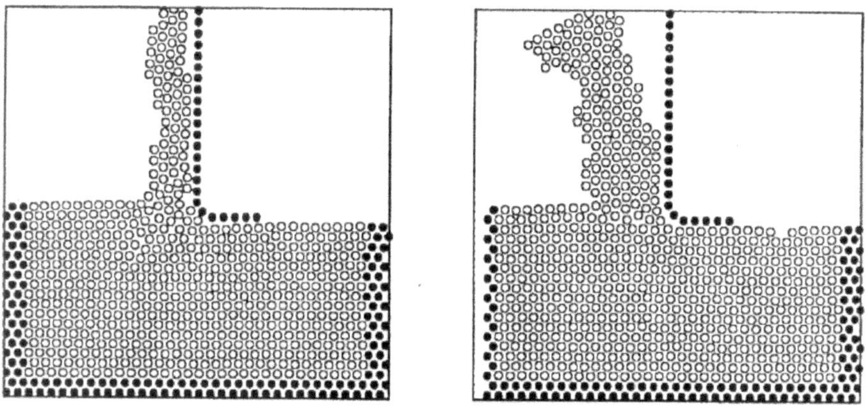

(a) (b)

Fig. 5. MD Simulation of the Orthogonal Cutting of Copper; (a) with cutting speed of 2500m/s and (b) with cutting speed of 500m/s (Belak & Stowers, 1990)

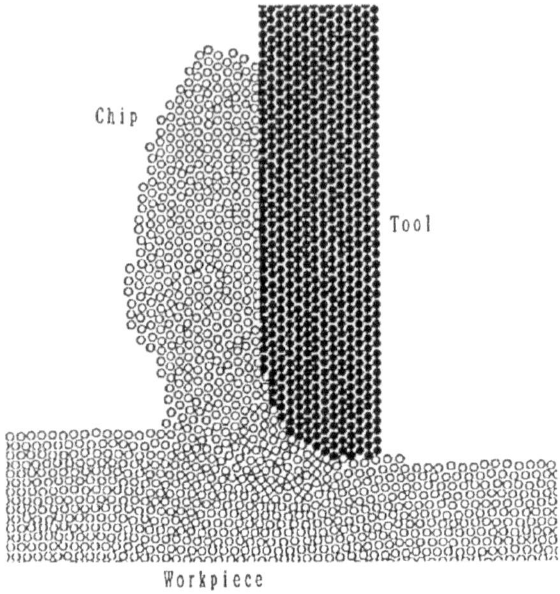

Fig. 6. MD Simulation of Cutting of Copper (Shimada et al., 1992)

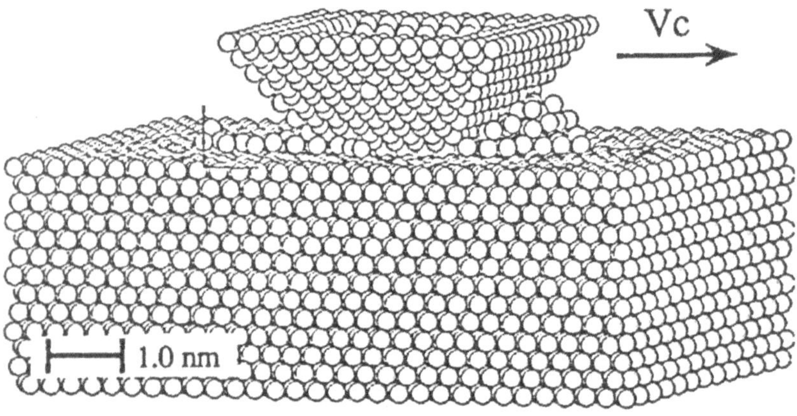

Fig. 7. Advanced MD Simulation with Straight Aligned Tool (Rentsch & Inasaki, 1994)

The MD simulation of nanometric cutting was carried out by Komanduri team with a range of negative-rake tools to simulate the Ultra-Precision Grinding (UPG) process (Komanduri et al., 1999). A copper work material and an infinitely hard tool (tungsten) were used in the simulations. A pairwise sum of Morse potential was used for the study, in which they concluded that simulation tests can facilitate a better understanding of the process without the need for expensive and time consuming machining or grinding experimental work.

The investigation of the fundamental atomistic processes in chemical mechanical polishing of copper was carried out by Ye et al (2002). They simulated the nanoscale polishing of a copper surface by a single abrasive particle, using the embedded-atom potential. The temperature was controlled by maintaining 1.2nm of the substrate at 300K and the rescaling of atom velocities was performed whenever the temperature deviated more than 10K from the specified value. This allowed the transfer of heat from the machined region to the bulk of the work-material. They focused on the mechanical abrasion aspect of material removal and found that dislocations and atomically rough planarized surface were formed. They also studied the nature of the material removal, chip formation, materials defects and frictional forces as functions of the cutting speed, depth of cut and abrasive geometry. They established that the material removal rate scales linearly with the depth of planarization and is directly proportional to the velocity of cut.

To be more realistic, (Rentsch & Inasaki, 2006) extended the MD material modelling to consider fluids, like coolants. They considered the impact of such fluids on the surface generation and the tribological contact conditions. The fluid-fluid interactions were calculated on the basis of the Lennard-Jones potential function and the embedded-atom potential function was used for the inner workpiece reactions (the internal tool dynamics were ignored). They observed an intensive self-diffusion of the fluid atoms, and these filled the whole free space above the workpiece. No impact on the stress distribution was observed, but the whole fluid-tool/workpiece contact was heated up in a narrow range (See Figures 8 (a) and (b).

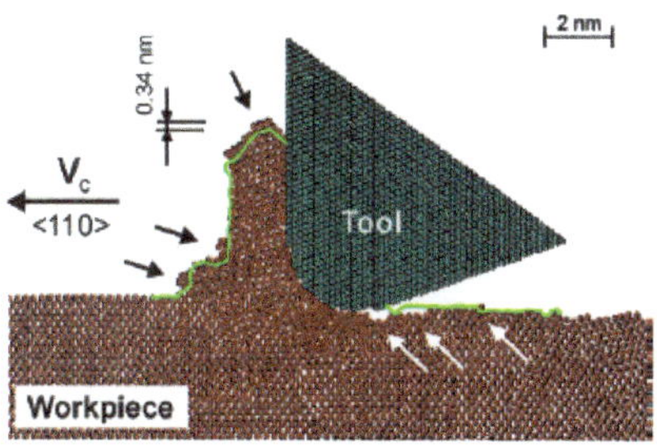

(a)

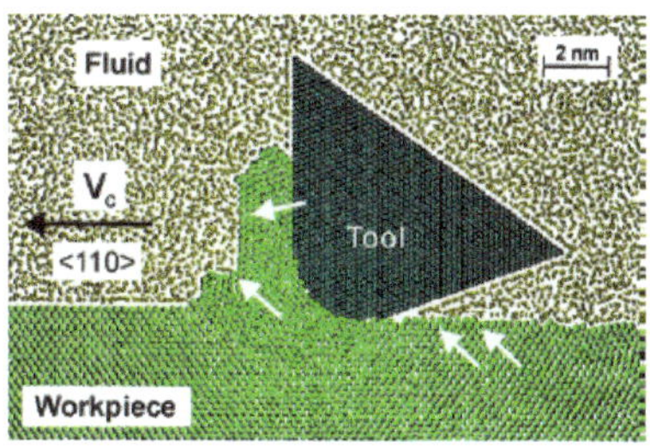

(b)

Fig. 8. MD Machining Simulation (a) without fluid (vacuum) after 230.7ps (b) with fluid after 230ps , (Rentsch & Inasaki, 2006)

Shimizu et al (2006) carried out MD simulations on the effect of vibration, velocity and acceleration in vibration assisted cutting. They used aluminium as the workpiece and diamond as the tool. Using Morse potentials for the interactions, they obtained that the effect of vibration on the plastic flow and cutting forces is more than the effect due to acceleration. See some snapshots of the simulation in Figure 9.

To extend the capability of MD studies, Pei et al., (2009), performed large scale MD simulations with model size of up to 10million atoms (Figures 10 (a), (b) and (c). Their results showed that as the cutting depth decreases, the tangential cutting force decreases and that size effects exists in nanometric cutting.

Oluwajobi & Chen, (2010a, 2011a, 2010b, 2011b, 2012) have conducted several MD simulations of nanometric machining. To show the effect of interatomic potentials on nanomachining, three popular potentials namely; EAM, Morse and the Lennand-Jones, were employed to model nanometric machining. The following shows the results;

Modelling with LJ Potential

The LJ parameters used for the atomic interactions are σ = 2.2277 Angstroms and ε = 0.415eV (Hwang et al 2004), which apply to both the Cu-Cu and the Cu-C interactions. The cutting forces are shown in Figure 11.

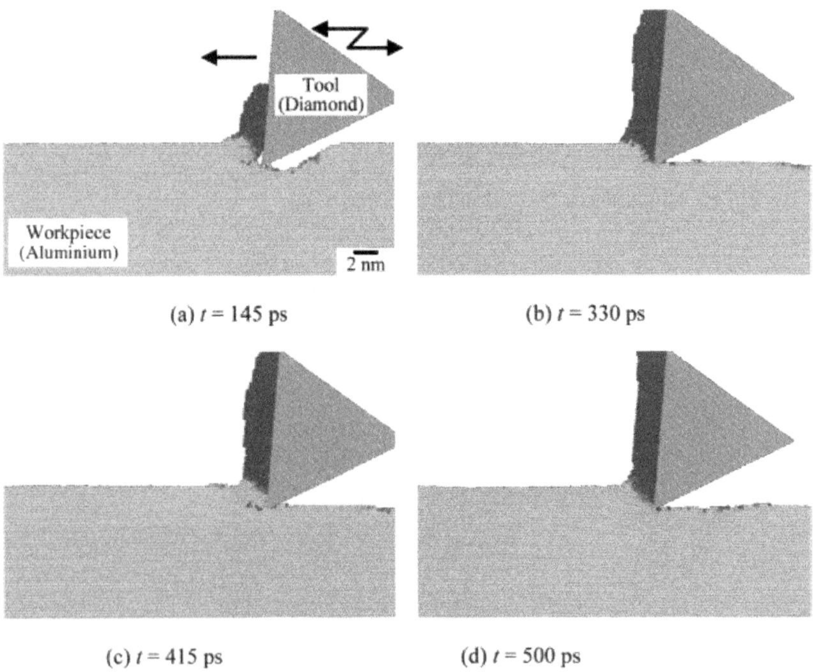

Fig. 9. Snapshots of Atomic Arrays in Vibration-assisted Cutting Process and Travelling Distance from Initial Arrays. Cutting speed, $V_c = 50m/s$, Amplitude, A=4nm and Frequency, f =4GHz (Shimizu et al., 2006).

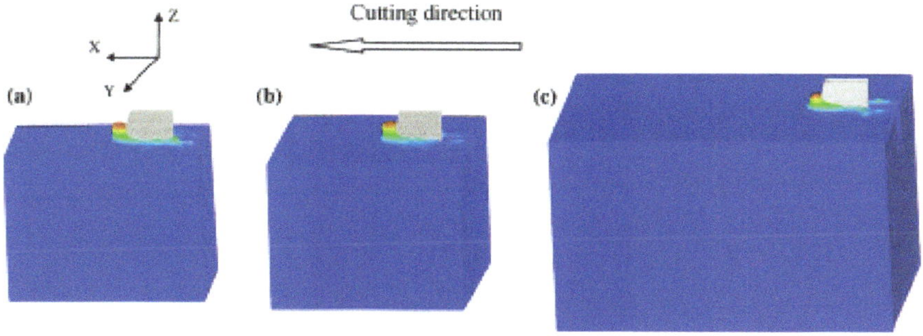

Fig. 10. MD Simulation Models with Number of the Atoms in the Workpiece around (a) 2 Million, (b) 4 Million and (c) 10 Million (Pei et al., 2009)

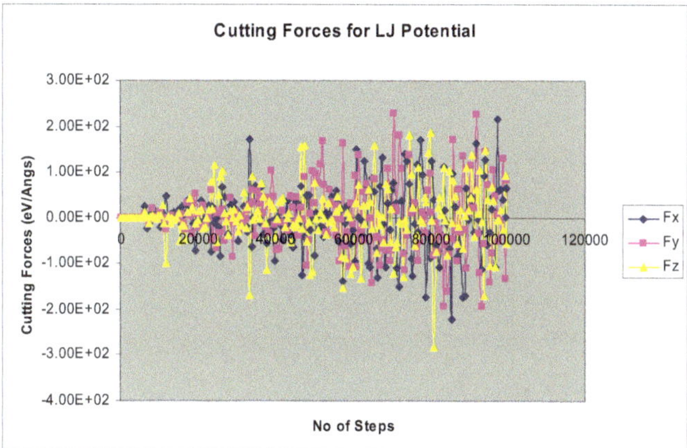

Fig. 11. Cutting Forces for LJ Potential (Oluwajobi & Chen, 2011a)

Modelling with Morse Potential

For Cu-Cu interactions: (Girifalco & Weizer, 1959, Pei et al., 2006)

$$D = 0.3429eV, \alpha = 0.13588(nm)^{-1}, \ r_e = 0.2866nm$$

For Cu-C interactions: (Hwang et al., 2004)

$$D = 0.087eV, \alpha = 0.17(nm)^{-1}, \ r_e = 0.22nm$$

The cut-off distance chosen was 6.4 Angstroms (that is, the interactions between atoms separated by more than this distance are neglected). The cutting forces are shown in Figure 12.

Modelling with EAM Potential

For the EAM potential, parameters used can be found in (Oluwajobi & Chen 2011a) and the cutting forces are shown in Figure 13.

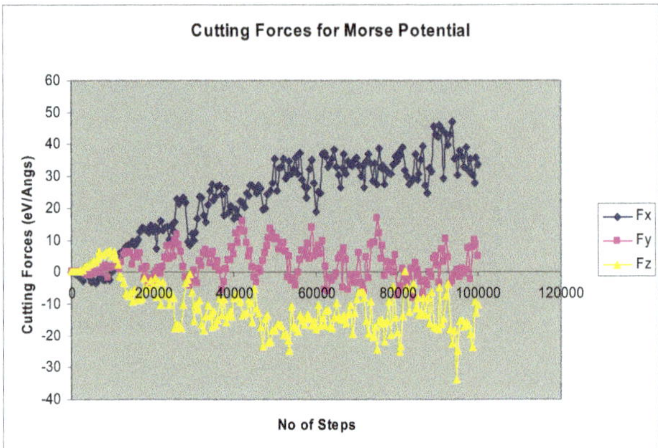

Fig. 12. Cutting Forces for Morse Potential (Oluwajobi & Chen, 2011a)

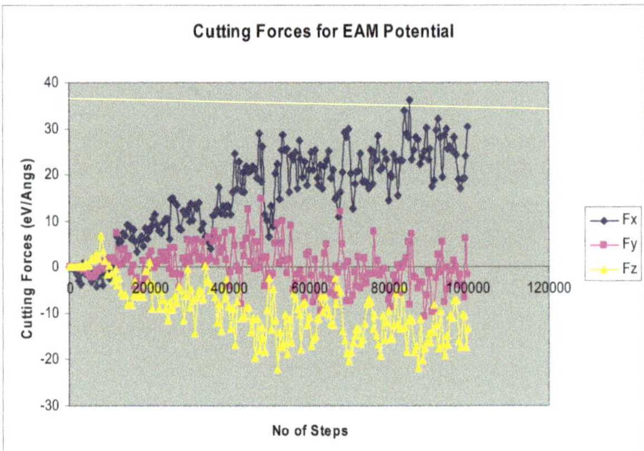

Fig. 13. Cutting Forces for EAM Potential (Oluwajobi & Chen, 2011a)

From Figures 11 - 13, it can be observed that the cutting forces variation is lowest for the EAM potential. The results of the EAM model are comparable with those of (Pei et al., 2006, Promyoo et al., 2008), and the EAM potential best describes the metallic bonding in the copper atoms. In contrast, the pair potentials (both LJ and Morse potentials), do not incorporate the many-body effects; they have no environmental dependence and they do not account for the directional nature of bonding in metals (Li et al., 2008). Through this investigation, it was identified that the EAM potential is the most appropriate of the 3 potentials commonly used for the modelling of nanomachining of copper with diamond tool. This is because the EAM potential provides the best description of the metallic bonding in the workpiece, also, the cutting forces variation is smallest; the potential and total energies are most stable for the depth of cut considered. The study recommended that the EAM potential should be used, rather than LJ and Morse potentials for the modelling of copper and other fcc metals in MD simulations of nanomachining (Oluwajobi & Chen, 2011a).

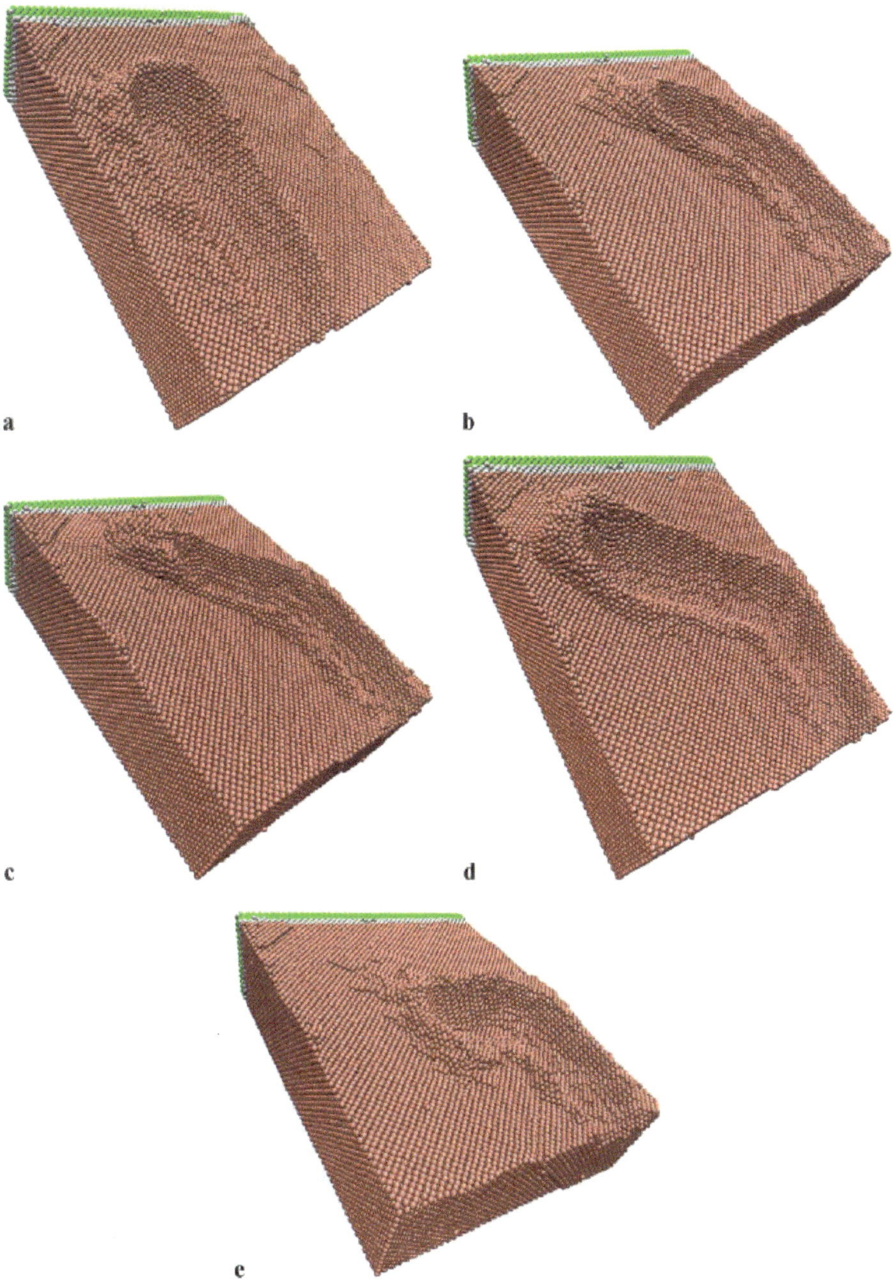

Fig. 14. Groove Profiles for Different Cutting Directions (a) 0deg (b) 30deg (c) 45deg (d) 60deg (e) 90deg (Zhang et al., 2009)

Many existing MD simulation studies on nanometric cutting have been limited to single pass or simple line-type groove. As an extension of the single pass studies, Zhang et al., (2009) modelled folder- line grooves for AFM-based nanometric cutting process of copper workpiece with diamond tool (Figure 14). They used the EAM potential for the copper-copper interactions and the Morse potential for the copper-diamond interactions. They treated the diamond tool as rigid and concluded that the normal, lateral and the resultant forces were almost symmetric with respect to the critical folder angle of 45°. Shi et al., (2011) investigated the multi-groove simulation of single-point tuning of copper workpiece with diamond tool. They used two diamond tools, offset by a fixed distance to simulate a two-groove cutting and modelled the copper-copper and the copper-diamond interactions by using the Morse potential (Figure 15). They also treated the tool as a rigid body and observed that the tool forces increase with increase in feed rate and depth of cut. In practice, most machining processes involve the use of multiple passes to create new surface patterns and the diamond tool is deformable and subjected to wear. To address the above problem, novel multi-pass nanometric machining MD simulations have been carried out, which show the consecutive passes of cut. See Figures 16 and 17 (Oluwajobi & Chen, 2011b).

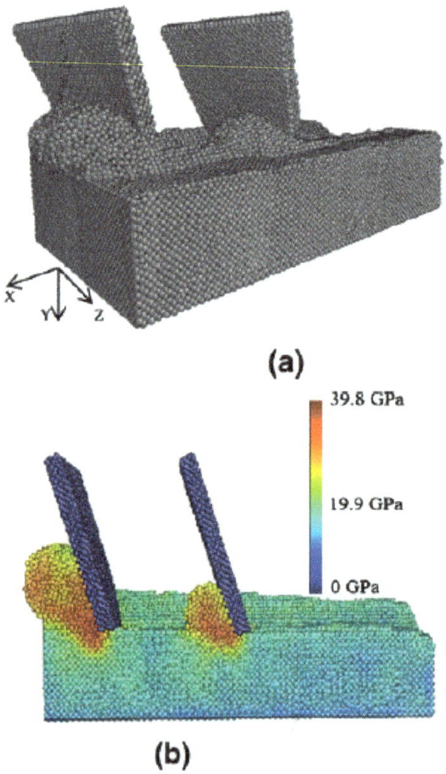

Fig. 15. MD Simulation of Multi-groove Cutting with Two Diamond Tools: (a) 3D perspective view (b) Stress Distribution (Shi et al., 2011)

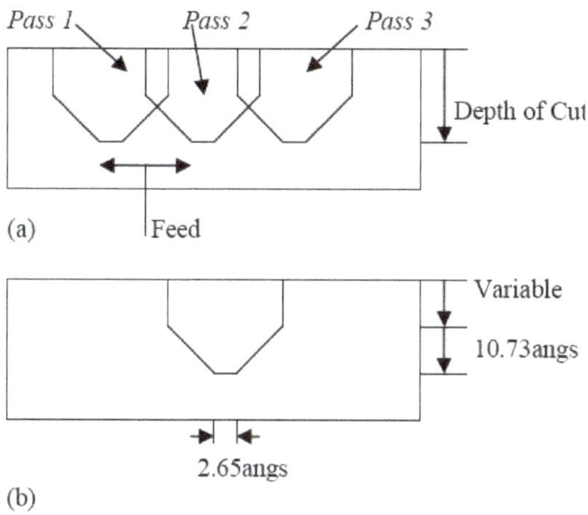

Fig. 16. a. Cross Section of the Machined Grooves with Passes 1-3 (direction of cut is perpendicular to the paper face) 16b: Tool Tip Dimensions (Oluwajobi & Chen, 2011b)

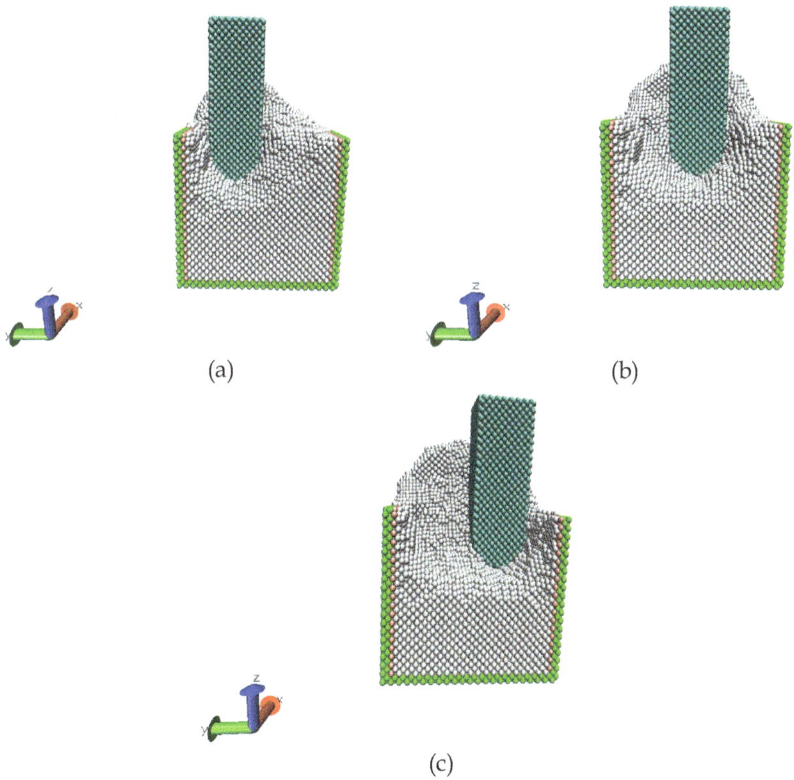

Fig. 17. MD Simulation of Multipass Nanometric Machining (a) pass 1 (b) pass 2 (c) pass 3 (Oluwajobi & Chen, 2011b)

From the simulations, it was observed that the magnitude of the tangential and the normal cutting force components increase with the increase in the depth of cut and they decrease with the consecutive passes. Also, the lateral force components are influenced by atomic vibrations and the cutting cross sectional area (Oluwajobi & Chen, 2011b, 2012).

5. Conclusions

The MD simulation of nanoscale machining has come a long way from inception; from the use of 'primitive' two dimensional systems to two and a half dimensional system and now to 'realistic' three dimensional systems. More developments have occurred from the first simulations where not many or no quantitative analyses were carried out, to now, where rigorous quantitative analyses for the evaluation of the machine process parameters are the norm. Also, the overall configuration have increased from a few thousands to millions atoms at present. The commonly used potentials for MD simulations of metals include LJ, Morse, EAM and MEAM. In many cases, the rationale or justification for their use are not given except for the reference to their use in earlier studies. Ideally, multi-body potentials (EAM and MEAM) should be used for metals, rather than pairwise potentials (LJ and Morse); because they more correctly model the metallic bonding. It is important to use appropriate interatomic potentials for MD simulations so as to obtain useful results. It is now feasible to carry out multipass nanometric machining MD simulations. From the results of such simulations, minimum surface roughness can be effectively evaluated, which is very critical in the industry.

At present, MD methods can only handle small length and time scales. This is far from what obtains during experimental studies. To tackle these problems, multiscale methods are been used. Spatial and temporal multiscale techniques would considerably extend the length and time scales in simulations respectively (Medyanik, 2007, Medyanik & Liu, 2008).

6. Nomenclature

a_i	*Acceleration of atom i,* $\frac{d^2r_i}{dt^2}$
a_{ij}	*Function that limit the range of interaction in Tersoff's potential*
A	*Material dependent constant*
b	*Third derivative of position,* $\frac{d^3r}{dt^3}$
b_{ij}	*A parameter that provides information for the direction and the length of bond (Bond strength)*
c	*Fourth derivative of position,* $\frac{d^4r}{dt^4}$
D	*Material dependent constant*
e	*Elementary Charge*
E	*Energy of interacting particles*
E_{tot}	*Total embedding energy*
f	*Material dependent constant*

f_A	*Attractive pair potential associated with bonding*
f_C	*A smooth cut-off function to limit the range of the potential*
f_R	*Repulsive pair potential*
F_i	*Force acting on atom i*
F_{wi}	*The force acting on the i-th workpiece atom*
F_{ti}	*The force acting on the i-th cutting tool atom*
F_{wtij}	*The interatomic forces acting on the i-th workpiece atom from the cutting tool atom j*
F_{wwij}	*The interatomic forces acting on the i-th workpiece atom from the workpiece atom j*
F_{twij}	*The interatomic forces acting on the i-th cutting tool atom from the workpiece atom j*
F_{ttij}	*The interatomic forces acting on the i-th cutting tool atom from the cutting tool atom j*
g	*Material dependent constant*
G_i	*Embedding energy for placing atom i into the electron density*
$G^{'}$	*First derivative of the embedding energy*
$G^{''}$	*Second derivative of the embedding energy*
h	*Planck constant (6.626069×10^{-34} J s)*
$\hbar$	*Reduced Planck constant* $\frac{h}{2\pi}$ *(1.054571×10^{-34} ms^{2}kg/s)*
m_i	*Mass of atom i*
M_1, M_2	*Nuclear Masses*
N	*Number of atoms*
N_e	*Number of electrons*
N_t	*Number of cutting tool atoms*
N_w	*Number of workpiece atoms*
r	*Distance between atoms*
r_e	*Equilibrium distance between atoms i and j*
r_i	*Position of atom or particle i*
r_{ij}	*Instantaneous distance between particles i and j*
r_o	*Material dependent constant*
t	*Time*
R	*Material dependent constant*
r_{wi}	*The displacement vector of the i-th workpiece atom*
r_{ti}	*The displacement vector of the i-th cutting tool atom*
r_{wtij}	*The displacement vector of the i-th workpiece atom and the interacting cutting tool atom j*
r_{wwij}	*The displacement vector of the i-th workpiece atom and the interacting workpiece atom j*

r_{twij}	*The displacement vector of the i-th cutting tool atom and the interacting workpiece atom j*
r_{ttij}	*The displacement vector of the i-th cutting tool atom and the interacting cutting tool atom j*
$U_{ij}(r)$	*Short range pair interaction representing the core-core repulsion*
v	*Velocity*
$V_{ij}, V(r_{ij})$	*Interatomic Pair Potential*
V_a	*Potential due to attractive forces between atoms i and j*
$V_e(r)$	*Electronic energy calculated by considering the two nuclei as fixed in space a distance r apart*
V_r	*Potential due to repulsive forces between atoms i and j*
$V_{i,j}$	*Short range pair interaction representing the core-core repulsion*
W	*Allowed energy levels*
α	*Material dependent constant*
ε	*Material dependent constant*
θ_{ijk}	*The bond angle between bonds ij and ik*
ζ_{ij}	*Effective co-ordination parameter (The count of the number of other bonds to atom i,besides the ij bond).*
$\rho(r)$	*Electron density*
$\rho_j^a(r)$	*Atomic electron density of atom j at the distance r_{ij} from the nucleus*
$\rho_{h,i}$	*Total electron density at atom i due to the rest of the atoms in the system*
$\bar{\rho}$	*Average electron density*
$\Psi(r,t)$	*Wave function (The probability of finding any particle at position r, at time t)*
Δt	*Finite time step*

7. References

Alder B.J.and T.E. Wainwright (1957), Phase Transition for Hard Sphere System, *Journal of Chemical Physics*, Vol. 27, pp 1208-1209

Alder B.J.and T.E. Wainwright (1959), Studies in Molecular Dynamics. I. General Method, *Journal of Chemical Physics*, Vol. 31 pp 459-466

Allen M.P. and D.J. Tildesley (1988), *Computer Simulation of Liquids*, (Oxford University Press) Baskes M.I. (1992), Modified Embedded-Atom Potentials for Cubic Materials and Impurities, *Physical Review B*, Vol. 46, pp 2727-2742

Beeman D. (1976), Some Multistep Methods for use in Molecular Dynamics Calculations, *Journal of Computational Physics* Vol. 20 pp 130-139

Belak, J. and Stowers I. F. (1990), A Molecular Dynamics Model of the Orthogonal Cutting Process, *Proceedings of the American Society of Precision Engineering*, pp 76-79.

Belak, J. and Stowers I. F. (1991), The Indentation and Scratching of a Metal Surface: A Molecular Dynamics Study, *Fundamentals of Friction: Macroscopic and Microscopic*, Singer, Pollock E 220, pp 1-10

Born M. and J.E. Mayer (1932), Zur Gittertheorie de Ionenkristalle, *Z. Physik*, Vol. 75, pp 1

Chamati H., N. Papanicolau, Y. Mishin & D.A. Papaconstantopoulos (2006), Embedded-Atom Potential for Fe and its Application to Self-Diffusion on Fe(100), *Surface Science*, Vol. 600, pp 1793-1803

Cheng K., X. Luo, R. Ward & R. Holt (2003), Modelling and Simulation of the Tool Wear in Nanometric Cutting, *Wear*, Vol. 255, pp 1427-1432

Cheong W.C.D, L. Zhang & H. Tanaka (2001), Some Essentials of Simulating Nano-Surface Processes Using the Molecular Dynamics Method, *Key Engineering Materials*, Vol. 196, pp 31-42

Cox H., R.L. Johnston & J.M. Murrel (1999), Empirical Potentials for Modelling Solids, Surfaces and Clusters, *Journal of Solid State Chemistry*, Vol. 145, pp 517-540

Ducobu F., E. Filippi & E. Rivière-Lorphèvre (2009), Chip Formation and Minimum Chip Thickness in Micro-milling, In *Proceedings of the 12th CIRP Conference on Modelling of Machining Operations*, pp 339-346.

Ercolessi F. (1997), A Molecular Dynamics Primer, *Spring College in Computational Physics, ICTP*, Trieste, June 1997, http://www.fisica.uniud.it/~ercolessi/md/md/ (Accessed in 2008)

Ercolessi F., E. Tosatti & M. Parrinello (1986), Au (100) Surface Reconstruction, *Physical Review Letters*, Vol. 57, pp 719-722

Ercolessi F. and J.B. Adams (1994), Interatomic Potentials from First Principles Calculations: The Force-Matching Method, *Europhysics Letters*, Vol. 26, No. 8, pp 583-588

Field M.J. (1999), *A Practical Introduction to the Simulation of Molecular Systems*, Cambridge University Press Finnis M.W. and J.E. Sinclair (1984), A Simple Empirical N-Body Potential for Transition Metals, *Philosophical Magazine A*, Vol. 50, Iss. 1, pp 45 -50

Foiles S.M. (1985), Application of the Embedded Atom Method to Liquid Transition Metals, *Physical Review B*, Vol. 32, No 6, pp 3409-3415

Foiles S.M., M.I. Baskes & M.S. Daw (1986), Emdedded-Atom-Method Functions for the FCC Metals Cu, Ag, Au, Ni, Pd, Pt, and their Alloys, *Physical Review B*, Vol. 33, No 12, pp 7983-7991

Frenkel D. and B. Smit (2001), *Understanding Molecular Simulations: From Algorithms to Applications*, Computational Science, Academic Press

Gear C.W. and K.W. Tu (1974), The Effect of Variable Mesh Size on the Stability of Multistep Methods, *SIAM Journal of Numerical Analysis*, Vol. 11, No 5, pp 1025-1043

Gear C.W. and D.S. Watanabe (1974), Stability and Convergence of Variable Order Multistep Methods, *SIAM Journal of Numerical Analysis*, Vol. 11, No 5, pp 1044-1058

Gibson J.B., A.N. Goland, M. Milgram & G.H. Vineyard (1960), Dynamics of Radiation Damage, *Physical Review*, Vol. 120, Iss. 4, pp 1229-1253

Girifalco L.A. and V.G. Weizer (1959), Application of the Morse Potential Function to Cubic Metals, *Physical Review*, Vol. 114, pp 687-680

Haile J.M. (1997), *Molecular Dynamics Simulation: Elementary Methods*, Wiley-Interscience

Hernandez E.R. (2008), Molecular Dynamics from Basic Techniques to Applications: A Molecular Dynamics Primer, *Frontiers in Contemporary Physics, AIP Conference Proceedings*, Vol. 1077, Iss.1, pp 95-123

Hwang H.J., O-K Kwon & J. W. Kang (2004), Copper Nanocluster Diffusion in Carbon Nanotube, *Solid State Communications*, Vol. 129, pp 687-690

Ikawa N., S. Shimada, H. Tanaka & G. Ohmori (1991), An Atomistic Analysis of Nanometric Chip Removal as Affected by Tool-Work Interaction in Diamond Turning, *Annals of the CIRP*, Vol. 40, Iss. 1, pp 551-554

Ikawa N., S. Shimada & H. Tanaka (1992), Minimum Thickness of Cut in Micromachining, *Nanotechnology* Vol. 3, pp 6-9

Jackson M.J. (2008), Micro and Nanomachining, In J.P. Davin – *Machining: Fundamentals and Recent Advances*, pp 271-297, Springer

Kenny S.D., D. Mulliah, C.F. Sanz-Navarro & R. Smith (2005), Molecular Dynamics Simulations of Nanoindentation and Nanotribology, *Philosophical Transactions of the Royal Society A.*, Vol. 363, No 1833, pp 1949-1959

Khukhryansky Y. P., E.A. Shunikov and V.V. Emelyanov (2004), Study of Cluster Forming at Growth of A3B5 Semiconductor Compounds from Liquid Phase, *Journal of Crystal Growth*, Vol. 269, pp 292-297

Komanduri R., and L.M. Raff (2001), A Review on the Molecular Dynamics Simulation of Machining at the Atomic Scale, *Proceedings of the Institution of Mechanical Engineers*, Vol. 215, Part B, pp 1639-1672

Kragelsky I.V., Dobychin M.N., & Kombalov V.S. (1982), *Friction and Wear-Calculation Methods*, Pergamon Press, New York

LAMMPS Manual, http://lammps.sandia.gov/doc/Manual.pdf, (Accessed in 2009)

Leach A. (2001), *Molecular Modelling: Principles and Applications*, Prentice Hall

Lee B, B.D.Wirth, J. Shim, J. Kwon, S.C. Kwon & J. Hong (2005), Modified Embedded-Atom Method Interatomic Potential for the Fe-Cu Alloy System and Cascade Simulations on Pure Fe and Fe-Cu Alloys, *Physical Review B*, Vol. 71, pp 184205 (1-15)

Lennard-Jones J.E. (1924), On the Forces between Atoms and Ions, *Proc. Royal Soc.* Vol. 109, pp 584-597

Li J.H., X.D. Dai, S.H. Liang, K.P. Tai, Y. Kong & B.X. Lin (2008), Interatomic Potentials of the Binary Transition Metal Systems and Some Applications in Materials Physics, *Physics Reports* ,Vol. 455, pp 1-134

Luo X., K. Cheng, X. Guo & R. Holt (2003), An Investigation on the Mechanics of Nanometric Cutting and the Development of its Test-bed, *International Journal of Production Research*, Vol.41, No.7, pp 1149-1165

Luo X. (2004), *'High Precision Surfaces Generation: Modelling, Simulation and Machining Verification'*, PhD Thesis, Leeds Metropolitan University

Medyanik S.N. (2007), Atomistic Simulation Methods and Multiscale Modelling, Accessed October 2011 http://multiscale.emsl.pnl.gov/docs/atomistic.pdf

Medyanik S.N. and W.K. Liu (2008), Multiple Time Scale Method for Atomistic Simulations, *Computational Mechanics*, Vol. 42, pp 569-577

Morse P.M. (1929), Diatomic Molecules according to Wave Mechanics II Vibrational Levels, *Physical Review*, Vol. 34, pp 57-64

Murrel J.N. and R.E. Mottram (1990), Potential Energy Functions for Atomic Solids, *Molecular Physics*, Vol. 69, pp 571-588

Oluwajobi A.O. and X. Chen (2010a), The Fundamentals of Modelling Abrasive Machining Using Molecular Dynamics, *International Journal of Abrasive Technology*, Vol. 3, No 4, pp. 354-381

Oluwajobi A.O. and X. Chen (2010b), On Minimum Depth Cut in Nanomachining, *In: Advances in Manufacturing Technology*: Durham University and Glasgow Caledonian University, *Proceedings of the 8th International Conference on Manufacturing Research*, University of Durham, Durham, pp 174-179

Oluwajobi A.O. and X.Chen (2011a), The Effect of Interatomic Potentials on the Molecular Dynamics Simulation of Nanometric Machining, *International Journal of Automation and Computing*, Vol. 8, No. 3, pp 326-332

Oluwajobi A.O. and X.Chen (2011b), The Effect of Depth of Cut on the Molecular Dynamics (MD) Simulation of Multi-Pass Nanometric Machining, *Proceedings of the 17th International Conference on Automation & Computing, University of Huddersfield, Huddersfield*, Sept. 2011, pp 40-45

Oluwajobi A. and X. Chen (2012), Multi-Pass Nanometric Machining Simulation Using the Molecular Dynamics (MD), *Key Engineering Materials*, Vol. 496, pp 241-246

Pei Q.X., C. Lu, F.Z. Fang & H. Wu (2006), Nanometric Cutting of Copper: A Molecular Dynamics Study, *Computational. Material Science*, Vol. 37, pp 434-441

Pei Q.X., C. Lu, H.P. Lee & Y.W. Zhang (2009), Study of Materials Deformation in Nanometric Cutting by Large-Scale Molecular Dynamics Simulation, *Nanoscale Reseaech Letters*, Vol. 4, pp 444-451

Plimpton S. J., (1995), Fast Parallel Algorithms for Short-Range Molecular Dynamics, *Journal of Computational Physics*, Vol. 117, pp 1-19 and www.lammps.sandia.gov

Potirniche G. P., M.F. Horstemeyer, G.J. Wagner & P.M. Gullett (2006), A Molecular Dynamics Study of Void Growth and Coalescence in Single Crystal Nickel, *International Journal of Plasticity*, Vol. 22, pp 257-278

Promyoo R. H. El-Mounayri and X. Yang (2008), Molecular Dynamics Simulation of Nanometric Machining under Realistic Cutting Conditions, *ASME International Conference on Manufacturing Science and Engineering* (MSEC2008), October 7-10, 2008, Evanston, IL.

Rafii-Tabar H. and A.P. Sutton (1991), Long Range Finnis-Sinclair Potentials for FCC Metallic Alloys, *Philosophical Magazine Letters*, Vol. 63, pp 217-224

Rahman A. (1964), Correlations in the Motions of Atoms in Liquid Argon, *Physical Review A*, Vol.136, pp 405-411

Rapaport D.C. (2004), *The Art of Molecular Dynamics Simulation*, Cambridge University Press

Rentsch R. and I. Inasaki (1994), Molecular Dynamics Simulation for Abrasive Processes, *Annals of the CIRP*, Vol. 43, No 1, pp 327-330

Rentsch R. and I. Inasaki (2006), Effects of Fluids on the Surface Generation in Material Removal Processes, *Annals of the CIRP*, Vol. 55, No. 1, pp 1-4

Rentsch R. (2008), Nanoscale Cutting, in *Nano and Micromachining*, edited by Davin .J.P. and M.J. Jackson, Wiley-ISTE, pp 1-24

Schlick T. (2002), *Molecular Modelling and Simulations: An Interdisciplinary Guide*, Springer

Shi J., Y. Shi and C.R. Liu, Evaluation of a Three-Dimensional Single-Point Turning at Atomistic Level by a Molecular Dynamics Simulation, *International Journal of Advanced Manufacturing Technology*, Vol. 45, No. 1-4, 2011, pp 161-171

Shimada S., N. Ikawa, G. Ohmori, H. Tanaka, & U. Uchikoshi (1992), Molecular Dynamics Analysis as Compared with Experimental Results of Micromachining, Annals of the CIRP Vol. 41, No 1, pp 117-120

Shimada S., N. Ikawa, H. Tanaka, G. Ohmori and J. Uchikoshi (1993), Feasibility Study on Ultimate Accuracy in Microcutting using Molecular Dynamics Simulation, *Annals of the CIRP*, Vol. 42, No.1, pp 91-94

Shimizu J. H. Eda, M. Yoritsune and E. Ohmura (1998), Molecular Dynamics Simulations of Friction on the Atomic Scale, *Nanotechnology*, Vol. 9, pp 118-123

Shimizu J., L. Zhou & H. Eda (2006), Molecular Dynamics Simulation of Vibration-Assisted Cutting: Influences of Vibration, Acceleration and Velocity, *International Journal of Nanomanufacturing*, Vol. 1, No. 1, pp 105-116

Smith R., D. Christopher & S. Kenny (2003), Defect Generation and Pileup of Atoms during Nanoindentation of Fe Single Crystals, *Physical Review B*, Vol. 67, pp 245405-10

Smith W. and I.T. Todorov (2005), The DL_POLY Molecular Dynamics Package,Z. *Kristallogra*, Vol. 220, pp 563-566

Stillinger F.H. and A. Rahman (1974), Improved Simulation of Liquid Water by Molecular Dynamics, *Journal of Chemical Physics*, Vol. 60, pp 1545-1557

Stillinger F.H. and T.A. Weber (1985), Computer Simulation of Local Order in Condensed Phases of Silicon, *Physical Review B*, Vol. 31, pp 5262-5271

Sutton A.P. and J. Chen (1990), Long-Range Finnis Sinclair Potentials, *Philosophical Magazine Letters*, Vol. 61, Iss. 3, pp 139-146

Tersoff J. (1988a), New Empirical Approach for the Structure and Energy of Covalent Systems, *Physical Review B*, Vol. 37 No 12, pp 6991-7000

Tersoff J. (1988b), Empirical Interatomic Potential for Silicon with Improved Elastic Properties, *Physical Review B*, Vol. 38 No 14, pp 9902-9905

Tuckerman M.E. and G.L. Martyna (2000), Understanding Modern Molecular Dynamics: Techniques and Applications, *Journal of Physical Chemistry*, Vol. 104, pp 159-178

Van Gunsteren W. F. and H.J.C. Berendsen (1990), Computer Simulation of Molecular Dynamics: Methodology, Applications, and Perspectives in Chemistry, *Anew. Chem. Int. Ed. Engl.*, Vol. 29, pp 992-1023

Verlet L. (1967), Computer Experiments on Classical Fluids I. Thermodynamics Properties of Lennard-Jones Molecules, *Physical Review D*, Vol.159, pp 98-103

Visual Molecular Dynamics (VMD), http://www.ks.uiuc.edu/Research/vmd/ (Accessed in 2010)

Wang C., T. Yu, W. Duan & L.Wang (1995), A First Principles Interatomic Potential and Application to the GrainBoundary in Ni, *Physics Letters A*, Vol. 197, Iss. 5-6, pp 449-457

Zhang J.T. , T. Sun, Y. Yan, Y. Liang & S. Dong, Molecular Dynamics Study of Groove Fabrication Process using AFM-Based Nanometric Cutting Technique, *Applied Physics A (Materials Science and Processing)*, Vol. 94, 2009, pp 593-600

Zhao K.J., C.Q. Chen, Y.P. Shen and T.J. Lu (2009), Molecular Dynamics Study on the Nano-void Growth in Face-centered Cubic Single Crystal Copper, *Computational Materials Science*, Vol. 46, pp 749-754

Permissions

The contributors of this book come from diverse backgrounds, making this book a truly international effort. This book will bring forth new frontiers with its revolutionizing research information and detailed analysis of the nascent developments around the world.

We would like to thank Lichang Wang, for lending her expertise to make the book truly unique. She has played a crucial role in the development of this book. Without her invaluable contribution this book wouldn't have been possible. She has made vital efforts to compile up to date information on the varied aspects of this subject to make this book a valuable addition to the collection of many professionals and students.

This book was conceptualized with the vision of imparting up-to-date information and advanced data in this field. To ensure the same, a matchless editorial board was set up. Every individual on the board went through rigorous rounds of assessment to prove their worth. After which they invested a large part of their time researching and compiling the most relevant data for our readers. Conferences and sessions were held from time to time between the editorial board and the contributing authors to present the data in the most comprehensible form. The editorial team has worked tirelessly to provide valuable and valid information to help people across the globe.

Every chapter published in this book has been scrutinized by our experts. Their significance has been extensively debated. The topics covered herein carry significant findings which will fuel the growth of the discipline. They may even be implemented as practical applications or may be referred to as a beginning point for another development. Chapters in this book were first published by InTech; hereby published with permission under the Creative Commons Attribution License or equivalent.

The editorial board has been involved in producing this book since its inception. They have spent rigorous hours researching and exploring the diverse topics which have resulted in the successful publishing of this book. They have passed on their knowledge of decades through this book. To expedite this challenging task, the publisher supported the team at every step. A small team of assistant editors was also appointed to further simplify the editing procedure and attain best results for the readers.

Our editorial team has been hand-picked from every corner of the world. Their multi-ethnicity adds dynamic inputs to the discussions which result in innovative outcomes. These outcomes are then further discussed with the researchers and contributors who give their valuable feedback and opinion regarding the same. The feedback is then collaborated with the researches and they are edited in a comprehensive manner to aid the understanding of the subject.

Apart from the editorial board, the designing team has also invested a significant amount of their time in understanding the subject and creating the most relevant covers. They scrutinized every image to scout for the most suitable representation of the subject and create an appropriate cover for the book.

The publishing team has been involved in this book since its early stages. They were actively engaged in every process, be it collecting the data, connecting with the contributors or procuring relevant information. The team has been an ardent support to the editorial, designing and production team. Their endless efforts to recruit the best for this project, has resulted in the accomplishment of this book. They are a veteran in the field of academics and their pool of knowledge is as vast as their experience in printing. Their expertise and guidance has proved useful at every step. Their uncompromising quality standards have made this book an exceptional effort. Their encouragement from time to time has been an inspiration for everyone.

The publisher and the editorial board hope that this book will prove to be a valuable piece of knowledge for researchers, students, practitioners and scholars across the globe.

List of Contributors

Martin Oliver Steinhauser
Research Group ShockWaves in Soft Biological Matter, Department of Composite Materials, Fraunhofer Institute for High-Speed Dynamics, Ernst-Mach-Institut, EMI, Eckerstrasse 4, Freiburg, Germany

Claudia Sandoval
Center for Bioinformatics and Integrative Biology (CBIB), Facultad de Ciencias Biológicas, Universidad Andres Bello, Santiago
Fraunhofer Chile Research Foundation - Center for Systems Biotechnology, FCR-CSB, Chile

Hossein Eslami and Nargess Mehdipour
Persian Gulf University, Iran

Wen-Jong Ma
Graduate Institute of Applied Physics, National Chengchi University, Taipei 11605, Taiwan

Wen-Jong Ma and Chin-Kun Hu
Institute of Physics, Academia Sinica, Nankang, Taipei 11529, Taiwan

Aatto Laaksonen and Alexander Lyubartsev
Stockholm University, Division of Physical Chemistry, Department of Materials and Environmental Chemistry, Arrhenius Laboratory, Stockholm, Sweden

Francesca Mocci
Università di Cagliari, Dipartimento di Scienze Chimiche, Cittadella Universitaria di Monserrato, Monserrato, Cagliari, Italy

Parul Sharma, Parvesh Singh and Krishna Bisetty
Department of Chemistry, Durban University of Technology, Steve Biko campus, Durban, South Africa

Juan J Perez
Department d' Enginyeria Quimica, UPC, ETS d'Enginyers Industrials, Barcelona, Spain

Hiromichi Tsurui
Juntendo University, Japan

Takuya Takahashi
Ritsumeikan University, Japan

Lucas J. Gutiérrez, Ricardo D. Enriz and Héctor A. Baldoni
Área de Química General e Inorgánica, Universidad Nacional de San Luis (UNSL), San Luis, Argentina

Lucas J. Gutiérrez and Ricardo D. Enriz
Instituto Multidisciplinario de Investigaciones Biológicas de San Luis(IMIBIO-SL, CONICET), San Luis, Argentina

Héctor A. Baldoni
Instituto de Matemática Aplicada San Luis (IMASL, CONICET), San Luis, Universidad Nacional de San Luis (UNSL), Argentina

Joseph M. Hayes
Institute of Organic & Pharmaceutical Chemistry, National Hellenic Research Foundation, Athens, Greece

Georgios Archontis
Department of Physics, University of Cyprus, Nicosia, Cyprus

Aurélien Perera and Bernarda Keži´c
Laboratoire de Physique Théorique de la Matière Condensée (UMR CNRS 7600), Université Pierre et Marie Curie, 4 Place Jussieu, F75252, Paris cedex 05, France

Bernarda Keži´c, Franjo Sokoli´c and Larisa Zorani´c
Department of Physics, Faculty of Sciences, University of Split, Nikole Tesle 12, 21000, Split, Croatia

Céline Durniak and Dmitry Samsonov
Department of Electrical Engineering and Electronics, University of Liverpool, United Kingdom

Koji Eriguchi
Graduate School of Engineering, Kyoto University, Japan

Yoshimichi Nakamura and Takahisa Ohno
National Institute for Materials Science, CREST-JST, Japan

Shih-Wei Hung and Pai-Yi Hsiao
Department of Engineering and System Science, National Tsing Hua University, Hsinchu, Taiwan

Ching-Chang Chieng
Department of Mechanical and Biomedical Engineering, City University of Hong Kong, Kowloon, Hong Kong

Mauro Chinappi
Department of Physics, Sapienza University of Rome, P.la Aldo Moro 5, 00185, Rome, Italy

Yeh Chun-Lang
Department of Aeronautical Engineering, National Formosa University, Taiwan, ROC

Robert O. J. Weinzierl
Imperial College London, United Kingdom

Akinjide Oluwajobi
Obafemi Awolowo University, Ile-Ife, Nigeria